Two Manuals Combined into One Book

Cavalry Troop

And

Crew-Served Machine Guns

Department of Defense

Pentagon
Publishing

Field Manual
No 17-97

HEADQUARTERS
DEPARTMENT OF THE ARMY
Washington, DC 3 October 1995

CAVALRY TROOP

TABLE OF CONTENTS

Page

DISTRIBUTION RESTRICTION: Approved for public release; distribution is unlimited.

*This publication supersedes FM 17-97, September 1988.

Preface

This publication serves as a doctrinal guide for commanders and subordinate leaders of both light (HMMWV equipped) and heavy (M1 and M3 equipped) cavalry troops. Heavy cavalry troops are found in Armored Cavalry Regiments (ACR), in Armor/Mechanized Divisions, and in Separate Armor/Mechanized Brigades. Light cavalry troops are found in Light Armored Cavalry Regiments (LACR), in Light Infantry Divisions, and in Separate Light Infantry Brigades.

Although FM 17-97 does not specifically address the TOE of the Light Division Cavalry Troop or the troops of a Separate Heavy or Light Brigade, the tactics, techniques, and procedures outlined in this publication still apply.

This publication lays out the organization, command and control, tactical employment, and service support of the cavalry troop in combat. It establishes the responsibilities and duties of key personnel in the troop during combat.

FM 17-97 is one of three publications in the cavalry troop family. It provides troop leaders with a doctrinal reference to link training and fighting. ARTEP 17-487-30-MTP, *Mission Training Plan for the Regimental Armored Cavalry Troop*, outlines how to train the troop. FKSM 17-97-3, *Cavalry Troop Common SOP*, describes troop operating procedures. This family of publications supports FM 17-15, *Tank Platoon*; FM 7-91, *Tactical Employment of Antiarmor Platoons, Companies, and Battalions*; and FM 17-98, *Scout Platoon*.

The information in this manual is based on the objective table of organization and equipment (TOE) for the ACR and LACR troops. Differences between actual unit organizations and equipment and those described in this manual are caused by differences in the modified table of equipment and organization (MTOE).

The proponent for this publication is HQ TRADOC. Submit changes for improving this publication on DA Form 2028 (Recommended Changes to Publications and Blank Forms), and forward it to Commandant, USAARMS, ATTN: ATSB-SBA-F, Fort Knox, KY 40121-5211.

Although this manual does not implement any particular international agreement, the material presented herein is in accordance with related international agreements. A list of related international agreements and other references can be found in the References section. The paragraph in Chapter 2, Section I, entitled Issue a Warning Order is in compliance with STANAG 2014. Chapter 7, Section III, Relief in Place, and Section IV, Battle Handover and Passage of Lines, are in compliance with STANAG 2082. Tables 7-2 and 7-4 are in compliance with STANAG 2083.

Unless this publication states otherwise, masculine nouns and pronouns do not refer exclusively to men.

Article I. Chapter 1

Section 1.01 Introduction

The role of the cavalry troop in Army operations remains unchanged from the traditional role of cavalry throughout the history of warfare. The troop is organized, equipped, and trained to protect and preserve the fighting ability of other combined arms forces. While its primary missions are reconnaissance and security, the cavalry troop may be called upon to execute attack, defend, and delay missions as part of squadron and regimental missions. *The troop accomplishes its missions by communicating, moving, and shooting in that order.*

The purpose of this chapter is—

- To depict organizations of the cavalry troop in—

  Heavy Troop:

  - The Armored Cavalry Regiment (ACR) (M1 and M3 equipped).
  - The Heavy Division Cavalry Squadron (M1 and M3 equipped).

  Light Troop:

  - The Light Armored Cavalry Regiment (LACR) (HMMWV equipped).

Contents

- To outline missions each troop performs and to highlight each troop's capabilities and limitations.
- To establish responsibilities of key personnel in combat.

(a) Section I. Organizations

(I) THE HEAVY CAVALRY TROOP

The heavy cavalry troop consists of 6 officers and 126 enlisted soldiers. The troop is organized into a headquarters section, two scout platoons, two tank platoons, a mortar section, and a maintenance section (see Figure 1-1).

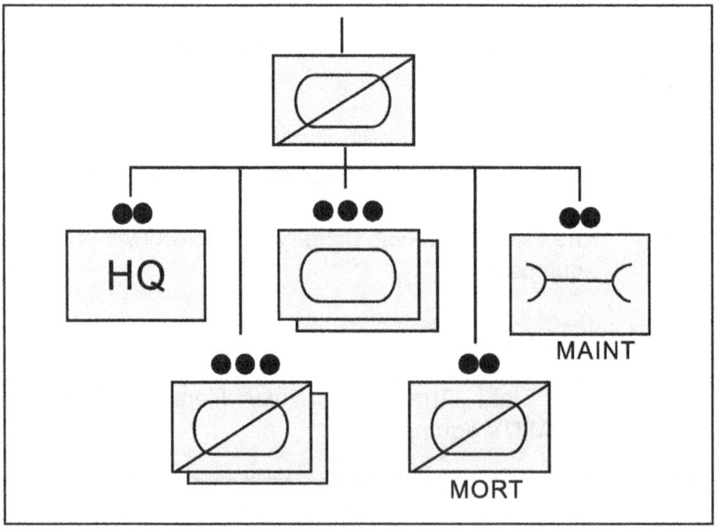

a) Figure 1-1. Heavy cavalry troop organization.

2) The Heavy Troop Scout Platoon

The scout platoon is organized and equipped to conduct reconnaissance and screening in support of its parent troop. However, when the parent unit is performing missions within an economy-of-force role, the scout platoon may conduct offensive, defensive, and retrograde operations in support of the troop mission. The platoon consists of 1 officer and 29 enlisted soldiers, and is equipped with six M3 cavalry fighting vehicles (CFV) organized into three scout sections (see Figure 1-2).

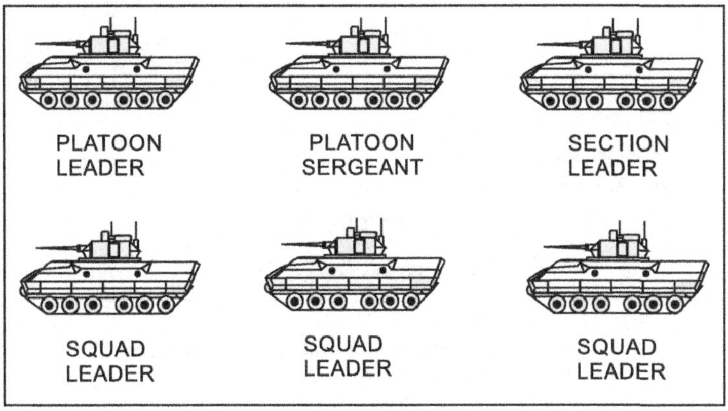

3)

a) Figure 1-2. Scout platoon organization.

4)

5) The Heavy Troop Tank Platoon

The tank platoon is organized and equipped to perform its three primary missions—attack, defend, and move. The platoon consists of 1 officer and 15 enlisted soldiers manning four M1A1 Abrams main battle tanks organized into two sections (see Figure 1-3).

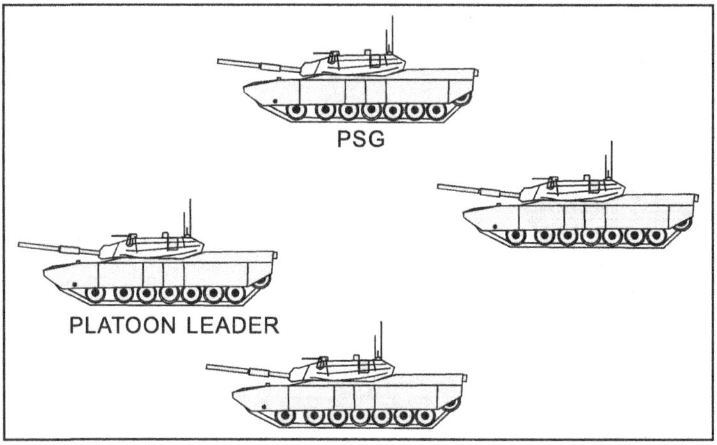

6)

a) Figure 1-3. Tank platoon organization.

7) The Heavy Troop Mortar Section

The mortar section is organized and equipped to provide immediate indirect fires in support of troop operations. Such supporting fires are usually suppression, screening, obscuration, or illumination. The section consists of nine enlisted soldiers. It is equipped with two 107-mm mortars mounted in two self-propelled mortar carriers (see Figure 1-4).

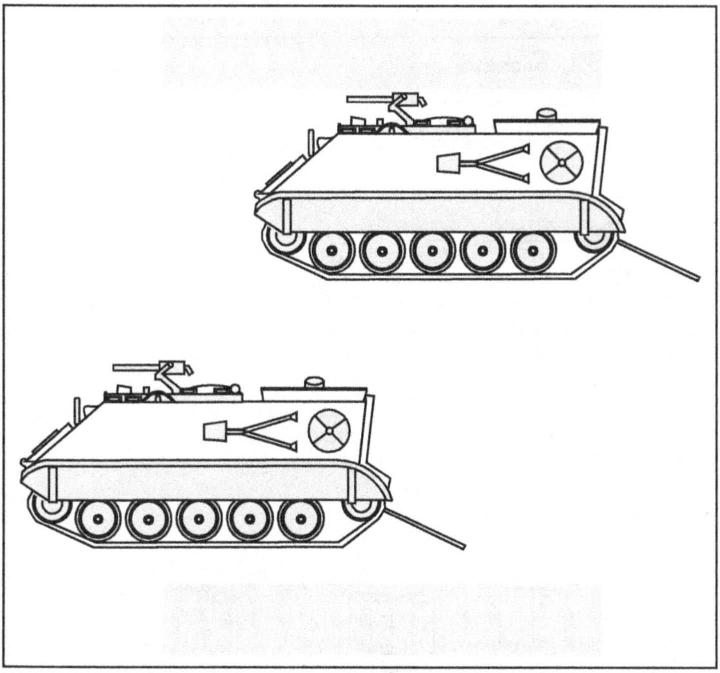

Figure 1-4. Mortar section organization.

8) The Heavy Troop Maintenance Section

This section is organized and equipped to diagnose and repair most equipment faults at troop level. It has the capability to recover all troop vehicles and to maintain the troop's equipment records. The section consists of 18 enlisted soldiers equipped with one armored personnel carrier (APC), one heavy recovery vehicle, one utility truck with cargo trailer, and two cargo trucks with cargo trailers (see Figure 1-5).

a) Figure 1-5. Maintenance section organization.

9) The Heavy Troop Headquarters Section

a)

The troop headquarters section is organized and equipped to perform command and control and logistical support functions for the troop. The section consists of 2 officers and 11 enlisted soldiers. It is equipped with one main battle tank, one command post (CP) carrier, one APC, one M3 CFV, one cargo truck with a 400-gallon water trailer, and two utility trucks with one cargo trailer (see Figure 1-6).

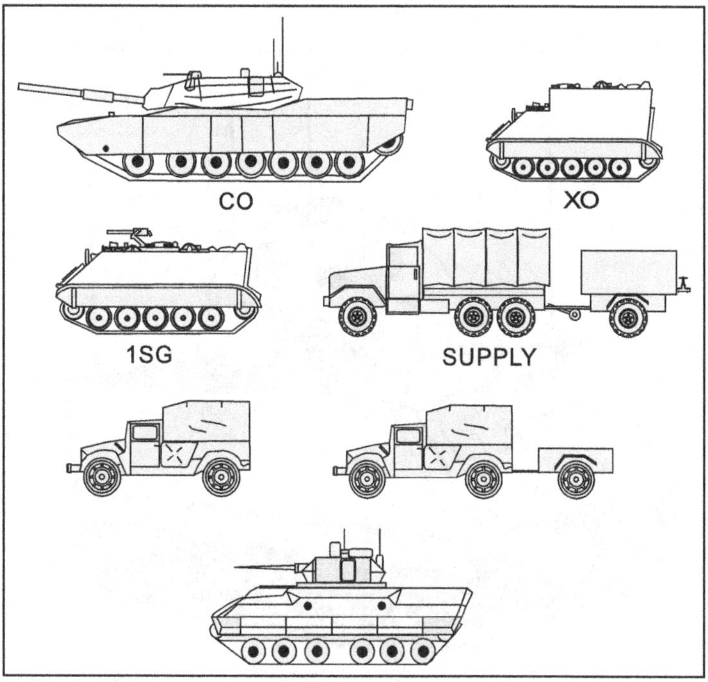

b) Figure 1-6. Headquarters section organization.

10) **The Fire Support Team (FIST)**

The FIST comes from the squadron howitzer battery in the heavy cavalry regiment and the division artillery in the division cavalry squadron. The FIST consists of one fire support officer (FSO), one NCO (team chief), one enlisted fire support specialist, and one enlisted radio operator. The FIST is responsible for coordinating indirect fires for the troop. In the heavy troop the team is mounted on an M981 FIST vehicle (see Figure 1-7).

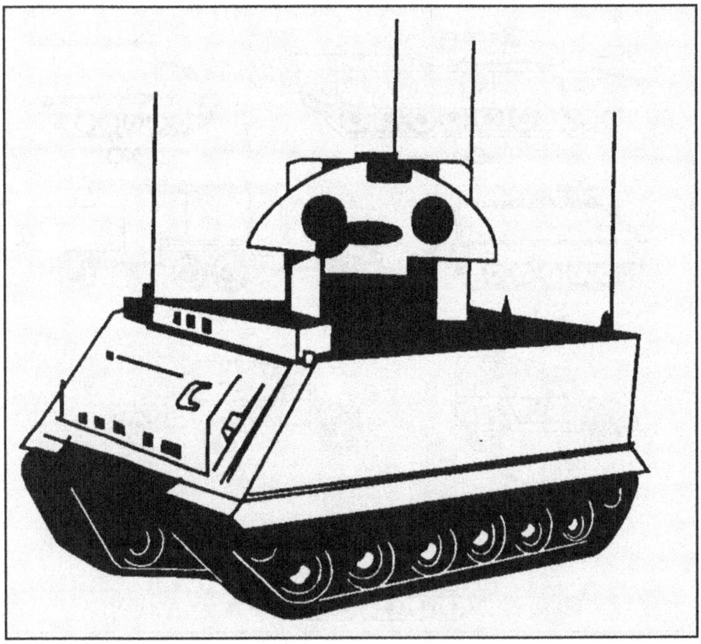

11)

a) Figure 1-7. FIST organization.

(II) THE LIGHT CAVALRY TROOP

The light cavalry troop consists of 6 officers and 107 enlisted soldiers. The troop is organized into a headquarters section, two scout platoons, two antitank (AT) platoons, a mortar section, and a maintenance section (see Figure 1-8).

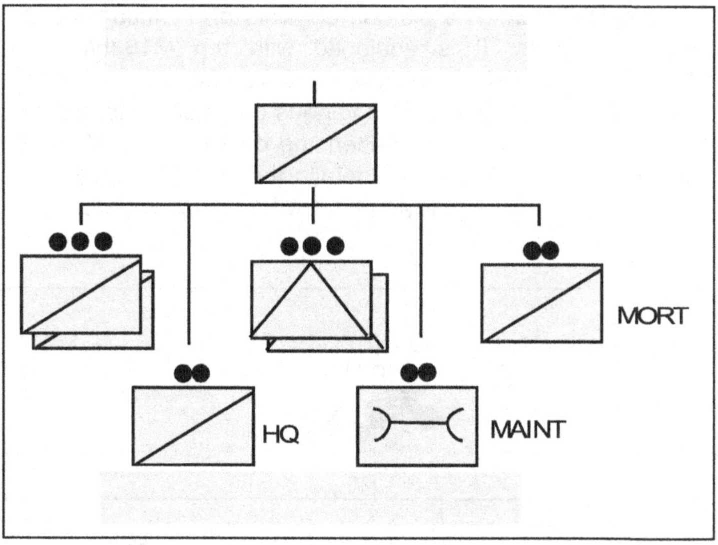

a) Figure 1-8. Light cavalry troop organization.

2) The Light Troop Scout Platoon

The scout platoon is organized and equipped to conduct reconnaissance and screening in support of its parent troop. However, when the parent unit is performing missions within an economy-of-force role, the scout platoon may conduct offensive, defensive, and retrograde operations in support of the troop mission. The platoon consists of 1 officer and 29 enlisted soldiers. It is equipped with ten M1025/M1026 HMMWVs (five MK-19 equipped and five cal .50 equipped). The platoon can organize into various configurations, usually from two to five sections, depending on factors of METT-T (see Figure 1-9). FM 17-98 details the tactics, techniques, and procedures for employing the HMMWV scout platoon.

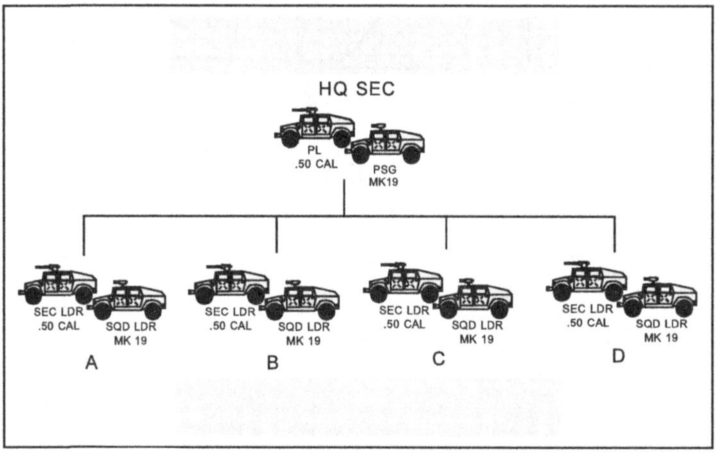

a) Figure 1-9. Light troop scout platoon organization.

3) The Light Troop Antitank Platoon

The AT platoon is organized and equipped to perform its three primary missions—attack, defend, and move. The platoon consists of 1 officer and 11 enlisted soldiers manning four M996 HMMWVs (TOW carriers) organized into two sections (see Figure 1-10).

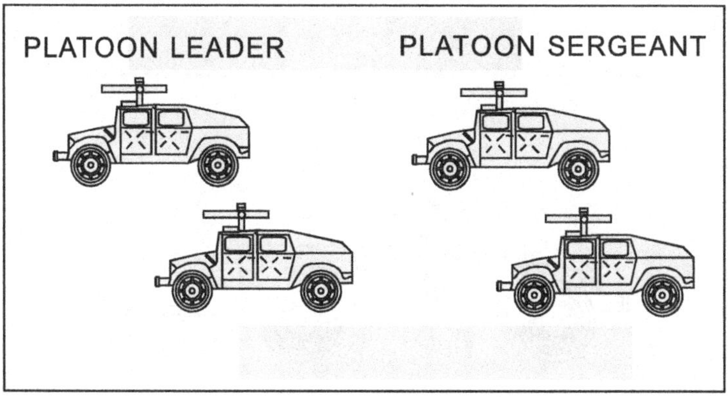

a) Figure 1-10. Light troop AT platoon organization.

4) The Light Troop Mortar Section

The mortar section is organized and equipped to provide immediate indirect fires to support troop operations. Such supporting fires are usually suppression, screening, obscuration, or illumination. The section consists of nine enlisted soldiers. It is equipped with two 120-mm mortars trailered by three M1025/M1026 HMMWVs (see Figure 1-11).

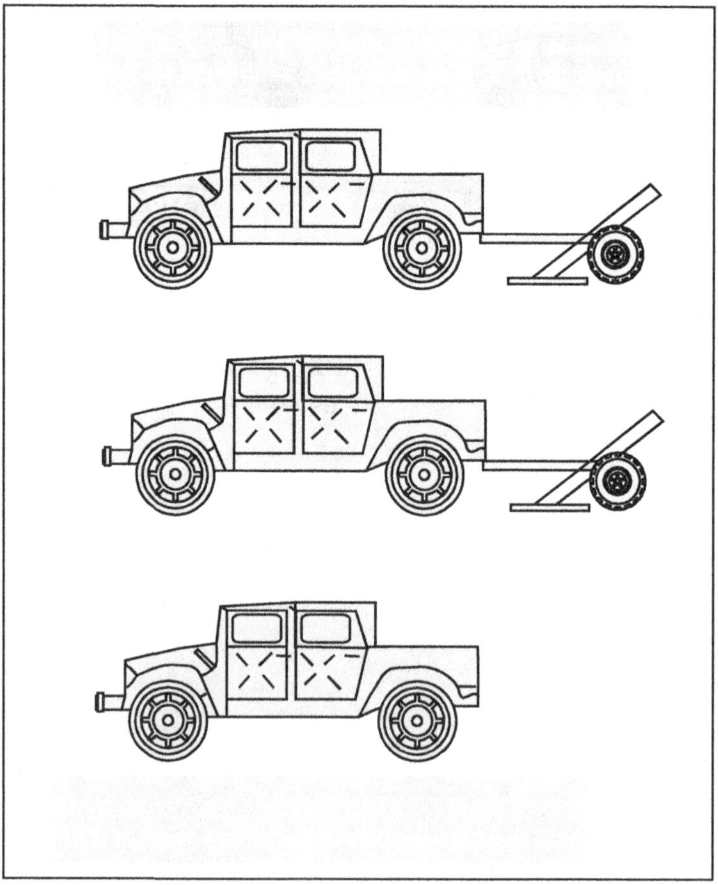

a) Figure 1-11. Light troop mortar section organization.

5) The Light Troop Maintenance Section

This section is organized and equipped to diagnose and repair most equipment faults at troop level. It has the capability to recover all troop vehicles and to maintain the troop's equipment records. The section consists of eight enlisted soldiers equipped with one M998 cargo/troop carrier HMMWV, one cargo truck with cargo trailer, and one 5-ton wrecker (see Figure 1-12).

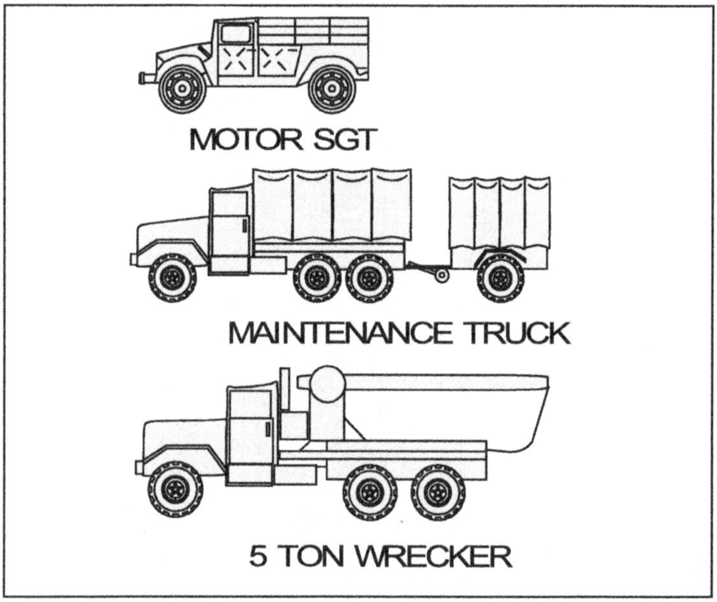

6)

a) Figure 1-12. Light troop maintenance section organization.

7) The Light Troop Headquarters Section

The troop headquarters section is organized and equipped to perform command and control and logistical support functions for the troop. The section consists of two officers and ten enlisted soldiers. It is equipped with one M1025 HMMWV, one M1037 HMMWV shelter carrier with S250 shelter, one M998 HMMWV with cargo trailer, and one 2 1/2-ton cargo truck with 400-gallon water trailer (see Figure 1-13).

a) Figure 1-13. Light troop headquarters section organization.

8) The Fire Support Team (FIST)

The FIST comes from the squadron howitzer battery in the light cavalry regiment. The FIST consists of one officer (FSO), one NCO (team chief), one enlisted fire support specialist, and one enlisted radio operator. The FIST is responsible for coordinating indirect fires for the troop. In the light troop the team is mounted on a HMMWV (see Figure 1-14).

(b)

a) Figure 1-14. Light troop FIST organization.

(c) Section II. Missions, Capabilities, and Limitations

(I) MISSIONS

Heavy and light cavalry troops perform reconnaissance and security missions to protect and preserve the fighting ability of the units to which they are assigned or attached. Both troops also conduct offensive, defensive, and retrograde operations in an economy-of-force role. Cavalry troops in general have limitations and capabilities associated with their TOEs and METT-T that must be considered when employing them in a specific mission role (see Figure 1-15).

	Hvy Trp	Lt Trp
Reconnaissance Missions		
Route Recon	✓	✓
Zone Recon	✓	✓
Area Recon	✓	✓
Security Missions		
Screen	✓	✓
Area Security	✱	✱
Convoy Security	x	x
Economy-of-Force Role		
Offensive Missions		
Hasty Attack	✓	✱
Attack	✓	✱
Movement to Contact	✓	✱
Defensive Missions		
Defend a Battle Position	✓	✱
Defend in Sector	✓	✱
Retrograde Missions		
Delay	✓	✱

✓ = fully capable
x = capable when reinforced
✱ = capable under permissive METT-T

a) Figure 1-15. Cavalry troop mission profiles.

(II) CAPABILITIES

a) The heavy cavalry troop is a unit that can be deployed by rail, sea, or both into a theater of operations. It can perform its missions under all visibility conditions and in any terrain that supports heavy armor movement. The integrated thermal sights on the M1A1 tank and M3 CFV along with ancillary passive night sights and night vision devices authorized to the troop provide an outstanding nighttime/reduced visibility acquisition and fighting capability. Additionally, the firepower and survivability organic to the heavy cavalry troop allow it to execute missions aggressively across the spectrum of warfare.

The light cavalry troop is a unit that can be rapidly deployed by air, rail, or sea into a theater of operations. It can perform its missions under all visibility conditions and in any terrain that supports wheeled vehicle movement. The ancillary thermal and night vision devices authorized to the troop provide an excellent reduced visibility acquisition capability. However, the light cavalry troop's ability to fight under reduced visibility conditions is limited because of its lack of integrated passive and thermal sights on stabilized weapons platforms.

(III)

(IV) LIMITATIONS

a) Given suitable terrain conditions, the ability of heavy and light cavalry troops to accomplish their assigned missions is mainly limited by the size and strength of threat forces encountered. Both troops are limited in close terrain, especially urban environments, due to their limited number of dismounted scouts. Also, the troops are dependent on their parent squadrons and regiments for additional combat support (CS) and combat service support (CSS) assets.

(d)
(e)

(f) Section III. Responsibilities

(I) THE TROOP COMMANDER

The troop commander is responsible to the squadron commander (SCO) for the discipline, combat readiness, and training of the troop, and for the maintenance of its equipment. He must be proficient in the tactical employment of the troop and its assigned and attached CS elements. He must also know the capabilities and limitations of the troop's personnel and equipment as well as those of CS elements attached to him.

a) The troop commander's responsibility in combat is twofold. He will—

- Accomplish all missions assigned to the troop in accordance with the SCO's intent.
- Preserve the fighting capability of the troop.

(II) THE EXECUTIVE OFFICER

In combat, the troop executive officer (XO) is second in command. He supervises the troop tactical operations center (TOC), where he stays abreast of the tactical situation in the troop's area of operations (AO). He manages the flow of combat information between the troop and the squadron from the troop TOC. With the assistance of the troop first sergeant (1SG), he plans and coordinates CSS for the troop.

(III) THE FIRST SERGEANT

The primary responsibility of the troop 1SG is sustaining the troop's ability to fight. He supervises the procurement and distribution of fuel, ammunition, food, water, clothing, equipment, replacements, and repair parts. He receives incoming personnel and assigns them to subordinate elements as needed. He is responsible for the medical evacuation of sick, injured, and wounded soldiers to the supporting medical treatment facility. He is also responsible for the evacuation of soldiers killed in action to the supporting graves registration collection point. He is also responsible for the recovery and evacuation of damaged combat equipment.

(IV) THE TROOP FIRE SUPPORT OFFICER

The troop FSO is responsible for the planning and coordination of the troop fire support plan. He relays troop requests for indirect fire and advises the commander on the employment of indirect-fire weapons in support of the troop's maneuver. The troop FSO may control and position the mortars during combat operations.

(V) THE PLATOON LEADER

The platoon leader is responsible to the troop commander for the discipline, combat readiness, and training of the platoon, and for the maintenance of its equipment. He must be proficient in the tactical employment of the platoon and know the capabilities and limitations of the platoon's personnel and equipment.

The platoon leader's responsibility in combat is twofold. He must—

- Accomplish all missions assigned to the platoon in accordance with the troop commander's intent.

- Preserve the fighting capability of the platoon.

(VI) THE PLATOON SERGEANT

The platoon sergeant (PSG) leads elements of the platoon as directed by the platoon leader, and assumes command of the platoon in his absence. The PSG assists the platoon leader in maintaining discipline, conducting training, and exercising control. He supervises platoon CSS, which includes supply and equipment maintenance.

(VII) THE MORTAR SECTION SERGEANT

The mortar section sergeant is responsible for providing responsive indirect fires to support the troop commander's concept of the operation.

(VIII) THE SUPPLY SERGEANT

The supply sergeant picks up, transports, and issues supplies and equipment to the troop. He works closely with the 1SG to accomplish these tasks. He also evacuates enemy prisoners of war and assists in the evacuation of soldiers who are killed in action to the graves registration collection point.

(IX) THE MAINTENANCE SERGEANT

The maintenance sergeant is responsible for the prompt battlefield repair or recovery of damaged or inoperable equipment. He works closely with the 1SG to accomplish these tasks.

(X) THE COMMUNICATIONS SERGEANT

The communications sergeant ensures the troop TOC and its crew are prepared for combat operations, and assists the XO in the TOC during combat operations. Within his capability, he repairs the communications equipment of subordinate elements.

(XI) THE NUCLEAR, BIOLOGICAL, CHEMICAL (NBC) NCO

The troop NBC NCO is responsible for troop NBC defense activities. He supervises radiological monitoring, chemical detection, and decontamination operations. He assists in maintaining NBC equipment and training NBC equipment operators and decontamination teams.

Chapter 2

Battle Command

Battle command is the art of battle decision making, leading and motivating soldiers and their organizations into action to accomplish missions. Battle command includes visualizing the current state and the future state, then formulating concepts of operations to get from one state to the other at least cost. Other functions of battle command include assigning missions; prioritizing and allocating resources; selecting the critical time and place to act; and knowing how and when to make adjustments during the fight.

Instantaneous response to orders is the hallmark of effective battle command in any cavalry operation. Leaders must quickly analyze information, make tactical decisions, and turn those decisions into successful battlefield actions.

Battle command of cavalry units is typically decentralized due to the size of the area of operations, vagueness of the enemy situation, and terrain unknowns. This places the burden of sound, timely decision making at the lowest levels. Leaders must develop a keen sense of situational awareness and constantly track the actions of subordinate units as well as those to the front, flank, and rear.

Contents

Effective battle command begins in the planning phase and continues through the execution phase of each mission. This chapter outlines the tools and techniques a troop commander needs to effectively command and control a cavalry troop in combat.

Section I. Troop Leading Procedures

The nine troop leading procedures are—

- Receive and analyze the mission.
- Issue a warning order.
- Make a tentative plan.
- Start necessary movement.
- Conduct a reconnaissance.
- Make final decisions and complete the plan.
- Issue an operations order.
- Rehearse.
- Supervise and refine.

RECEIVE AND ANALYZE THE MISSION

The troop commander will receive missions from squadron in the form of written, briefed, or radioed operation orders (OPORD) and fragmentary orders (FRAGO). Upon receipt of a mission, he will conduct a mission analysis to determine the *who*, *what*, *when*, *where*, and *why* elements of the mission and how much time is available until mission execution.

Mission. The mission analysis must identify the following:

- Specified tasks.
- Implied tasks.
- Essential tasks.
- Intent of the higher commander.
- Any constraints or limitations.

The items for analysis are derived from the squadron order as noted below.

- Paragraph 2, MISSION.
  - Mission Statement.

- Paragraph 3, EXECUTION.
  - Concept of the Operation.
  - Commander's Intent.
  - Maneuver.
  - Specific Instructions.
  - Coordinating Instructions.

- Execution matrixes.

- Thorough map reconnaissance.

List all specified and implied tasks not covered by unit SOP in a mission list. Identify those tasks that are vital to the squadron accomplishing its mission. These are essential tasks. Put the essential tasks in the form of a restated mission for the troop. The restated mission statement should be short and simple, and should cover *who, what, when, where,* and *why.*

One final note on this—keep the mission list at hand. It makes the job easier when preparing a course of action and assigning missions or tasks to platoons and sections.

Time. Because time is usually the most limited resource available, it must be used as efficiently as possible. Figure out what has to be done and what can be done in the time available. Use the backwards planning process to determine when critical events in the planning process must occur and stick to the timeline developed. Get the timeline out to subordinates in the warning order. Strive to use no more than one-third of the time available in the planning process at troop level. The remaining two-thirds is for subordinate leaders to plan and prepare.

Consider the following in terms of time when conducting mission analysis and preparing the troop warning order:

- Combat service support.

- Precombat inspections.

- Route reconnaissance (time/distance factors):
  - Assembly area to route start point.
  - Start point to command posts.
  - Command posts to release point.
  - Release point to line of departure.

- Subordinate's planning time.

- Platoon rehearsals:
  - Battle drill rehearsals.
  - Actions on contact.
  - Obstacle breaching/emplacement.
  - Movement formations.

Note. The preparatory actions above are triggered by receipt of a warning order at platoon level.

- Time to issue troop OPORD.

Note. One-third of the available time includes the time it takes to issue the troop OPORD.

- Troop rehearsals:
  - Backbriefs.
  - Rock drills.
  - Radio rehearsals.

- Time to move from—
  - Assembly area to line of departure.
  - Phase line to phase line.
  - Line of departure to objective.

- Effect of weather on movement.

Once the mission and time available have been analyzed, move to the next step in the trooping leading procedures.

ISSUE A WARNING ORDER

Issue the warning order promptly. Elements of the warning order include—

- Situation.
- Mission.
- Earliest time of movement.
- Time and place of OPORD.
- Attachments and detachments.
- Actions triggered:
  - Security.
  - Reconnaissance.
  - Combat service support actions.
  - Precombat inspections.
  - Sleep plans.
  - Rehearsals.

MAKE A TENTATIVE PLAN

Develop a course of action (COA). Consider the factors of METT-T when developing two, or better yet, three separate COAs.

- **Mission.** See *Receive and Analyze the Mission* paragraph on page 2-2.

- **Time.** In addition to the factors mentioned in the *Receive and Analyze the Mission* paragraph on page 2-2, the commander should consider the following details in relation to time when developing his tentative plan.

  - Time to move from—

    Line of departure to phase lines.
    Phase line to phase line.
    Line of departure to objective.

- Time for enemy to—
  - Attack.
  - Counterattack.
  - Reinforce.
- Effect of weather on movement and identification of friend or foe.
- Effect of NBC operations on movement.

• **Enemy.** Conduct intelligence preparation of the battlefield (IPB). IPB integrates enemy doctrine with weather and terrain to determine and evaluate enemy capabilities, vulnerabilities, and probable courses of action. IPB is a key part of preparing for battle. Much of the IPB workup can be found in Paragraph 1 of the squadron OPORD.

• **Terrain.** Analyze terrain in the area of operations. Consider the factors of OCOKA. (Arranged here in a more logical, systematic approach, based on IPB procedures and preparation of the modified combined obstacle overlay [MCOO].)

■ ***Obstacles.*** Identify all obstacles and restricted and severely restricted terrain throughout the sector or zone (a terrain sketch may be beneficial). Consider the following as obstacles:

- Water.
- Power lines.
- Bridges.
- Defiles.
- Slopes.
- Towns.
- Embankments.
- Railroads.
- Existing engineer obstacles.
- Wooded areas.

Terrain that flows around, or in some instances through or over, the obstacles identified is the primary maneuver corridor, or avenue of approach.

■ ***Avenues of Approach.*** Consider the following for each avenue of approach:

- What size force does it support? In what formation?
- Does it support long range observation and fields of fire? From where?
- Does it provide any cover and concealment? Where?
- Where does it lead?
  Through sector/zone?
  Enters/exits a flank?

The answers to these questions will help determine what terrain, if any, is key terrain.

■ ***Key Terrain.*** Key terrain is that terrain which control over would provide a clear advantage over another force. Seizing, securing, or even avoiding key terrain in sector/zone will figure prominently in the final plan. Identify what terrain is key and evaluate it in terms of observation and fields of fire along each avenue of approach.

■ ***Observation/Fields of Fire.*** Evaluate all terrain along the avenues of approach in terms of observation and fields of fire. Determine the following:

- What terrain provides the best observation of critical areas or along avenues of approach?
  At night?
  When visibility is obscured by smoke or fog?
- What terrain offers effective range of friendly and enemy weapon systems?
- What terrain provides the best fields of fire? For what size force?

■ ***Cover and Concealment.*** Evaluate all terrain along the avenues of approach in terms of cover and concealment. Determine the following:

- What terrain affords the best cover and concealment to the enemy and to the friendly forces?
- Where are likely hide positions that support potential enemy and friendly battle positions?
- What routes would a force use to move from hide positions to battle positions?

• **Troops Available.** Analyze the troops available to execute the mission. Determine the following:

■ ***Soldiers.***

- What is the strength of the platoons in the troop?
- Can all weapon systems be manned?
- How is the morale of the troop?
- What is the experience level of the troop?
- What training and skills are required? Are specialists such as engineers needed?
- What is the general health status and fatigue level of the soldiers?

■ ***Equipment.***

- How many weapon systems are operational and employable?
- Is special equipment or additional equipment needed to accomplish the mission?

START NECESSARY MOVEMENT

While developing his tentative plan, the commander may decide to reposition some of his forces before the operation starts so that he is prepared to meet mission requirements. If the entire troop needs to move, he will often execute the movement in accordance with a squadron plan. The amount of time he has to reposition his forces is determined by the readiness condition (REDCON) in effect.

Note. See Chapter 7 for a discussion of REDCON.

CONDUCT A RECONNAISSANCE

If sufficient time is available after determining his tentative course of action, the commander should make every effort to get out into his AO and look at the terrain. Ground reconnaissance is the norm; however, air transportation may be available. Staying on the ground gives a better feel for the terrain than does reconnoitering by helicopter, but it is slower. Although faster, aerial reconnaissance does not provide an appreciation of the terrain from a vehicle commander's point of view in terms of fields of fire and cover and concealment. If circumstances permit, subordinate leaders should accompany the commander to help perform the reconnaissance. It gives them an opportunity to become familiar with the area.

MAKE FINAL DECISIONS AND COMPLETE THE PLAN

Based on the new information gathered by his personal reconnaissance, the commander makes final adjustments to his plan and nails down the details. Once the plan is complete, he prepares the order. Troop orders need not be written and handed to subordinate leaders. Speak face to face with them if possible. Put the plan in the form of a standard five-paragraph field order. Jot it down in a notebook so every detail can be recalled.

ISSUE THE OPERATIONS ORDER

Assemble the orders group. Make sure everyone in the orders group is present. Issue necessary graphics to subordinate leaders and give them time to prepare their maps. Ensure the overlays are neat and accurate. Messy overlays with broad pen strokes cause confusion and waste valuable time. Once the subordinate's overlays are prepared, check them for accuracy. (A technique for ensuring accuracy is to include in the order a list of grid coordinates for key locations that might be ambiguous in the operational graphics. Even neat graphics, once copied from squadron down to platoon, may often deviate by as much as 500 meters.) Arrange the orders group in a semicircle from left to right in the order in which they will be addressed when giving specific instructions.

Issue the order. Use notes when issuing the order; do not rely on memory. If possible, issue the order while overlooking the AO. Keep in mind that the order will be translated a couple of times before the privates receive it, and it must make sense to them. Talk in their language. The order should be short and simple, and must be logical and easy to follow; be clear, organized, and concise so that no one becomes confused. Misunderstanding the commander's intent most often occurs in the transmission of orders to the platoon leader and his subordinates.

Note. See Appendix C for the operations order format and sample OPORD.

Conduct the confirmation brief. Immediately after the order is issued, call for questions from the subordinate leaders. After all questions concerning the order have been answered, begin the confirmation brief by each of the troop's subordinate leaders.

The confirmation brief is used by the commander to confirm that his intent and guidance for the conduct of the operation are clearly understood by everyone in the orders group before they are dismissed to begin their planning. The confirmation brief adjourns only when the commander is confident his subordinates understand their mission, his and the higher commander's intent, his concept of the operation, the scheme of maneuver, the time plan, and the type and location of the rehearsal.

REHEARSE

Rehearsals are of paramount importance before executing any plan. Rehearsals help in the following ways.

- Clarify the commander's intent.
- Expose combat, combat support, combat service support, or disconnected activities in the plan.
- Reinforce the scheme of maneuver and fire support plan.
- Focus on actions and decision points critical to mission accomplishment.
- Ensure subordinates explicitly understand their missions, how their missions relate to each other, and how each mission relates to the commander's plan.
- Outline conditions that, when present, would necessitate execution of branch plans.
- Provide feedback to the commander.

Rehearsals instill confidence in participants by—

- Giving participants faith in the success of their own plan as well as in their commander's plan.
- Providing subordinates with purpose, direction, and motivation.
- Enabling leaders to execute missions with speed, flexibility, and audacity.

There are generally seven types of rehearsals—full, key leader, terrain model, sketch map, map, radio, and backbrief. Preparation time and resources for each range from extensive to minimal. At troop level, the most common rehearsal types are backbrief, map, sketch map, terrain model, and radio.

Backbrief. During the backbrief, each subordinate leader briefs the commander of how he intends to accomplish his mission before he issues his OPORD to his respective unit. By having subordinates explain their intent and concept of operation, the commander can ensure their plans support his own. Flaws or potential problems with the plan may be revealed at this time.

Do not confuse the backbrief rehearsal with the confirmation brief the commander uses immediately after he issues an OPORD to determine how well his subordinate leaders understand the order.

Map/Sketch Map/Terrain Model. Using either a map and overlay of the same scale as used to plan and execute the operation, a large scale sketch map, or a terrain model, the commander and subordinate leaders move unit markers in a sequential (either by phase, event, or time), interactive, verbal execution of the operation. These type rehearsals are essentially conducted like a war game to show the planned sequence of action-reaction-counteraction to critical events or phases of the operation. This is a good opportunity to coordinate not only actions of the unit but also critical locations such as contact points, checkpoints, boundaries, battle positions, and hide positions among subordinate leaders.

Radio. Using existing communications networks (either FM or wire), the commander and subordinate leaders verbally and interactively execute critical portions of the operation. This technique can have obvious communications security disadvantages; if so, then only the essential, most critical portions of the operation are rehearsed. Radio rehearsals require few resources and little time and are best used in conjunction with other methods or to further refine the plan.

SUPERVISE AND REFINE

The last step in troop leading procedures must not be neglected. Failure to supervise the execution of the order, or to refine the plan as the situation changes, is the road to ruin. The importance of this step cannot be overemphasized. Actions during the final step may include, but are not limited to, higher commander and his staff being present in the subordinate unit staff planning process.

Section II. Situational Awareness

Situational awareness is always keeping a clear picture of the tactical situation, both mentally and graphically. This picture includes both the friendly and enemy situation and an understanding of the relevant terrain. Since the troop normally operates dispersed over wide frontages, it is essential that all leaders maintain situational awareness so they can make sound, quick tactical decisions. Situational awareness also permits the leaders to anticipate events and relate separate pieces of information to form logical conclusions. One of the critical outcomes of situational awareness observed by all leaders is a reduction of fratricide incidents.

BATTLEFIELD FRAMEWORK

How the commander has structured the battlefield impacts the troop commander's ability to maintain good situational awareness. A commander will structure the battlefield based on the conditions of METT-T. The framework of the battlefield can vary from a very rigid extreme with an obvious front and rear boundary and closely tied adjacent units, to a very dispersed and decentralized structure with no secure areas, unit boundaries, or definable front. Between these extremes is an unlimited number of possible variations. Maintaining situational awareness will become more difficult the more unstructured the battlefield is. Modern, highly mobile operations with small forces lend themselves to a less rigid framework which will challenge the ability to maintain a good picture of the battlefield.

PICTURING THE BATTLEFIELD

To have a good picture of the battlefield, all leaders must have virtually perfect knowledge of the friendly situation one level higher than their own. This means that all principal subordinate leaders must know the troop situation and the troop commander must know the squadron situation. All leaders must have accurate knowledge of the terrain, and they must know as much as possible about the enemy. The requirement to maintain a real-time picture of the battlefield one level higher does not relieve leaders of the requirement to understand the situation two levels higher. The difference is that a leader's understanding of the situation two levels higher than his own does not have to be as specific or in real time.

Most of the information passed between elements is in the form of reports over FM radio. The troop commander receives many reports as a result of his troop graphics. Good graphics require that the subordinate elements report periodically as they accomplish tasks. The troop commander must be aware of when subordinates report so he will know how current his visualization of the situation is. If an element does not report in a timely manner, the commander must quickly determine the situation of the overdue element.

Although many reports may not be addressed specifically to him, particularly on the squadron net, the commander must monitor them by eavesdropping on the nets as traffic is sent. How effectively he can accomplish this is, to some degree, experience dependent; however, there are techniques he can use to relate the information to his map and thereby track the tactical situation. The troop XO plays an important role in assisting the troop commander and platoon leaders in maintaining situational awareness. The troop XO provides leaders with periodic updates of friendly units to the front, flanks, and rear, based on traffic from the squadron operations and intelligence (OI) net.

The commander's map is the key to maintaining situational awareness (see Figure 2-1). He should plot all friendly position reports up to one level higher than his own. Information from spot reports should also be plotted. Using different colors for friendly and enemy elements allows quick distinction. To avoid cluttering the map, he should place a dot or symbol with a number on his map where the element is located. The same number is then written in the map margin (or beyond the area of operations) with the complete spot report or unit identification next to it. This notation should also include the time. As positions or reports are updated, the old symbol is crossed off and a new one with a corresponding notation is added. This system allows the commander to easily track and monitor the tactical situation. This system is augmented by a formal operations log kept in the troop TOC by the XO or NCOIC.

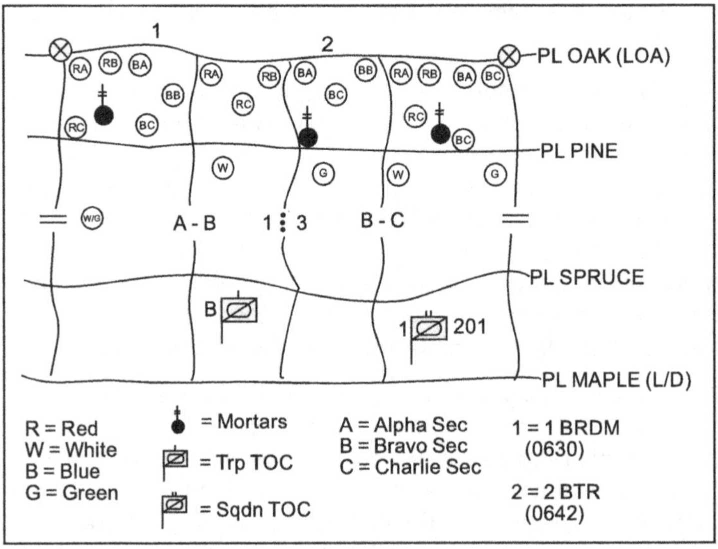

Figure 2-1. Sample situation map (troop commander).

BATTLE SPACE

Situational awareness, as previously discussed, is critical to successful reconnaissance and security missions; however, the troop commander's primary attention must be on his battle space. Battle space for the cavalry troop is determined by the location of individual platoons and sections, the range of direct-fire weapons, observation, sensors, and the terrain on which they are applied. This space is the immediate area over which the troop has influence. The troop commander must be aware of the general situation, but he must actively direct and manage all activity within his battle space. Most of the troop commander's command and control efforts focus on what is happening within his battle space and shaping his battle space to make it more efficient.

FRATRICIDE

Fratricide is a significant danger to all forces operating on a mobile battlefield where weapon system lethality is significantly greater than friend or foe identification ability. For this reason, situational awareness on the part of all leaders, particularly the commander, is critical not only to mission success but also to survival.

Under these types of conditions, it is critical that the commander know where other friendly elements are operating. With this knowledge of the situation, he must anticipate dangerous situations and take steps to either avoid or mitigate them. The commander must constantly be vigilant to changes and developments in the situation that may place his elements in danger. He must also ensure all subordinate unit positions are constantly sent to higher headquarters so all other friendly elements are aware of where they are and what they are doing. When the commander perceives a potentially dangerous situation, he must personally use the squadron command net to coordinate directly with the friendly element involved.

Section III. Communications

NETS

Several troop radio nets are used to circulate the volume of information in any combat operation. A troop command net links the troop commander with his subordinate units (see Figure 2-2). A troop fire support net links subordinate units with the fire support team (FIST) and mortar section (see Figure 2-3). Each net has an assigned radio frequency, but not necessarily a dedicated radio.

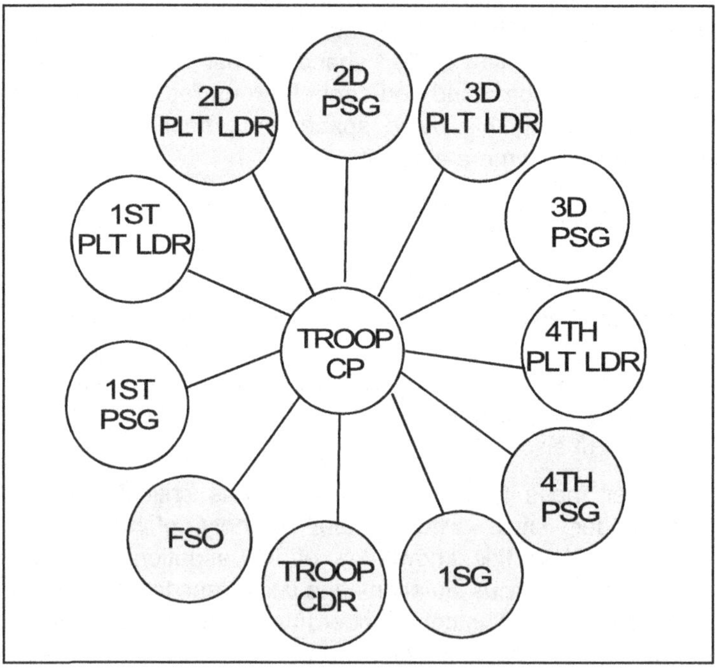

Figure 2-2. Troop command net.

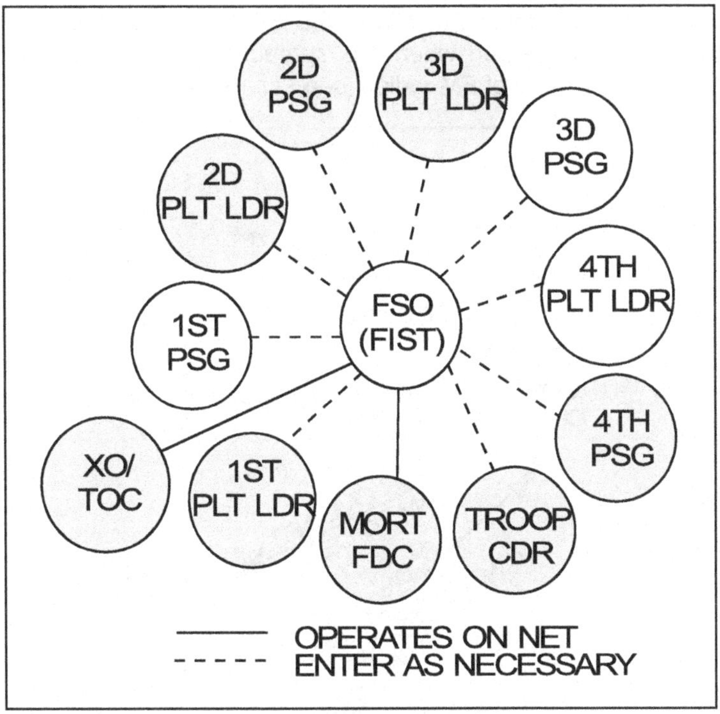

Figure 2-3. Troop fire support net.

In the troop CP, three additional radio nets are established to link the troop with the SCO and his headquarters. The squadron operations and intelligence (OI) net is used primarily to transfer enemy and friendly information by FM radio. All routine and recurring reports from the troop to squadron are transmitted on this net. Squadron headquarters uses this net to transmit information about current enemy and friendly situations within the squadron's AO. It is often used to issue FRAGOs. The troop commander and the XO are linked to the SCO by an FM squadron command net. This net is reserved strictly for use by the SCO, XO, S3, and CPs; and by subordinate troop, company, and battery commanders. In addition, an AM squadron command net allows the SCO to command and control the squadron when it is spread over distances that exceed the range of FM radios (see Figure 2-4).

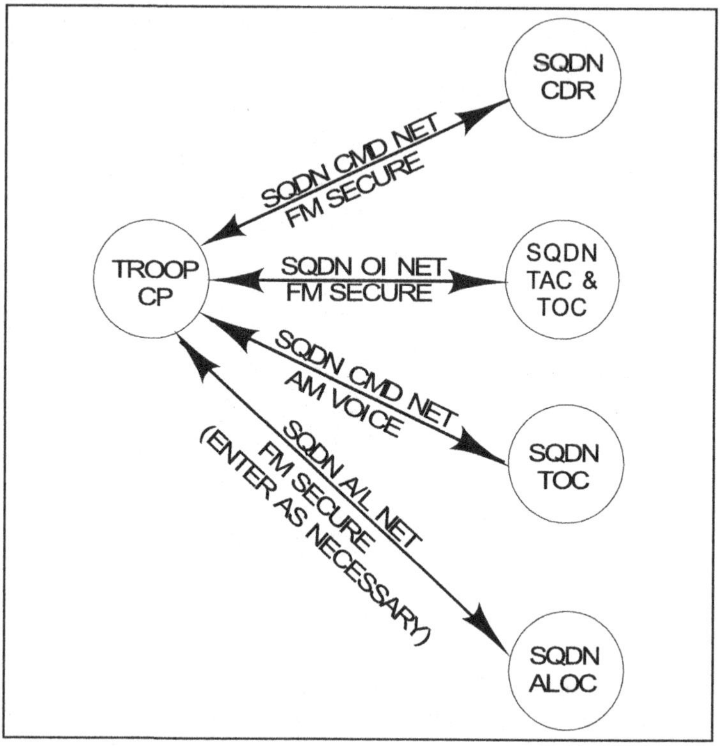

Figure 2-4. Troop command post external nets.

EAVESDROP

This tool is used by all cavalrymen. It is designed to give leaders time to think before they act. Most leaders have a hard time making decisions while they are talking on the radio. Decisions are usually better if the commander listens closely to what is said, and if he has uninterrupted time to consider information received. Eavesdropping is a method of communicating based on that principle. Everyone simply listens to the information sent from one leader to another, and unless necessary, stays off the radio net. In the troop, for example, reports from scout platoons are transmitted to the troop CP, where the XO reviews and records them. All others listen in and record the information themselves. Everyone stays off the net yet remains informed. If all or part of a transmission is missed, enter the net and quickly get the needed information from the troop CP.

FIXED CALL SIGNS

Fixed call signs are tools that establish immediate identity of subunits and leaders, thereby reducing the length of transmissions on the radio. These call signs eliminate the confusion often caused by the ever-changing signal operation instructions (SOI). Each platoon is referred to by a color, and leaders are referred to by numbers. These call signs do not change. An example of fixed call signs for a cavalry troop is outlined below.

Fixed Call Signs

Element	Call Sign
TROOP HEADQUARTERS	BLACK
FIST	Black 1
Medics	Black 2
TOC	Black 3
Supply	Black 4
XO	Black 5
Commander	Black 6
1SG	Black 7
Maintenance	Black 8
Mortars	Black 9
NBC	Black 10

1ST PLATOON (Scout)	RED
Plt Ldr 11	Red 1
Sec Ldr 12	Red 2
Sqd Ldr 13	Red 3
PSG 14	Red 4
Sec Ldr 15	Red 5
Sqd Ldr 16	Red 6
*Sec Ldr 17	Red 7
*Sqd Ldr 18	Red 8
*Sec Ldr 19	Red 9
*Sqd Ldr 10	Red 0
2D PLATOON (Tank or AT)	WHITE
Plt Ldr 21	White 1
TC 22	White 2
TC 23	White 3
PSG 24	White 4
3D PLATOON (Scout)	BLUE
Plt Ldr 31	Blue 1
Sec Ldr 32	Blue 2
Sqd Ldr 33	Blue 3
PSG 34	Blue 4
Sec Ldr 35	Blue 5
Sqd Ldr 36	Blue 6
*Sec Ldr 37	Blue 7
*Sqd Ldr 38	Blue 8
*Sec Ldr 39	Blue 9
*Sqd Ldr 30	Blue 0
4TH PLATOON (Tank or AT)	GREEN
Plt Ldr 41	Green 1
TC 42	Green 2
TC 43	Green 3
PSG 44	Green 4

*Light troop only

Section IV. Techniques of Tactical Control

CHAIN OF COMMAND

The troop commander cannot personally supervise the execution of all combat tasks he assigns. He must depend on a trusted team of subordinate leaders. The chain of command in the troop extends from the commander to the platoon leaders and section leaders. Within the platoons, it extends from platoon leaders and their platoon sergeants to section sergeants to vehicle/crew commanders. The troop XO and the first sergeant serve as the commander's principal assistants. They help coordinate the actions of subordinate units to ensure unity of effort. They also ensure the troop has what it needs to accomplish its assigned tasks.

ORDERS GROUP

The troop orders group consists of all subordinate leaders required to receive and execute the troop commander's orders. As a minimum, the orders group should include the XO, the first sergeant, the platoon leaders, the mortar section sergeant, the fire support officer (FSO), and the leaders of any units that are attached or under the troop's operational control (OPCON).

STANDING OPERATING PROCEDURES

A troop SOP is a written collection of standing orders. It establishes procedures for performing routine combat tasks. For example, the SOP tells subordinates what procedures to follow when the troop occupies an assembly area, breaches an obstacle, conducts a passage of lines, or reacts to air attack. It lists reports that are routinely sent to troop and squadron headquarters, and shows their correct formats. It also states procedures for communicating within the troop, obtaining indirect fire, or getting additional fuel or ammunition. The SOP is a valuable time saver for troop leaders. Most important, the SOP keeps everyone in step, reduces confusion, and provides the backbone of combat proficiency. Disciplined execution of the troop SOP is the hallmark of outstanding units.

TERRAIN INDEX REFERENCE SYSTEM

Terrain index reference system (TIRS) is used to maneuver the troop, rapidly report friendly locations without fear of compromise, and pass out control measures. TIRS is not used to report enemy locations.

Description

Each TIRS point is shown by a tick mark located on a grid line intersection. Each point is given a designator of one letter and two numbers, such as X56, placed in the upper right quadrant of the tick mark. TIRS point designation is SOP, and units determine which letters they will use. They may designate specific letters for specific unit sectors or AO. The TIRS point is then recorded as X56, located at PA2450 (four-digit grid).

TIRS point-to-point references designate kilometers. For example, 500 meters is given as "POINT FIVE," 1,000 meters as "ONE," and 3,500 meters as "THREE POINT FIVE."

For shifts from the TIRS point, use cardinal directions instead of left, right, up, or down. Using TIRS, "FROM X-RAY FIVE SIX—EAST ONE POINT SEVEN—NORTH POINT SEVEN" translates to "From the tick mark for TIRS point X56, shift east 1,700 meters and north 700 meters." When each TIRS point is placed on a four-digit grid intersection, the use of shifts makes the TIRS as accurate as the six-digit grid system. For an example of a TIRS point in use, refer to Figure 2-5.

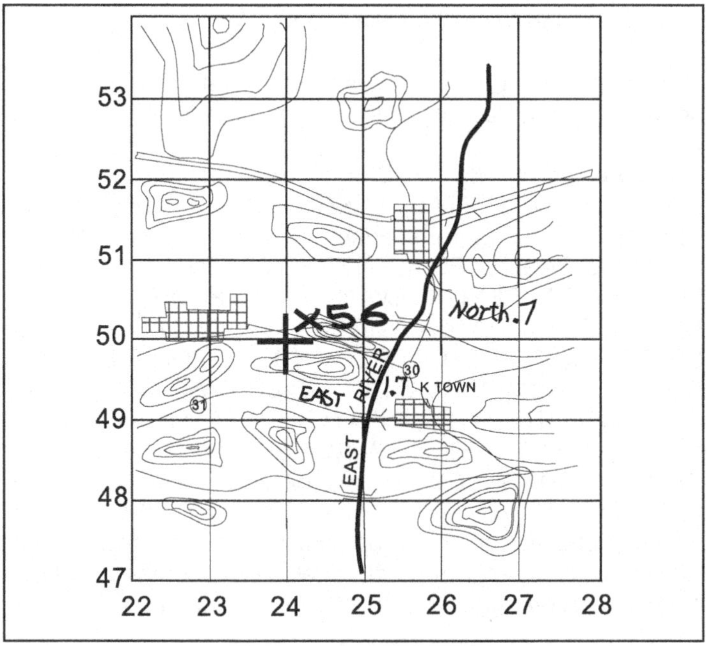

Figure 2-5. Placing TIRS on the map.

Application

The squadron normally issues the TIRS to be used for the operation as early as possible, perhaps with the warning order. The TIRS list should be issued to elements as an annex to a written OPORD.

The squadron should designate four to six TIRS points in each 10-kilometer square. The squadron TIRS should normally be sufficient for the troop to operate; however, if the troop designates additional TIRS, it should always ensure only squadron TIRS are referenced in its communications with higher headquarters.

TIRS should be used routinely to control combat operations. Use them—

- To identify BPs and to pass out control measures (such as LDs, PLs, and boundaries) quickly.
- To report friendly unit locations.

Passing control measures, such as BPs, sector boundaries, and PLs, are quick and accurate using TIRS.

Figure 2-6 shows BP B22. A troop commander could report "TROOP ESTABLISHED VICINITY ALFA ZERO SEVEN." To be more precise, he could report "TROOP BRAVO ESTABLISHED ALFA ZERO SEVEN WITH RED AT FROM ALFA ZERO SEVEN EAST POINT THREE—NORTH POINT FIVE—WHITE AT FROM ALFA ZERO SEVEN WEST POINT THREE—NORTH POINT EIGHT—BLUE AT FROM ALFA ZERO SEVEN EAST ONE POINT NINE—NORTH ONE POINT NINE—TO FROM ALFA ZERO SEVEN WEST POINT NINE—NORTH ONE POINT SEVEN—AND GREEN AT FROM ALFA ZERO SEVEN EAST ONE POINT ZERO—NORTH POINT FIVE." In the second transmission, the troop commander gives the precise location of the center of mass for all his platoons: red, white, blue, and green represent the 1st, 2d, 3d, and 4th platoons.

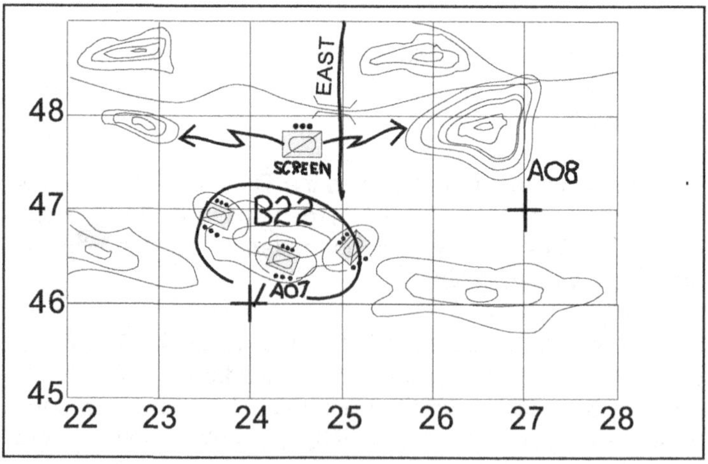

Figure 2-6. Reporting location using TIRS.

Using this transmission, the SCO orders Troop B to occupy a BP that is not on the troop commander's map: "BRAVO SIX SIX OCCUPY A BP EXTENDING FROM ALFA ZERO EIGHT WEST POINT SEVEN—SOUTH ONE POINT ZERO—TO FROM ALFA ZERO EIGHT EAST POINT SIX—SOUTH POINT FIVE—ORIENT NORTHEAST."

Figure 2-7 shows how the SCO could quickly establish new phase lines and boundaries between troops.

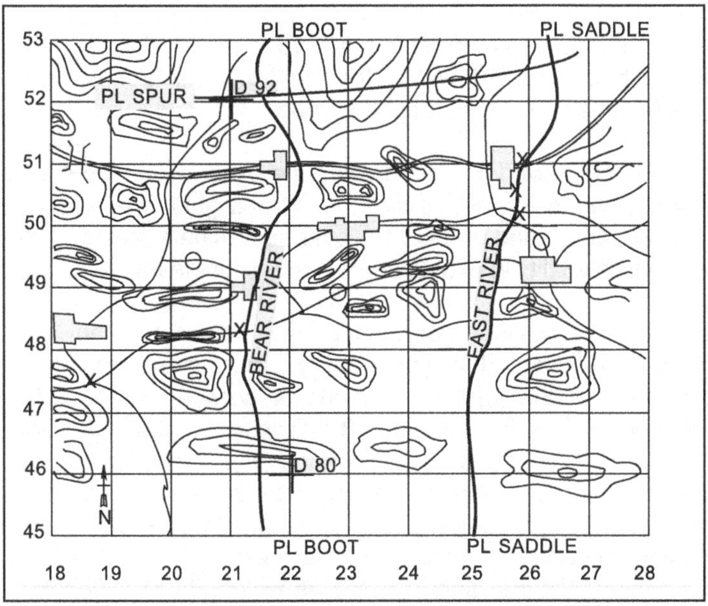

Figure 2-7. Establish graphic control measures using TIRS.

The SCO could transmit a new boundary between troops B and C by radio: "BRAVO SIX SIX AND CHARLIE SIX SIX BOUNDARY IS FROM DELTA EIGHT ZERO EAST POINT NINE—NORTH POINT EIGHT—TO FROM DELTA NINE TWO EAST TWO—SOUTH THREE—TO FROM DELTA NINE TWO EAST ONE POINT NINE—SOUTH POINT ONE."

PL SPUR could be sent as "PL SPUR IS FROM DELTA NINE TWO WEST POINT NINE—TO FROM DELTA NINE TWO EAST FIVE POINT THREE—NORTH POINT SEVEN."

Section V. Command Guidance and Organizational Control

TROOP COMMANDER

The troop commander is mounted in an M1A1 tank in the heavy troop or a HMMWV in the light troop. His first priority is the command and control of the troop.

The commander must provide the troop with clear intent and guidance for the mission. Intent and guidance facilitate positive command and control over subordinate platoons yet provide flexibility in execution for subordinate leaders.

The commander positions himself where he can best command and control the troop. He may maneuver along with one of the tank/AT platoons or between the tank/AT platoons and scout platoons. It is imperative that the troop commander not become decisively engaged with fighting his vehicle. He must retain the ability to control the actions of his troop. The troop commander operates on both troop and squadron command radio nets. He must have the capability to communicate to subordinates as well as to the squadron commander.

TROOP TOC

The purpose of the troop tactical operations center (TOC) is to report information to the higher headquarters. The troop TOC is the critical link between squadron and troop. One of the primary purposes of the cavalry troop is to report, and the troop TOC is central to that function. The troop TOC operates under the direction of the troop XO who is responsible for assisting the commander in the command

and control of the troop. The troop TOC takes reports from subordinate units in the troop, processes the information, and reports the information to the higher headquarters. The TOC tracks the battle at the troop and squadron level and relays information to the commander and subordinate platoons pertaining to the friendly/enemy situation. The TOC is the net control station for the troop command net and operates on the squadron command and OI net. The troop TOC also monitors and operates as necessary on the squadron A/L and the troop fire direction net.

In the heavy troop the TOC vehicle is an M577. An M3 CFV moves with the heavy troop TOC to provide local security for the M577 and to serve as a command and control vehicle for the XO if the commander is out of action. The XO should not conduct TOC operations out of the CFV because his experience is needed in the TOC vehicle to handle the information flow between troop and squadron over the OI and command nets. If the commander is out of action, the XO should avoid, if possible, rushing forward to assume command of the troop because of the importance of the activities in the TOC. The XO should assume command of the troop from the TOC until the battle is over. However, if the XO does need to move forward to command and control the troop, a qualified replacement must be placed in the TOC to assume his duties. This individual may be the FIST chief, one of the platoon leaders, or the first sergeant. The correct replacement for the XO in the TOC should be based on the depth available at any one of the above mentioned positions.

The light troop TOC is a cargo HMMWV with a shelter.

The primary concern when positioning the troop TOC is based on the TOC's ability to communicate with the squadron and the subordinate elements of the troop. The TOC position should also be an area that provides good cover and concealment from enemy and limited access to civilian population. During reconnaissance or offensive operations, maneuver the TOC at least one terrain feature behind troop combat elements. During security or defensive operations, the TOC should be positioned in sufficient depth to avoid contact with the enemy yet maintain communications with the forward scout platoons. The TOC and the troop trains may collocate for increased local security.

FIST

The FIST is the critical link with the supporting artillery and responsible for coordinating indirect fires (both mortar and FA) for the troop. The team processes calls for fire from the platoons and allocates the appropriate indirect-fire system based on the commander's guidance for fire support. The FIST can also assist the squadron ALO with the employment of close air support.

The FIST operates on three radio nets: troop command, troop fire direction, and squadron FSE digital/voice. The FIST monitors at least one of the following nets: squadron command, squadron OI, and howitzer battery (supporting artillery headquarters in the heavy and light division).

The FIST vehicle also may serve as the alternate troop TOC. The FSO has ready access to the higher level situation and the radio systems to replicate the troop TOC if it becomes damaged or destroyed.

Command guidance to the FIST should include the following:

- **Purpose of indirect fires.** How does the commander intend to use field artillery and mortar fires to support his maneuver?
  - Screening.
  - Suppression.
  - Development of the situation.
  - Disengagement.

- **Engagement/attack criteria.** How many rounds and of what type and mix will be fired at a particular target? Which targets will be engaged with artillery and which with mortars?

- **Control of troop mortars.** If the FIST controls movement of troop mortars, how far forward of the scouts will the mortars be able to range? Where are the mortars going to move? When are the mortars going to move?

- **FIST movement.** The primary considerations when positioning the FIST are security of the team and the ability to communicate with the squadron FSE, howitzer battery, or the DS artillery. The FIST is not the forward observer team for the troop; the troop has 19Ds who act as forward observers. The five techniques to maneuvering the FIST are—

  - Maneuvers with the commander.
  - Maneuvers with or near the mortar section.
  - Maneuvers with the TOC.
  - Maneuvers alone to maintain communications.
  - Maneuvers with the scouts or tanks to directly control fires or to use the ground laser designator (GLD).

Note. See Chapter 8 for a more in-depth discussion of troop fire support command and control techniques.

TROOP MORTARS

The troop mortars provide responsive indirect-fire support to the troop. In the light troop they are equipped with three cargo HMMWVs and two 120-mm towed mortars. The heavy troop is equipped with two M106 mortar carriers with 107-mm mortars. The mortars operate on the troop command and fire direction nets.

Command guidance for the mortars should include the following:

- **Purpose of mortar fires.** How does the commander intend to use mortar fires to support his maneuver?

  - Screening.
  - Suppression.
  - Development of the situation.
  - Disengagement.

- **Engagement/attack criteria.** How many rounds and of what type and mix will be fired at a particular target?

- **Positioning.** How far forward of the scouts will the mortars be able to range?

- **Movement.** Where will the mortars move and what will the mortars orient their movement on, for example, checkpoints, a route, mortar firing positions? Who will control the movement of mortars—the mortar section sergeant, the FIST, or the troop XO?

SCOUT PLATOONS

The scout platoons, HMMWV or Bradley equipped, perform reconnaissance and security missions for the troop. They may also perform missions within an offensive or defensive role in certain METT-T situations. To enhance command and control and to provide scout platoons the freedom of action required in cavalry operations, the troop commander should give the following guidance to scouts, regardless of what missions they are performing.

- Engagement, disengagement, destruction, and bypass criteria.

- Tempo of the operation. Tempo is dictated by the higher commander's intent for the operation, and is a function of the enemy situation as it relates to the survivability and firepower of the systems conducting the operations (METT-T). Tempo for the scout platoon determines the rate at which they conduct their mission and develop the situation. High tempo operations, for example, could be characterized by very aggressive reconnaissance focusing on mounted avenues of approach and likely enemy defensive positions. Engagement criteria in high tempo operations would be unrestricted, allowing the scouts to rapidly develop the situation by fire and maneuver. Low tempo operations could be characterized by scouts conducting a very detailed reconnaissance and maximizing the use of stealth to develop the situation.

The scout platoon leader and platoon sergeant operate on both the troop command and platoon command nets. As required, they will operate on the troop fire direction net. For platoon organization for combat, see FM 17-98, Chapter 2, Section VI.

TANK/AT PLATOONS

The tank/AT platoons are the principal killers in the troop. They assist the troop commander in the rapid development of the situation during operations by fire and maneuver. The troop commander should give the platoons the following guidance to facilitate their command and control during battle.

- Location in the troop formation and distance behind the scout platoon.

- Engagement, disengagement, and destruction criteria. When and where will they be employed and against what size force? Will they be used to suppress in support of another maneuver force, or destroy by fire and maneuver?

- Tempo of the operation. (Same as for the scout platoons above.)

The platoon leader and platoon sergeant operate on troop command and platoon command nets. As required, they will operate on the troop fire direction net.

TROOP TRAINS

The troop trains provide logistical support and casualty evacuation to the troop. The trains are under the direction of the first sergeant and consist of the troop combat and field trains. The first sergeant monitors the troop command and squadron A/L nets. He directs resupply, treatment and casualty evacuation, and maintenance/recovery operations for the troop.

The troop trains may be consolidated or echeloned. Either way, the trains should move one terrain feature behind the trail combat elements of the troop. If echeloned, the field trains move with the squadron field trains under the direction of the squadron HHT commander. The troop combat trains consist of the first sergeant in his vehicle, a maintenance contact team with recovery vehicle, and a medical evacuation vehicle. The field trains are made up of the supply sergeant and the remaining members of the maintenance team.

Command guidance to the troop trains should include the following:

- Trains organization. Should they move consolidated or echeloned?

- Movement guidance. Where do the trains move in the troop formation and what is their orientation for movement? Use checkpoints, routes, or trains locations to orient the movement of the trains.

Note. See Chapter 9 for a more in-depth discussion of the TTP for combat service support.

Section VI. Tactical Movement Formations

During tactical operations, the troop normally moves in one of four movement formations: troop column, troop line, troop vee, or troop split-vee.

TROOP COLUMN

Use the troop column when moving on a designated route, when speed is essential, and when contact with the enemy is not expected. This formation moves the troop quickly and efficiently from one place to another. Control of the troop while moving in column is eased, but the troop is vulnerable to enemy air or ground attack.

The troop usually moves at a designated speed and with a set distance between vehicles when moving in column formation. The march speed and distance between vehicles are designated by the troop SOP, based on the situation.

Note. See Chapter 7 for a discussion of column formation.

TROOP LINE

The troop line formation can be used when conducting defensive or retrograde operations in the heavy and light troops (see Figure 2-8). The formation may also be used by the light troop during reconnaissance operations when the troop is scrambled.

Figure 2-8. Troop line formation.

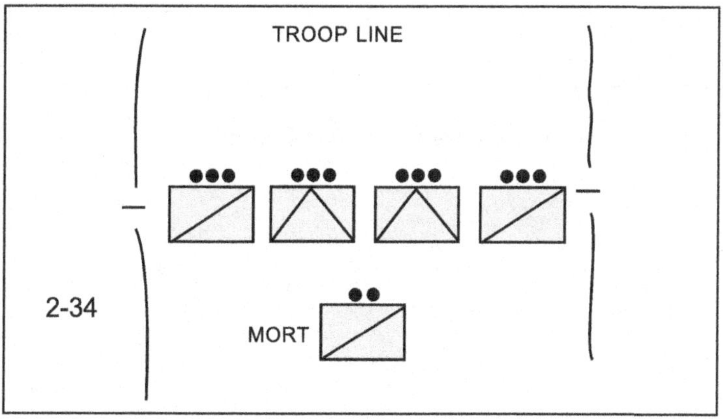

TROOP VEE

The troop vee formation is used to conduct reconnaissance or security missions (see Figure 2-9). The purpose of the vee formation is to keep maximum reconnaissance forward; the two tank or AT platoons are positioned in depth, ready to attack or defend any where in the troop zone/sector. This formation allows the troop to mass combat power at the decisive point (see Figure 2-10).

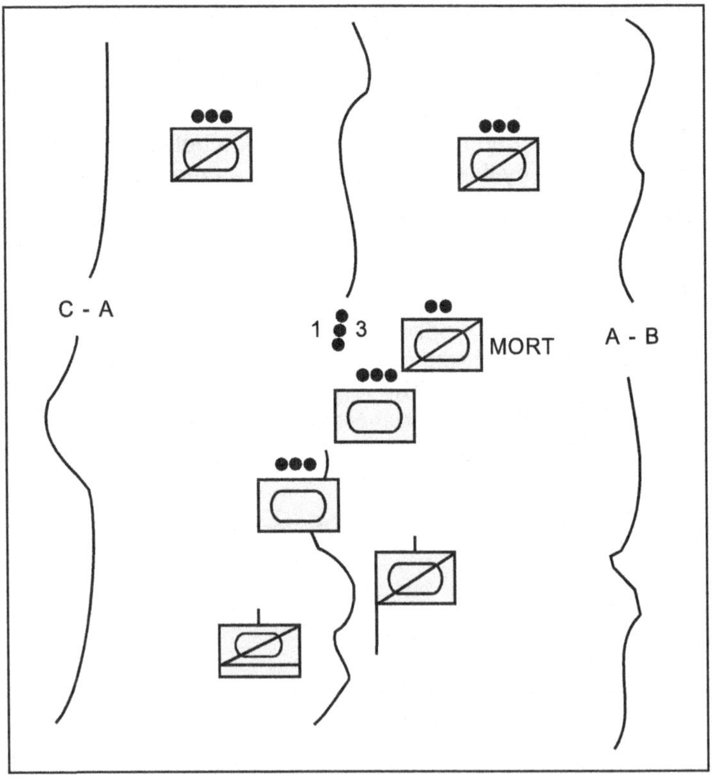

Figure 2-9. Troop vee formation.

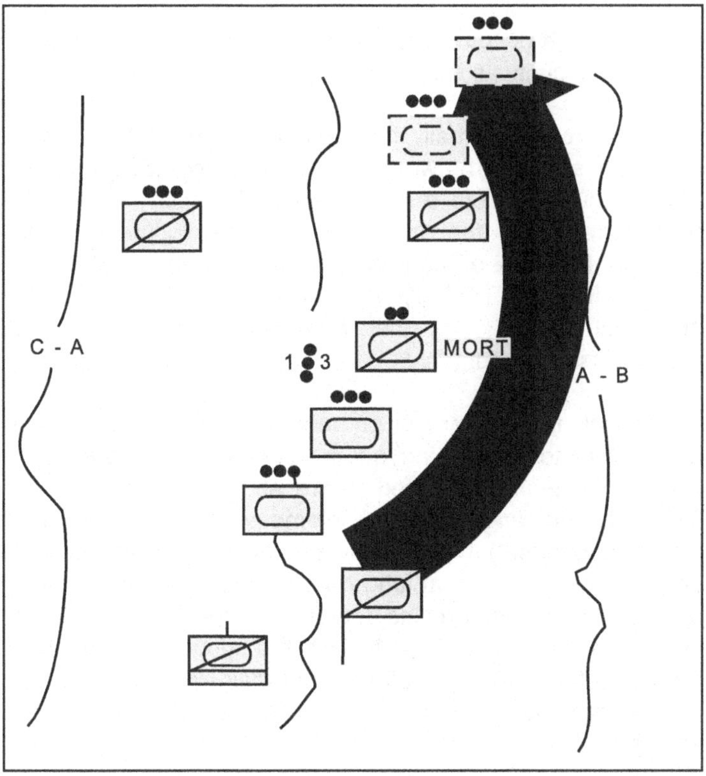

Figure 2-10. Troop commander maneuvers his tank platoons (troop vee).

In the troop vee formation, the two scout platoons work abreast in their respective zone/sector. The two tank/AT platoons are collocated in depth of the scouts, and they maneuver in the troop zone/sector.

Heavy Troop. Normally the distance between the scouts and tanks varies from 1 kilometer to 4 kilometers. The tanks should be close enough to respond quickly, but far enough back to respond to enemy contact based on the commander's guidance for engagement. Generally, do not position the tank platoons to overwatch the scout platoons. When the scouts make contact with the enemy, the tanks should be far enough in depth to maintain their ability to maneuver. The scouts are usually responsible for providing their own security as they move.

Light Troop. Normally the distance between the scouts and TOWs varies between 500 meters to 1,500 meters. TOW vehicles should respond quickly to enemy contact in order to assist in the scout platoon's development of the situation. Because of the reduced fire power in the HMMWV scout platoon, the TOW systems may provide the scouts with additional security in the form of overwatch during reconnaissance operations.

Position the mortars in or near the center of the troop zone/sector to provide indirect-fire support across the troop front. Keep them positioned to fire about one third to two-thirds of their range (about 3 kilometers to 3.5 kilometers, terrain dependent) beyond the scouts, so the scouts can engage enemy forces at long range with indirect fires. (Line of sight is the determining factor in range forward of the scouts. In densely wooded terrain, the scout's visibility may be only 100 to 200 meters forward of their front line trace, so mortar range need not always be 3,000 meters forward of the scouts.)

TROOP SPLIT-VEE

The troop split-vee is a variation of the troop vee, and can be used in reconnaissance and security missions (see Figure 2-11). In the split-vee formation, the scout platoons work abreast and forward of the troop. The two tank/AT platoons work in depth of the scout platoons, but each tank/AT platoon initially follows a single scout platoon in its zone/sector. The scout platoons conduct their reconnaissance or screening mission within their zone/sector, and the tank/AT platoons key their movement on the progress of the scout platoons. Give the tank/AT platoon leaders command guidance on where they should move in relation to the scout platoons. Tank/AT platoon leaders eavesdrop on the troop command net to monitor the progress of the scouts and to move their platoons to maintain their orientation with the scouts based on the troop commander's guidance.

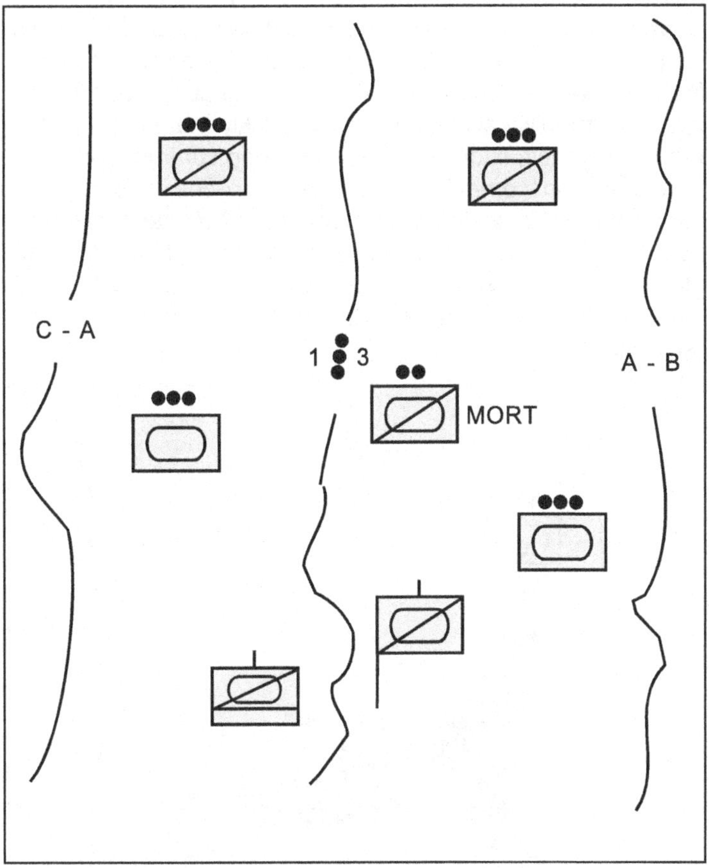

Figure 2-11. Troop split-vee formation.

Develop a habitual relationship between a scout platoon and a tank/AT platoon. First and second platoons work together, and third and fourth platoons work together. This builds teamwork between the scout and tank/AT units, and gives them confidence in each other.

All four platoons still work for the troop commander in this formation. The purposes of having a tank/AT platoon work with a scout platoon are to have combat power spread across the entire troop front, to keep the tanks/TOWs in a position to respond rapidly in support of a scout platoon, and to have scouts available to reconnoiter routes and positions for use by the tanks/TOWs. The troop split-vee is often used in close or restrictive terrain that reduces the tank/AT platoon's mobility so that it cannot quickly maneuver across the troop zone/sector. Generally the troop commander is the only one who commits the tank/AT platoons to conduct a hasty attack, to counterattack, or to move into overwatch positions. If he must mass the combat power of both tank/AT platoons, the troop commander can still shift the tank/AT platoons across the troop zone/sector during the mission (see Figure 2-12).

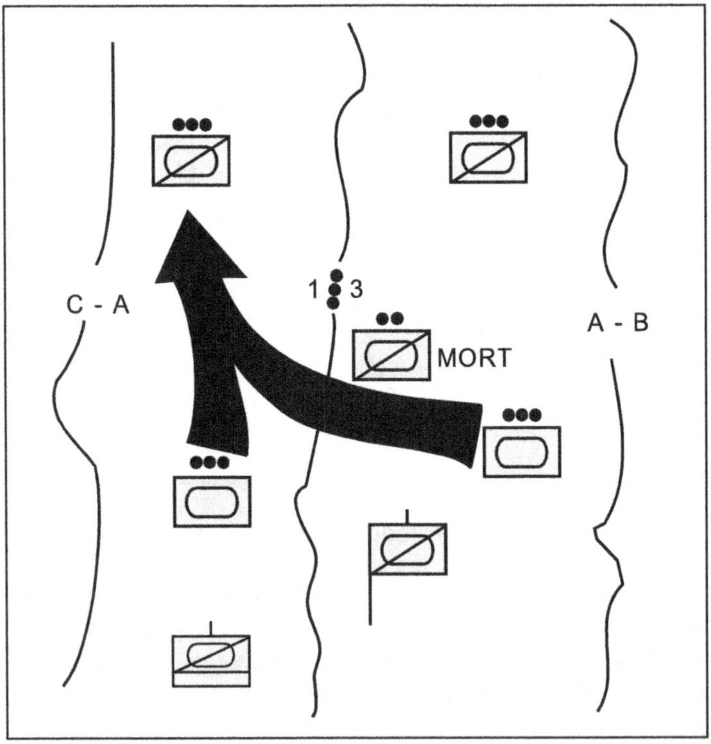

Figure 2-12. Troop commander maneuvers his tank platoons (troop split-vee).

LIGHT TROOP SCRAMBLE

The light troop scramble is a task organization that allows the troop commander to use a mix of weapon systems to enhance acquisition and destruction of the enemy under reduced visibility conditions and in restrictive terrain, where massed TOW fires are not possible. This task organization can also be used to respond to multiple but separate taskings of a troop usually associated with operations other than war, for example, manning checkpoints and area security missions. The troop commander may maneuver his troop in a vee, split-vee, column, or line formation when task organized into a scramble.

The troop is task organized into four like platoons of seven vehicles each. Five scouts and two AT vehicles are in each platoon (see Figure 2-13). The platoon may task organize and operate two ways:

- TOWs integrated with the scout sections (see Figure 2-14).
- TOWs maneuvering as a section to overwatch the movement of the scouts (see Figure 2-15).

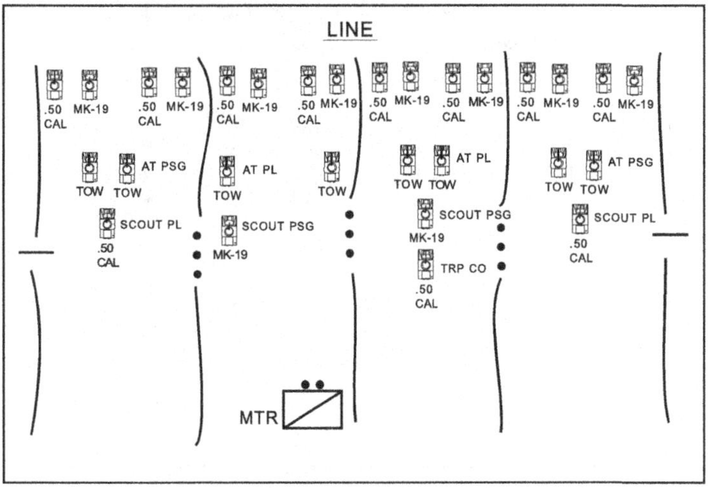

Figure 2-13. Light troop scramble.

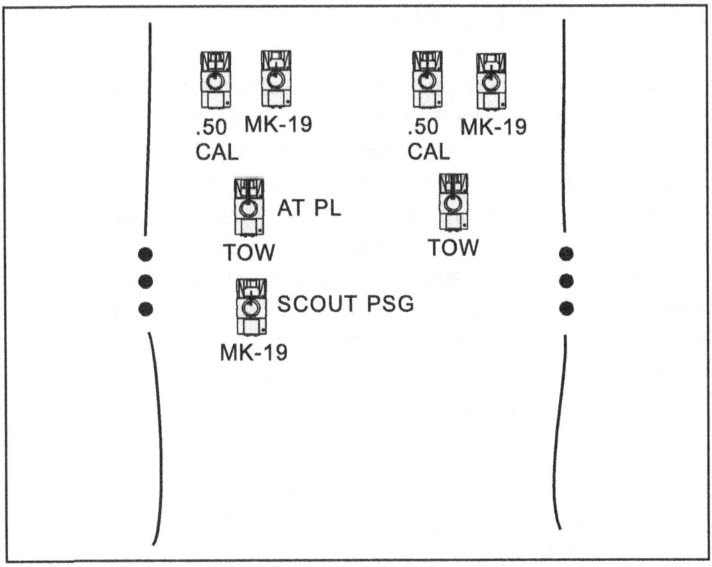

Figure 2-14. TOWs integrated with scout section.

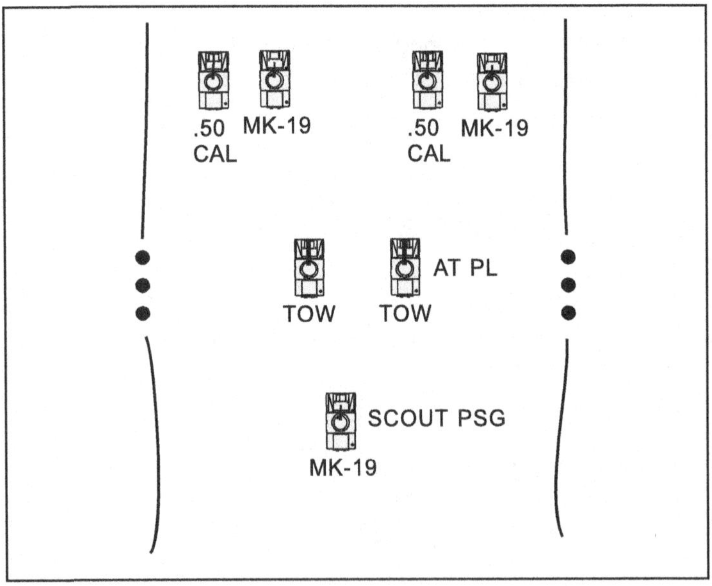

Figure 2-15. TOWs maneuvering as a section
to provide overwatch.

Chapter 3

Reconnaissance

Reconnaissance is a mission undertaken to obtain information about the activities and resources of an enemy or about the meteorologic, hydrographic, or geographic characteristics of a particular area. Reconnaissance produces combat information. Combat information is a by-product of all operations acquired as they are in progress. Reconnaissance, however, is a focused collection effort. It is performed before and during other combat operations to provide information used by the commander to confirm or modify his concept.

Cavalry is the corps or division commander's principal reconnaissance organization. Cavalry troops perform reconnaissance using a combination of mounted and dismounted techniques. The cavalry troop is uniquely organized, trained, and equipped to perform the crucial task of reconnaissance for other combined arms forces. Reconnaissance is the cavalryman's specialty.

CONTENTS

Page

Section I. Purpose, Fundamentals, and Capabilities

PURPOSE

Cavalry troops conduct reconnaissance forward of another friendly force to provide information about the terrain and enemy within the area. The reconnaissance allows the follow-on force an opportunity to maneuver freely and rapidly to its objective. Reconnaissance keeps the follow-on force from being surprised or interrupted, and protects it against losing soldiers and equipment on the way to the objective. The cavalry troop performs three types of reconnaissance—route, zone, and area.

FUNDAMENTALS

Successful reconnaissance operations are planned and performed with six fundamentals in mind.

- Maximum reconnaissance force forward.
- Orient on the location or movement of the reconnaissance objective.
- Report all information rapidly and accurately.
- Retain freedom to maneuver.
- Gain and maintain enemy contact.
- Develop the situation rapidly.

Maximum Reconnaissance Force Forward. In reconnaissance, every pair of eyes makes a difference. Do not keep scouts in reserve. This does not mean that every scout should be forward in a strictly linear sense, but actively employed in the conduct of the reconnaissance.

Orient on the Location or Movement of the Reconnaissance Objective. From the IPB (intelligence preparation of the battlefield) the S2 will identify gaps in the squadron's knowledge of the enemy and terrain it will be operating in. These gaps together with the commander's guidance form the PIR (priority intelligence requirements) and direct the reconnaissance efforts of the troop. Focus the efforts of the troop in selecting a course of action and a scheme of maneuver that continuously orients the troop on its reconnaissance objective based on the commander's guidance. The troop's focus may be a terrain feature, a specific area, or an enemy force.

Report All Information Rapidly and Accurately. Higher commanders base their decisions and plans on the battlefield information cavalry troops gather during reconnaissance. Combat information loses value as it ages, so the fresher the better. Scouts report exactly what they see. Troop commanders and executive officers may summarize or clarify reports to the squadron, but must never delete raw information reported by the scouts. Information that seems unimportant to the troop may be extremely valuable at higher echelons.

Retain Freedom to Maneuver. Cavalry troops must be able to maneuver on the battlefield to continue the reconnaissance mission. If the troop becomes decisively engaged, the reconnaissance stops. Use of proper movement techniques and overwatch helps prevent decisive engagement. IPB provides the troop commander information that allows him to anticipate battlefield events and to retain freedom of maneuver.

Gain and Maintain Enemy Contact. Contact is any condition ranging from a surveillance sighting to engaging in close combat. Surveillance is often sufficient and is the preferred method of maintaining contact. When necessary or required, troops will use fire and maneuver to maintain contact with an enemy force. Once contact is gained, however, it is not lost unless ordered by higher headquarters.

Develop the Situation Rapidly. During reconnaissance operations cavalry troops frequently and repeatedly encounter situations that require action to determine what the troop must face. These situations may be terrain oriented, obstacles, or enemy. Terrain or obstacle situations require close reconnaissance, bypass, hasty breach, if necessary, and marking. If an enemy force is encountered, the troop will determine the size, composition (What is the enemy force made up of—tanks, personnel carriers?), disposition (Is the enemy force in an offensive or defensive posture? Is he dug in? What is his orientation?), and activity. IPB provides the threat situational information that guides the effort. Reconnaissance techniques, often in the form of drills, are used while developing the situation.

Troops must develop the situation rapidly to get inside the enemy's decision cycle and force him to react to the actions of the troop rather than the troop reacting to the enemy. This requires the troop to rapidly execute development of the situation at the platoon level through the execution of rehearsed battle drills.

CAPABILITIES

The ability of a cavalry troop to conduct reconnaissance is a function of the enemy situation, the terrain the troop is operating in, and the type of troop conducting the reconnaissance.

- Heavy Troop
  - Can reconnoiter up to a 10-kilometer wide zone.
  - Can reconnoiter up to two routes simultaneously.
  - Can conduct reconnaissance at the rate of about 1 kilometer per hour, depending on the terrain.
  - When faced with a heavy-equipped threat, will conduct either aggressive or stealthy reconnaissance, depending on the higher commander's guidance.
- Light Troop
  - Can reconnoiter up to a 10-kilometer wide zone.
  - Can reconnoiter up to two routes simultaneously.
  - Can conduct reconnaissance at the rate of about 1 kilometer per hour, depending on the terrain.

- When faced with a heavy-equipped threat, will conduct stealthy reconnaissance and aggressive reconnaissance when reinforced with heavy forces.
- When faced with a light-equipped threat, will conduct either aggressive or stealthy reconnaissance, dependent on the commander's guidance.

Section II. Planning Considerations, Methods, and Techniques

The purpose of this section is to outline the planning considerations, methods, and procedures a cavalry troop uses to execute reconnaissance missions.

PLANNING CONSIDERATIONS

When planning a reconnaissance mission, the troop commander must take into account his unit's capabilities and limitations and consider the following:

- Time available from mission receipt to completion.
- Threat size, composition, disposition, and will to fight.
- Terrain and weather effects on the troop's ability to maneuver.
- Tempo of the operation.
- The squadron commander's (SCO) intent and guidance, as follows:
  - What are the SCO's focus and desired endstate of the reconnaissance?
  - What triggers the squadron's employment of the tank/AT company?
  - What does the SCO want the troop to destroy, fix, and bypass?
- Task organization or reinforcements.
- Critical tasks to be accomplished by the troop. Specifically, identify which critical tasks may be deleted during the reconnaissance.

Based on the considerations above, the troop commander determines the following:

- What is the focus of the reconnaissance? What critical tasks must be accomplished within the constraints of time and terrain?
- What specified or implied missions are associated with the squadron endstate?

Note. If given a limit of advance (LOA) for the reconnaissance mission, the troop commander should plan to screen along the LOA (see Chapter 4).

- How will the troop deal with enemy contact? What are the troop's criteria for engagement, destruction, and bypass?
- Under what situations does the commander see employment of his tank (heavy troop) or antitank (AT) (light troop) platoons?
- How will the commander use indirect fires from mortars and artillery to support his maneuver?
- Who controls the troop's attachments and how are they integrated into the reconnaissance?

The troop may receive attachments from higher headquarters. These assets may be maintained under troop control or tasked down to platoons for their use in the execution of the platoon's specified tasks. Examples include attached engineers, ground surveillance radar (GSR), or a chemical reconnaissance element.

- **Engineers.** If an engineer platoon is attached to the troop, the commander may elect to keep them under his control and treat them as a maneuver platoon. He may assign them to conduct the reconnaissance of the route, while the scout platoons move just ahead and reconnoiter terrain on either side. If the troop receives a squad or section, the troop commander may elect to task it to the scout platoon conducting the reconnaissance of the route.

- **GSR**. If a GSR squad or section is attached, the troop commander may elect to task organize them to a platoon or keep them under troop control. During reconnaissance operations, GSRs may be focused on flank avenues of approach into the troop zone with their movement controlled by the scout platoon leader or commander. In either case the commander should give clear guidance for positioning and orientation of the reconnaissance.
- **Chemical reconnaissance element.** If a chemical reconnaissance element (squad, section, or platoon) is task organized to the troop, the commander may maintain the element in reserve to reinforce a scout platoon that comes in contact with a contaminated area, or he may task organize the element down to one of the scout platoons.

METHODS OF RECONNAISSANCE

There are three methods of conducting reconnaissance at the cavalry troop level: *dismounted, mounted, and reconnaissance by fire.* The troop commander may use any method or combination of methods to accomplish the reconnaissance mission under the restrictions placed on him by METT-T (mission, enemy, terrain, troops, and time available) and the higher commander's intent and guidance. A fourth method of reconnaissance, *aerial reconnaissance*, may be executed by an air cavalry troop conducting a coordinated reconnaissance forward of the ground troop.

Dismounted Reconnaissance

The troop commander may direct scouts to conduct dismounted reconnaissance when—

- Time is not a limiting factor.
- Detailed information is required.
- Stealth is required.
- Enemy contact is expected or has been achieved through visual means.

- Scout vehicles cannot move through an area because of terrain.
- Security is the primary concern.

Dismounted reconnaissance permits the cavalry troop to collect the most detailed information about the terrain and enemy within a given zone, area, or route. However, dismounted reconnaissance is also the most time-consuming of all the reconnaissance methods. The cavalry troop, heavy or light, is limited in the number of dismounted scouts it can employ at any time.

Mounted Reconnaissance

The troop commander directs scouts to conduct mounted reconnaissance when—

- Time is limited.
- Detailed reconnaissance is not required, or mounted method affords the same opportunity to collect information as the dismounted method.
- Enemy locations are known.
- Enemy contact is not likely.
- An air cavalry troop is conducting a coordinated air reconnaissance.

Reconnaissance by Fire

When conducting reconnaissance by fire, the troop places direct and/or indirect fire on positions the enemy is suspected of occupying. This action causes the enemy to disclose his presence by moving or by returning fire. The troop commander may use reconnaissance by fire when—

- Time is critical.
- Natural or man-made obstacles that could be overwatched by an enemy force are encountered.
- A suspected enemy position fits the situational template.
- Bunker complexes that may or may not be occupied are encountered.
- Enemy locations are known.

The disadvantage of the reconnaissance-by-fire method is that the troop will lose any element of surprise it may have had. However, reconnaissance by fire may reduce the chance of some portion of the troop being caught in an enemy kill zone. Reconnaissance by fire may not always provide the desired effect. A well-disciplined force will resist the inclination to move when probed by weapon fires.

When indirect-fire situations exist, the troop commander ensures scouts are in a position to observe the target area. Once the decision is made to use reconnaissance by fire, weapons should be used in the following priority.

- Indirect-fire systems.
- Machine gun.
- 25-mm chain gun or MK-19.
- TOW or tank cannon fire.

Reconnaissance by fire does not mean the indiscriminate use of direct and indirect fires at all woodlines and hilltops in the hopes of causing the enemy to react. Not only will the enemy recognize this ploy for what it is and not react to it, but also it wastes valuable ammunition.

Aerial Reconnaissance

Air cavalry troops perform aerial reconnaissance when time is critical. Aerial reconnaissance is often coordinated closely with a ground reconnaissance troop. The air and ground forces complement each other. The air element can move forward of the ground unit and reconnoiter key pieces of terrain or restrictive terrain, allowing the ground troop to concentrate its efforts in other areas or to increase the tempo of its reconnaissance. The air troop provides the ground troop with added security by clearing the ground forward of the ground unit, thereby facilitating movement of the ground force and quickening the pace of the operation. The ground troop can move rapidly mounted to the areas of interest within its area or zone of operations and have the time to dismount and collect detailed information (see Figure 3-1).

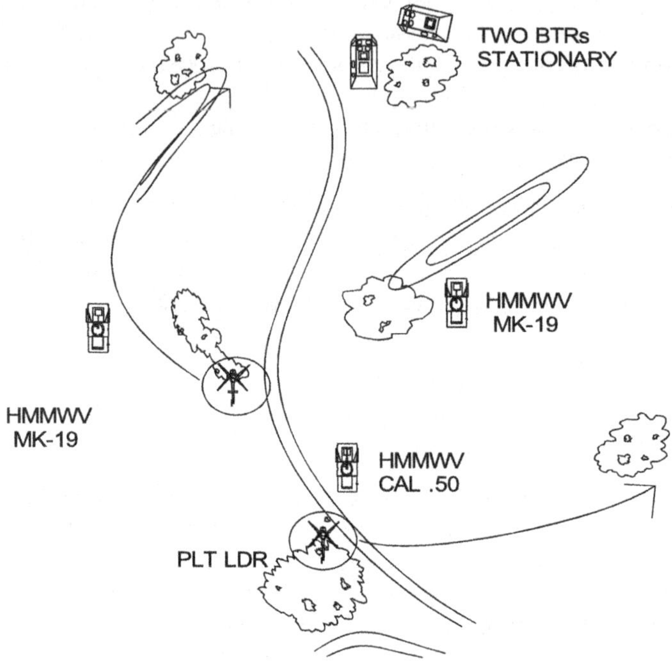

- Air cavalry platoon identifies two stationary BTRs along the woodline.
- Air cavalry platoon leader sends spot report to air cavalry troop commander. Troop commander tells platoon leader to develop situation and get in contact with ground platoon leader in zone.
- Air cavalry platoon leader sends information to ground platoon leader and assists ground scout section in developing situation.
- Air cavalry troop commander and ground troop commander choose a course of action.
- This coordination of air-ground supports squadron commander's intent, which increases troop survivability and squadron operating tempo of reconnaissance.

Figure 3-1. Air scouts conduct coordinated reconnaissance of zone with ground troop.

Aggressive Versus Stealthy Reconnaissance

The cavalry troop uses various combinations of reconnaissance methods. The method or combination of methods chosen by the troop and its scout platoons is based on the higher commander's guidance and METT-T. The method or methods chosen will characterize the reconnaissance as either aggressive or stealthy.

The stealthy approach to reconnaissance is more time-consuming. It emphasizes avoiding detection and engagement by the enemy. To be effective, a stealthy approach must rely primarily on dismounted reconnaissance and maximum use of covered and concealed terrain.

Aggressive reconnaissance is characterized by the speed and manner in which the reconnaissance force develops the situation once contact is made with an enemy force. A troop conducting aggressive reconnaissance uses firepower from direct- or indirect-fire systems and maneuvers to rapidly develop the situation. The troop will primarily use mounted reconnaissance and reconnaissance by fire when conducting aggressive reconnaissance.

Cavalry troops must be trained to conduct both aggressive and stealthy reconnaissance. The troop commander, scout platoon leader, and squad leaders will most likely use many different methods over the course of a reconnaissance operation. The constraints of commander's guidance, the opposing threat, the capabilities of the troop, the terrain the troop is operating in, and the time available to complete the mission (METT-T) all have an impact on the method or methods chosen by the troop leadership.

TECHNIQUES

Actions on Contact

Actions on enemy contact are a series of steps the troop takes when it encounters an enemy force or situation that warrants/demands action. Actions on contact are important because they allow the troop to maintain its tempo of operation by rapidly developing the situation and taking action before the enemy can gain the initiative and force the troop to react. At platoon level, actions on enemy contact consist of four steps.

- Deploy and report.
- Develop the situation.
- Choose a course of action.
- Recommend or execute a course of action.

Note. See FM 17-98 for an in-depth discussion of platoon actions on contact during reconnaissance.

While the platoon that makes contact executes actions on contact, the commander must continue to maneuver the remainder of the troop to ensure a clear picture of the enemy situation across the entire troop front. The following steps demonstrate the actions taken by the platoon in contact and the corresponding actions at the troop level.

Deploy and Report

Platoon Action. The elements of the scout platoon that make initial contact with the enemy immediately deploy to terrain that affords them both cover and good observation. If necessary, scouts return fire to suppress the enemy, and then deploy to their positions. The scout making contact sends a contact report to his platoon leader. The platoon leader forwards the report to the troop commander. Once the scout in contact is in a good covered and concealed position, he sends his initial spot report, which is forwarded through the platoon leader to the troop commander (see Figure 3-2).

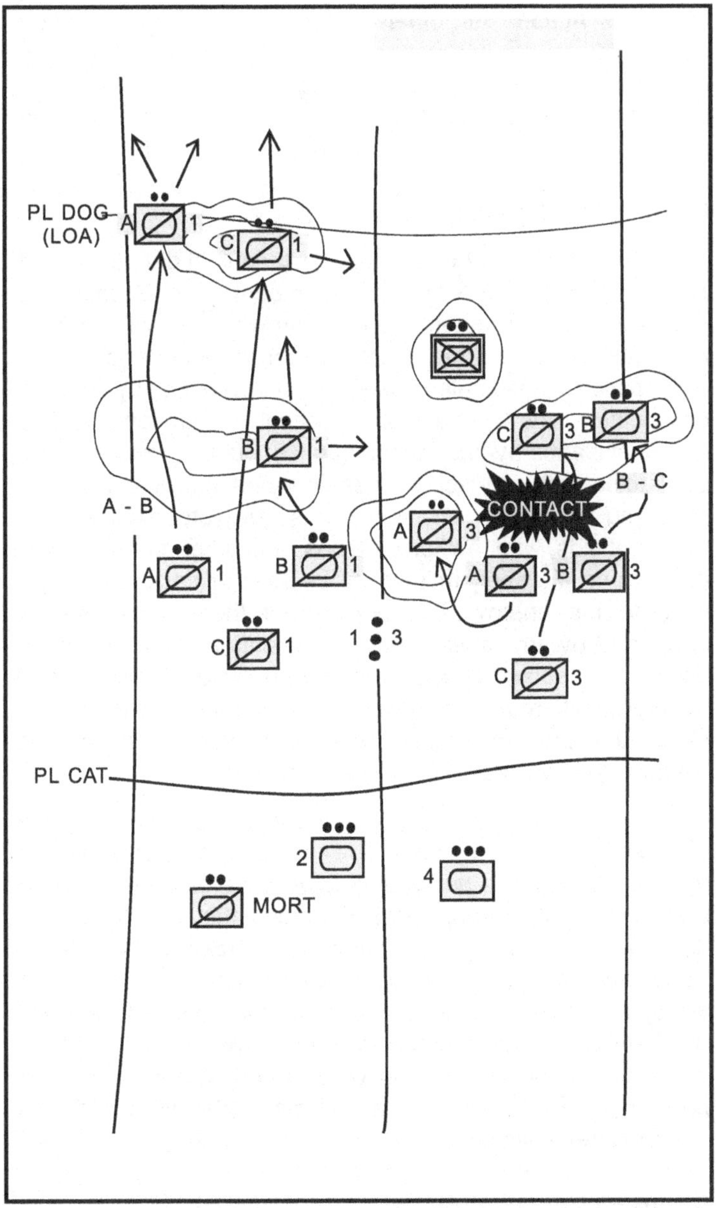

Figure 3-2. Scouts deploy and report then begin to develop the situation.

Troop Action. All other platoon leaders and platoon sergeants monitor the contact report. The troop commander assesses the information and moves, if necessary, to a position where he can observe the action. However, the troop must not lose focus of the reconnaissance mission.

Develop the Situation

Platoon Action. Next, the platoon in contact defines what it is up against. Scouts use dismounted and mounted reconnaissance to determine the enemy's size, composition, and orientation, and the exact location of weapon systems. The platoon may also use reconnaissance by fire to determine the enemy's tactical intentions. The reconnaissance-by-fire technique should, however, be conducted with indirect-fire assets when possible to avoid revealing the scouts' position. The scout platoon will search for minefields, wire, antitank ditches, and other obstacles that could force a friendly unit into a fire sack or cause it to flank itself to the enemy. To determine if the enemy can be supported by any other forces, the scouts search for enemy flanks and scour all adjacent terrain. They identify good counterattack routes into the flanks or rear of the enemy. Once the platoon leader determines the extent of the situation, he forwards a follow-up spot report.

Troop Action. The troop commander will most likely tell the scout platoon not in contact to continue its reconnaissance to a designated LOA to develop the situation across the entire troop front. By doing this, the troop can determine if there are any other enemy forces with which the troop must be concerned. The scout platoon not in contact will establish hasty observation posts along the LOA oriented on likely enemy locations or avenues of approach. The tank platoons will continue to monitor the troop command net and prepare to act based on the commander's intent/guidance and the enemy situation.

Choose a Course of Action

Platoon Action. Now that the platoon leader knows what he is up against, he considers two or three possible courses of action and selects the one that best meets the commander's intent/concept of the operation, is within his capabilities, and allows the troop to resume its reconnaissance mission as soon as possible. The possible courses of action open to him, based on the commander's intent/concept, might be hasty attack, bypass, hasty defense/screen, or support of a hasty attack by another platoon(s). Some courses of action will quickly be ruled out because they do not meet the commander's intent and guidance for the operation.

- **Hasty Attack.** The platoon leader can conduct a hasty attack if he has enough combat power to defeat the enemy quickly. In most cases, the scout platoon does not have the capability to mass enough combat power to defeat an enemy in prepared positions. In addition, the scout platoon leader may not want to risk battle losses that would reduce his effectiveness and his ability to complete the mission.

- **Bypass.** If the platoon does not have enough combat power to conduct a hasty attack, or if it wants to remain undetected and continue the reconnaissance mission, the scout platoon can bypass the enemy. The platoon leader must receive the troop commander's permission to bypass. If he has permission to bypass, the platoon leader must leave scouts in contact with the enemy force unless ordered to break contact by the troop commander. The platoon leader will know from the troop commander's guidance in the operations order if this is a viable course of action.

- **Hasty Defense/Screen.** If the platoon cannot conduct a hasty attack and cannot bypass, it establishes a hasty defense or screen. The platoon will conduct a hasty defense if it can defend against an enemy force. If the enemy contact exceeds the platoon's capability to conduct a hasty defense, it may elect to establish a screen and maintain contact through observation. The platoon concentrates on maintaining contact with the enemy and fixing it in place with indirect or possibly direct fire until additional support comes from the troop.

- **Support a Hasty Attack by Another Platoon(s).** The platoon in contact may become the support element for a hasty attack by a tank/AT platoon(s).

Troop Action. The troop commander continues to maneuver the troop and assists the platoon leader as required. He maintains situational awareness.

Recommend/Execute a Course of Action

Platoon Action. If the course of action the platoon leader chooses meets the commander's guidance for actions on contact, the platoon leader executes the course of action. If the situation dictates that the platoon execute a hasty defense/screen, or support a hasty attack by another platoon, the platoon leader reports this information to the troop. The report is brief. He updates his spot report (Blue 1) with any additional information, tells the troop commander what he is doing about the situation, and recommends the course of action that he thinks best suits the situation. The platoon in contact has then completed its actions on contact.

Troop Action. The commander must approve or disapprove the recommended course of action, based on its effect on the troop and squadron mission.

If the scout platoon is required to establish a hasty defense/screen, the troop commander assumes responsibility for continuing to develop the situation. The commander can do this best in a position from which he can influence troop actions. Based on what the scouts have found and reported and what the commander can see, he must weigh the advantages and disadvantages of each course of action, and choose the one that best meets the troop's mission requirements.

- **Hasty Attack**. The troop can conduct a hasty attack, which throws the weight of one or both tank/AT platoons against the enemy and destroys him quickly, then continue its reconnaissance. The commander may coordinate the efforts of the scout platoon and tank/AT platoon(s) in the hasty attack or delegate the responsibility to the scout platoon in contact. The hasty attack at troop level should have two elements: the *assault/attack-by-fire* element and the *support* element.

  Steps to the execution of a hasty attack are as follows:

  - The scouts determine that the enemy force encountered cannot be bypassed based on the enemy disposition and composition. The enemy encountered meets the commander's intent for destruction.

  - The scout platoon leader recommends hasty attack to the troop commander and identifies a good attack-by-fire or assault position for the tank/AT platoons.

  - The troop commander approves the scout platoon leader's recommended course of action and issues orders to execute a hasty attack.

  - The troop commander decides to use either attack by fire or assault as his method of destruction. Heavy troops equipped with tanks can perform attack by fire or assault. Because of the defensive nature of the TOW, light troops tend to use attack by fire when conducting hasty attack.

  - The scout squad or section moves to link up with the tank/AT platoons and guides them to an assault/attack position. The assault element reports when set in assault positions, and if time is available, reconnoiters the attack-by-fire position or the attack axis.

- The troop commander moves to the scouts who are positioned in overwatch of the enemy contact. He also positions the FIST in overwatch to assist in controlling indirect fires.

- The troop commander designates the scouts in overwatch as the support element. He establishes direction of fire and methods of control for direct-fire weapon systems; for example, left and right limits for fires, and signals for the initiation of direct fires and the shifting of fires as the assault element begins its attack.

- The troop commander directs the FIST to fire for effect on enemy positions when given the signal (AT MY COMMAND). The FIST reports when guns are ready.

- The troop commander issues the fire command for the support force and indirect-fire systems and directs the attack-by-fire or the assault element to execute its movement. The troop commander lifts and shifts direct and indirect fires as necessary to cover the movement of the assault element and to seal off the enemy withdrawal (see Figure 3-3).

- **Bypass**. The troop may bypass the enemy force and continue the reconnaissance to further develop the enemy situation throughout the depth of the zone while maintaining an element of surprise. The decision to bypass is based on the higher commander's intent for the operation. The troop should leave an element in contact with the enemy force; however, the more elements in contact with the bypassed enemy mean the less reconnaissance will be conducted.

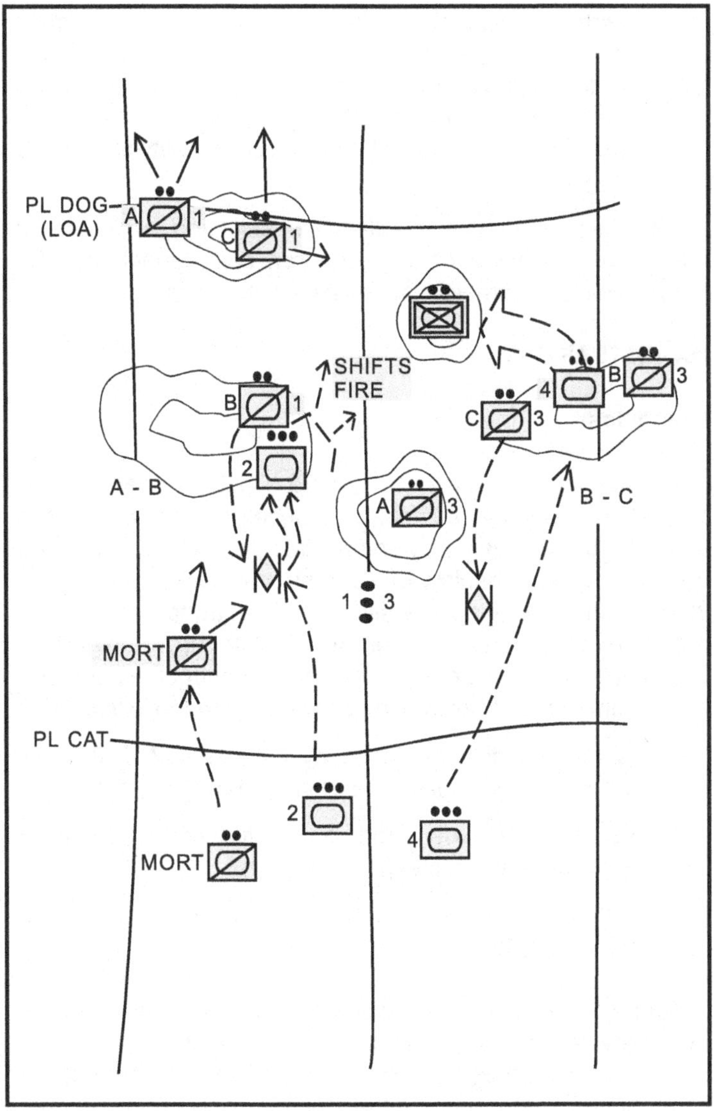

Figure 3-3. Troop positions for hasty attack.

- **Hasty Defense**. If the troop cannot conduct a hasty attack or bypass the enemy, it may establish a hasty defense to fix the enemy while waiting for further orders. If this is the troop's course of action, the squadron commander has the responsibility to develop the situation further.

The overriding considerations in selecting a course of action are the intent of the squadron commander and the troop's ability to complete the mission with minimum losses.

DANGER AREAS

Obstacles

Obstacles encountered during reconnaissance are treated as enemy contact, because all obstacles are assumed to be covered by enemy fire. Upon encountering an obstacle, scouts deploy to covered positions, report, and then begin to develop the situation. The troop commander may move his tank platoon(s)/AT platoon(s) forward to overwatch the obstacle while the scouts conduct dismounted and mounted reconnaissance to determine the following:

- Is the obstacle defended by the enemy? If so, how many enemy soldiers are there and where are they?
- What are the extent and composition of this obstacle? How deep, wide, steep, or long is it?
- Can the obstacle be bypassed, or will breaching be necessary?

Once the scouts have completed the reconnaissance of the obstacle and recommended a course of action (either bypass, hasty attack/breach, or hasty defense), the troop commander must choose the course of action that best meets the commander's intent and guidance for the mission. If the commander's decision is to bypass or breach the obstacle, the troop must mark the obstacle and any lane created through the obstacle (see Figure 3-4).

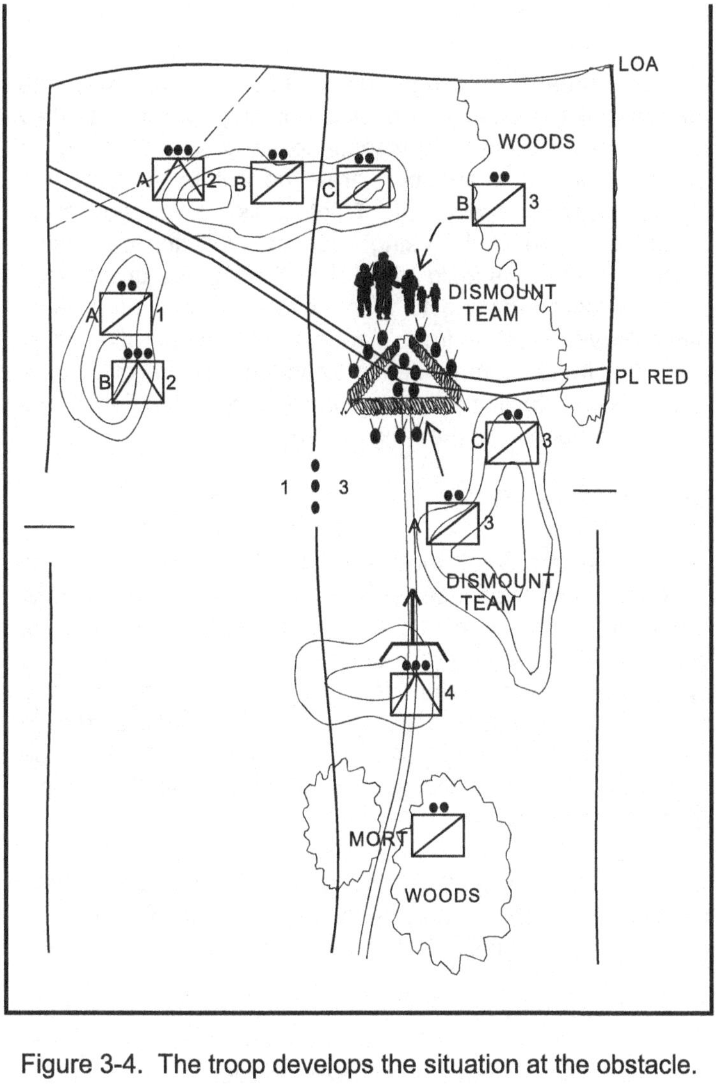

Figure 3-4. The troop develops the situation at the obstacle.

Open Areas

Open areas are dangerous for cavalry troops because they permit the enemy to observe or engage units as they move through them. When reconnoitering a route, zone, or area that includes an open area, do not try to force the reconnaissance through it. Treat it like a danger area. Reconnoiter the flanks using good covered and concealed routes. The troop should try to clear the far side of the open area before attempting to reconnoiter the open area itself. Open areas within the troop's zone or area of operations may force the troop commander to abandon the use of platoon boundaries to facilitate the reconnaissance to the flanks and rear of the open area (see Figure 3-5).

Built-up Areas

In most areas of the world the troop can expect to conduct reconnaissance of built-up areas (BUA). BUAs range from small hamlets to large towns, even portions of cities. BUAs are very dangerous for mounted units and should be avoided when possible. If the situation permits, the troop should conduct a reconnaissance of BUAs from a distance, and then bypass them. However, it may be necessary to move through a town as part of the troop's reconnaissance mission.

The troop's ability to conduct reconnaissance of a BUA is constrained by the lack of dismounted scouts. The troop is capable of finding the following in BUAs:

- Enemy vehicles.
- Enemy command and control facilities.
- Obstacles such as rubble, blown bridges, and craters.
- Logistics elements.
- Bypasses within the BUA.

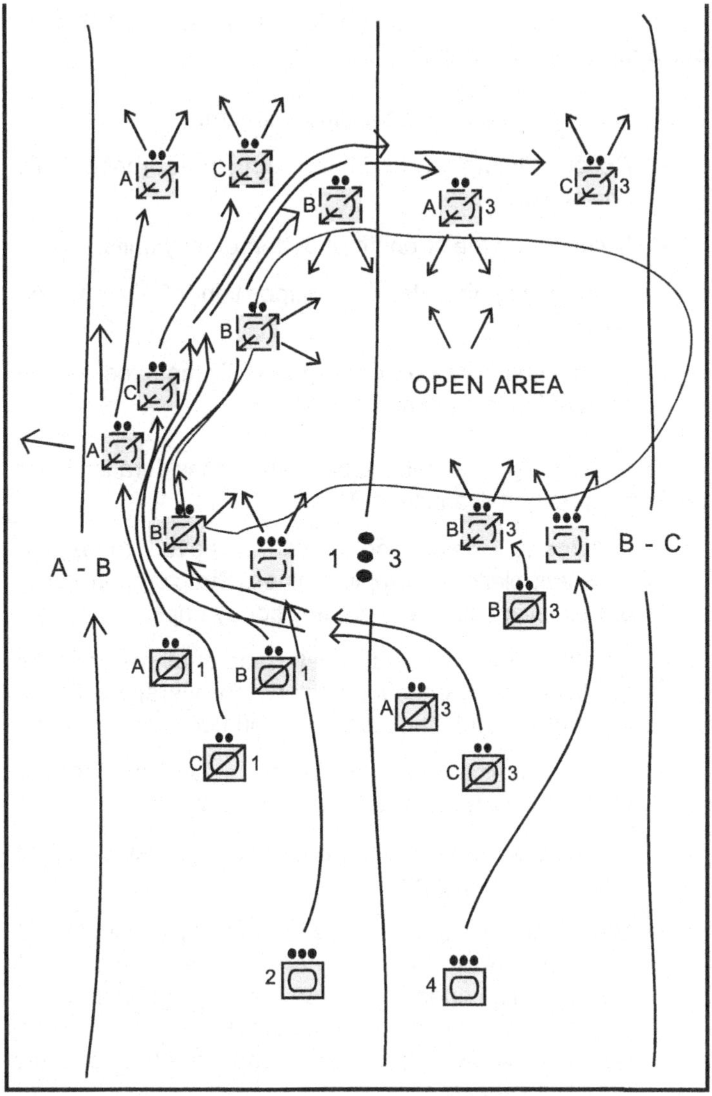

Figure 3-5. The troop reconnoiters to the flanks and rear of an open area.

If the troop is not augmented, it is not capable of performing the following in BUAs:

- Reconnoitering each building and street.
- Reconnoitering underground structures (basements and sewers).
- Determining the strength of dug-in enemy units.
- Determining the detailed disposition of dismounted enemy units.
- Conducting hasty attacks against dismounted enemy units of larger than platoon size.

The following are techniques and considerations for performing reconnaissance in BUAs:

- Dirt roads, alleys, and curves in paved roads are excellent places to employ mines. Structures in BUAs are excellent places to employ booby traps.
- Ask the S2 for detailed street maps of all major BUAs the troop might encounter during an operation. These are more useful than standard 1:50,000 military maps.
- Clearly define platoon zones. Do not divide responsibility for a street between platoons.
- Observe BUAs from outside for signs of enemy activity prior to entering them.
- Main roads through towns are normally accompanied by open areas such as parks, traffic circles, and medians. The enemy may use these same open areas for logistics elements, concentrations of armored vehicles, artillery positions, and combat support vehicle locations.
- Collateral damage constraints may limit the ability to employ some types of weapon systems.

LIGHT TROOP EXAMPLE

The squadron is conducting a zone reconnaissance. The troop zone contains a large town, 2 to 3 kilometers square, and the squadron main supply route, ROUTE GOLD, passes through the town. The troop will have to reconnoiter enough of the town to ensure there are no enemy who can disrupt movement along ROUTE GOLD.

On approaching the town, the troop has options on how to enter it. The techniques used will largely depend on the amount of time available for the mission. If time is limited, the scouts will halt short of the BUA, just off ROUTE GOLD, and observe the town from multiple vantage points before entering. If more time is available, the scouts will move well off the route and observe the BUA before entering. In any case, remember if enemy is present, he will probably cover the main road with fires. Once the platoon is inside the BUA, the main road may provide the best axis for the reconnaissance.

As the scout platoon observes the town, they look for any signs of enemy presence. The antitank platoons move into overwatch positions. The mortars establish a firing position in support of the scouts. Scouts reconnoiter to the flanks and rear of the BUA to find a bypass and establish flank and far side security (see Figure 3-6).

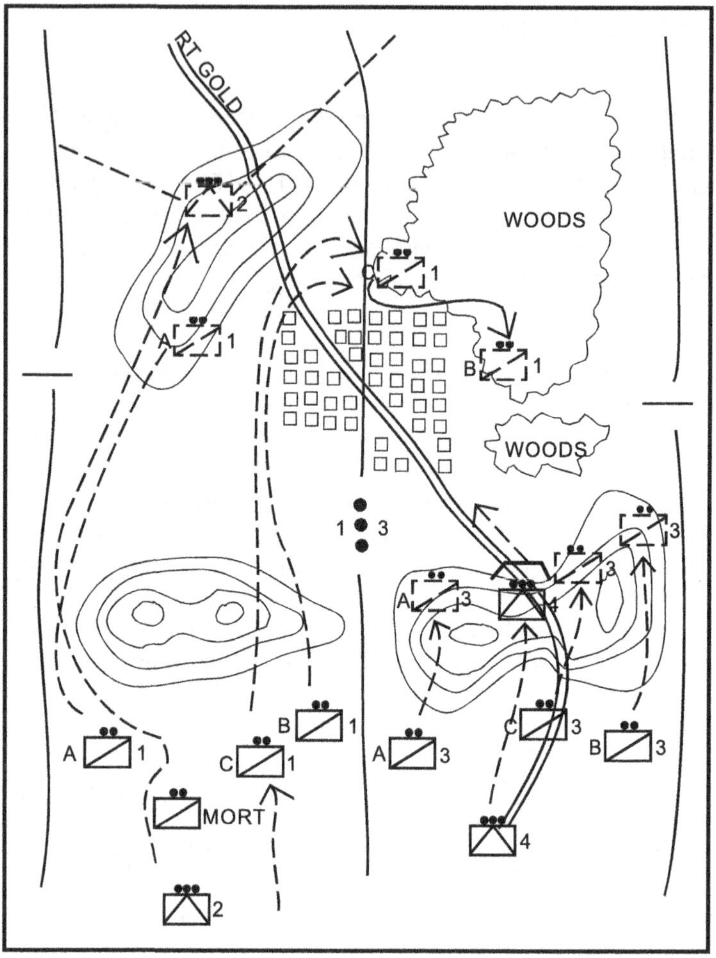

Figure 3-6. The troop securing the BUA with one scout platoon, ATs in overwatch, and one scout platoon preparing to reconnoiter the BUA itself.

Once the area surrounding the BUA is secured, the scouts move mounted to the last concealed position and dismount, searching for suitable entry points into the BUA. *If there is enemy contact*, the overwatching vehicles suppress with direct and indirect fires while the scouts move to a covered and concealed position. At this point the troop commander must decide to maintain contact and bypass, or continue to probe for entry points into the BUA. The determining factors for the commander's decision are the intent of the higher commander and the level of threat in the BUA. In most instances, cavalry troops maintain contact and bypass BUAs occupied by enemy forces. The enemy forces are handed off to follow-on forces for destruction.

If the troop does not encounter enemy contact on its initial reconnaissance of the outskirts of the BUA, the scout platoons continue their search for suitable entry points into the town. Once the scouts determine those entry points into the town are clear, they move into the BUA and reconnoiter nearby buildings (moving through side and back yards, alleys, and other off-street paths) and surrounding areas. The dismounted teams from the scout platoon do not enter or clear each building, but look for obvious signs of enemy presence. HMMWVs may be brought forward to overwatch the dismounted team's movement with caliber .50 machine guns or MK-19s (see Figure 3-7). The AT platoons should continue to overwatch from outside the BUA. After the scouts complete the reconnaissance of the BUA, the troop continues its reconnaissance in zone, ensuring that bypasses around the BUA are marked.

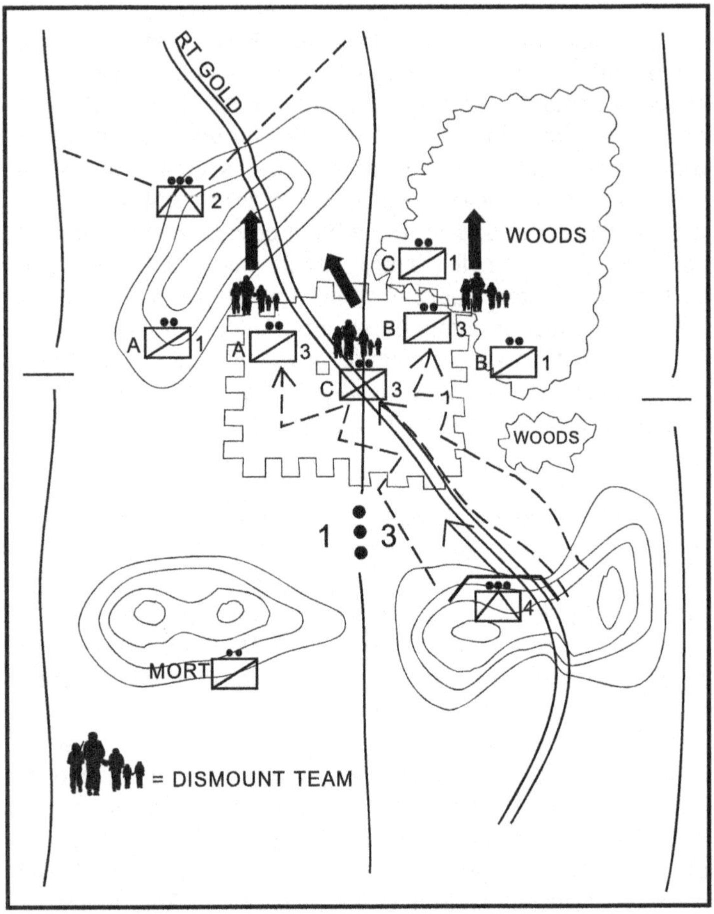

Figure 3-7. Scout section reconnoitering in a built-up area.

Section III. Route Reconnaissance

Route reconnaissance is a directed effort to gain detailed information about a specific route and the terrain on either side of the route that the enemy could use to influence movement along the route. Depending on the terrain surrounding the route, a route reconnaissance mission may be accomplished by a single scout platoon, or it may require the entire troop to cover the terrain and accomplish the required tasks.

A cavalry troop may be assigned the mission to conduct a zone reconnaissance (see Section IV of this chapter), but also have the task of reconnoitering a route that is within its zone. In this case, the troop mission would be zone reconnaissance. However, one of the subordinate units in the troop, scout platoon, or attached engineers may be given the route reconnaissance mission by the troop commander.

CRITICAL TASKS

During a route reconnaissance, the following critical tasks must be accomplished unless the SCO directs the troop to do otherwise:

- Reconnoiter and determine trafficability of the route.
- Reconnoiter all terrain the enemy can use to dominate movement along the route.
- Reconnoiter all BUAs along the route.
- Reconnoiter all lateral routes in the area of responsibility.
- Inspect and classify all bridges along the route.
- Locate fords or crossing sites near all bridges along the route.
- Inspect and classify all overpasses, underpasses, and culverts.

- Reconnoiter all defiles along the route; clear all defiles of enemy and obstacles within its capability, or locate a bypass.

- Locate mines, obstacles, and barriers, and within its capability, clear the route.

- Locate a bypass around BUAs, obstacles, and contaminated areas.

- Report route information.

- Find and report all enemy that can influence movement along the route.

TECHNIQUES

The SCO directs the troop to conduct a route reconnaissance as a mission or as a specific task in another mission. This section discusses route reconnaissance in the context of an assigned troop mission.

The troop commander considers several factors in formulating his concept.

- Start point, release point, and designation of the route.

- Mission to be performed at the start point and after reaching the release point.

- Time the mission is to start, and if required, be completed.

- Critical points along the route identified as checkpoints.

- Any constraints or restrictions.

- IPB information on the route, to include current enemy situation.

IPB provides critical information on the enemy and terrain. Enemy threats may be encountered in two basic forms.

- Ambushes along the route in close or restricted terrain or even tied to obstacles along the route.
- Attack by long-range direct or indirect fires from dominating terrain along the route.

Analysis of the terrain provides an indication of danger areas and the nature of the potential threat. From this information, the commander determines how much terrain on each flank of the route must be reconnoitered and the organization for combat. Any constraints or restrictions may also influence how much terrain is reconnoitered.

The troop commander adds control measures to provide an adequate framework for the mission. He places a boundary on both sides of the route far enough out to provide reconnaissance of the dominating terrain. A line of departure is placed perpendicular to the route short of the start point, allowing sufficient space to deploy into a tactical formation. An LOA is placed far enough beyond the release point to enclose dominating terrain that overwatches the release point. The commander uses additional phase lines to maintain a coordinated effort throughout the troop. Boundaries and phase lines are drawn along recognizable terrain features. The terrain features should be visible from both ground and air to facilitate air-ground integration. The commander uses other control measures as necessary for flexibility in the maneuver and execution of the mission (see Figure 3-8).

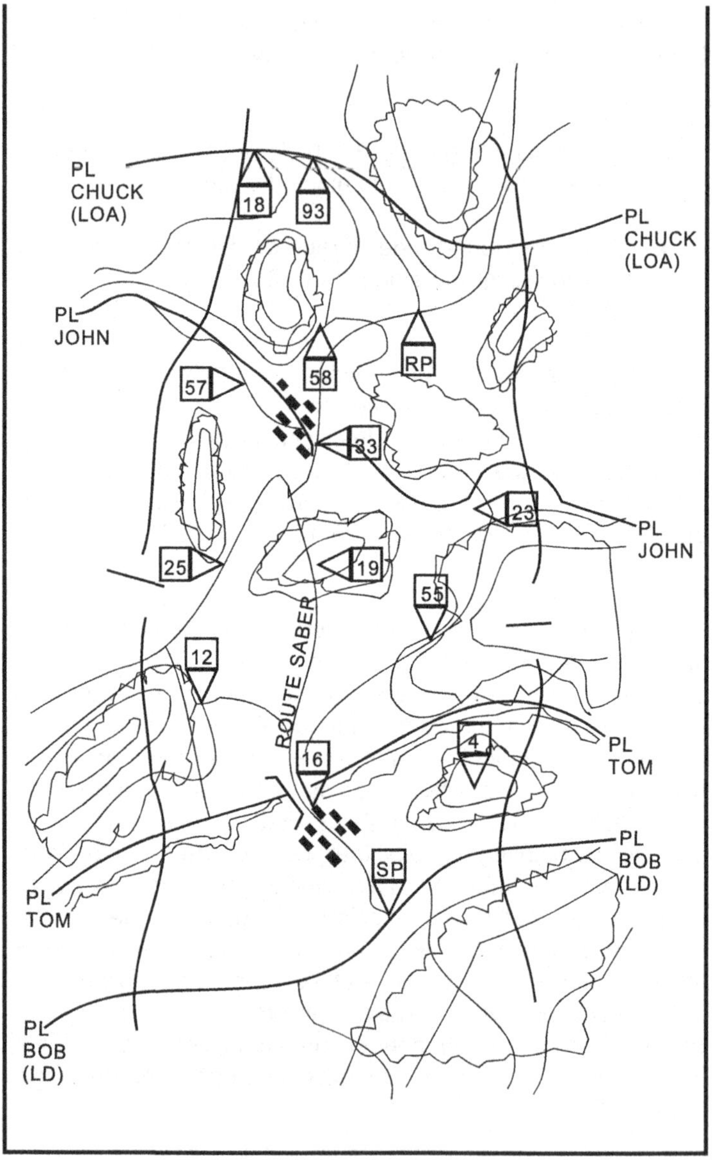

Figure 3-8. Troop graphic control measures for a route reconnaissance.

The cavalry troop normally performs a tactical road march to the line of departure and deploys to execute the reconnaissance of the route. Based on the amount of intelligence known about the enemy, the commander will determine how much security is required for the move forward to the line of departure. Also the commander should consider the effect his final disposition of forces will have on the troop's follow-on mission.

One technique for accomplishing all tasks as rapidly and safely as possible is as follows: Draw a 2.5-kilometer to 3-kilometer boundary on either side of the route. Near the start point, draw a line of departure from one boundary to the other. Next, draw an LOA about 3 kilometers forward of the release point from one boundary to the other. Add phase lines between the line of departure and LOA, as needed, to help control the progress of the troop. To control movement on the route, establish checkpoints. Use checkpoints to identify specific areas or locations to be reconnoitered. These checkpoints will serve as references to orient reconnaissance efforts.

One scout platoon moves out early across the line of departure to reconnoiter the terrain or BUAs on either side of the route. The other scout platoon reconnoiters the route itself, trailing 1 kilometer to 2 kilometers behind the lead scout platoon. This provides a good measure of security for the troop and the platoon working to reconnoiter and classify the route. The mortar section follows the lead scout platoon and remains positioned to range from 3,000 to 3,500 meters forward of the lead scout elements. The mortar section avoids moving on the route and stays on the flanks near the route. The tank/AT platoons follow and support behind the second scout platoon. They key their movement on the advance of the scout platoons. The tank/AT platoons should be kept back in depth to retain flexibility, because they must be ready to react anywhere in the troop zone. The troop trains follow about 2 kilometers behind the tank/AT platoons, and bound from one covered and concealed position to another. The troop CP displaces generally along the route using terrain that affords effective and continuous communications with troop elements and the squadron. The troop commander positions himself to observe the actions of the scout platoons.

EXAMPLE OF A TROOP ROUTE RECONNAISSANCE

The 1st platoon (scout) crosses the LD (PL BOB) using the two-section organization, and reconnoiters the terrain on either side of ROUTE SABER up to PL TOM. The platoon looks for the threat around the outskirts of the village, and explores routes into the village that intersect ROUTE SABER. The scouts also search the woods near checkpoint 4 for the enemy. The platoon scans the terrain north of the stream, and then moves up. One section locates a fording site west of the bridge at checkpoint 16, and the other section swims the stream on the east. The platoon makes a quick visual inspection of the bridge, then searches for the threat in the forested terrain near checkpoints 12 and 55. The mortar section crosses the LD next, about 2 kilometers behind the 1st platoon (scout), and establishes a firing position on the outskirts of the village. The 3d platoon (scout) crosses the SP when the 1st platoon (scout) crosses PL TOM.

The 3d platoon (scout) reconnoiters and classifies the route through the village, then moves up and classifies the bridge just north of town. The troop commander bounds closely behind the 3d platoon (scout) and observes their actions. The 2d and 4th platoons (tank) follow along the flank of ROUTE SABER about 1 kilometer behind the 3d platoon (scout), then move into the outskirts of the village. They observe the dominating terrain on the far side of the stream while the 3d platoon (scout) reconnoiters the bridge. The troop CP and trains remain in concealed positions south of the LD (see Figure 3-9).

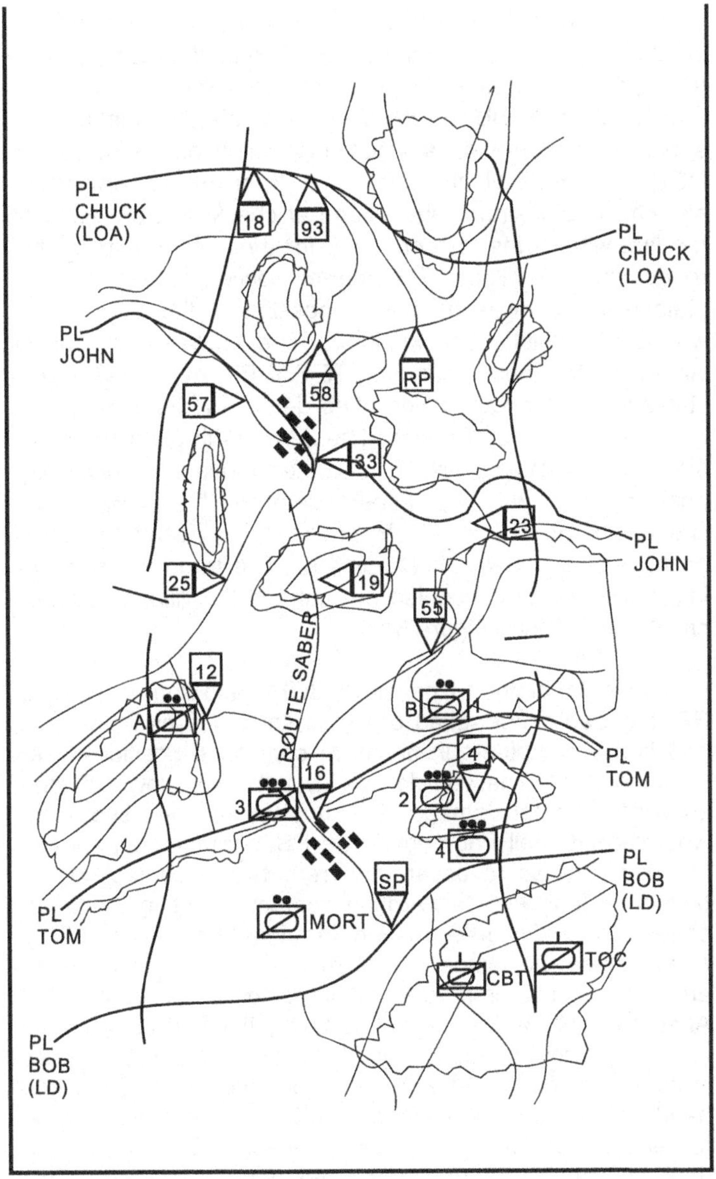

Figure 3-9. Route reconnaissance (part one).

The 1st platoon continues to reconnoiter in zone up to PL JOHN. The platoon searches the wooded areas for the threat, and explores the high-speed lateral routes to the east and west of ROUTE SABER. The platoon clears the shoulders and makes a visual inspection of the defile on ROUTE SABER at checkpoint 19, then moves forward to search the outskirts of the village at checkpoint 33. As with the first village, the platoon looks for threat forces along the roads into the village. The mortar section follows the 1st platoon and establishes a firing position on the edge of the woods southwest of checkpoint 19. The 3d platoon continues classifying the route along ROUTE SABER. At the defile near checkpoint 19, scouts using mine detectors check the route to ensure it is free of mines. The 2d and 4th platoons cross the bridge at checkpoint 16, and then move into concealed positions in the woods on either side of ROUTE SABER near checkpoint 19. The troop CP moves to a concealed position on high ground near checkpoint 12. The troop trains move into covered and concealed positions in the village south of checkpoint 16 (see Figure 3-10).

As the 1st platoon crosses PL JOHN and approaches the RP, the lead scout squad of Bravo section is hit by cannon and heavy machine-gun fire from concealed positions near checkpoint 93. The vehicle has its left track blown off and sustains external damage. The gunner and one scout are wounded by shell fragments. The PSG immediately returns fire and deploys to cover then reports the contact to the platoon leader who forwards a contact report to the troop commander. Once the PSG determines the location of the enemy force, he sends a spot report to the platoon leader and a request for immediate suppression to the troop FSO. After the call for fire is sent forward, the PSG directs his remaining scout squad to search for additional threat forces in the woods north of the RP and along the high-speed route heading northeast (see Figure 3-11). As mortar fire falls on the enemy position, the 1st platoon leader quickly moves up near checkpoint 58 to observe the threat. He directs his section to search the woods north of checkpoint 57, and to move around to the west flank of the threat and find out

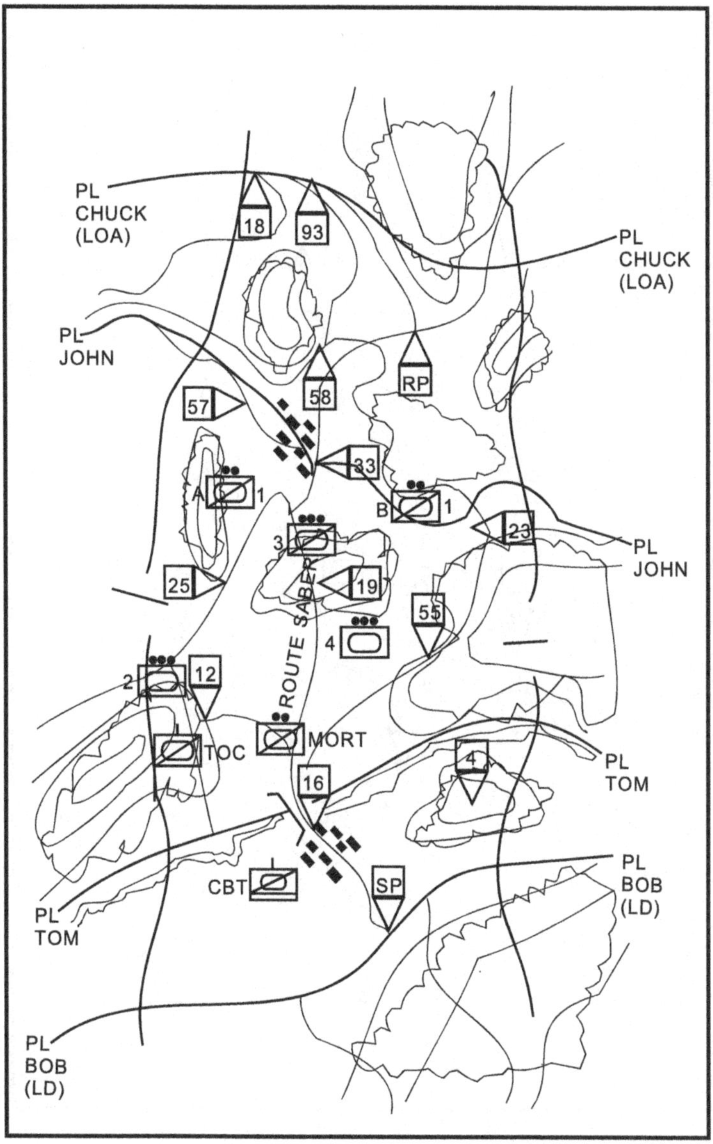

Figure 3-10. Route reconnaissance (part two).

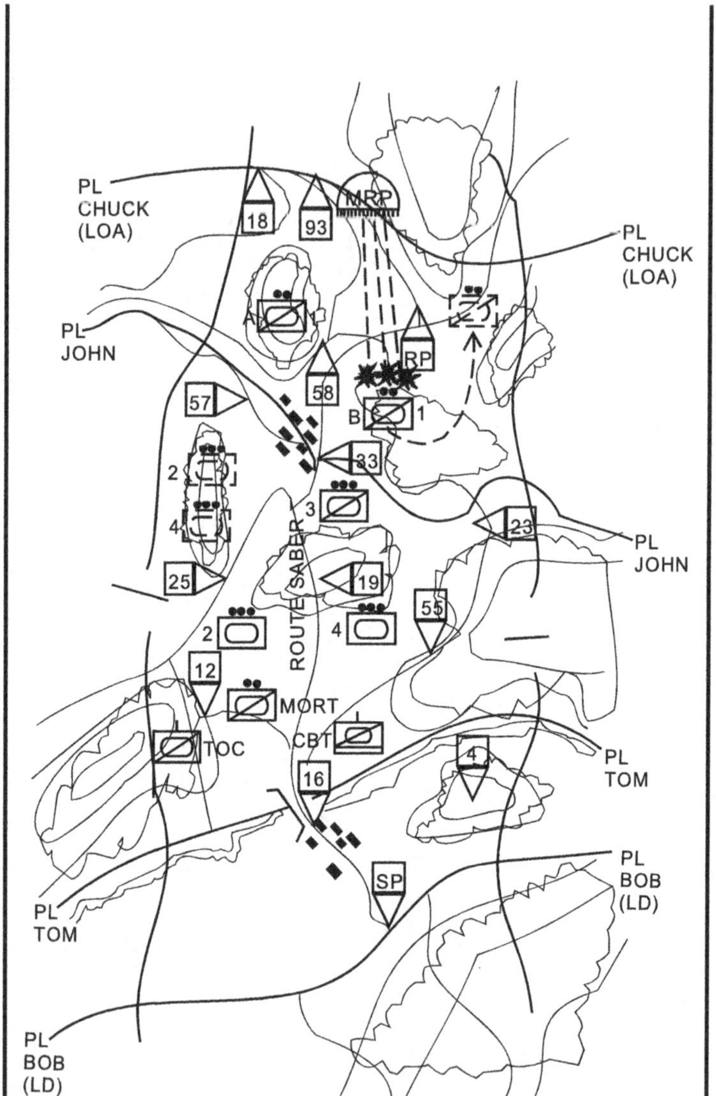

Figure 3-11. Route reconnaissance (part three).

what the platoon is up against and whether or not the threat force is supported by other elements in the village at checkpoint 18. He then reports to the troop commander. The troop commander immediately moves forward to join the 1st platoon leader and to see the situation for himself. En route, the commander tells the 2d and 4th platoons to move into concealed positions in the village at checkpoint 33, and to prepare to counterattack. The 3d platoon continues to reconnoiter the route through the village at checkpoint 33, and then moves into covered positions along ROUTE SABER just south of checkpoint 33. The troop CP holds in place and reports the situation to squadron. The 1SG quickly leads the M88 and the medic vehicle forward to treat and evacuate the injured scouts and their damaged vehicle.

Dismounted scouts from the 1st platoon determine that the threat force consists of three BMPs and one T-64 in prepared positions. Threat crews remain mounted and continue to slew turrets, scanning for more friendly elements. The threat reinforced platoon is oriented southeast toward the route. Their western flank is free of obstacles and unobserved. The troop commander decides to conduct a hasty attack (see Figure 3-12).

When the tank platoons complete their hasty attack, the troop commander gets permission to advance beyond the LOA and reconnoiter the enemy position. The 3d platoon returns to complete the reconnaissance of ROUTE SABER. As the platoon checks for mines along the route to the RP, they discover the route is mined just short of the RP. They find a bypass around the obstacle and forward a bypass report (Blue 10) to the troop XO. The platoon then organizes into two sections and moves out to occupy OPs overlooking the lateral routes at the east and west flanks of the troop near PL CHUCK. The troop CP moves into position on high ground near checkpoint 58. The 1SG directs activities of the troop trains, to include evacuation of injured soldiers and inoperable vehicles, security and evacuation of injured prisoners, requests for casualty replacements and resupply.

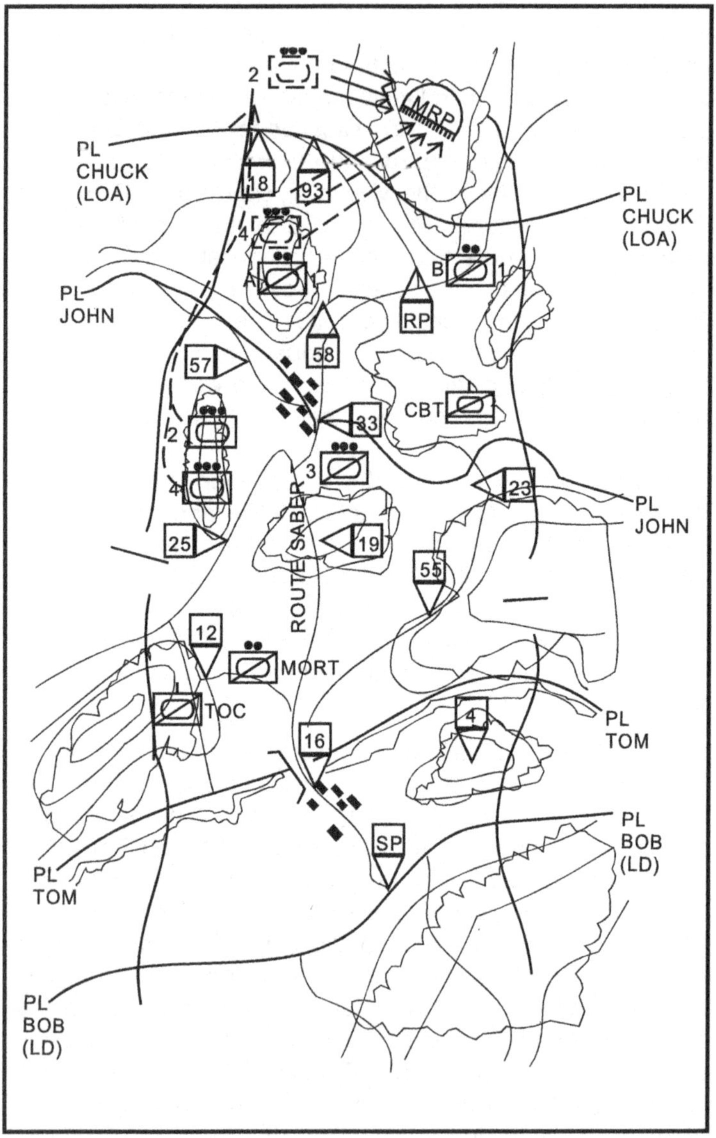

Figure 3-12. Route reconnaissance (part four).

Section IV. Zone Reconnaissance

Zone reconnaissance is the directed effort to obtain detailed information concerning all routes, obstacles, terrain, and enemy forces within a zone defined by boundaries. Obstacles include chemical and radiological contamination. A zone reconnaissance is assigned when the enemy situation is vague or when information concerning cross-country trafficability is desired. It is appropriate when previous knowledge of the terrain is limited or when combat operations have altered the terrain.

CRITICAL TASKS

Zone reconnaissance is a deliberate, time-consuming process. During a zone reconnaissance, the troop accomplishes the following critical tasks unless specifically directed otherwise by the higher commander:

- Reconnoiter all terrain within the zone.
- Inspect and classify all bridges within the zone.
- Locate fords or crossing sites near all bridges in the zone.
- Inspect and classify all overpasses, underpasses, and culverts.
- Locate and clear all mines, obstacles, and barriers in the zone within its capability.
- Locate a bypass around BUAs, obstacles, and contaminated areas.
- Find and report all enemy forces within the zone.
- Report reconnaissance information.

TECHNIQUES

The cavalry troop is usually assigned a zone reconnaissance as part of a squadron zone reconnaissance mission.

A zone reconnaissance mission is very time-consuming. Any commander who orders this mission must remember the number and complexity of the tasks to be accomplished. If he wants a faster tempo, he must prioritize reconnaissance tasks for the troop.

The cavalry troop can effectively reconnoiter a zone from 6 to 10 kilometers wide. If stretched any farther, the troop quickly exceeds its ability to accomplish the critical tasks.

Because the enemy situation is vague and knowledge of the terrain is limited to what is shown on maps or aerial photos, platoons will face unexpected situations around every corner; therefore, the course of action selected must also provide a good measure of protection for the troop as it executes the mission. When considering techniques of conducting a zone reconnaissance, the scheme of maneuver has to be flexible. The troop commander must convey his intent to subordinates so they can act quickly and without orders. He should employ forces, tank or AT platoons, in depth to give the troop flexibility and responsiveness.

When the troop receives a zone reconnaissance mission, the zone is usually identified by lateral boundaries. The line of departure and a reconnaissance objective or LOA are specified. The commander should divide the troop zone into two platoon zones for the scout platoons. Use caution when drawing the boundary. Make sure it is on easily identifiable terrain that is not a high speed avenue of approach. The number of critical tasks within the troop zone, and terrain restrictions should guide the commander in determining the location of the scout platoon boundaries. Doctrinal distances are not always the best solution if one platoon will be overtasked.

Add phase lines every 5 to 8 kilometers on easily identifiable terrain to control progress of the troop through the zone. Place contact points near the intersection of the boundary and all phase lines. Ensure the contact points provide at a minimum the potential for concealment from enemy observation to allow the exchange of information between flank units. Use checkpoints to focus the efforts of scout platoons and to assist in maneuvering the troop (see Figure 3-13). Use TIRS on maps as described in Chapter 2.

Use a troop vee or split-vee to conduct zone reconnaissance. As scout platoons reconnoiter the zone, the tank/AT platoons follow and support behind the scouts, keying their movement off the scouts' forward progress. The tank/AT platoons are kept far enough in depth to retain flexibility, and yet remain responsive to situations developed by the scout platoons. The distance from the scouts is determined by the terrain and enemy situation. Therefore, the distance the tanks/ATs move from the scouts may vary during the course of the operation. Control tank/AT platoon movement by one of the following methods:

- Give movement guidance to the platoon leaders. Have them key movement off the scouts.
- Move the platoons by CPs, battle positions, and hide positions.

Move the mortar section center of the troop zone to range forward of the scouts. The range forward of the scouts is determined by the scouts' ability to acquire the enemy. The mortars may move under the control of the FIST or eavesdrop on the troop net and key their movement off the progress of the scout platoons. Regardless of who controls their movement, the TOC and the FIST track the mortar position constantly to ensure they can support the scouts.

Note. If the mortar section is busy firing missions and computing fire data, they may not be able to stay abreast of the friendly situation. Therefore, the FIST or XO should be ready to control the mortars displacement.

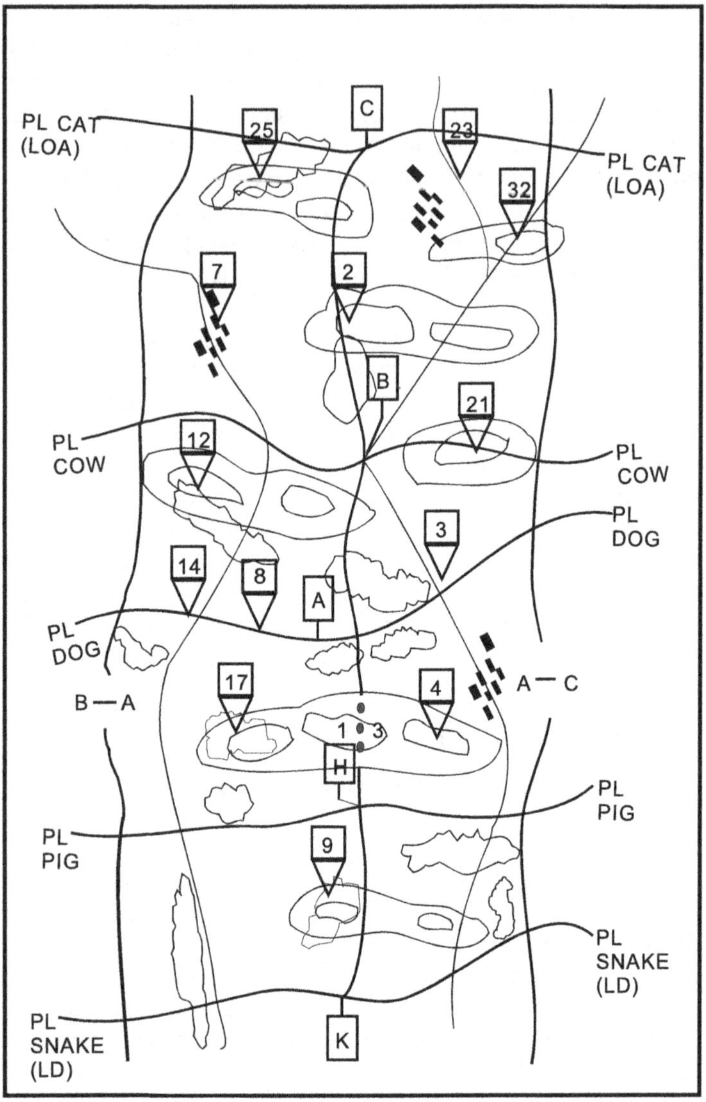

Figure 3-13. Zone reconnaissance graphic control measures.

The XO displaces the troop CP through the zone using terrain that affords effective and continuous communications with troop elements and the squadron. The 1SG moves the troop trains about 2 kilometers behind the tank platoons and bounds the trains from one covered and concealed position to another. The troop commander positions himself well forward to observe the actions of one scout platoon or the other. His position usually depends on where he expects initial enemy contact or problems to occur.

LIGHT TROOP EXAMPLE

The troop commander deploys the troop in a split-vee formation. The 1st and 3d platoons (scout) cross the LD, PL UTAH, abreast and reconnoiter the terrain in troop zone up to PL OHIO. The platoons determine both the open terrain and the trafficability of the route within their zone. Some hill slopes are greater than thirty percent and dense woods inhibit armor movement. Except for a marshy area north of checkpoint 9, the open terrain is dry and firm. Both routes will support heavy armor movement. The troop commander follows the 1st platoon. The mortar section eavesdrops on the troop command net to follow the progress of the scout platoons, and coordinates its moves with the troop FSO accordingly. As the scout platoons cross PL OHIO, the 2d and 4th platoons (antitank) cross the LD. They eavesdrop on the troop command net and key their movement on the progress of the scout platoons. The troop CP and the trains hold in place.

The scout platoons continue reconnaissance up to the river (PL BAMA). They take a close look at the dominant terrain near checkpoint 17 and at the village near checkpoint 4. As the platoons approach the river, scouts move into covered and concealed positions, dismount, and visually search the dominant terrain on the north side of the river. The 1st platoon inspects the bridge at checkpoint 14 and determines that it will support up to 40 tons. Scouts also verify that a good fording site exists near checkpoint 8. The 3d platoon confirms there is a fording site with a rock bottom just east of the blown bridge at checkpoint 3. It will support heavy armored traffic (see Figure 3-14).

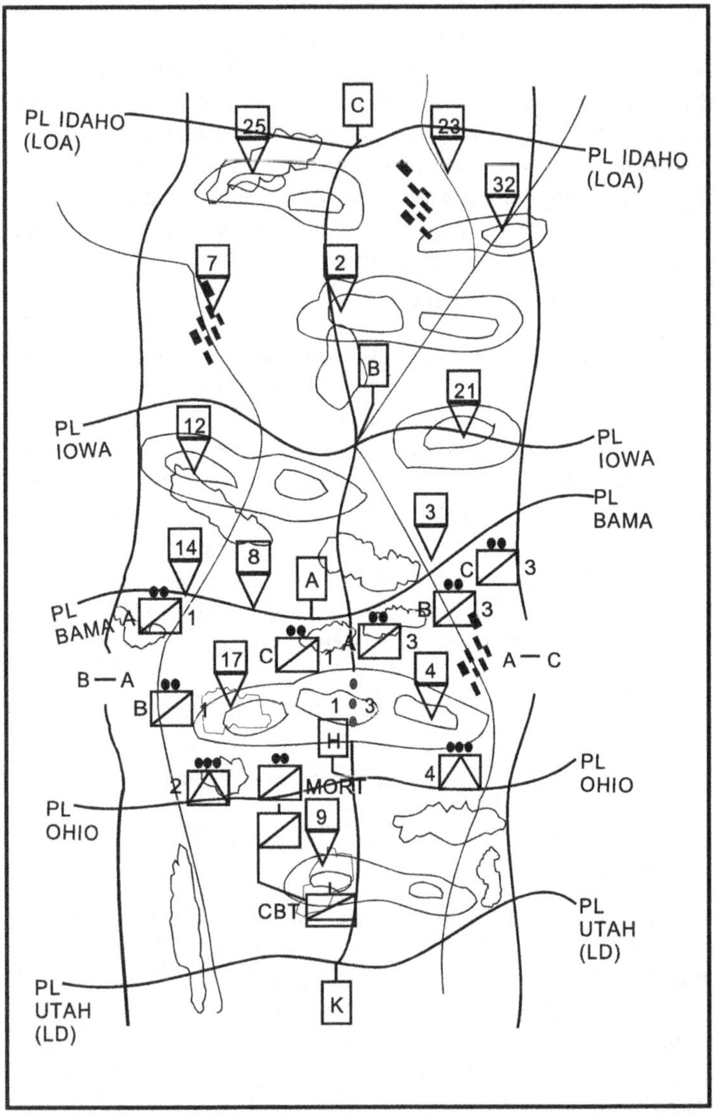

Figure 3-14. Zone reconnaissance (part one).

As the scout platoons continue reconnaissance towards PL IOWA, the 1st platoon observes two stationary BRDM-2s on the high ground near checkpoint 12. They look like a security outpost. The platoon engages with immediate suppression from the troop's 120-mm mortars. As the rounds impact, the outpost withdraws behind the ridgeline to the north. The 3d platoon determines the dominant terrain near checkpoint 21 is clear of threat forces. The mortar section crosses the river at checkpoint 8. The 2d platoon follows, crossing at checkpoint 8, while the 4th platoon uses the ford near checkpoint 3. The troop CP bounds forward to high ground near checkpoint 17. The troop trains cross the river and move into the woods on the northern bank.

The scout platoons continue reconnaissance while moving toward the troop LOA. As the 3d platoon crosses the ridge (PL IOWA), Bravo section receives heavy machine-gun fire from concealed positions south of checkpoint 2. The platoon immediately deploys, reports contact to the troop commander, and calls for indirect fire to suppress the threat (see Figure 3-15).

Under this protection, the platoon, using primarily a dismounted technique, reconnoiters to the flanks and rear to develop the situation. The troop commander acknowledges the report, and moves immediately to link up with the 3d platoon leader. The troop commander orders the 1st platoon to continue its reconnaissance to locate the enemy's flank and coordinate with the FIST to switch indirect fires to the squadron howitzer battery. The 3d platoon is organized in a three-section, three-vehicle configuration; the platoon leader is with the Charlie section. The platoon leader sends his Bravo section forward on the enemy's left and his Alpha section on the right to determine if the enemy is mutually supported by other threat forces from the flanks or rear. The Charlie section sergeant and a two-man dismount team move in closer and determine that the enemy consists of an MRP with three BTR 70s and one BMP-2. The Alpha scout section and scouts from 1st platoon identify the right flank of the threat and discover the flank is unprotected by obstacles and is exposed to direct fires.

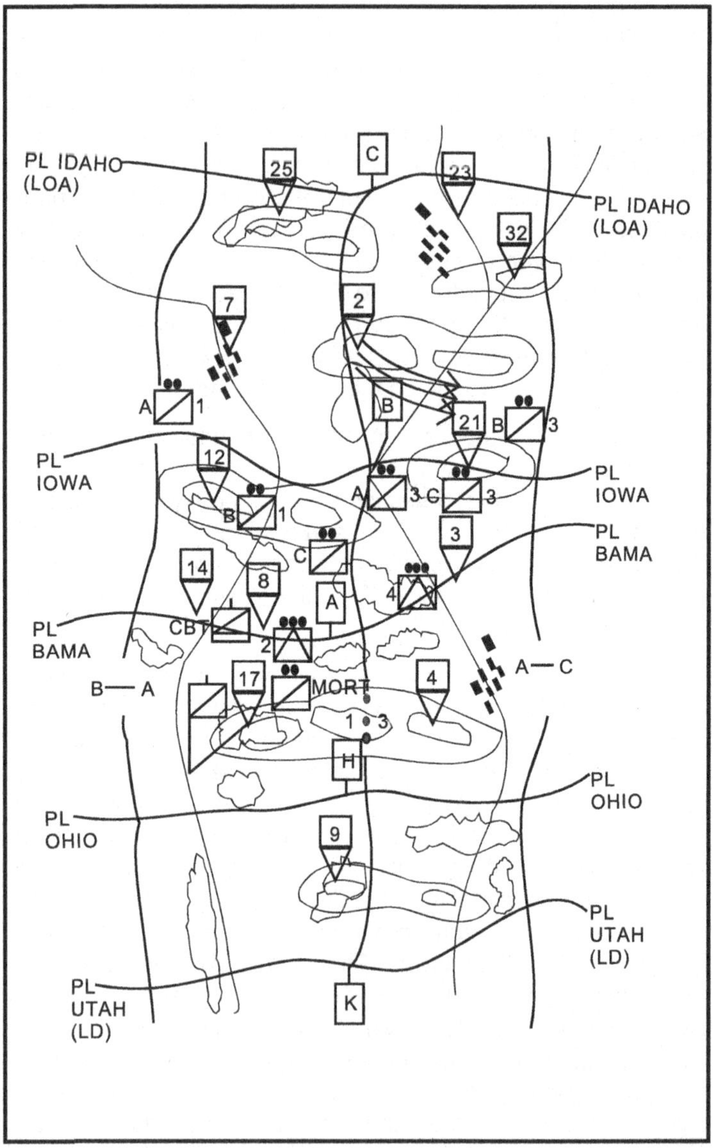

Figure 3-15. Zone reconnaissance (part two).

While en route to the 3d platoon's zone, the troop commander tells the 1st platoon to continue reconnaissance forward to PL IDAHO and determine if other enemy forces are providing mutual support to the enemy in the 3d platoon's zone. At PL IDAHO, he tells the platoon to establish a screen. The troop commander tells the 4th platoon to move over and join the 3d platoon Charlie section in overwatch of the enemy MRP, and prepare to overwatch with fires. The troop commander tells 2d platoon to prepare to conduct a hasty attack of the enemy MRP. The 1st platoon scout squad moves to checkpoint 12, links up with the 2d platoon, and guides them into position to attack the threat platoon.

The 2d platoon reports they are set in the assault position. The 4th platoon reports they are set in the overwatch position with the 3d platoon Charlie section. The 3d platoon leader reports that his dismount teams from his Alpha and Bravo sections are out of the target area and set. The 1st platoon leader reports he is set in a screen along PL IDAHO. The troop commander orders the FSO to fire for effect on the enemy position at his command. The commander issues orders and guidance for limits of fire and signals for the initiation, shifting, and lifting of fires. The troop commander issues the fire command for indirect fire over the command net and follows with a fire command for direct fire for the 4th platoon and the Charlie section of 3d platoon. The indirect fire from the squadron's howitzer battery and the troop mortars impact on the enemy position and is immediately followed by the fire from four TOWs, MK-19s, and M2 machine guns. The commander then issues the orders for 2d platoon to initiate its move to the attack-by-fire position followed by the order for the support element (4th platoon, Charlie section of 3d platoon) to shift fires. The troop FSO shifts fires to the rear of the enemy position. The 2d platoon moves forward and engages the enemy target from the flank to complete the destruction of the enemy force (see Figure 3-16).

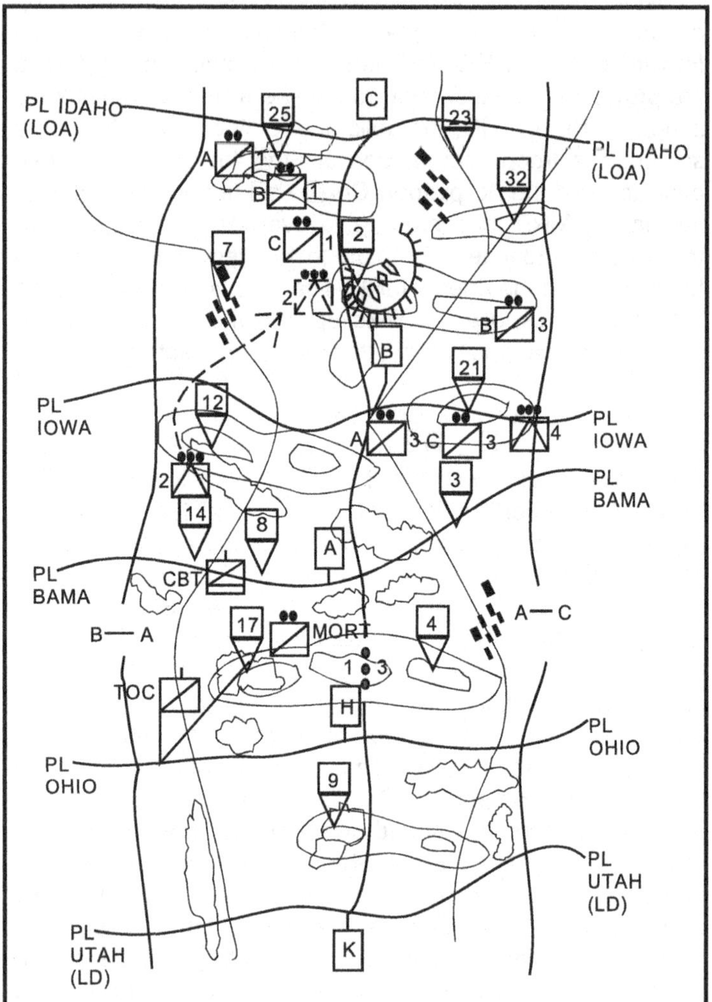

Figure 3-16. Zone reconnaissance (part three).

The antitank platoons consolidate in covered and concealed positions west of checkpoint 2. Scout dismount teams from the 3d platoon move in quickly, capture two wounded prisoners, and search the vehicles and personnel. The 3d platoon leader orders one scout squad to stay and secure the prisoners, and then completes his reconnaissance up to the LOA. The 1SG leads the medics to the location of the 3d platoon scout squad and the EPWs. The XO collects and transmits final reconnaissance reports to the squadron.

AIR-GROUND COORDINATION

Cavalry troops will often conduct a coordinated reconnaissance with an air cavalry troop. Because of the air cavalry troop's ability to maneuver faster and look deeper than ground cavalry, they are often placed to the flanks or forward of ground cavalry. The air cavalry troop should focus its reconnaissance efforts on areas that impede ground cavalry movement.

The air cavalry troop quickens the pace or tempo of the ground reconnaissance because it provides the ground unit with added security during movement and increased situational awareness once contact is made. The air cavalry troop can gain initial contact with the enemy or a critical piece of terrain. Once the air troop develops the situation with his assets, he can pass off the information/contact to the ground troop (see Figure 3-17).

- Ground troop and air troop coordinate zone reconnaissance operations.
- Air troop divides the zone into platoon zones.
- Air troop will use checkpoints to identify critical areas. Checkpoints should enhance ground troop movement throughout the zone.
- Platoon on the right reconnoiters forward of ground troop, identifies platoon ford site, and conducts route reconnaissance.
- Platoon on the left reconnoiters forward of ground troop, supports the other platoon in conducting route sweep, checks lateral routes in the zone, and looks for sites along the obstacle (river/road).

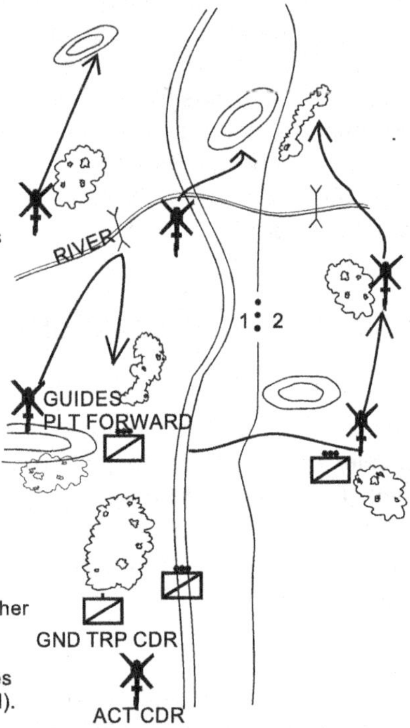

- ACT commander and ground troop commander identify critical areas where air-ground coordination is required or needed.
- ACT commander positions himself close to the ground troop commander (if needed).
- If the ford site or bridge site is blocked, ACT elements reconnoiter the site forward of the location. They also find a bypass to the obstacle and lead ground troops to bypass the location. The ACT will start the screen forward of ground elements.

Figure 3-17. Air-Ground troop coordinated reconnaissance.

Section V. Area Reconnaissance

An area reconnaissance is a specialized form of zone reconnaissance. It is a mission conducted to gain detailed information about terrain features and enemy forces within a specified area or point that other forces intend to occupy, pass through, or avoid. A commander usually calls for area reconnaissance before he sends his forces into or near an area to avoid being surprised by actual terrain conditions or unexpected enemy forces.

CRITICAL TASKS

During an area reconnaissance, the following critical tasks must be accomplished unless the SCO directs the troop to do otherwise:

- Reconnoiter all terrain within the area.
- Inspect and classify all bridges within the area.
- Locate fords or crossing sites near all bridges within the area.
- Inspect and classify all overpasses, underpasses, and culverts.
- Locate and clear all mines, obstacles, and barriers in the area within its capability.
- Locate a bypass around BUAs, obstacles, and contaminated areas.
- Find and report all enemy within the area.
- Report reconnaissance information.

TECHNIQUES

An area reconnaissance is conducted like a zone reconnaissance. When the troop receives an area reconnaissance mission, the assigned area is identified as the terrain inside a solid, continuous boundary. Planning the movement to the area is the first step. Select the route(s), establish a march order on each route, and specify a start point, checkpoints, and a release point. Use a movement technique that keeps the troop moving quickly and securely. If possible, avoid contact with the enemy while en route. Report and bypass. Be careful when approaching the area to begin the reconnaissance. Use common sense. Avoid known enemy forces outside of the area where reconnaissance will be conducted; start somewhere else.

As in a zone reconnaissance, enclose the area within a troop zone. Draw a line of departure, an LOA, and lateral boundaries. Divide the troop zone into two platoon zones. Add phase lines along identifiable terrain to control movement through the area. Place contact points at the intersection of the platoon boundary and all phase lines. Place TIRS on the map. To identify specific areas or features, use checkpoints for reference (see Figure 3-18).

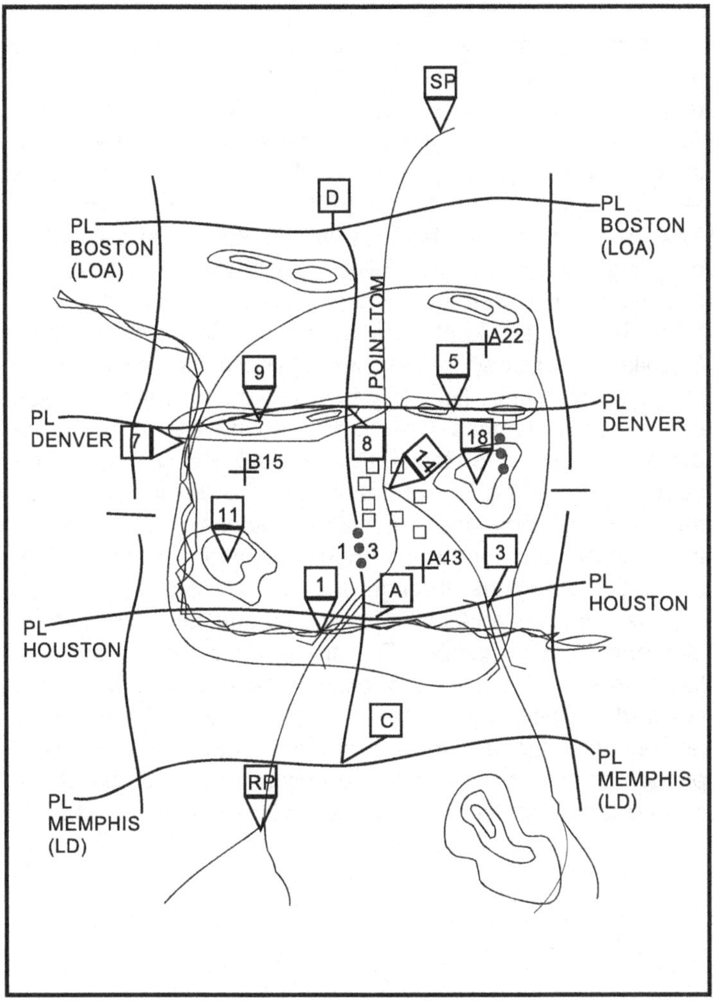

Figure 3-18. Area reconnaissance.

Use a troop vee or split-vee to conduct an area reconnaissance. Scout platoons deploy abreast from the release points to accomplish all the reconnaissance tasks. They move out across the line of departure first. The mortar section moves through the center of the troop zone, remaining in position to range from 3 to 3.5 kilometers forward of the lead elements in the scout platoons. The mortar section could also follow one of the scout platoons, depending on where initial enemy contact is expected. The tank platoons are kept back far enough in depth to retain flexibility, but remain responsive to situations developed by the scout platoons. They key their advance on the progress of the scout platoons. The tank platoons are told how far behind the scout platoons they should stay as the troop deploys across the line of departure. The distance between the tank platoons and the scout platoons will usually be adjusted as the terrain or enemy situation changes. The troop CP displaces through the zone using terrain that affords effective and continuous communication with troop elements and squadron. The troop trains follow about 2 kilometers behind the tank platoons and bound from one covered position to another. The troop commander positions himself well forward to observe the action of one scout platoon or the other. The location usually depends on where initial enemy contact or problem situations are expected.

Chapter 4

Security

Security is an essential part of all offensive and defensive operations. Cavalry provides security for the commander along an exposed front, flank, or rear of the main body where a threat may exist. Surveillance is continuous during security operations. Even during security missions that involve fighting the enemy, the scouts' primary task remains gathering information. Scouts do this by establishing OPs, conducting patrols, and performing reconnaissance.

Counterreconnaissance is an inherent task in all security operations. Counterreconnaissance is not a mission. It is the sum of actions taken at all echelons to counter enemy reconnaissance and surveillance efforts through the depth of the area of operations. Counterreconnaissance denies the enemy information about friendly units. It is both active and passive and includes combat action to destroy or repel enemy reconnaissance elements.

CONTENTS

Section I. Purpose, Fundamentals, and Capabilities

Security includes screening operations, guard operations, covering force operations, area security operations, convoy security operations, and route security operations. The squadron performs screen, guard, and route security missions. Covering force operations are normally the mission of a cavalry regiment. Separate brigades or task organized divisional brigades may perform cover operations as well.

Both the heavy and light cavalry troops perform two security missions (screen and convoy security) independently or as part of their parent squadron. Both also participate in guard, cover, and route security missions as part of their parent squadron or regiment. Either troop will normally perform reconnaissance, screen, defend, delay, attack, or a combination of these missions in support of their parent squadron or regiment (see the applicable section of Chapters 3, 5, and 6 for specifics).

PURPOSE

Security operations are designed to obtain information about the enemy and to provide reaction time, maneuver space, and protection to the main body. Security operations are characterized by conducting continuous reconnaissance to reduce terrain and enemy unknowns, gaining and maintaining contact with the enemy to ensure continuous information, and providing early and accurate reporting of information to the protected force.

FUNDAMENTALS

Five fundamentals are common to all cavalry security operations.

Orient on the Main Body. As a security force, the troop will be operating at a specified distance from a main body, between it and a known or suspected enemy force. If the main body moves, the troop also moves. The troop commander must know how the main body commander intends to maneuver his forces and where he wants the troop in relation to his movement. The troop commander maneuvers his troop to positions where he can provide the needed security.

Perform Continuous Reconnaissance. The troop's security and the security of the main body come in large measure from knowing everything about the terrain and the enemy within the troop's area of operations (AO). Reconnaissance and continuous patrolling go hand in hand with security operations. Cover all the ground in the AO. Determine what the terrain will allow the troop *and* the enemy to do.

Provide Early and Accurate Warning. Early and accurate warning of enemy approach is the cornerstone of security operations. The main body commander needs as much time as possible to shift and concentrate his forces to meet and defeat an unexpected enemy attack. Put observers in positions that afford long-range observation of expected enemy avenues of approach. Use aeroscouts and ground surveillance radar (GSR), if available, to enhance their ability to see. Place remote sensors in the ground to monitor avenues of approach that cannot be easily observed. If possible, send dismounted or mounted patrols forward of OPs to extend their ability to see, providing additional reaction time for the main body commander.

Provide Reaction Time and Maneuver Space. All security operations are designed to provide reaction time and maneuver space for the main body so it can deal effectively with an unexpected enemy attack. The troop provides early warning to the main body commander and sometimes has to fight hard to buy time and space so the main body's combat power can be concentrated on defeating the enemy.

Maintain Enemy Contact. Once visual or physical contact with the enemy is gained, do not allow him to break the contact. Maintain enemy contact and continue to report his activities until told to stop.

CAPABILITIES

Capabilities of the heavy troop include—

- Screen up to a ten-kilometer-wide sector.
- Maintain continuous surveillance of up to six battalion-size avenues of approach.*
- Can establish up to 12 short-duration OPs.

Capabilities of the light troop include—

- Screen up to a ten-kilometer-wide sector.
- Maintain continuous surveillance of up to six battalion-size avenues of approach.*
- Can establish up to 16 short-duration OPs.

*The maximum six long-duration OPs either cavalry troop can occupy is a function of personnel required to perform the following tasks at each OP:

- Man the actual OP.
- Maintain radio communications with the OP and with the platoon leader.
- Provide local security for the vehicles.
- Conduct dismounted patrols as required.
- Conduct resupply.
- Perform maintenance.
- Sleep/rest.

The ability to perform the above tasks simultaneously for periods in excess of 12 hours requires at least 9 to 10 personnel (collocation of two scout squads for M3-mounted scouts, or three scout squads for HMMWV-mounted scouts).

Section II. Screen

Screen is the most common security mission heavy and light cavalry troops conduct. Both troops conduct screen missions for their parent squadrons or other forces to—

- Provide early warning of enemy approach.
- Provide real-time information, reaction time, and maneuver space to the protected force.
- Destroy enemy reconnaissance elements within their capability (perform counterreconnaissance).
- Impede and harass the enemy.

The screen mission provides the least amount of protection of any security mission, and is appropriate when operations have created extended flanks, when gaps between forces exist and cannot be secured in force, or when required to provide early warning over gaps that are not considered critical enough to require security in greater strength. A commander normally assigns cavalry this mission when he needs time to respond to an unexpected enemy attack, and cannot afford to commit other forces to the task.

The screen mission is defensive in nature. As such, both heavy and light cavalry troops screen the front, flanks, and rear of a stationary force but only to the flanks or rear of a moving force. Screening operations are not performed forward of a moving force. Zone reconnaissance or movement to contact is the appropriate cavalry troop mission suited to the requirements of the offensive force.

CRITICAL TASKS

A screen mission has certain critical tasks that guide planning. To achieve the intent of a screen mission, the troop must accomplish the following critical tasks:

- Maintain continuous surveillance of all battalion-size avenues of approach into the troop sector under all visibility conditions.

- Destroy or repel all enemy reconnaissance patrols (if within the unit's capability and directed by the higher commander).

- Locate lead elements of enemy order of battle and determine the direction of movement of each.

- Maintain contact with, report the activity of, and impede and harass the enemy while displacing.

EXTENDED SCREENS

Both the heavy cavalry troop and the light cavalry troop normally screen a sector up to ten kilometers in width. However, METT-T may dictate an extended screen across frontages in excess of the norm. Either troop's ability to accomplish its critical tasks, or its ability to screen in depth, can diminish rapidly as frontages increase.

Examples of extended screens are described below.

- Heavy cavalry troop screens 20 kilometers of southern bank of unfordable river crossed by four bridges in sector.

- Light cavalry troop screens 25 kilometers of desert terrain, from dominant ridge.

DEPTH

Depth is also important in a screen. The term *"screen line"* is descriptive only of the forward trace along which security is provided. Depth allows an enemy contact to be passed from one element to another without requiring displacement. Depth is advantageous to—

- Destroy an enemy reconnaissance patrol without compromising critical OPs.
- Prevent an enemy from penetrating the screen line too easily.
- Prevent gaps from occurring when OPs displace or are lost.
- Maintain contact with moving enemy without compromising OPs.
- Prevent enemy templating of the screen line.

Depth is achieved primarily by positioning OPs, particularly where there are limited avenues of approach. Tank platoons, AT platoons, the mortar section, and attached elements positioned behind the screen line establish local security and provide surveillance. The degree to which depth can be attained is a function of many factors, which include—

- Higher commander's intent and concept as expressed in—
  - Graphical trace of the screen line (LOA).
  - Engagement criteria.
  - Destruction criteria.
  - Displacement/disengagement criteria.
- Width of the sector.
- Depth of the troop sector.

- Terrain and avenues of approach it will support.
- Attachments and detachments.

Screening is largely accomplished by establishing a series of OPs and conducting patrols to ensure adequate surveillance of the assigned sector. Screens are active operations. Stationary OPs are only one part of the mission. Employing patrols (mounted and dismounted), aerial reconnaissance, ground-based sensors, intelligence from space-based sensor systems, and OPs relocated on an extended screen ensure that continuous overlapping surveillance occurs. Inactivity in an immobile screen promotes complacency.

PLANNING CONSIDERATIONS/GUIDANCE TO SUBORDINATES

The enemy situation is often vague when planning a screen. The troop should develop plans that are flexible enough to react to any enemy course of action, particularly the worst case. Planning considerations for a screen should include a detailed description of how contact with the enemy reconnaissance will be gained then how and where it will be destroyed. Planning should also cover the method of displacement once the main body of the enemy force has been identified and how that force will be handed off to the main body in the main defensive belt. Because of the need for flexibility, screen operations will often begin to inherit the characteristics of defense or delay missions. To cover the displacement of scouts, some elements of the troop may be required to execute missions such as delay or defend.

Screen operations at troop level usually occur in four phases.

- Movement to and occupation of the screen line.
- Surveillance and counterreconnaissance.

- Gaining and maintaining contact with the enemy main body displacement of screen.
- Rearward passage of lines.

Higher command guidance should address each phase of the operation and cover at least the following:

- Location/orientation/width of the screen.
- Duration of the screen.
- Method of movement to and occupation of the screen line.
- Location and disposition of the friendly force being screened.
- Engagement/destruction criteria.
- Displacement/disengagement criteria.
- Follow-on missions.
- Positioning and orientation guidance for GSRs (if attached).

The following must be considered when developing and completing the plan and executing the screen mission:

Time Screen Must Be Established. The time the screen must be set and active will influence the troop's method of deploying to and occupying the screen line.

Movement to the Screen Line. If the screen mission is the result of a previous tactical maneuver such as zone reconnaissance, the troop will essentially be postured to begin screening from present positions. This situation occurs frequently, and may be the result of a FRAGO to halt at a specified phase line.

If the troop is not currently sited on the screen line, obviously, deployment to the screen line must occur before actually beginning the screen mission. Time determines the method of occupying the screen line. Thorough analysis of METT-T will determine which deployment technique or combination of techniques best meets mission requirements.

Trace and Orientation of Screen Line. The initial screen line is depicted as a phase line and often represents the forward line of own troops (FLOT). As such, the screen line may be a restrictive control measure for movement (limit of advance); coordination/permission would be necessary to move beyond the line to establish OPs or to perform reconnaissance. When occupied, OPs are sited on or behind the phase line. OPs should be given specific orientation and observation guidance.

Initial OP Locations. The squadron or troop commander may determine tentative initial OP locations to ensure effective surveillance of the sector and designated named area(s) of interest (NAI). At a minimum, the troop commander designates a primary orientation for the scouts during the conduct of the screen. Scouts, once set on the screen line, will report their location to the troop TOC and verify they are in compliance with the commander's orientation guidance. The scouts who occupy each OP always retain the responsibility to modify the location to achieve the commander's intent and guidance for orientation. The OPs are positioned along or behind the screen line.

OPs are generally categorized based on their expected duration of employment as either—

- Short-duration (less than 12 hours)

  or
- Long-duration (more than 12 hours).

OPs may be either mounted or dismounted. Mounted OPs maximize use of vehicular optics, weapon systems, and speed of displacement, but are more readily detected by the enemy. Dismounted OPs provide maximum stealth at the expense of speed of displacement, and vehicle-mounted optics and weapons.

A heavy cavalry troop can occupy up to 12 short-duration OPs (one per scout squad, six per scout platoon). For extended periods of time, the heavy troop can occupy six OPs (one per scout section, three per scout platoon).

A light cavalry troop can occupy up to 16 short-duration OPs (one per scout squad, eight per scout platoon). For extended periods of time, the light troop can occupy six OPs (one per scout section, three per scout platoon).

Width and Depth of the Screened Sector. The troop sector is defined by lateral boundaries extending out to a limit of advance (the initial screen line), forward of a rear boundary. The troop sector is established by the squadron or unit being screened. The troop rear boundary may be a squadron phase line and may serve as a battle handover line (BHL) to control passing of responsibility for the enemy to the protected force. The troop's ability to gain depth decreases as screened frontage increases.

Locations of Subsequent Screen Lines. The squadron or troop commander uses additional phase lines to control the operation. These phase lines may serve as subsequent screen lines. Displacement to the subsequent screen lines is event driven.

Scout Platoon Sectors. Assign clear responsibility of identified avenues of approach and designated NAIs. The nature of a screen normally requires both scout platoons to deploy abreast.

Tank/AT Platoon Sectors. Position the tank (heavy troop) or AT (light troop) platoons in the scout platoons' sectors. They may occupy hide or battle positions along avenues of approach. The tank/AT platoons remain responsive to the troop commander. They are the primary direct-fire killing asset.

Force to be Screened. The troop must orient on the force it is securing. If the main body is moving, the troop must move to maintain the screen's position relative to the main body.

Reinforcements. Any unique requirement posed by the mission may require assets not organic to the troop. GSR and engineers are common attachments at troop level.

- **GSR.** During screen operations, GSR is used to augment scout OPs and to add depth to the screen. GSRs should be attached to scout platoons, and the commander should provide the scout platoon leader with positioning and orientation guidance.
- **Engineers.** If engineers are attached to the troop, the troop commander should assign them with priority of mission and priority of effort in support of commander's guidance. During screen operations, engineers will normally dig survivability positions for scouts and tanks. Also engineers may emplace obstacles in support of the counterreconnaissance battle or assist the troop with displacement of the screen once contact has been established with the enemy body.

Special Requirements or Constraints. Specify all requirements for observing any NAI identified during the intelligence preparation of the battlefield (IPB). Task subordinate platoons as required. Specify the following in the troop OPORD:

- Engagement criteria.
  - What size force will scouts engage and destroy (if any)?
  - Where will this action occur?
  - What size force will the tank or AT platoons engage and destroy?
  - Where will this action occur?

- Disengagement criteria.
  - What event will cause scouts to displace from the initial screen line?
  - How will scouts maintain contact with the enemy while displacing?
  - What event will cause the tank or AT platoons to displace to subsequent or alternate positions?

Indirect Fire Planning. Fire planning integrates artillery and mortar fires. Position the troop mortars to fire up to two-thirds of their maximum range forward of the initial screen line. A wide sector may require the troop commander to position them to provide effective coverage of the most likely avenue of approach determined by IPB. The troop FSO plans artillery fires to adequately cover any gaps in mortar coverage.

Direct Fire Planning. Based on his analysis of the terrain, the troop commander determines where to engage the enemy (engagement areas). He also determines the location of battle positions that provide observation, fields of fire, and cover and concealment that support each engagement area.

Positioning of C2, CS, and CSS Assets. The troop commander positions himself to observe the most dangerous enemy avenue of approach. The troop TOC positions itself in depth to provide continuous control and reporting during initial movements. After the screen line has been reestablished in depth following displacement from the initial screen line, the TOC can reposition. Combat trains position behind masking terrain close enough for rapid response. They are best sited along routes providing good mobility laterally and in depth.

Patrol Requirements. Patrols may be required to cover gaps between OPs. The troop commander tasks the scout platoon leaders to perform specific patrols.

Coordination. The troop commander coordinates his concept with air cavalry troop (ACT) commanders who may be operating the same ground, flank troop commanders, tank company (heavy cavalry) and AT company (light cavalry) commanders, and other unit commanders as appropriate.

TECHNIQUES

Graphics

The following is a list of common graphic control measures used to control screen missions:

- Boundaries.
- Phase lines.
- Checkpoints.
- Contact points.
- NAIs.
- OPs.
- Mortar firing positions.
- Battle positions.
- Hide positions.
- TIRS.
- Unit symbols.

Movement to the Screen Line

In deploying to the screen line, the troop commander must deal with the competing requirements to establish the screen quickly to meet mission requirements and to provide the necessary level of security for the troop in doing so. The troop moves to the screen line using one of three basic methods—a tactical road march, zone reconnaissance, or movement to contact.

Tactical Road March. The troop conducts a tactical road march to a release point behind the screen line. From the release point, platoons deploy to occupy initial positions. This method of deploying to the screen line is the fastest, but least secure. It is appropriate when enemy contact is not expected and time is critical, or when an air cavalry troop is conducting zone reconnaissance forward of the ground troop.

Movement to Contact. The troop conducts a movement to contact from a line of departure to the initial screen line. This method is slower than a tactical road march, but more secure. It is appropriate when enemy contact is likely, time is limited, or when an air cavalry troop is conducting zone reconnaissance forward of the ground troop.

Zone Reconnaissance. The troop conducts a zone reconnaissance from a line of departure to the initial screen line. Given adequate time, this method is preferred as the troop can clear the zone of any enemy and enables platoons to become thoroughly familiar with the terrain. The troop can reconnoiter potential subsequent OP locations, battle and hide positions, and mortar firing positions, for example, as they move to the screen line. A zone reconnaissance is appropriate when time is available and information about the enemy or terrain is unknown.

Security Drill

A security drill is a series of rehearsed actions (battle drills) a scout platoon or cavalry troop takes to maintain contact with the main body of an advancing enemy force. It is used when collapsing the screen line to subsequent OP positions or when transitioning from a screen mission to a delay or defend mission.

At platoon level, OPs gain contact with the enemy main body, then report and prepare to displace to a subsequent position. When the enemy force reaches the OP's break point (point where the OP must displace or his position/movement will compromise him to the enemy), the OP passes off the responsibility to track the enemy to another OP in depth. The platoon displaces its OPs to subsequent positions in depth while maintaining contact with the enemy.

At troop level, the security drill combines the collapse of the initial screen line with the actions of organic tank or AT platoons. Scout platoons may perform platoon security drills initially, consolidating some or all of their combat power at a battle position to aid execution of a troop-level engagement.

At platoon and troop levels, conduct of security drills is tempered by the commander's overall concept, intent, and scheme of maneuver. Enemy actions (events) drive security drill execution (response) (see Figure 4-1).

STATIONARY SCREEN

Take a close look at the high-speed avenues of approach into the sector. Divide the sector into two platoon sectors. Make sure the platoon boundary is on easily identifiable terrain. Do not split avenues of approach with a platoon boundary or place the boundary on a road. Place NAIs, TIRS, or checkpoints beyond the screen line to focus surveillance from OPs. If needed, add additional phase lines to control withdrawal of the troop at 5- to 8-kilometer intervals. Place contact points at the intersection of the platoon boundary and all phase lines. Place TIRS on the map or overlay as described in Chapter 2.

Deploy the scout platoons abreast and establish a series of OPs along or behind the initial screen line, as terrain allows, but never forward of it without permission. Make it clear to the scout platoon leaders which avenues of approach (depicted as NAIs or checkpoints) they are to observe. If unobservable areas between OPs need to be routinely checked, have the scout platoon leaders prepare patrol plans for approval and subsequent execution.

Position the mortar section to fire 3 to 4 kilometers forward of the initial screen line, oriented on the expected enemy avenue of approach. Establish subsequent firing positions for the mortar section back through the sector. Position each tank (heavy) or AT (light) platoon in one of the scout platoon sectors, or have the tank/AT platoons consolidate. Position the tank/AT platoons in hide positions or battle positions in depth behind the scout platoons, oriented on the expected enemy avenue of approach. Establish subsequent positions for the tank/AT platoons back through the sector to support the scheme of maneuver.

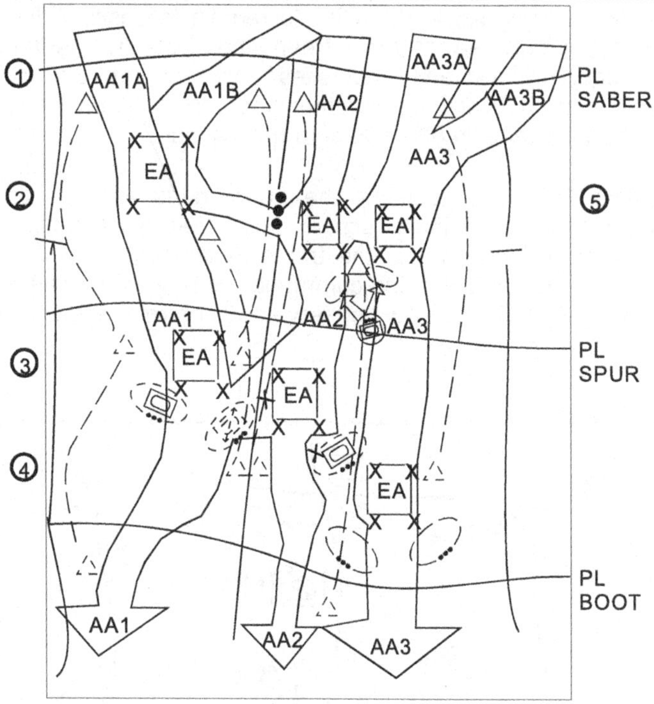

INDEX:

1. Initial screen line.
2. Scout platoon engagement area dependent on commander's intent.
3. Subsequent OPs.
4. Troop consolidates fire power of tanks and scouts to execute engagement while maintaining contact.
5. Scouts maintain contact and execute engagement with tank platoon.

Figure 4-1. Scout platoon/cavalry troop security drill.

Have the XO position the command post on terrain that affords good FM radio communications with troop elements and squadron headquarters. If possible, the command post should be positioned behind the subsequent screen positions. This allows the TOC to remain in position during the initial collapse of the screen line. Establish tentative subsequent command post sites back through the troop sector. The XO retains the authority to adjust the actual TOC location to maintain effective communications.

The 1SG first positions the troop trains within 5 to 8 kilometers of the initial screen line, then he establishes subsequent locations for the trains to bound back through the troop sector. The troop commander positions himself well forward where he can best observe and control the actions of the troop (see Figure 4-2).

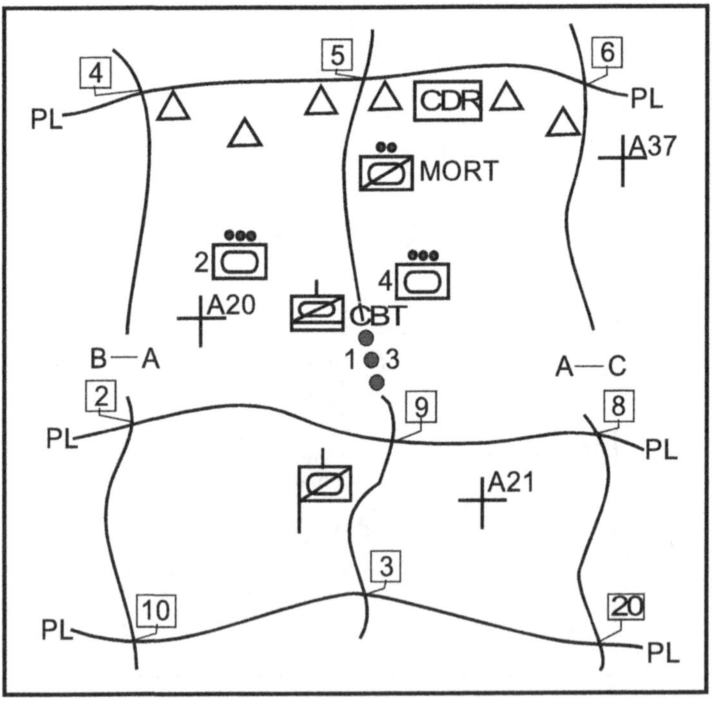

Figure 4-2. Troop screen positioning.

MOVING FLANK SCREEN

The troop may conduct a moving flank screen by itself or as part of the squadron. It may be tasked to screen the exposed flank of the squadron while the squadron conducts a movement to contact, a hasty attack, or a zone reconnaissance (see Figure 4-3). The troop may also participate in a squadron mission to screen or guard another combined arms force (see Figure 4-4).

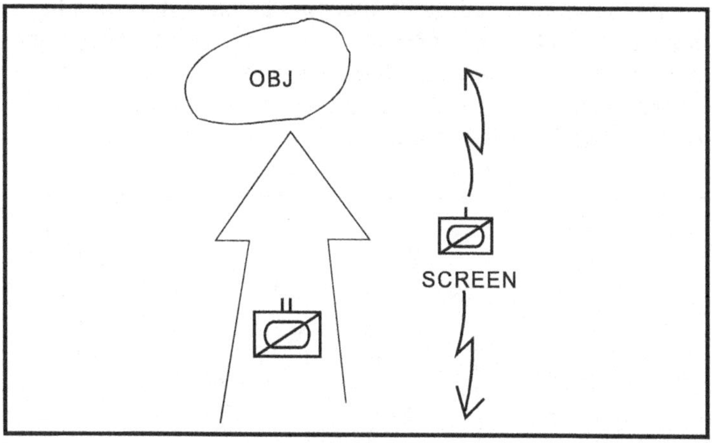

Figure 4-3. Squadron movement to contact; troop flank screen.

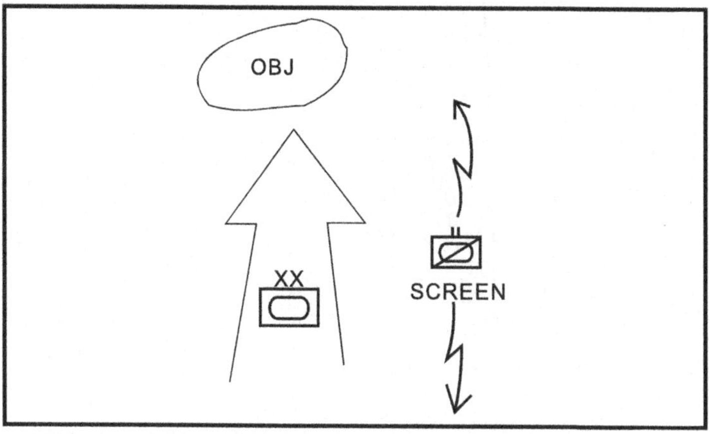

Figure 4-4. Squadron flank screen.

A moving flank screen uses the same techniques as when screening a stationary unit. Position the scout platoons to maintain continuous surveillance on the avenues of approach, the mortar section to cover likely avenues of approach with indirect fire, and the tank/AT platoons in depth to destroy or repel enemy reconnaissance units.

Applying Graphic Control Measures

Because of the inherent dual orientation of a moving flank screen (direction of movement versus orientation of the screen line), control of the operation poses numerous challenges. Control measures must facilitate both orientations (see Figure 4-5).

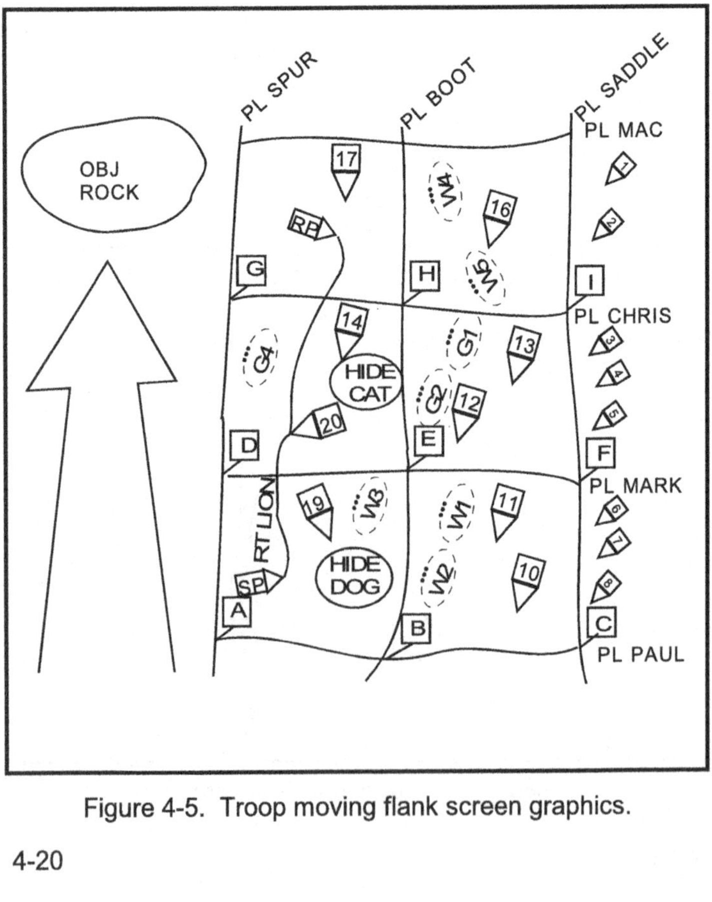

Figure 4-5. Troop moving flank screen graphics.

Follow the procedures below when applying graphic control measures to a moving flank screen.

- Use phase lines to control the scout platoon's movement (placed perpendicular to the screen line). Plan to use these phase lines as on-order boundaries for subordinate platoons if enemy contact is gained. Place phase lines no more than 5 kilometers apart (corresponding to the width of a scout platoon's screen frontage). Do not divide avenues of approach with them.

- Use additional phase lines rearward of (parallel to) the initial screen line to control retrograde movement (toward the protected force). Plan to use these phase lines as subsequent (on-order) screen lines.

- Use objectives, checkpoints, or axes of advance to control the movement of the tank/AT platoons for movement. Position and plan to use these objectives or checkpoints as battle positions for the tank/AT platoons if the scouts make enemy contact at the screen line. Plan subsequent (on-order) battle positions between the screen line and the protected force.

- Use mortar firing positions or checkpoints to control movement of the mortar section. Position these firing positions rearward of the screen line where they allow the mortars to fire two-thirds maximum range forward or cover likely avenues of approach. Plan subsequent (on-order) mortar firing positions between the screen line and the protected force.

While this number of graphic control measures and required planning may seem excessive, they provide maximum flexibility in terms of mission execution. The troop commander can issue simple FRAGOs to adjust the plan to the enemy situation.

Repositioning the Screen

The troop must reposition to stay oriented on the force it is securing. Movement along the screen line is determined by the speed of the protected force. Movement is conducted by one of three techniques—continuous marching, bounding by platoons (alternately or successively), and bounding by OPs (alternately or successively).

Continuous Marching. This technique is appropriate when the protected force is moving quickly and contact is not likely. It is the least secure movement technique.

Deploy both scout platoons abreast with the two tank or AT platoons, the mortar section, and the remainder of the troop in depth (between the screen line and the protected force). The trace of the screen line is essentially the route of advance for the two scout platoons in column. Have the tank/AT platoons and the remainder of the troop move along a designated route or axis of advance (see Figure 4-6).

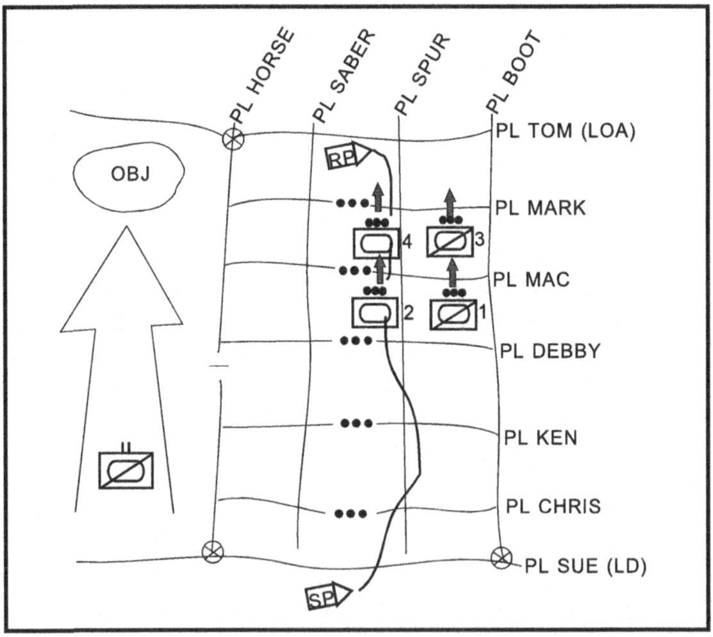

Figure 4-6. Troop moves by continuous marching.

Bounding by Platoons. This technique is appropriate when the protected force requires greater protection than afforded by continuous marching, is not moving quickly, or knows enemy contact is possible. Bounding platoons alternately may leave temporary gaps in the screen line as they move. Bounding platoons successively is more secure but slower than bounding platoons alternately.

Deploy both scout platoons abreast with the two tank/AT platoons, the mortar section, and the remainder of the troop in depth (between the screen line and the protected force). Alternately bound one scout platoon around (to the rear of) the other to assume new positions along the screen line (see Figure 4-7), or successively bound the scout platoons along the screen line (see Figure 4-8). Have the tank/AT platoons occupy designated positions sequentially or alternately.

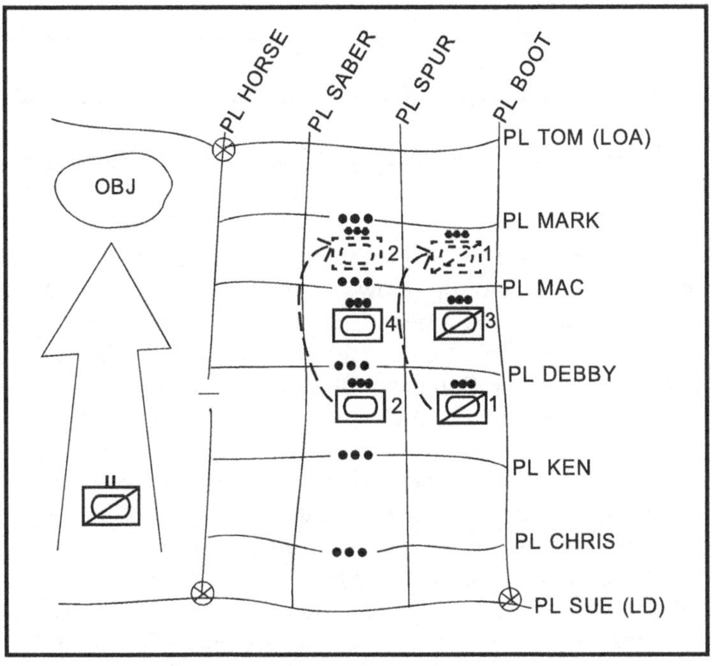

Figure 4-7. Troop moves by alternately bounding platoons.

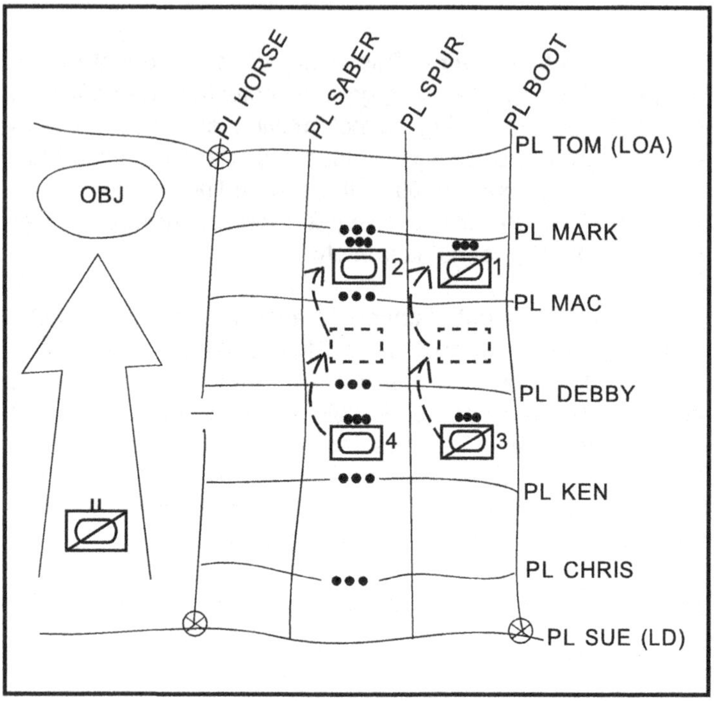

Figure 4-8. Troop moves by successively bounding platoons.

Bounding by OPs. This technique is appropriate when the main body is moving slowly, contact is possible, and maximum security is required. Bounding OPs alternately will disrupt the integrity of the scout platoons as OPs bound to their next position. Bounding OPs successively is easier for the scout platoons to control.

Deploy both scout platoons abreast with the two tank/AT platoons, the mortar section, and the remainder of the troop in depth (between the screen line and the protected force). Alternately bound the rearmost OP around (to the rear of) the other OPs (both platoons) to assume a new position along the screen line (see Figure 4-9), or successively bound the OPs along the screen line (see Figure 4-10). The number of OPs on the screen line may be reduced, as two or more may be bounding at any given time. The rate of advance of the protected force will determine this. Have the tank/AT platoons occupy designated positions sequentially or alternately.

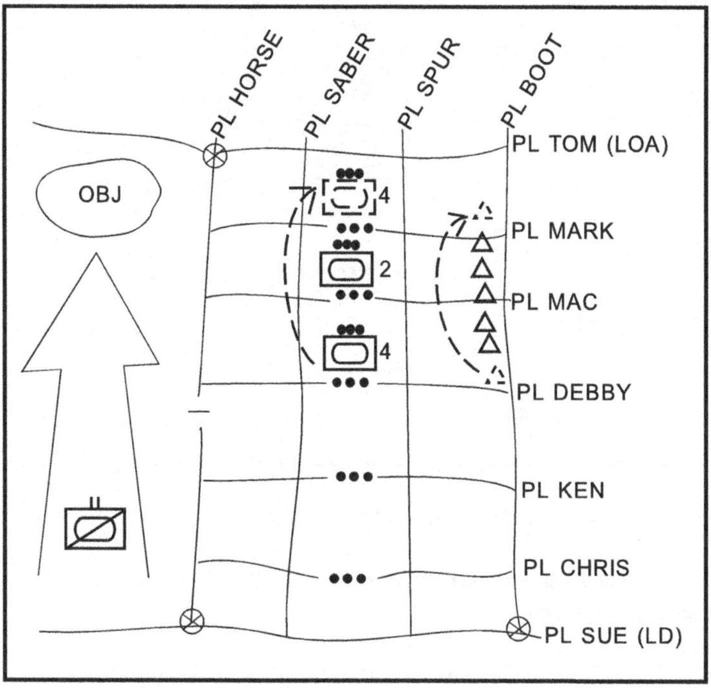

Figure 4-9. Troop moves by alternately bounding OPs.

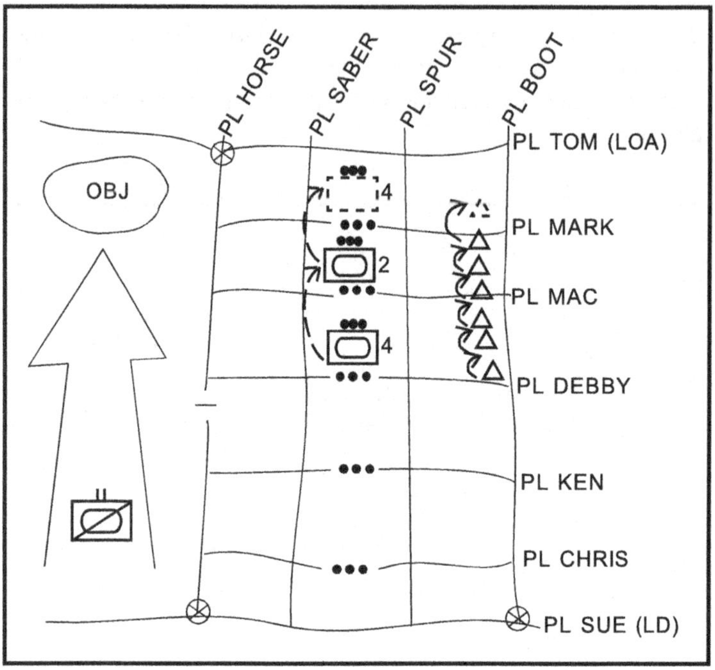

Figure 4-10. Troop moves by successively bounding OPs.

Example of a Screen Forward of Stationary Main Body (Heavy Troop)

The 1st and 3d platoons deploy abreast along PL BOOT, the initial screen line, and establish OPs to maintain continuous surveillance of the high-speed avenues of approach into the troop sector. Route 220 in the 1st platoon sector is the most likely route of enemy approach. Scout platoons execute their approved patrol plans between OPs, and periodically check the areas that cannot be observed. The mortar section lays its guns in a centrally located position about 3 kilometers behind the initial screen line and orients on Route 220. The 2d platoon occupies a hide position behind the 1st platoon. The 4th platoon occupies a hide position north of checkpoint F. Both positions provide good lateral movement and several maneuver options. The troop CP positions south of TIRS A21. The 1SG positions the troop trains in the woods just north of contact point 8, which is near a good lateral road network. The troop commander and FIST collocate well forward with the 1st platoon leader, overlooking Route 220 (see Figure 4-11).

The 1st platoon leader reports two BMPs and one T-64 bounding across the open terrain, moving southwest astride Route 220 about 2,000 meters east of PL BOOT. He maintains contact with the patrol as it crosses PL BOOT and enters the troop sector. The troop commander orders 2d platoon to attack the patrol by fire. The 2d platoon moves into hull-down positions and ambushes the patrol. All three vehicles are hit and damaged. The 1st platoon leader sends his PSG to capture the surviving BMP crew members and to search the vehicles for vehicle unit symbols, maps, sketches, and any other items of intelligence value. The PSG links up with the 1SG, who has moved forward. He hands over the prisoners and material, and returns to his position on the screen line. The 1SG evacuates the prisoners and captured material to the squadron combat trains. The 3d platoon reports no contact in its sector. The XO reports the situation to the squadron.

The troop commander tells the troop to be alert for other divisional reconnaissance patrols in the sector, and to expect regimental reconnaissance within the next hour. The XO eavesdrops on the squadron OI net and monitors reports from Troop B that indicate it has encountered several divisional reconnaissance patrols. The XO passes this information to the troop commander.

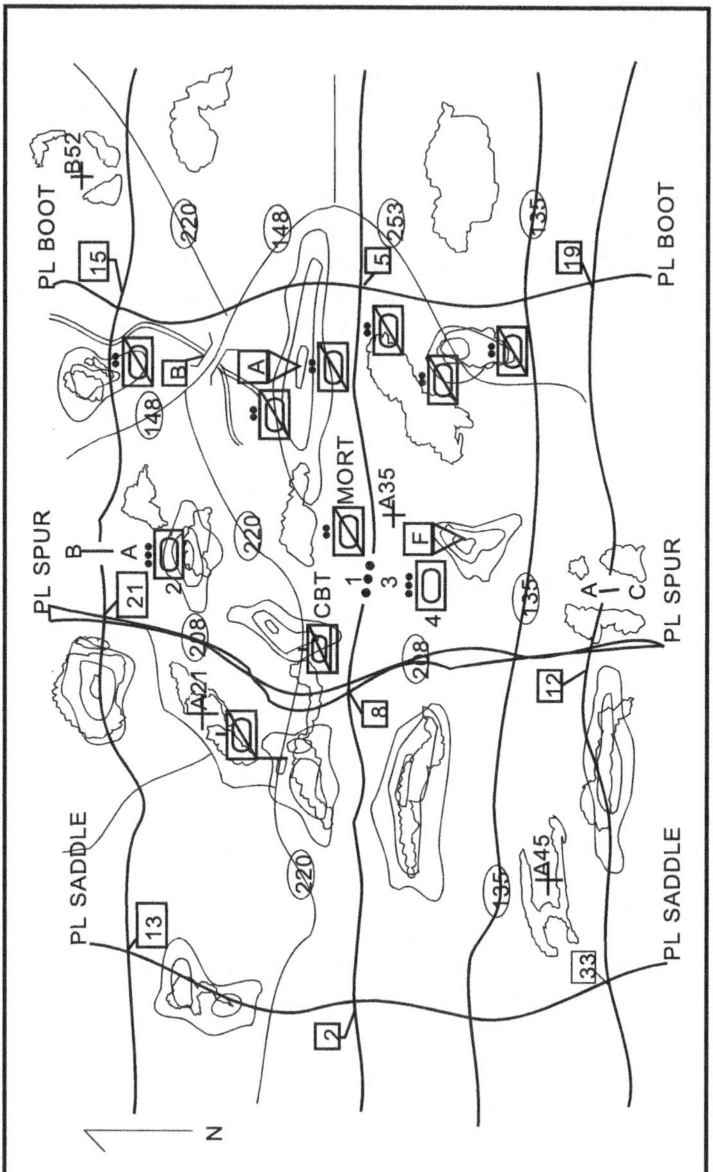

Figure 4-11. Troop screen.

Within the next 45 minutes, both scout platoons report reconnaissance patrols of two BMPs each in their sectors. The troop commander maneuvers a tank platoon into position to ambush each patrol. Three of the four BMPs are quickly destroyed. One BMP in the 1st platoon's sector is damaged, but is able to move into a covered and concealed position in the wooded area west of checkpoint A. Artillery begins to fall on the dominant terrain east of PL SPUR. The XO reports the situation to squadron.

The troop commander orders the 1st platoon to regain contact with the damaged BMP. The 1st platoon leader sends Bravo section to locate the damaged BMP. Bravo section moves into the wooded area and quickly finds and destroys the BMP.

The troop commander orders the scout platoons to move one of their scout sections back in sector in anticipation of the arrival of enemy reconnaissance elements. Both scout platoons reposition a scout section back towards PL SPUR while the troop commander shifts his tank platoons, mortar section, and trains.

The 1st platoon later reports a reinforced MRP moving southwest along Route 220. The 1st platoon continues to observe the enemy reconnaissance elements and reports their progress to the troop. The FSO engages the enemy with mortar high explosive (HE) and smoke rounds to disrupt their movement and keep them buttoned up. The troop commander tells the 4th platoon to move north and be prepared to counterattack the enemy by fire. He quickly outlines its route using TIRS. The enemy reconnaissance elements rapidly move in an attack formation toward the terrain previously occupied by the 2d platoon. It appears that the enemy elements are attempting to secure the dominant terrain in the 1st platoon sector and fix the tank platoon so the lead company of the enemy advance guard can maneuver around it.

The FSO continues to suppress the enemy with smoke and HE while it attempts to maneuver toward its objective. The troop commander directs 2d platoon into position to establish a base of fire and fix enemy elements while 4th platoon counterattacks by fire from the south. The FSO coordinates mortar and artillery fires to suppress and isolate enemy elements as the tank platoons execute their counterattack.

While the tank platoons complete the destruction of the enemy, heavy artillery begins to impact along and east of PL SPUR. The 1st platoon scouts report a tank platoon followed by six BTR-60s moving west along Route 220. The troop commander reports this information directly to the squadron commander. He orders the troops to fall back to PL SPUR, and to begin coordination for battle handover and passage of lines. The troop commander orders the 3d platoon to begin to bound back to PL SPUR. He tells the 1st platoon to maintain contact with the enemy lead company as it falls back to PL SPUR. The troop commander maneuvers the tank platoons west of PL SPUR.

The 1st platoon leader bounds his scout sections back to PL SPUR as he maintains contact with the lead company. The FSO engages the MRC with mortar and artillery fires to slow and disrupt its advance. The XO keeps squadron informed of the situation in the troop sector.

The troop XO moves back to the contact point to begin coordination for battle handover and passage of lines. The troop continues to maintain contact with the lead MRC as it maneuvers back in sector, while the FSO engages the MRC with mortar and artillery fires.

Section III. Guard

Cavalry squadrons conduct guard missions to provide another unit early warning of enemy approach, and to prevent enemy ground maneuver forces from coming within direct-fire range of the protected unit as it performs other missions or tasks. A guard is conducted to the front, flanks, or rear of a stationary or moving force. Many of the considerations for conducting a moving flank screen apply to conducting a moving flank guard.

A cavalry troop mission unique to moving flank guard operations is assigned to the lead troop of the squadron. This three-fold mission consists of the following tasks:

- Maintain contact with the protected force.
- Reconnoiter the zone between the protected force and the squadron route of advance.
- Reconnoiter the squadron route of advance.

The lead troop accomplishes these tasks by performing zone reconnaissance. The speed of the protected force determines how thoroughly the reconnaissance is performed. Assistance is required if the zone is too wide for the lead troop. An air cavalry troop may maintain contact with the protected force, or a following troop may perform route reconnaissance along the squadron's route of advance.

Note. For a detailed description of techniques for performing a squadron guard mission, see Chapter 4, FM 17-95.

The cavalry troop does not conduct guard operations independently. It normally performs reconnaissance, screen, defend, delay, and attack missions as part of a squadron guard operation. Review the appropriate chapter in this manual for specific mission details. See Chapter 3, Sections III and IV, Route and Zone Reconnaissance; Chapter 5, Section III, Hasty Attack; and Chapter 6, Sections II, III, and IV, Defend From a Troop Battle Position and Defend/Delay in Troop Sector.

Section IV. Cover

A covering force operates apart from the main body to develop the situation early and deceives, disorganizes, and destroys enemy forces. It accomplishes all the tasks of screening and guard forces. Unlike screening or guard forces, a covering force is a tactically self-contained force (that is, it is organized with sufficient combat support and combat service support assets to operate independently of the main body). Covering force operations can be either offensive or defensive in nature and can be conducted to the front, flanks, or rear of a stationary or moving force.

Cover operations are performed by the regiment (ACR or LACR) for the corps commander to provide the time and space he needs to make decisions and to achieve his operational objective.

Defensive cover operations, whether forward, flank, or rear, prevent the enemy from attacking the corps main body at the time and place and with the combat strength he chooses. Defensive cover operations are generally intended to destroy the enemy's initiative and set him up for defeat. Both heavy and light cavalry troops participate in defensive cover operations as part of their parent squadrons. Depending on the squadron commander's scheme of maneuver, they either screen, defend, delay, or attack.

Offensive cover operations are conducted by the regiment to seize and retain the initiative for the corps commander, and to allow him to attack decisively with the main body of the corps at the time and place he chooses. In offensive cover operations, the troop usually conducts movement to contact or zone reconnaissance missions. See the appropriate chapters in this manual for mission details.

Section V. Route Security

Cavalry squadrons and regiments conduct route security missions to prevent enemy ground maneuver forces from coming within direct-fire range of the protected route. A route security force operates on and to the flanks of a designated route. Route security operations are defensive in nature and, unlike guard operations, are terrain oriented. A route security force conducts reconnaissance, screens, attacks, defends, and occupies key locations along the route to prevent an enemy force from impeding, harassing, containing, seizing, or destroying traffic along the route.

Route security operations are not conducted independently by a cavalry troop. Both the heavy and light cavalry troops participate in route security operations as part of their parent squadrons and may conduct reconnaissance, screen, defend, delay, attack, or convoy security missions throughout the area of operations.

Section VI. Area Security

An area security force neutralizes or defeats enemy operations in a specified area. It operates in an area delineated by the headquarters assigning the area security mission. It screens, reconnoiters, attacks, defends, and delays as necessary to accomplish its mission. Area security operations may be offensive or defensive in nature and focus on the enemy, the force being protected, or a combination of the two.

Area security operations are conducted to deny the enemy the ability to influence friendly actions in a specific area or to deny the enemy use of an area for his own purposes. This may entail occupying and securing an area before the enemy can, or taking actions to destroy enemy forces already present.

The area to be secured may range from specific points (bridges, defiles) to areas such as terrain features (ridgelines, hills) to large population centers and adjacent areas. The factors of METT-T and unit capability will determine specific unit missions.

Area security missions are conducted by cavalry troops, squadrons, and regiments who employ the techniques of screen, guard, offense, and defense, depending on the nature and purpose of the mission.

Section VII. Convoy Security

Convoy security operations are performed as a minimum by a cavalry troop or a company team. Both heavy and light cavalry troops are suited to the requirements of protecting a convoy due to their organic reconnaissance capability and combat power. Both the cavalry troop and company team should be reinforced with engineers. METT-T considerations, such as restrictive terrain and limited time, may dictate a coordinated effort with air cavalry assets.

Convoy security operations are conducted when insufficient friendly forces are available to continuously secure lines of communication in an area of operations. They may also be conducted in conjunction with route security operations. A convoy security force operates to the front, flanks, and rear of a convoy element moving along a designated route. Convoy security operations are offensive in nature and orient on the force being protected.

A convoy security mission has certain critical tasks that guide planning and execution. To protect a convoy, the security force must accomplish the following critical tasks:

- Reconnoiter the route the convoy will travel.
- Clear the route of obstacles or positions from which the enemy could influence movement along the route.

- Provide early warning and prevent the enemy from impeding, harassing, containing, seizing, or destroying the convoy.

The convoy security force is organized into three or four elements to accomplish the following (see Figure 4-12):

- **Reconnaissance element.** The reconnaissance element performs tasks associated with zone and route reconnaissance forward of the convoy.
- **Screen element.** The screen element provides early warning and security to the convoy's flanks and rear.
- **Escort element.** The escort element provides close-in protection to the convoy. May also provide a reaction force to assist in repelling or destroying enemy contact.
- **Reaction force.** Provides firepower and support to the elements above in order to assist in developing the situation or conducting a hasty attack. May also perform duties of the escort element.

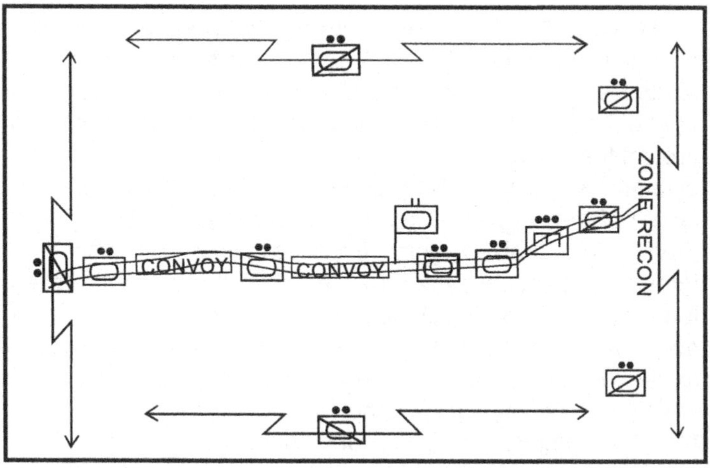

Figure 4-12. Cavalry troop conducts convoy security.

TECHNIQUES

The troop commander organizes and coordinates the efforts of his unit to fulfill the critical tasks associated with the convoy security mission.

Tasks of the reconnaissance element can usually be fulfilled by a single scout platoon. The troop commander ensures the reconnaissance element focuses on trafficability of the route and enemy forces that may influence movement along the route. METT-T may dictate the use of engineers to assist in reconnoitering and clearing the route. Convoy speed is determined by the pace of reconnaissance (METT-T). As a guide, the reconnaissance element should operate a minimum of 3 to 5 kilometers ahead of the main body of the convoy.

Tasks of the screen element can usually be fulfilled by a single scout platoon also. The troop commander ensures the screen element establishes OPs to provide early warning on critical portions of the route or key avenues of approach to the route. OPs have a limited ability to destroy enemy forces; therefore, their primary purpose is to acquire the enemy and direct reaction forces or indirect fire to destroy it.

Tasks of the escort element are best performed by tank platoons in a heavy troop. A light troop should consider a scramble of platoons (MK-19/M2/TOW—see Chapter 2, Figure 2-13) to gain the suppressive firepower necessary to protect the convoy. The troop commander ensures the escort element is positioned to provide security throughout the length of the convoy. This requires elements of the two tank (heavy troop)/mixed (light troop) platoons be dispersed throughout the convoy order of march. If there is no reaction force available or designated, a task of the escort element may be to provide reaction forces that respond to enemy forces identified by the reconnaissance or screening elements. Depending on the length of the convoy and METT-T considerations, the troop commander may keep one tank/mixed platoon consolidated or specifically designated as the reaction force.

Chapter 5

Offense

Offense is the decisive form of war. Brigades, battalion/task forces, and company teams are the principal offensive force for the corps or division. Cavalry units normally perform reconnaissance and security missions in support of corps and division offensive operations. Cavalry troops may perform certain offensive missions as part of a squadron, regiment, or other combined arms force. These offensive missions are normally performed during reconnaissance or security operations.

If required, cavalry troops may perform offensive operations within an economy-of-force role for a higher headquarters.

CONTENTS

Section I. Purpose and Fundamentals

PURPOSE

The main purpose of the offense is to defeat, destroy, or neutralize the enemy force. Offensive operations are also undertaken to secure decisive terrain, to deprive the enemy of resources, to gain information, to deceive and divert the enemy, to hold the enemy in position, to disrupt his attack, and to set up conditions for future successful operations.

FUNDAMENTALS

Successful offensive operations have four fundamentals in common.

- **Surprise.** Strike the enemy at the time and place or in a manner that he least expects.
- **Concentration.** Mass available forces, strive for overwhelming superiority in men, weapons, and firepower. With concentration, however, vulnerability becomes a factor. A force that is dispersed is much more survivable. The commander must maintain a high sense of situational awareness to anticipate the conditions of battle that will allow him to mass at the critical point, kill the enemy, and quickly disperse to survive.
- **Tempo.** Tempo is the rate of speed of military action. Controlling or altering the rate is essential for maintaining the initiative. Tempo can be fast or slow dependent on the capabilities of the troop relative to that of the enemy. Commanders must adjust tempo to ensure synchronization.
- **Audacity.** Boldness in the plan's execution is key to success in offensive operations. Commanders should understand when and where they are taking risks, but must not become tentative when executing their plan.

Section II. Movement to Contact

A movement to contact is a mission undertaken to gain or regain contact with the enemy. Unlike a zone reconnaissance, which is focused on reporting detailed information on the terrain and enemy within a given zone, movement to contact is focused on finding the enemy. The critical tasks are geared for achieving fast movement and rapid location of enemy forces.

The cavalry troop normally conducts a movement-to-contact mission as the lead element of a squadron or combined arms force hasty attack. The troop may also conduct the mission as part of a squadron movement to contact when the squadron is leading the advance of another combined arms force, such as a brigade or division.

A movement to contact is characterized by rapid aggressive action. The troop commander must rapidly develop the situation and may be permitted to bypass enemy forces to maintain momentum.

CRITICAL TASKS

During a movement to contact, certain critical tasks will be accomplished. Unless the squadron commander gives guidance otherwise, the troop will—

- Reconnoiter and determine the trafficability of all high-speed routes within the zone.
- Inspect and classify all bridges, culverts, overpasses, and underpasses along high-speed routes. Identify all bypasses and fords that cannot support rapid heavy-armor movement.
- Clear all high-speed routes of mines, obstacles, and barriers within its capability, or locate a bypass.
- Find and report all enemy forces within the zone and determine their size, composition, and activity.

TECHNIQUES

The cavalry troop can perform movement to contact in a zone up to 10 kilometers wide. The size of the zone depends on the terrain and enemy situation. Scout squads, and in some cases sections, must be mutually supporting during movement based on the enemy situation. Mutual support between squads and sections is gained by reducing the size of the zone based on the terrain. The more restrictive the terrain the less mutual support becomes available.

When considering techniques for conducting a movement to contact, remember the basics.

- Always retain the ability to maneuver.
- Make contact with the smallest force possible.
- Employ forces in depth; stay flexible.
- Develop the situation rapidly once contact is made with the enemy force. Operate at a tempo that forces him to react to you, not you to him.

When the troop receives a movement-to-contact mission, the zone of action is normally identified by lateral boundaries. A line of departure and an objective or limit of advance are specified to orient the troop's movement. The objective is usually a terrain feature placed deep enough to ensure contact with the enemy. Divide the troop zone into two platoon zones. Think about where to draw the boundary. Make sure it does not split enemy avenues of approach and that it is on easily identifiable terrain. Do not use a road for a boundary. To control progress through the zone, add phase lines every 5 to 8 kilometers on easily identifiable terrain. Place contact points near the intersection of the platoon boundary and all phase lines. If specific areas need to be checked, identify them using checkpoints for reference. Place TIRS on the map.

Use a troop vee or split vee formation when conducting a movement to contact. Deploy scout platoons abreast and have them accomplish all the related reconnaissance tasks. Move them out first across the line of departure. The mission of the scouts is to find the enemy, develop the situation, and recommend to the troop commander the best course of action for the employment of the tank/AT platoons.

The mortar section, under control of the troop fire support officer or mortar section sergeant, follows about 2 kilometers behind the scouts, centered in the zone, to provide continuous coverage forward of the scouts. The troop command post, under control of the XO, displaces in the zone, positioning on terrain that affords effective and continuous communication with troop elements and the squadron. The troop trains, under control of the 1SG, follow about 2 kilometers behind the tank/AT platoons, and bound from one covered and concealed position to another. The troop commander positions himself well forward to observe one scout platoon or the other. His position is usually located where he expects initial enemy contact or difficult situations.

Heavy Troop

Due to the survivability and firepower of scout platforms (M3) in the heavy troop, the tank platoons may be kept in considerable depth (1,000 to 3,000 meters) behind the scouts until enemy contact is made. Once contact is made and the scouts have developed the situation/recommended a course of action, the troop commander may then employ his tanks, using firepower and maneuverability to fix the enemy or destroy it by hasty attack.

Light Troop

In the light cavalry troop, the AT platoon(s) should be kept close (500 to 1,500 meters) to the scout platoons. Upon contact, the HMMWV scouts will require more time to develop the situation than Bradley-mounted scouts. The AT platoon(s) can assist the scouts in developing the situation by overwatching the scouts. Also the TOW takes time to employ, so reaction time is slower than with the tank. During movement to contact, the light troop commander must be able to execute his actions on contact very quickly to maintain the momentum of the operation. A technique the light troop might use is outlined below.

- Scouts make contact with an enemy force.
- The troop commander immediately moves one AT platoon forward to overwatch the scouts as they develop the situation.

- If the situation dictates, the AT platoon suppresses with direct fires to allow the scouts to maneuver.

- Once the situation is developed, the troop positions to either fix the enemy for the follow-on force or destroy it by a hasty attack.

EXAMPLE OF MOVEMENT TO CONTACT (HEAVY TROOP)

The troop is ordered to reestablish contact with elements of a withdrawing enemy mechanized infantry platoon. As the troop deploys in a split-vee formation, the 1st and 3d platoons cross the line of departure abreast, north and south respectively, and begin to reconnoiter in troop zone up to PL BOOT. The scout platoons determine the trafficability of high-speed routes within their zones, and search for the enemy in areas that dominate the routes. The bridge across the creek at checkpoint B will support only 20 tons, so the 1st platoon finds a solid-based fording site 100 meters south of the bridge. Thus far, the high-speed routes are not damaged or blocked. The 3d platoon reports finding tracks of what appears to be three BMPs heading northwest from Hill 244. The 1st platoon reports tracks of an enemy company-size force heading west along Route 220. Track patterns indicate BMPs and T-64s/T-72s. The mortar section eavesdrops on the troop command net, and moves with the scout platoons, staying back about 1.5 kilometers, centered in the troop zone. The 2d and 4th platoons are paired with their sister scout platoons and key their movement on the scout platoon's progress, staying back about 1 kilometer and using terrain that affords good cover and concealment. The troop CP and trains hold in place (see Figure 5-1).

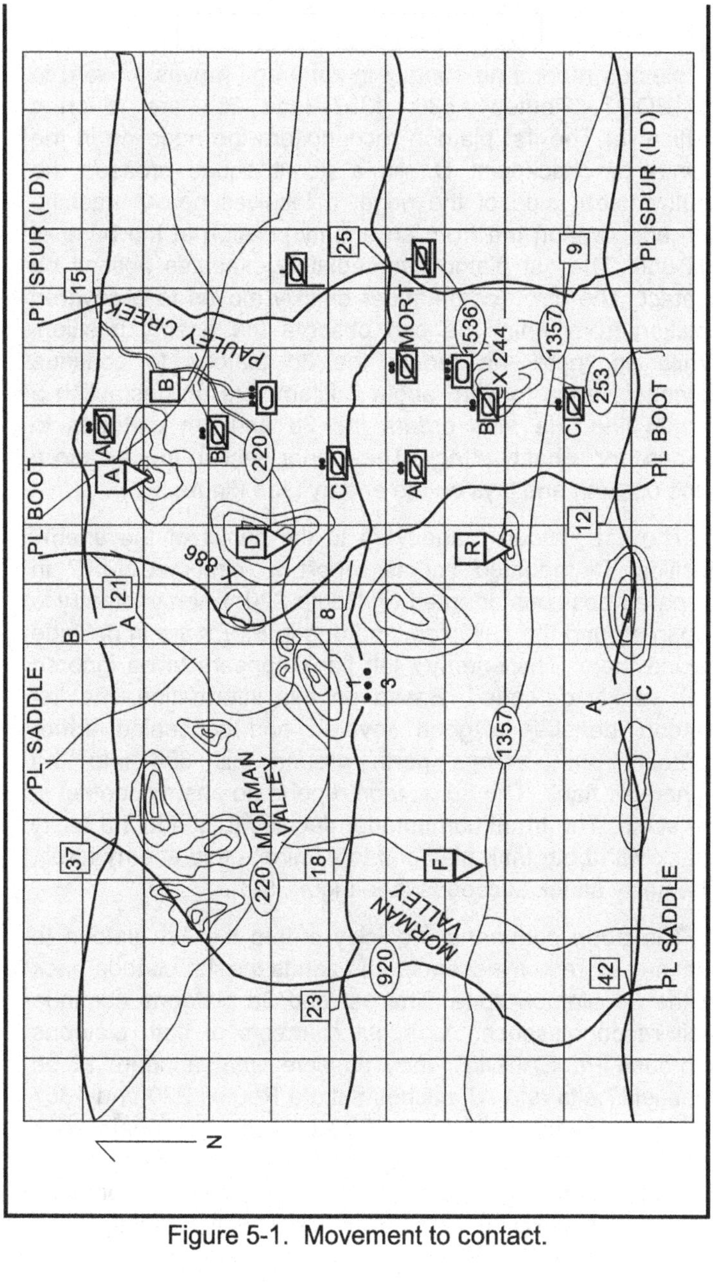

Figure 5-1. Movement to contact.

As the 1st platoon approaches PL BOOT, it reports two BRDM-2s withdrawing west at high speed on Route 220. The 3d platoon reports no contact in zone and moves forward to PL BOOT. Routes 1536, 1357, and 253 are reported trafficable. The 1st platoon reconnoiters the ridgeline in the vicinity of checkpoint D. As a scout squad crosses the southwestern side of the ridge, it receives heavy machine gun and cannon fire from an enemy position at the base of Hill 886. The 1st platoon immediately executes actions on contact. The troop commander quickly moves to a covered position from which he can observe the enemy position. While en route, he orders the 3d platoon to continue reconnaissance forward about 2 kilometers and establish a screen line. He also orders the 2d and 4th platoons to prepare for a hasty attack. The mortar section moves into a firing position and lays on the enemy (see Figure 5-2).

The 1st platoon maneuvers to the flanks of the enemy position. Dismounted scouts report a reinforced MRP in prepared positions oriented on Route 220. Enemy infantry is dug in around the buildings, and dug-in BMPs are in defilade behind them. The enemy's left flank appears to be ignored and weakly defended. Based on this information, the 1st platoon identifies a good covered and concealed armor approach that swings north around Hill 886 into the vulnerable flank. The 3d platoon reports no enemy contact in the south. The troop commander decides to conduct a hasty attack with both tank platoons to eliminate this enemy force. The hasty attack succeeds (see Figure 5-3).

The troop commander quickly orders the 1st platoon to continue its reconnaissance, and sends the 4th platoon back to the 3d platoon zone. The 1st and 3d platoons continue their reconnaissance. As scout elements of both platoons approach PL SADDLE, they observe what appears to be minefield belts and AT ditches astride Routes 220 and 1357 in the Morman Valley. Dismounted reconnaissance determines the extent of the obstacles, and reveals heavy concentrations of enemy forces in several company strongpoints along the dominating ridgeline by Highway 920.

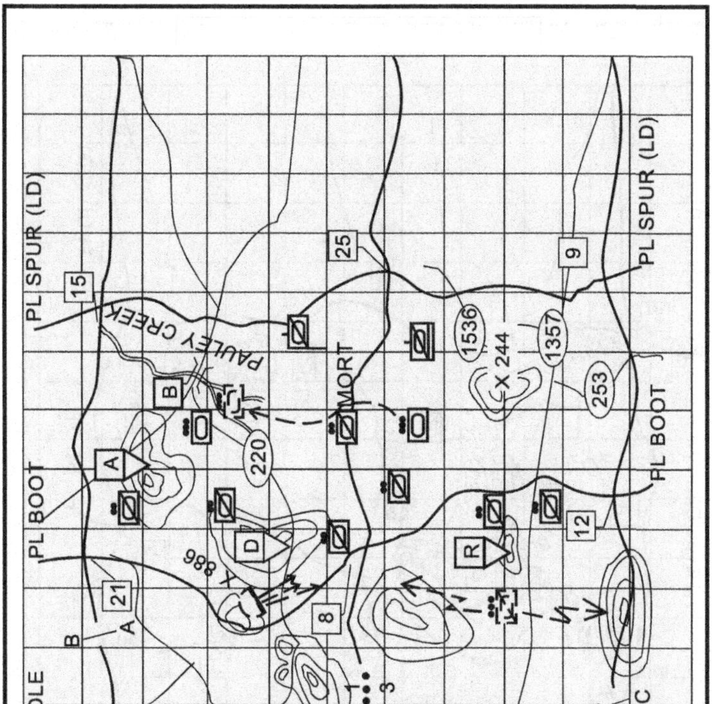

Figure 5-2. Actions on contact.

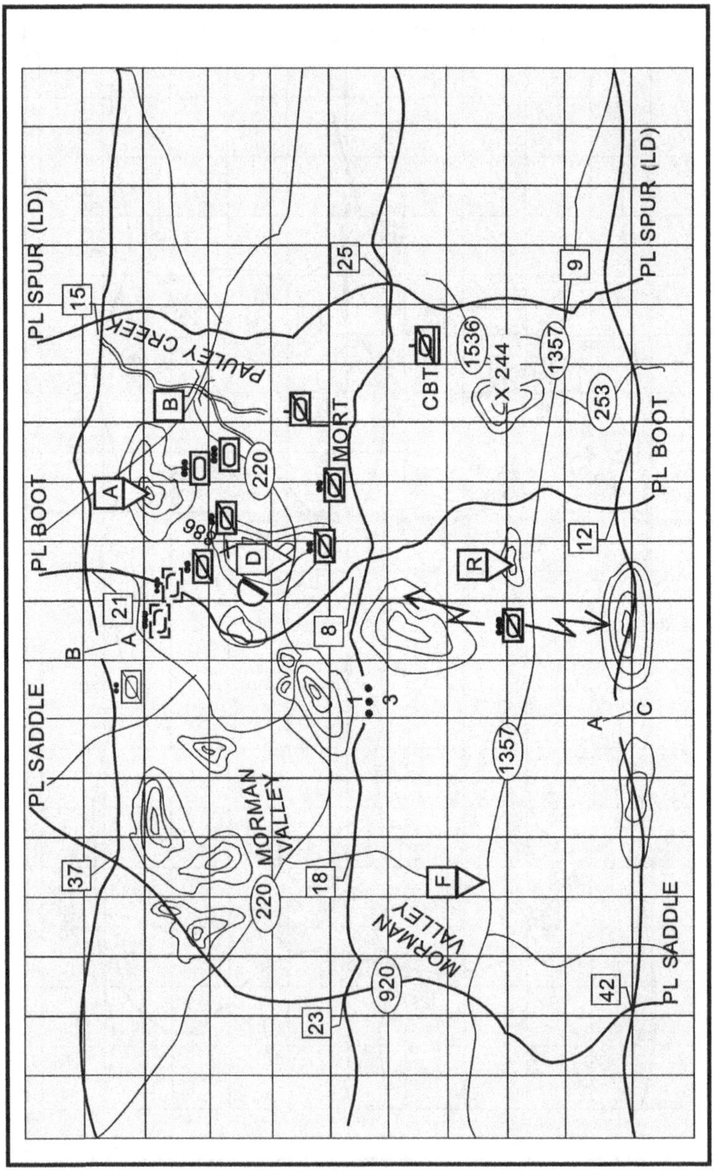

Figure 5-3. Hasty attack.

The troop commander takes a look. En route, he orders both scout platoons to determine if any gaps or boundaries exist between the enemy positions. After seeing the situation, the troop commander calls his squadron commander and recommends the troop establish a hasty defense and prepare to assist in battle handover and forward passage of following attack forces (see Figure 5-4).

Section III. Hasty Attack

A hasty attack is conducted with a minimum of preparation to defeat an enemy force that is not prepared or deployed to fight. It is a course of action routinely employed in cavalry operations to seize or retain the initiative, or to sustain the tempo of operations. A hasty attack can be executed while the troop is engaged with a zone reconnaissance mission or movement to contact. (See Chapter 3, Section II, and Section II of this chapter for other examples of hasty attack.)

CRITICAL TASKS

To execute a hasty attack, the following critical tasks must be accomplished:

- Reconnoiter and determine the size, composition, and orientation of the enemy force.
- Determine if the objective enemy force is supported by other units nearby.
- Find a high-speed covered and concealed approach into the enemy's flank(s).
- Establish a maneuver element (usually one or both tank platoons) to move to a position of advantage over the enemy and attack him by fire.

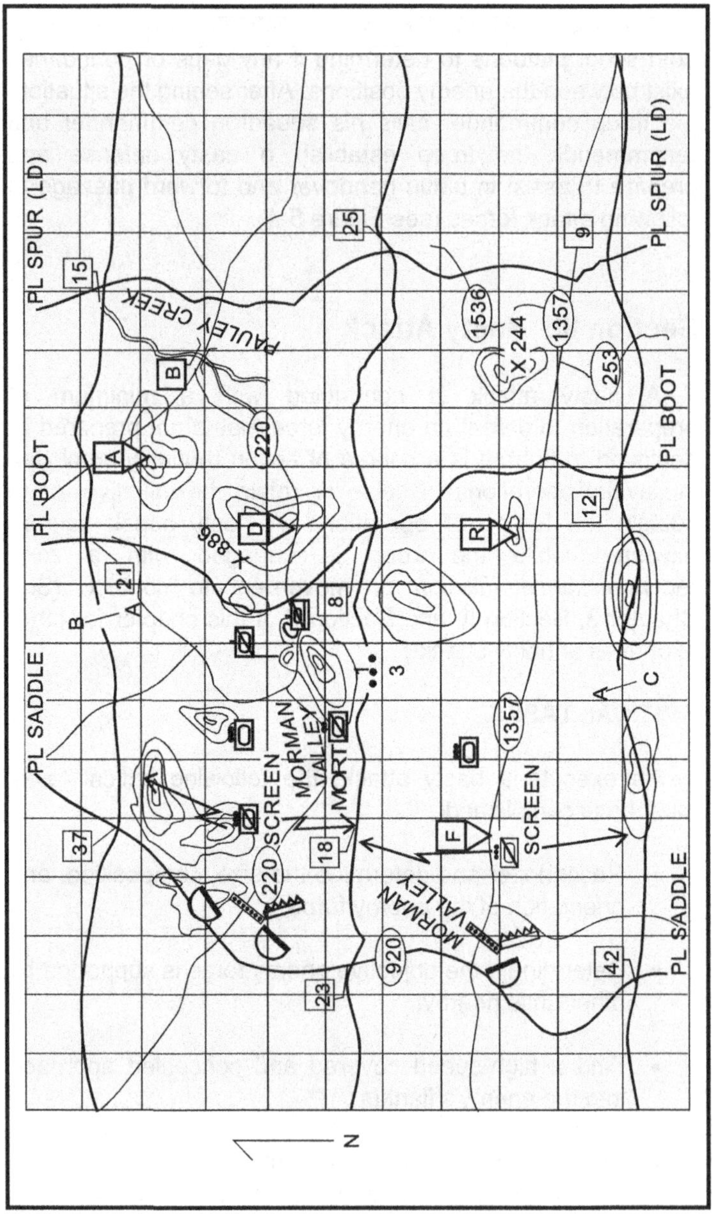

Figure 5-4. Hasty defense.

- Establish a base-of-fire element (usually one or both scout platoons) to defeat or suppress all observed enemy AT weapons with long-range direct and indirect fires before the maneuver force deploys into its attack.

- Isolate the objective enemy force from other mutually supporting units with indirect fires (usually with smoke and HE ammunition).

- Attack the enemy by fire or by fire and movement, and defeat him.

- Once the attack is completed, immediately establish hasty defensive positions and OPs on high-speed avenues of approach into the troop position.

TECHNIQUES

Each critical task has a time at which it will be accomplished in relation to all other critical tasks. A good hasty attack depends on the commander's sense of timing and on his ability to employ his forces to accomplish the tasks in the proper sequence. The commander has to synchronize—to concentrate and apply different forms of combat power against the enemy at the right times and places. The decision to conduct a hasty attack is usually made after an enemy force's reconnaissance and dispositions show that winning requires a quick strike, with little preparation. Tactics for conducting a hasty attack have three common features.

- Known or suspected enemy AT weapons are suppressed and destroyed with direct and/or indirect fires before the maneuver force is committed.

- The enemy is forced to fight in two directions.

- The enemy is suppressed and unable to react.

SETTING UP THE TROOP FOR A HASTY ATTACK

While conducting other missions, scouts will often make contact with an enemy force. In developing the situation, a scout platoon may recommend hasty attack as a course of action to the troop commander, who decides to execute the recommended course of action. The troop commander issues FRAGOs that will position forces to execute an attack simply and effectively.

The scout platoon in contact continues to reconnoiter the enemy's position and to accomplish its reconnaissance tasks. One section of the scout platoon remains in contact with the enemy. The other scout platoon continues its reconnaissance up to a limit of advance established by the troop commander. Both scout platoons continue to develop the situation further by looking for the presence of other enemy units, to the flanks or rear, supporting the enemy contact.

The FIST moves to a good position to see the battlefield and to control the indirect fires. The FSO places the mortar section on terrain where it establishes a firing position and prepares to suppress the enemy position.

The size and strength of the enemy may require the use of one or both tank/AT platoons. The commander may choose to do one of the following options:

- Move one tank/AT platoon into an overwatch position and attack by fire with one.
- Keep one in reserve behind the scout platoon not in contact and attack by fire with one.
- Attack by fire with both platoons.

The scouts determine a good attack position and attack-by-fire position for the tank/AT platoon(s). A scout squad from the scout platoon in contact moves to a link-up point with one or both tank/AT platoon(s) to guide them into the attack position.

The troop commander moves to collocate with the scout platoon in overwatch. The 1SG moves medics and recovery vehicles close to the battlefield. The XO assists the commander in control of the troop and keeps the higher commander informed.

INDIRECT FIRES IN SUPPORT OF A HASTY ATTACK

Indirect fires should complement the troop's scheme of maneuver. In the light troop, dependent on availability of supporting indirect-fire systems, indirect fires may prove to be the best weapon of destruction. The troop can acquire and engage the enemy with indirect fire from positions offering good protection from enemy direct/indirect fire. Because the light troop is in covered and concealed positions, the survivability of the troop is enhanced. Indirect fires must also be controlled to prevent fratricide. The troop commander must determine the following:

- Who will control the indirect fires during the hasty attack?
- Who will initiate indirect fires onto the objective?
- Who will lift and shift the indirect fires to subsequent targets?
- What will the signal be for lifting and shifting indirect fires?
- Are there any restrictive fire measures or restrictive fire areas?

There are many different answers to these questions. METT-T will determine which answer works best. Under most conditions the commander or the FSO will be in the best position to control the engagement of indirect-fire systems.

The troop commander should use available indirect fires from mortars and supporting artillery to—

- Suppress the enemy while scouts are maneuvering to develop the situation.
- Screen enemy observation of scouts or assault element during the conduct of the hasty attack.
- Isolate the enemy contact by firing HE and smoke between the enemy force and any possible supporting positions.
- Shift indirect fires off the objective to block enemy withdrawal routes.

DIRECT FIRES DURING HASTY ATTACK

When executing the hasty attack the troop must combine indirect fires with direct fires. The direct-fire systems available to the heavy troop commander are highly destructive. The high rate of fire from the M3 and M1 allows the commander to quickly place effective fires into the objective area while stationary or moving.

The systems available to the light troop commander can be just as destructive when used in combinations and against a comparably equipped enemy. The light troop commander should strive to engage the enemy with combinations of weapons. Using the bounding technique and move-set drills by platoons, the troop commander can place effective suppressive and destructive fires into the objective area. (The high rates of fire from the M2 caliber .50 and MK-19 should be used to suppress the enemy and cover the TOW systems as they engage.) See Figure 5-5.

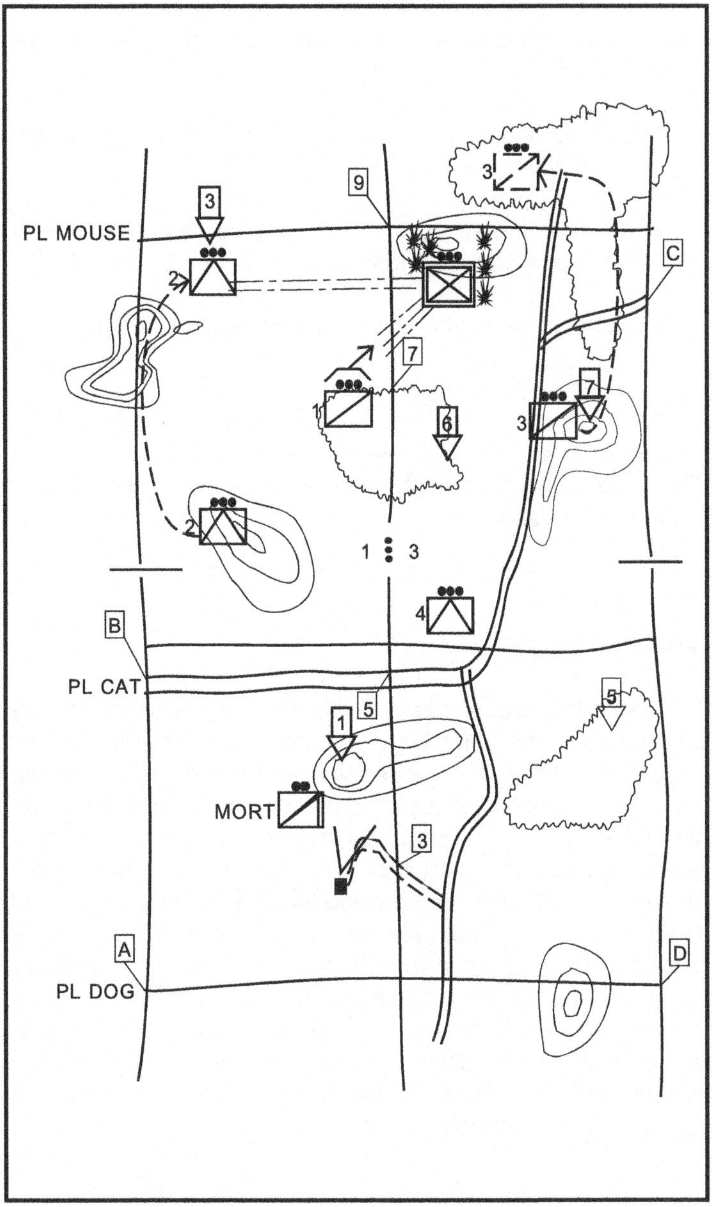

Figure 5-5. Light cavalry troop using fire and maneuver.

Like indirect fires, direct fires must be controlled. The commander must determine the following before executing a hasty attack:

- Who will initiate direct fires into the objective area and from where will they be initiated?
- What is the aim point for the support element? When do they shift fires and where do they shift fires to?
- What is the limit of advance for the assault force?

To determine the answers to these questions the commander must look at the enemy he is facing, the terrain he is operating in, and the forces he has to bring to bear against the enemy.

THE OBJECTIVE

Once the troop has initiated its attack and enemy resistance in the objective area has ceased, the troop may begin consolidating in preparation of continuing its mission. Have one or both scout platoons dismount scouts to sweep the area quickly for prisoners and other items of tactical value. Immediately establish a hasty defense with tank/AT platoons oriented on high-speed approaches into the troop position, and with scout platoons oriented on expected high-speed enemy approaches at OPs out from the defensive position. Position the mortar section and have the guns lay on the most likely enemy avenue of approach. Redistribute ammunition and balance crews as time allows. Replace key leaders quickly. If time is available and enemy counterattack unlikely, begin resupply operations.

EXAMPLE OF A HASTY ATTACK

A troop is conducting a zone reconnaissance moving toward its limit of advance (LOA). As the 3d platoon crosses the ridge (PL SABER) in its zone, Charlie scout section is hit by enemy tank cannon and heavy machine-gun fire from concealed positions near checkpoint 2. The platoon immediately takes cover and reports contact to the troop commander. The troop commander acknowledges and immediately moves to link up with the platoon leader. The platoon sergeant calls for indirect fire to suppress the enemy force. Under this protection, the platoon leader orders dismounted reconnaissance through the woods on the enemy's right to find his flank. He then sends his Bravo scout section forward on the enemy's left to find his other flank and to determine if the enemy is mutually supported by other enemy forces near checkpoint 2. The platoon leader and two scouts dismount, move in closer, and determine the enemy force consists of a reinforced MRP, two tanks, and three BMPs, in a hasty defensive position. Scouts report what appear to be minefields forward and to the flanks of the enemy position. The Alpha section identifies the right flank of the enemy position and discovers the flank is not protected by obstacles and not covered by direct-fire weapons (see Figure 5-6).

En route to the 3d platoon's zone, the troop commander tells the 1st platoon to continue reconnaissance forward to the LOA and find out if other enemy forces are providing mutual support to the enemy platoon in the 3d platoon's zone. He also tells the platoon leader to establish a screen along the LOA in his platoon zone. He then tells the 4th platoon to join the 2d platoon near checkpoint 12, using the ridgeline to cover its move, and prepare to conduct a hasty attack. Using TIRS, he outlines a route that runs from checkpoint 12 around the western edge of the village to checkpoint 7, then to a position from which it can engage the enemy on his most vulnerable flank (see Figure 5-7).

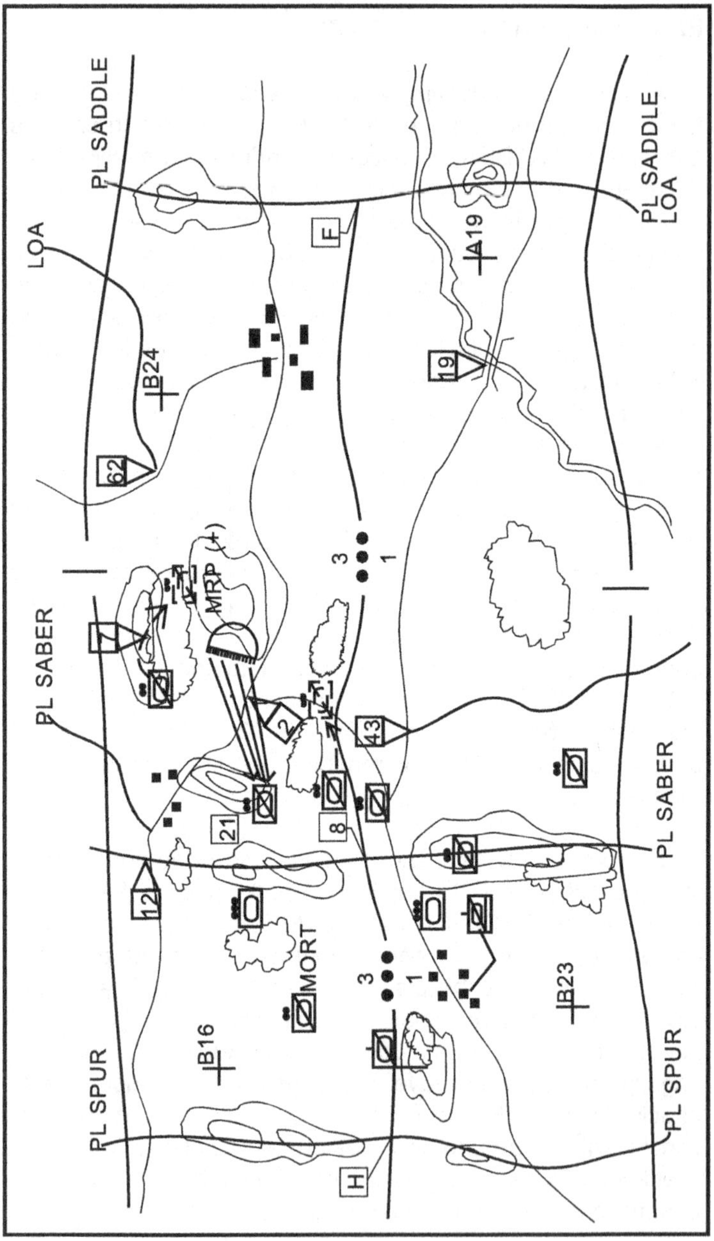

Figure 5-6. Develop the situation.

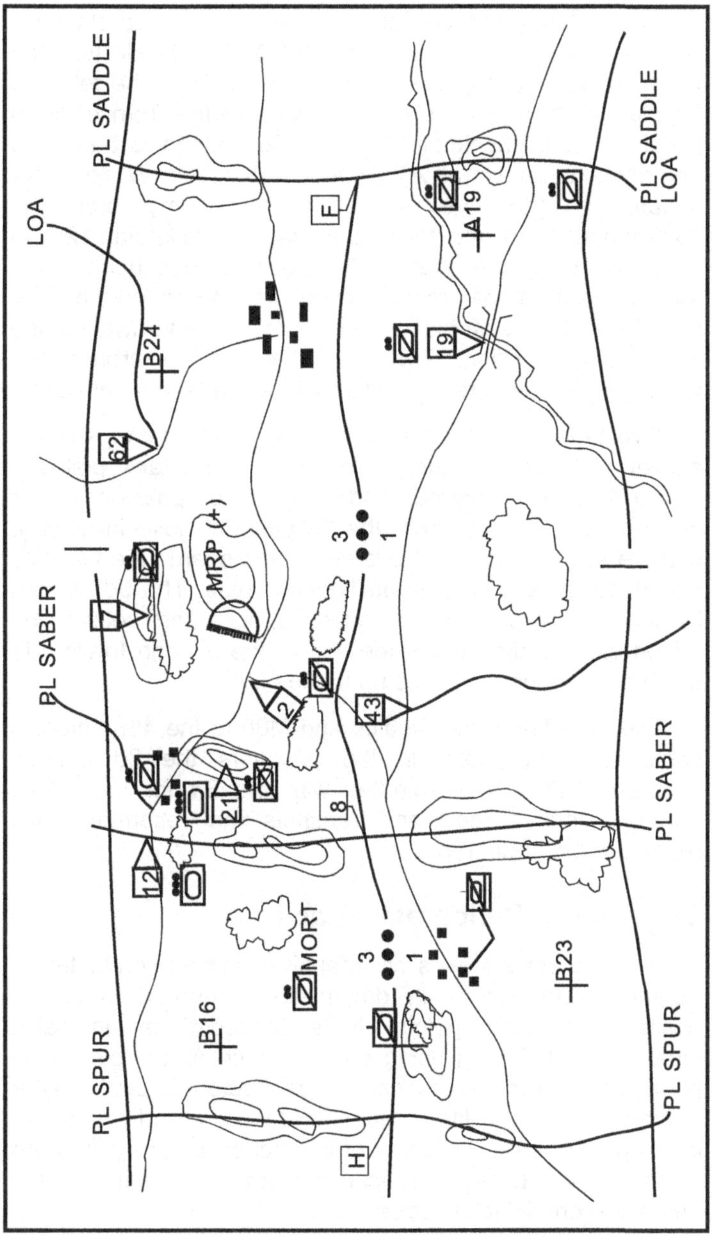

Figure 5-7. Set for hasty attack.

After a first-hand look at the situation in the 3d platoon's zone, the troop commander decides to attack. He tells the FSO to continue suppressing the enemy. Next, he tells the 3d platoon leader to establish a base of fire from covered positions near checkpoint 21. The 3d platoon is tasked to destroy or suppress all AT weapons observed. While heavy suppressive fires are raining on the enemy force, the commander joins the tank platoons at checkpoint 12 and leads them into the attack. After passing checkpoint 7, the tank platoons begin maneuvering. The troop commander tells the FSO to shift his fires on routes of withdrawal behind the enemy position. The tank platoons go around the woodline, deploy abreast, and attack by fire (see Figure 5-8).

Two tanks in the 4th platoon are hit and damaged; one soldier is killed and three are injured. The tank platoons consolidate in covered and concealed positions near checkpoint 7. Scouts from the 3d platoon move in quickly, capture several wounded prisoners, and search the vehicles and dead for anything of intelligence value. The 3d platoon leader orders one scout squad to stay and secure the prisoners, and then takes the rest of the platoon forward to finish zone reconnaissance up to the LOA.

The 1SG leads the medics and M88 to the 4th platoon's casualties and picks up the prisoners the 3d platoon secured. The XO coordinates with the S4 for immediate resupply, then collects and transmits final reconnaissance reports to the squadron.

Section IV. Deliberate Attack

A deliberate attack is an offensive mission conducted to defeat enemy forces in dug-in and prepared defensive positions or strongpoints. It is preceded by intensive intelligence gathering using sources such as patrols, aerial photographs, reconnaissance by fire, radio intercepts, radar surveillance, satellite reconnaissance, and prisoner interrogation. Large volumes of indirect preparatory fires are normally used. Detailed tactical planning and rehearsal of the attack are crucial to success.

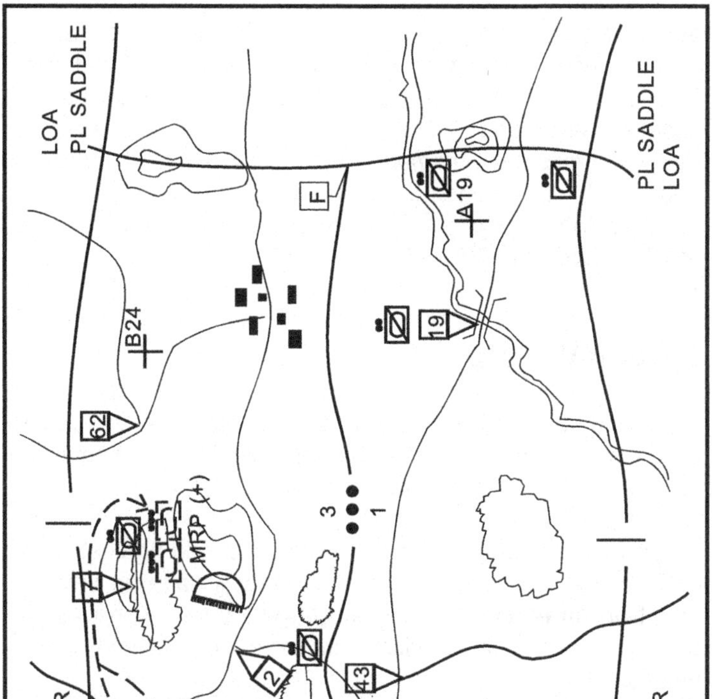

Figure 5-8. Tank platoons move into position for the hasty attack.

The cavalry troop alone cannot conduct a deliberate attack against prepared defensive positions. It usually participates with its parent squadron or other combined arms force as one of the three major elements in a breaching and assault operation. The troop can serve as the *overwatch/support element,* the *breaching element* (if reinforced with engineers), or the *assault element*. Doctrine for attacking a complex obstacle or a strongpoint is found in FM 71-1 and FM 17-95-10.

Section V. Raid

A raid is an attack into enemy territory for a specific purpose, with no intent to gain or hold terrain. The troop returns to friendly lines after the attack.

The cavalry troop normally conducts a raid against clear, well-defined targets whose destruction will have an adverse effect on an enemy force. Good targets for a troop raid are enemy artillery, rear service units, or command and control headquarters.

CRITICAL TASKS

During a raid, the following critical tasks must be accomplished:

- Coordinate and establish a restricted fire area around the raid area.
- Reconnoiter the troop's direction of attack from the point of departure to the objective.
- Establish visual contact with the enemy. Reconnoiter and determine the best place to initiate the attack.
- Establish a base of fire element to suppress the enemy force.

- Establish a maneuver element (usually one or both tank platoons) to move to a position of advantage over the enemy and attack them by fire.
- Suppress the enemy force with long-range direct fires while the maneuver element moves up to begin its attack.
- Establish surveillance on enemy avenues of approach to the objective, where follow-on enemy maneuver units would most likely appear.
- Attack the enemy by fire or by fire and movement to defeat them. Finish quickly.
- Establish security to protect the withdrawal of the remainder of the troop.
- Return rapidly to friendly lines and enter at the designated passage point.

TECHNIQUES

A raid achieves its purpose through speed, violent execution, and surprise. The troop's maneuver should produce that effect. There will probably not be a chance to reconnoiter before the attack; speed and surprise generate the security needed. Consequently, the troop might not always take the mortar section, combat trains, or command post. Vehicles disabled on a raid may be left behind in the interests of speed and security for the entire troop. When the troop receives a FRAGO to conduct a raid, the commander should receive the following graphic control measures from the S3 or squadron commander: a point of departure along the line of departure or line of contact, a direction of attack, an objective, a route of withdrawal, and a passage point to enter friendly lines. Analyze the terrain and the enemy situation, then issue the troop FRAGO.

The FSO should coordinate with the squadron FSO to establish a restricted fire area around the raid area. The 1SG should coordinate for immediate resupply of Class III and Class V when the troop returns. Have one scout platoon take the lead and reconnoiter the troop direction of attack from the point of departure on the line of departure to the objective. The lead platoon should establish visual contact with the raid objective to confirm the troop's scheme of maneuver and establish surveillance of the objective. The other scout platoon may act as the base-of-fire element, and follow about 1 kilometer behind the lead platoon. Have the lead platoon guide it into an overwatch position to deliver long-range suppressive fires on the enemy. Consolidate the tank platoons, move forward to join them, then follow about 1 kilometer behind the second scout platoon.

Once the base-of-fire element is set, have the platoon suppress the enemy with direct and indirect fires. Guide the maneuver element into a position to attack by fire. Finish quickly and move out.

Tell the scout platoon performing reconnaissance and surveillance tasks to screen and protect the withdrawal of the remainder of the troop back to friendly lines. Have the other scout platoon lead the way back on the designated route of withdrawal, and have the tank platoons follow. Stay back and supervise the conduct of the rear screen until all troop elements have entered friendly lines. Consolidate and resupply the troop with fuel and ammunition as soon as possible. Get ready to fight again.

EXAMPLE OF A RAID (HEAVY TROOP)

The squadron is defending in sector astride a likely Threat regimental zone of attack. The three armored cavalry troops are deployed abreast with the tank company in reserve. The squadron frontage is about 10 kilometers wide. Threat forces attack from the march. Squadron elements in the northern half of the sector are softened with long-range preparatory fires delivered by rocket and artillery forces. As artillery fires increase in intensity, the fire support element and the advance guard battalion of the regiment deploy from the march into prebattle formation and then attack. The advance guard battalion is quickly defeated by the concentrated fires of the northern cavalry troop and a surprise counterattack by the tank company. The squadron rapidly repositions to meet the assault of the regimental main body. Within one hour, the regimental main body approaches the line of contact with two MRBs abreast. Their axis of advance indicates most of their combat power will be concentrated in the northern half of the squadron's sector. Artillery suppressive fires increase as the Threat battalions deploy into attack formation. Aeroscouts report two 122-mm artillery battalions of the regimental artillery group moving into firing positions about 4 kilometers behind the attacking combat maneuver units. In this situation, with good defensible terrain, the squadron commander decides to concentrate the combat power of the troops, the tank company, and his available artillery in the north to defeat the advancing maneuver forces. He orders the southern cavalry troop to conduct a raid behind enemy lines to destroy the regimental artillery group simultaneously.

The troop commander quickly analyzes the terrain, and then issues a FRAGO to the troop. He tells the 1st platoon, which is currently on the screen line, to take the lead and reconnoiter the troop's direction of attack from the point of departure on the LD to the objective area. The platoon is tasked to establish visual contact with the Threat artillery units and to locate the best covered and concealed approach into their flanks or rear. The troop commander tells the 3d platoon to follow about 1 kilometer behind the 1st platoon and establish a base of fire to suppress and fix the artillery units. He orders the 2d and 4th platoons to consolidate rapidly near the point of departure. He tells the mortar section to remain in the troop sector. Priority of fires goes to the 1st platoon. As the 1st platoon approaches the point of departure, the troop FSO coordinates with the squadron FSO to establish a restricted fire area around the raid area. Once it is established and confirmed by the FSO, the troop commander orders the 1st platoon forward (see Figure 5-9).

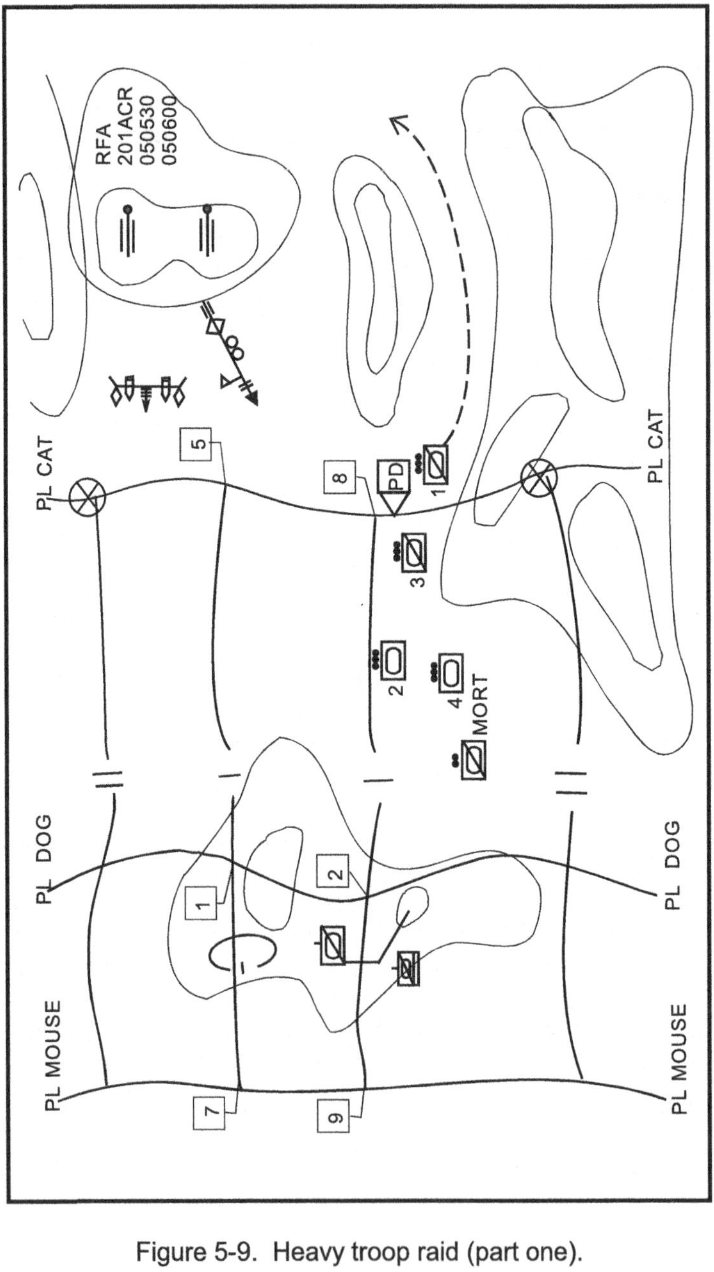

Figure 5-9. Heavy troop raid (part one).

The 1st platoon makes visual contact in about 10 minutes, and continues to reconnoiter for a good axis of attack. The 3d platoon moves into an overwatch position about 1,800 meters away and lays down a curtain of suppressive fire with chain guns, TOW missiles, and mortar fire. While suppressive fires are delivered, the tank platoons maneuver into the left rear of the artillery units, then deploy abreast and attack the Threat artillery battalions by fire (see Figure 5-10).

After silencing the battalions, the troop commander immediately orders the 3d platoon to lead the troop back to friendly lines along the designated route of withdrawal. The 2d and 4th platoons follow while the 1st platoon screens the withdrawal of the troop. The troop commander joins the 1st platoon and supervises the conduct of the rear screen until all troop elements have reentered friendly lines (see Figure 5-11). The troop returns to defensive positions, and the 1SG begins resupply actions.

EXAMPLE OF A RAID (LIGHT TROOP)

The light troop should consider the raid target when planning its operation. A combination of weapon systems may best suit the destruction of the raid objective rather than operating in pure platoons. A mix of weapon systems may also assist in security of the objective during the raid. The troop scramble should be considered as a possible task organization when planning a raid.

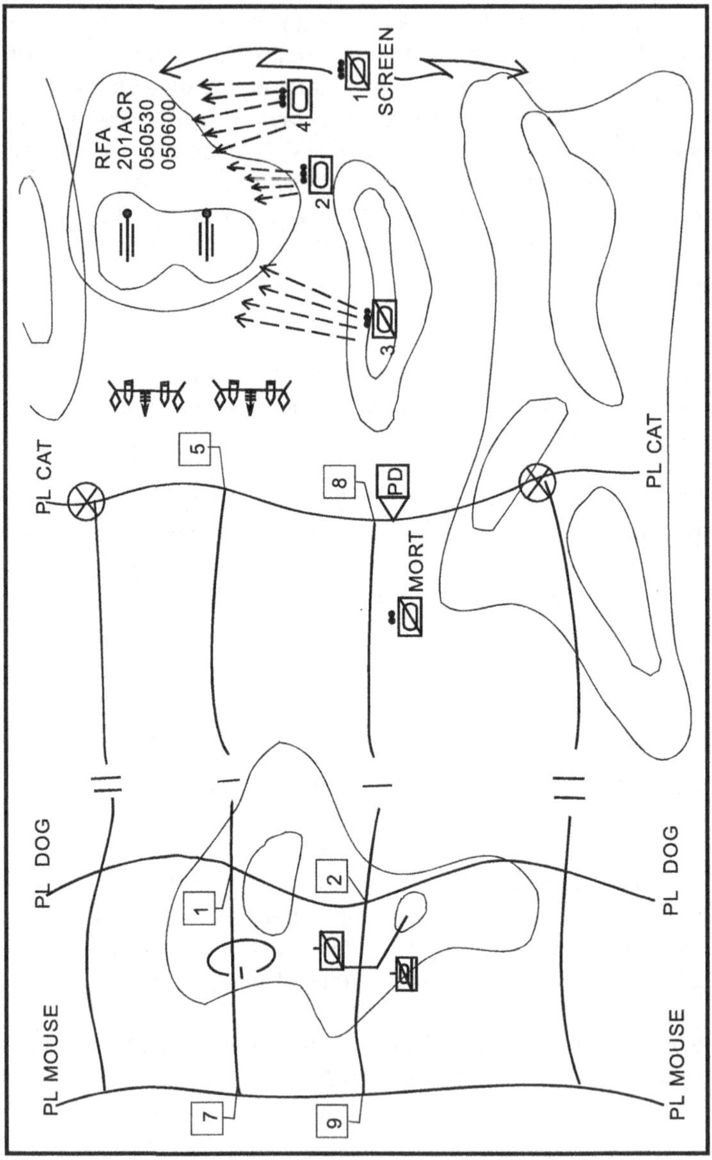

Figure 5-10. Heavy troop raid (part two).

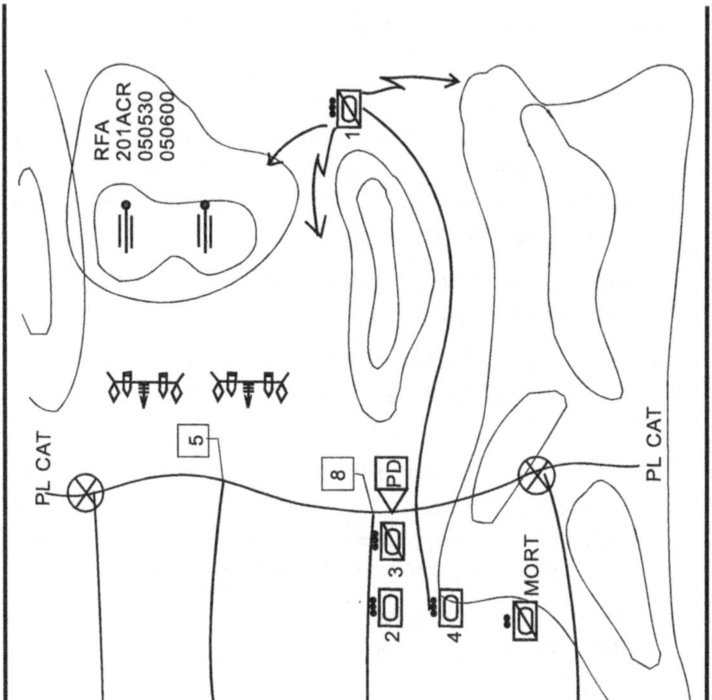

Figure 5-11. Heavy troop raid (part three).

Example steps of a light troop conducting a raid are as follows:

- Troop moves in column, platoons in a vee. Troop is a scramble organization with platoons integrating the AT vehicles into the scout sections. Each scout platoon is running two sections of three vehicles plus the platoon leader. See Figure 5-12.
- Lead platoon (1st platoon) conducts reconnaissance forward of the troop.
- 1st platoon establishes contact with the objective, then secures the flanks and conducts surveillance of the objective.
- 2d platoon, second in the order of movement, moves through 1st platoon, establishes security on the far side of the objective, orienting on the mounted avenues of approach into the objective area.
- 3d platoon establishes an attack position vicinity of BP1. 4th platoon establishes an attack position vicinity of BP2.
- Troop commander initiates the attack with indirect fires. 3d and 4th platoons move and attack by fire from BP1 and BP2.
- Troop engages enemy position using TOWs to destroy hard targets and fires from caliber .50 and MK-19s for soft targets and suppression of enemy defenses.
- Troop uses indirect fires to cover their withdrawal.
- 1st platoon covers the troop withdrawal.
- 4th platoon displaces first, followed by 3d platoon, then 2d, then 1st.
- Troop moves in column, platoons in vee with 1st platoon providing rear security. 4th platoon conducts a rapid zone reconnaissance forward of the troop as it moves toward friendly lines. See Figure 5-13.

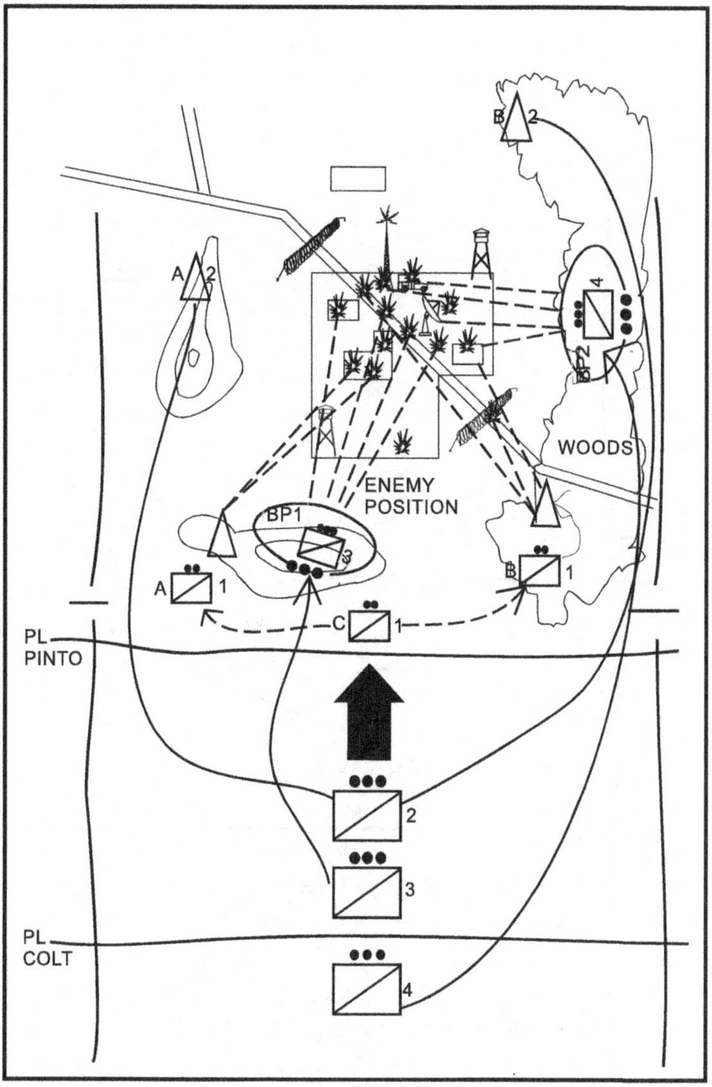

Figure 5-12. Light cavalry troop conducts a raid using a troop scramble.

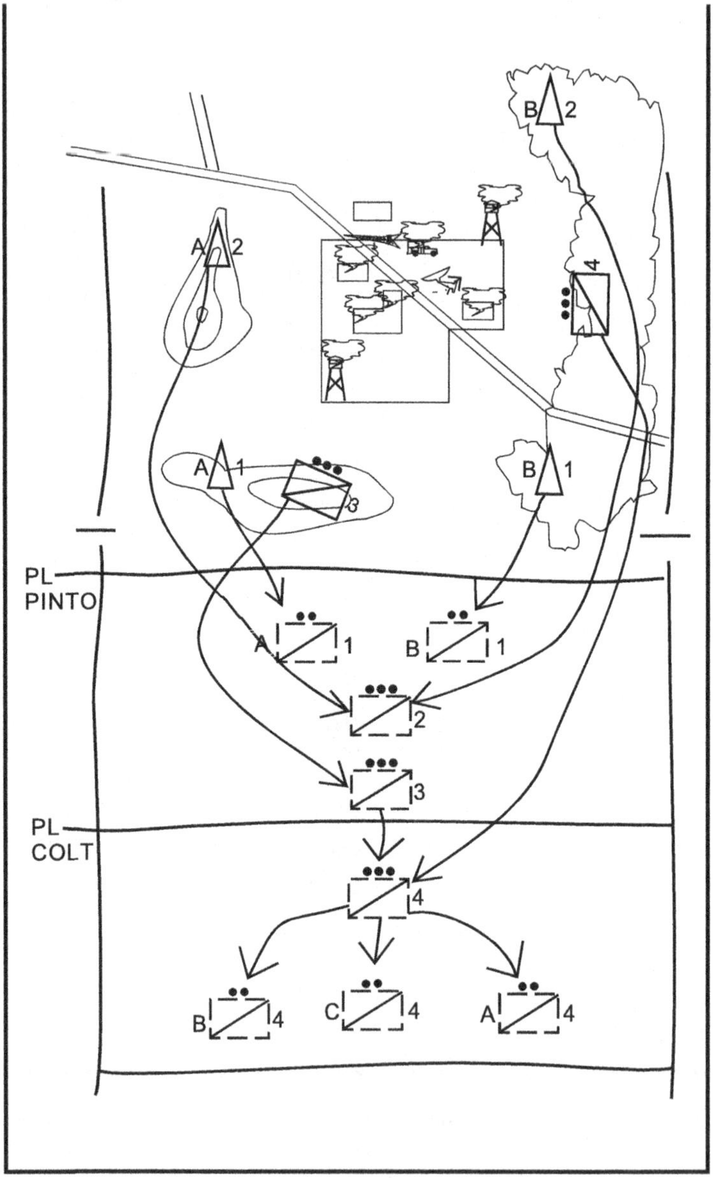

Figure 5-13. Light cavalry troop completes the raid and returns to friendly lines.

Chapter 6

Defense

Defense is not the decisive form of war. While defense can deny success to the enemy, it seldom assures victory. The defense is, however, the stronger form of war due to the inherent advantages of the defender. Army operations recognize the strength of the defense, but emphasize the necessity to transition to the offense quickly.

Brigades, battalion task forces, and company teams are the principal defensive forces for the corps or division. Cavalry units normally perform security missions for the defense or reconnaissance missions to support attacks. Cavalry units frequently perform defensive operations as a part of security missions, or when required to defend in an economy-of-force role for a higher headquarters.

Contents

Section I. Purpose, Fundamentals, and Schemes of Maneuver

PURPOSE

The immediate purpose of defensive operations is to defeat an enemy attack and create opportunities to go on the offense. For both the heavy and light cavalry troops, defensive operations may also achieve one or more of the following:

- Gain time.
- Concentrate forces elsewhere.
- Attrit enemy forces as a prelude to offensive operations.
- Control key or decisive terrain.
- Retain tactical, strategic, or political objectives.

In some cases, the troop may be required to defend because it cannot muster enough combat power to attack.

FUNDAMENTALS

Eight fundamentals are common to cavalry troop defensive operations.

Depth. Position platoons in depth, and place obstacles in depth. Depth allows the troop to—

- Gain enemy contact early.
- Perform counterreconnaissance tasks.
- Ascertain enemy direction of attack/intentions.
- Develop the situation, providing reaction time and maneuver space to concentrate combat power when and where it is needed.

Dispersion. Deploy subordinate elements as far apart as possible without losing their ability to concentrate (mass) firepower against the enemy. The more dispersed the troop, the harder it is for the enemy to mass fires against it as a whole. However, do not allow the enemy to concentrate its forces or fires against isolated elements.

Security. The troop may employ passive or active measures, or a combination of techniques. All must be considered in the defensive plan. Examples of passive and active security measures follow:

Passive	**Active**
Disperse vehicles and platoons.	Screen/Establish OPs.
Use camouflage/ cover and concealment.	Perform mounted/ dismounted patrols.
Impose radio listening silence.	Establish GSR posts.
Use hide positions.	Establish M8 chemical alarm net.
Enforce noise and light discipline.	
Minimize movement.	
Do not position in likely target areas.	

Maximize Terrain Advantages. Study the terrain. Reconnoiter it from both the troop commander's and the enemy's view, if possible, to determine—

- Avenues of approach.
- Reconnaissance avenues of approach.
- Restricted/severely restricted areas.
- Defiles (canalizing terrain).

- Engagement areas.
- Battle positions.
- Subsequent and alternate battle positions.
- Hide positions to support battle positions.
- OP positions forward of defensive positions.
- Subsequent and alternate OP positions.
- Positions where obstacles can be tied in with natural obstacles to turn, disrupt, or block the enemy.
- Positions that facilitate counterattacking by fire or by fire and maneuver into the flanks and throughout the depth of the enemy.
- Routes to and from each position.

Stop Enemy Rate of Advance. Offense is based on two principles—speed and mass. Develop a defensive plan that blunts the momentum of the enemy attack through the use of obstacles and fires (direct and indirect). If most of his combat power is killed, the enemy will be forced to dig in and establish a hasty defense or withdraw from combat.

Mass Combat Power at the Right Place and Time. In order to defeat a massed attack, the troop must mass fires against the enemy where and when he is least able to escape its effects. The ability of the troop to mass combat power when and where it is needed is a function of—

- Early warning/reaction time.
- Responsive/rapid maneuver by subordinate elements.

Force the Enemy to Fight in Two Directions. When engaging the attacking force, maneuver platoons into positions that force the enemy to turn and fight in two or more directions. This will force him to split his fires, preventing him from concentrating fires, and to expose his vulnerable flanks.

Counterattack. Draw the enemy into structured engagement areas and attack him *en masse* with overwhelming firepower to destroy him quickly and decisively. Wrest the initiative from him. Maneuver forces to exploit the situation. Attack by fire and by fire and maneuver into his vulnerable flanks and throughout the depth of his formation.

DEFENSIVE SCHEMES OF MANEUVER

There are three basic schemes of maneuver the commander can use in designating a course of action for a defensive mission. These schemes of maneuver center on the use of battle positions and sectors for subordinate platoons, or a combination of the two.

Defend from a Troop Battle Position. Both heavy and light cavalry troops may defend from a battle position. This method is be used when the enemy situation is clear, when there is only one avenue of approach, or when the troop commander must coordinate subordinate unit fires. In this scheme of maneuver the troop commander retains most of the authority for fighting the battle. The troop commander must understand his squadron commander's intent and concept to prevent holding the troop in place and risking its destruction.

Defend in Troop Sector. Both the heavy and light cavalry troops may defend in troop sector. This method may be used when the enemy situation is vague, when there is more than one avenue of approach, or when subordinate platoons require more freedom or action. In this scheme of maneuver, the troop commander delegates much of the responsibility for fighting the battle to his subordinate platoon leaders. He maintains control through the use of effective control measures and a clearly understood intent and concept.

Defend by Combination of Battle Positions and Sectors. Both heavy and light cavalry troops may defend using a combination of battle positions and sectors. This is the most common method of defense for a cavalry troop as it provides the troop commander maximum flexibility to maneuver and mass fires throughout the depth of the sector when changes in METT-T dictate.

Section II. Defend From a Troop Battle Position

Either the heavy or light cavalry troop may defend from a troop battle position. This mission is normally assigned when the squadron commander elects to concentrate the direct fires of the troop or squadron within an engagement area. The troop cannot maneuver outside the position without the squadron commander's permission. Within the battle position, the troop commander positions his platoons to concentrate all direct fires where the squadron has specified. The troop fights to retain the position unless ordered by the squadron commander to counterattack or withdraw. The troop may still retain the task to screen/maintain contact with the enemy forward of the battle position, depending on the squadron commander's intent. If so, the tank/AT platoons will initially be positioned within the troop battle position, and the scout platoons may assume adjacent positions within the battle position following execution of a security drill (see Security Drill paragraph in Chapter 4, page 4-15).

CRITICAL TASKS

To defend a troop battle position, the following critical tasks must be accomplished:

- Decide where the enemy will be killed and designate the engagement area.
- Establish OPs oriented forward and to the flanks of the BP to gain contact with the enemy force and provide early warning.
- Establish primary and alternate platoon BPs to concentrate direct fires within the engagement area as directed by the squadron commander.

- Designate supplementary platoon BPs to cover other routes of enemy approach.
- Establish sectors of fire for each platoon.
- Reconnoiter and establish platoon routes from hide positions to platoon BP, and for withdrawal to subsequent platoon/troop BPs.

TECHNIQUES

The squadron commander assigns troop battle positions when he can mass the fires of two or more troops in a squadron engagement area. The size of a troop battle position can vary, but it should provide enough depth and maneuver space for platoons to maneuver into alternate/supplementary positions and execute local counterattacks.

The troop commander conducts a thorough terrain study before the troop occupies the position, keeping in mind where the squadron commander wants the troop to concentrate its fires.

- Designate primary and alternate positions for each platoon.
- Position platoons to achieve flanking fires along the avenue of approach. Consider the effective range of each platoon's weapon systems.
- Position platoons to mass direct fires within the engagement area and to provide mutual support.
- Position platoons to cover any dead space in the engagement area. If not possible, plan indirect fires to cover the dead space.
- Use TRPs to control fires and orient weapon systems for each platoon.
- Occupy the BPs from the rear. Establish OPs to support the squadron plan. Allow platoon leaders time to reconnoiter and position vehicles to place effective direct fires within their sectors of fire. If better

defensive terrain lies outside the BP, call the S3 or commander for approval to adjust the boundaries.

- Once platoons are set, inspect them to make sure each platoon is properly oriented and has good fields of fire. When satisfied, have the platoons reconnoiter routes to subsequent positions and select firing positions for each vehicle. If time and engineer support are available, dig prepared positions for vehicles. Have the platoon leaders prepare platoon fire plans for approval.
- After preparing the BP, have all elements except OPs move to hide positions to reduce the risk of enemy observation and to decrease their vulnerability to enemy fires.
- Maintain security. Position the command post where it has FM communications with subordinate elements and the squadron. Position the troop trains behind good cover that is out of direct-fire range and allows quick access to each platoon position. The troop commander is positioned where he can observe the engagement area and control the troop. The FSO should be nearby to ensure coordinated fire support.
- Identify trigger points/lines.
  - Upon the enemy reaching what terrain feature does the troop initiate indirect fires? Direct fires?
  - Upon the enemy reaching what terrain feature does the troop displace to prevent decisive engagement?
- If the troop has to disengage and displace under fire to a subsequent position, bound the troop back by platoon(s), consistent with the squadron scheme of maneuver. If the troop disengagement and displacement are covered by another element, the troop may move as a whole to a subsequent position.

See Figure 6-1 for an illustration of the techniques described above.

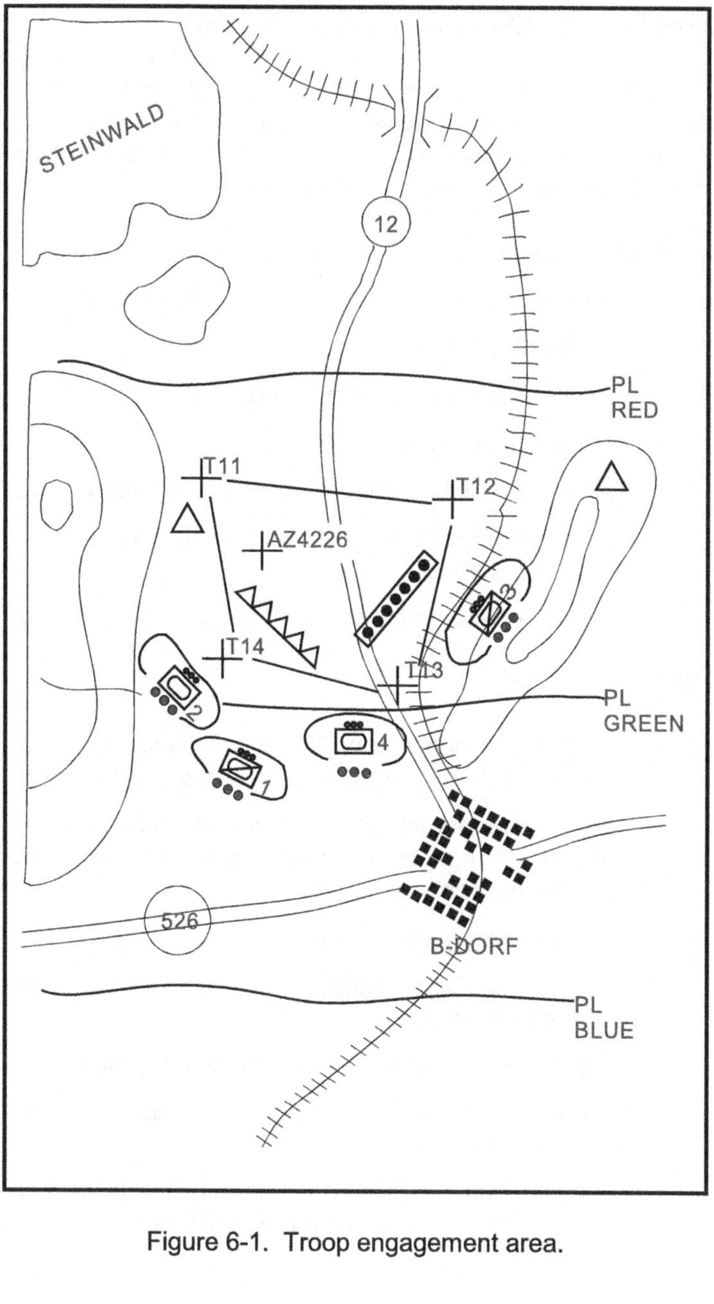

Figure 6-1. Troop engagement area.

Section III. Defend in Troop Sector

Both heavy and light cavalry troops may defend in sector. METT-T considerations determine optimal troop sector width; however, the troop is normally allocated a sector oriented on a single battalion-size avenue of approach.

Either troop may defend in sector when—

- The squadron cannot concentrate its fires due to—
  - extended frontages.
  - defending along a cross compartment.
  - multiple avenues of approach.
- Retention of specific terrain features is not necessary.
- The troop may use the depth of the sector to dissipate the enemy's attack.
- Maximum flexibility to maneuver is desired.

CRITICAL TASKS

- Maintain continuous surveillance of high-speed routes or avenues of approach into the troop sector (screen).
- Destroy or repel all enemy reconnaissance elements forward of the troop's initial defensive positions (counterreconnaissance).
- Structure engagement areas.
- Position platoon battle positions to support engagement areas.
- Engage the enemy from more than one direction.
- Determine criteria for initiating fires, counterattack, and disengagement.
- Prevent the enemy from penetrating the troop rear boundary or designated NPL (no penetration line).

TECHNIQUES

When given the order to defend in sector, the squadron will usually provide the following graphic control measures: *troop boundaries, an initial screen line, a rear boundary, phase lines, contact points between troops, TIRS, and TRPs that support any squadron engagement areas.*

Study the terrain in sector. Identify the terrain near the initial screen line from which OPs can maintain continuous long-range surveillance of enemy avenues of approach.

Determine where platoons can be positioned astride or on the enemy avenues of approach. Look for positions that provide good observation and fields of fire into the avenue of approach, and good cover and concealment for hide and defilade positions. Take a look at proposed platoon battle positions, and determine where troop fires can be massed on the avenue of approach. Use this portion of the avenue of approach to structure a troop engagement area(s).

The engagement area is where the troop will destroy an advancing enemy force. Establish a series of TRPs and use them to assign sectors of fire to each platoon. This allows the commander to control the fires of the troop and to achieve overlapping platoon fires.

With the assistance of supporting engineers, plan obstacles within the sector to support the defensive plan. Reinforce existing obstacles within engagement areas, and plan more obstacles to slow, canalize, or turn the enemy. Obstacles can buy the troop time to engage the enemy, and increase the effectiveness of indirect fires in the engagement area by compressing threat formations, slowing them down, and detaining them in the engagement area. Obstacles can give the commander time to maneuver platoons to counterattack or to move to subsequent positions. Plan obstacles in depth so the enemy gets bottled up in the engagement area and is confronted with a series of breaching operations. Make sure the troop can observe and place fires on all obstacles in the sector. Place the obstacles to achieve different effects. Obstacles emplaced on the reverse side of a hill or depression will cause the enemy to pile into them before he sees them. Strategically emplaced obstacles will cause the enemy to turn when he sees them, exposing his flanks to direct fires.

Give the FSO planning guidance so he can develop the troop indirect-fire support plan for the mission. Plan fires to support the scouts on the screen line forward of the troop. The scouts need indirect-fire support to engage enemy reconnaissance forces, to disrupt enemy lead echelon formations, and to attack follow-on forces. Plan indirect fires to engage enemy forces in the engagement area, when they are slowed by obstacles. These fires suppress, disrupt, and confuse the enemy and allow platoons to set up the direct-fire engagement. Plan indirect fires behind the engagement area to isolate the enemy. In addition, plan fires forward of the troop's positions to help disengage from the enemy in case the troop cannot stop him from initial positions. Identify rally points in sector behind the battle positions. Crews or troop elements that become separated or disorganized during battle move to these identified rally points to reassemble or reorganize.

Position platoons to maximize their weapons' effectiveness and crew/vehicle survivability based on the given terrain and the capabilities of the enemy. Platoons positioned at the base and along one or both flanks of the engagement area will force the enemy to fight in two or more directions.

Determine how to employ the scout platoons. Although the primary role of scout platoons is to conduct reconnaissance and screening in support of the troop, the troop commander may need to use their firepower to support troop defensive missions. Depending on sector width and number of avenues of approach, one scout platoon may be employed in a screen mission forward of the troop during a mission to defend in sector. It may fall back to a battle position after identifying the attacking enemy force (security drill), or it may stay forward of the troop, continuing to screen to identify follow-on forces. The other scout platoon may fight the attacking force from a battle position. The placement of the scout platoon's BP depends on the role the commander wants the platoon to play in the troop fight.

After making final adjustments to initial battle positions with the platoon leaders, plan alternate positions and subsequent positions in depth. Give platoon leaders time to reconnoiter covered and concealed withdrawal routes to their alternate subsequent positions.

Position the troop mortars where they can support the scouts on the screen line. Ensure they can engage targets from 3 to 3.5 kilometers beyond the screen line, or as far as the scouts can observe. Plan other mortar positions so that they can support the troop fight as the threat enters the engagement area. Also, plan positions through the sector.

The first sergeant positions the troop trains behind the initial troop battle positions, where they are responsive to troop needs but not vulnerable to direct fires. The XO positions the command post behind the initial platoon BPs on terrain that affords good FM radio communications with the troop elements and squadron headquarters. If possible, the XO positions the command post behind the subsequent troop positions; this reduces its vulnerability to fires, and allows it to remain stationary and maintain good FM communications while the troop displaces to other positions. The XO and first sergeant plan subsequent positions through the sector.

Position the FIST where it can maintain good digital FM communications with the supporting artillery unit. If possible, keep the FIST where it can use the laser designator to designate high-priority targets in the engagement area for Copperhead or other laser-guided munitions.

The troop commander must be in position to see the battlefield. To control the troop fires, he must understand which areas can and cannot be engaged by platoon battle positions.

The first critical task for a defend mission is to destroy or repel the threat reconnaissance. The scout platoon(s) on the screen line is responsible for identifying enemy reconnaissance forces, engaging them with indirect fire, and defeating them if possible. The other platoons may need to assist the scout platoon(s) in defeating the reconnaissance forces after the scouts identify them.

After destroying or repelling enemy reconnaissance forces, the troop is prepared to take on the lead echelon of the enemy force. Remember the scheme of maneuver. Let the enemy enter the engagement area and then mass the direct fires of the troop to strike a decisive blow. Depending on how the obstacles are set up, the troop commander may want to strike the enemy just before he reaches the obstacles; then, as the enemy deploys in reaction to troop fires, he hits the minefields and tank ditches. The commander may wish to wait until the enemy gets into the obstacles, and strike when he is confused and his formations are compressed.

Continue the fight by maneuvering platoons into alternate or supplementary positions while counterattacking by fire to complete the destruction of the lead echelon forces. If the troop is unable to defeat the enemy in the initial engagement area, it must be prepared to displace to subsequent positions. Do not allow the troop to become decisively engaged. Use the disengagement criteria from the scheme of maneuver to ensure adequate time to bound the troop back to subsequent positions by platoon(s). Keep one or two platoons in contact with the enemy, engaging him with direct and/or indirect fires. The other platoons move back in sector to subsequent positions. Once they are set, they engage the enemy with indirect fires or long-range TOW missile shots or tank main gun fire so the remainder of the troop can break contact and move to its subsequent positions.

Once the platoons are in their subsequent positions, make any needed adjustments to their positions or orientations based on what the enemy is attempting to do. Finish off the enemy from these positions, and then move forward and reoccupy initial or alternate positions, if possible.

EXAMPLE - DEFEND IN TROOP SECTOR (HEAVY TROOP)

The troop is given a mission to defend in sector and hold the threat forward of PL SABER. The troop commander completes his plan and issues the order. The 1st platoon moves forward and establishes a screen along PL SPUR. The 3d platoon prepares BP 20 at the base of the initial engagement area and occupies a hide position to the rear. BP 20 is about 2,500 meters away from the base of EA BEAR. The platoon's fires are concentrated in the direction of TRP 25. The 2d and 4th platoons occupy BP 30 and BP 40, respectively, along the eastern flank of EA BEAR. The 2d platoon orients on TRP 21, and the 4th platoon orients on TRP 27. The western flank is protected by a steep ridgeline. The troop command post is on high ground about 2 to 3 kilometers behind the initial defensive positions. The troop trains are just south of BP 30. The troop commander positions between BP 20 and BP 30, where he has a good view of all platoon positions and the engagement areas (see Figure 6-2).

A threat reconnaissance patrol of two BMPs approaches the screen line moving along the flanks of the high-speed route into the troop sector. As the patrol crosses PL SPUR, a 1st platoon scout element unmasks and quickly ambushes the patrol with cannon fire. One scout squad immediately searches the threat vehicles for intelligence information and captures two wounded prisoners. The troop commander tells the 1st platoon leader to expect a CRP within an hour. The rest of the troop stays in hide positions away from terrain most likely targeted by threat artillery units (see Figure 6-3).

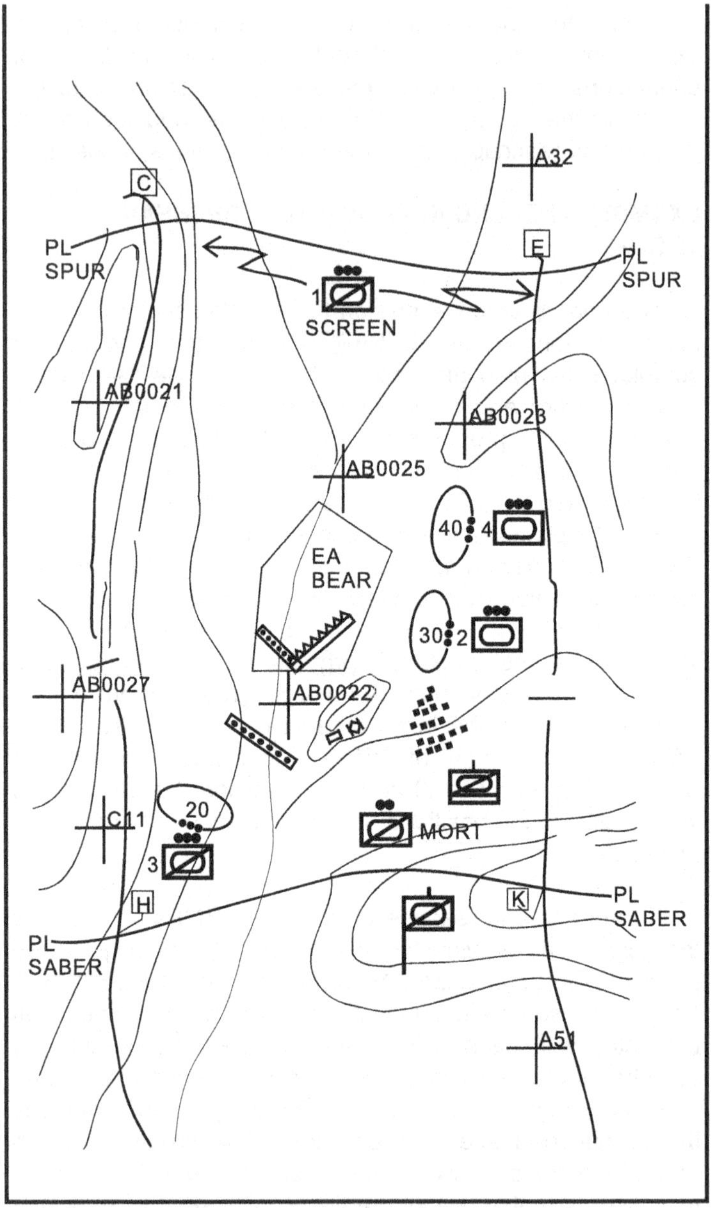

Figure 6-2. Defend in sector (part one).

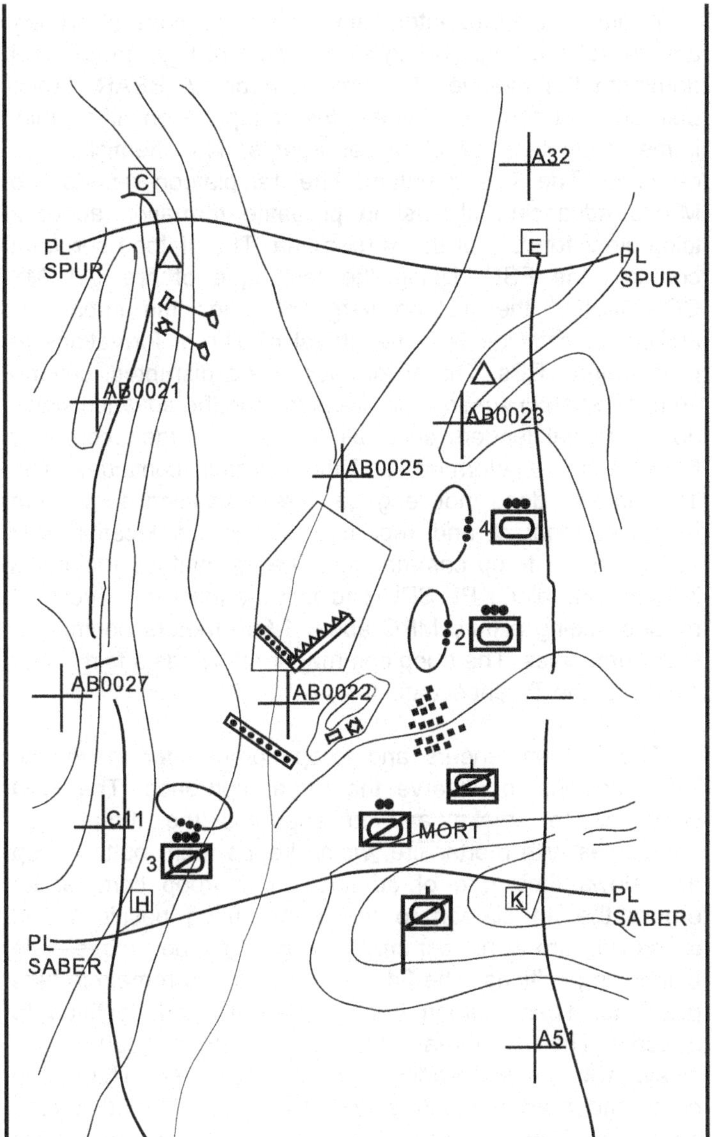

Figure 6-3. Defend in sector (part two).

About 45 minutes later, heavy concentrations of artillery and rocket fire begin falling along areas of high ground that dominate the avenue of approach near EA BEAR. Troop elements button up. When the suppression lifts, they immediately test for chemical agents. No chemicals are detected. The TCs unbutton. The 1st platoon reports two MRCs advancing abreast in prebattle formation about 2 kilometers forward of the screen line. The platoon sergeant contacts the FSO. Using the technique of fire "AT MY COMMAND," the platoon sergeant times the impact of artillery to coincide with the arrival of Threat formations at preplanned TRPs. Threat formations are disrupted, several vehicles sustain suspension damage, and the advance slows down. Threat leaders scramble to restore order, company formations are reformed, and the advance continues. The 1st platoon does not engage the advancing force, but maintains contact and reports the Threat's location and activity to the troop commander. The 1st platoon maintains its positions along PL SPUR to identify follow-on forces. It reports seeing a third MRC about 1,500 meters behind the lead companies. The troop commander now has a fairly clear picture of the Threat situation.

The platoon leaders and troop commander move into hide positions to observe the threat approach. The FSO continues to smoke and to suppress the threat lead companies with mortar fire, which keeps them buttoned up and slows their rate of advance. The troop commander orders the 3d platoon to move into firing positions and prepare to fire at his command. The 3d platoon moves into hull-down positions. The 2d and 4th platoons remain in hide positions. Their platoon leaders stay up and continue to observe. The lead threat MRCs appear about 2,800 meters away, with tanks leading platoon columns. The troop commander orders the 3d platoon to engage. The 3d platoon sends six TOW missiles down range, targeting mine-roller tanks and the threat platoon leader's BMPs (see Figure 6-4). The threat force detects the antitank guided missile (ATGM) fire and moves toward the 3d platoon, which quickly backs into defilade.

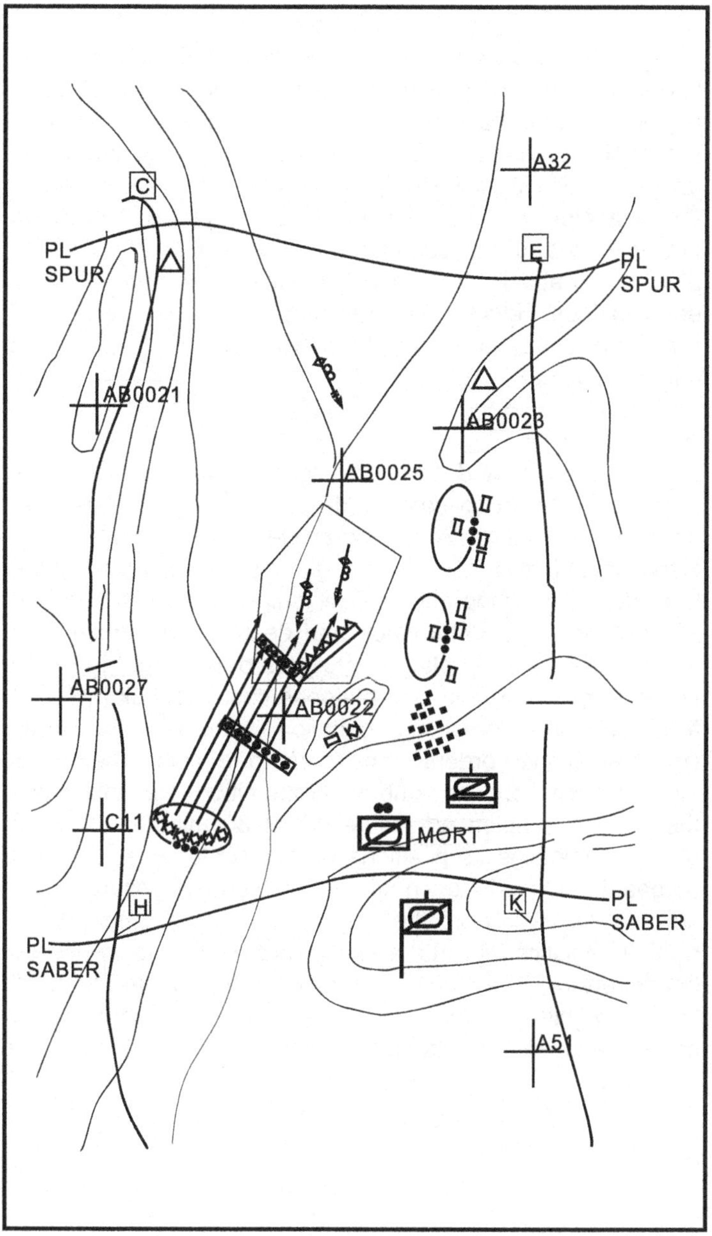

Figure 6-4. Defend in sector (part three).

The threat continues to advance through EA BEAR, piling into a minefield and tank ditch hidden on the reverse slope of a long, shallow draw. Several threat vehicles are caught in the obstacle. As the FSO observes this development, he concentrates all available fires on group targets that cover the obstacle. With the advance disrupted and stalled at the obstacle, the troop commander orders the 2d and 4th platoons to attack by fire into the flanks of the threat's lead echelon (see Figure 6-5). The concentrated cannon fire of both tank platoons quickly destroys most of the two lead companies.

The threat's third company comes into view in attack formation. The troop commander orders the 3d platoon to engage this force with long-range TOW missile fires. Survivors of this battalion begin to withdraw from the battlefield. The troop commander quickly orders the 2d and 4th platoons to counterattack and destroy the remnants of the battalion. The 3d platoon provides overwatch. Moving to alternate positions, the tank platoons counterattack by fire to finish off the remaining battalion vehicles. The troop commander then orders all platoons to reoccupy their initial hide positions and redistribute ammunition. He anticipates the second echelon will arrive within 30 minutes. The 1st platoon maintains its positions on the screen line. The first sergeant moves to each of the platoons to resupply. He collects the PWs from the 1st platoon. The XO collects all routine logistics reports and forwards them to squadron headquarters. The troop commander checks the status of leaders within the troop and designates replacements and cross-levels within the troop as necessary.

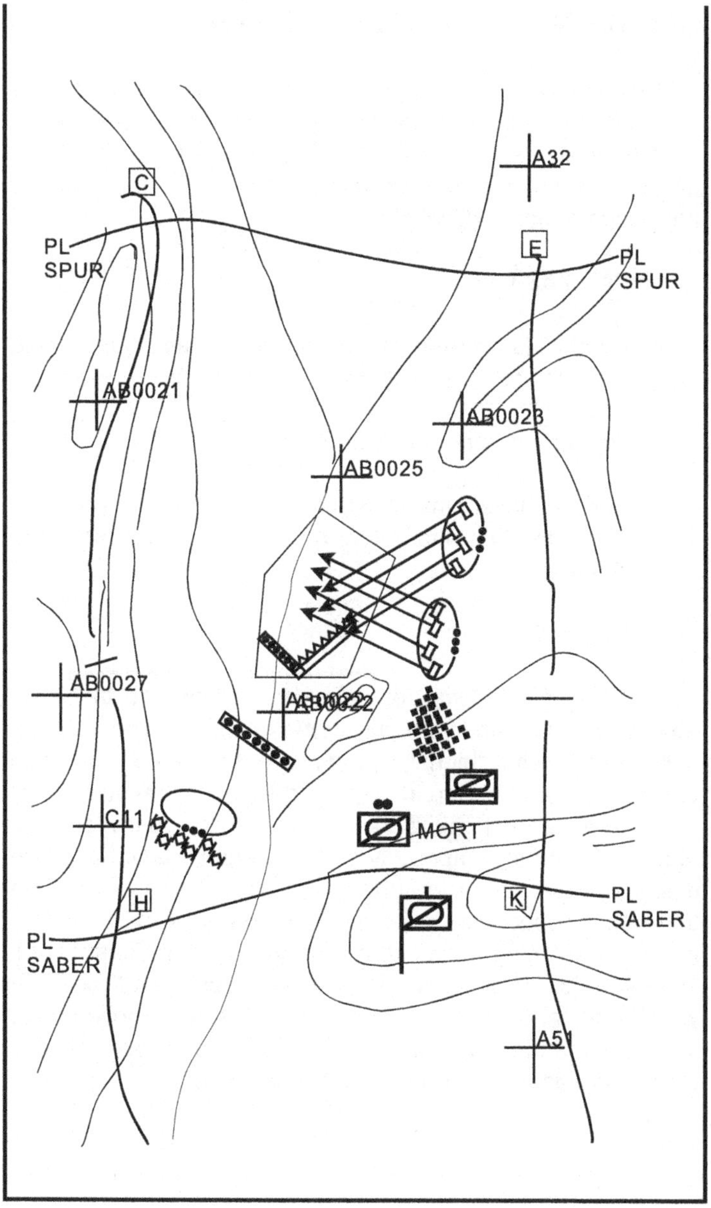

Figure 6-5. Defend in sector (part four).

Section IV. Delay in Troop Sector

Delay is a continuous series of defensive actions over successive positions in depth that trades the enemy space for time while retaining freedom of action. It is an economy-of-force operation that buys time to permit something else to happen at a more critical place on the battlefield.

CRITICAL TASKS

The critical tasks for delay include all the tasks associated with defend in sector as well as—

- Preserve freedom to maneuver.
- Cause the enemy to deploy from march or prebattle formation into attack formation as the troop moves to the rear.

TECHNIQUES

Planning and tactics for delay are identical to those for defend in sector, and vary only in their purpose. The flow of a delay resembles a "hit hard, then move" technique. The troop commander and subordinate platoon leaders must be very aware of disengagement criteria. The troop must mass the effects of fires to temporarily stop the enemy advance, then disengage and move to subsequent positions in depth. The troop cannot become decisively engaged. It must maintain a mobility advantage over the enemy. This means taking advantage of terrain, being familiar with high-speed routes of withdrawal, and rehearsing engagements and movements. The commander will probably have to use one or two platoons to assist disengagement of the others.

Chapter 7

Other Tactical Operations

Several combat operations are routinely associated with successfully accomplishing the missions described in Chapters 3 through 6. These operations require special planning and training considerations and techniques because of their complexity. At troop level, these operations are based on standing operating procedures (SOP) to ensure they can be conducted quickly and efficiently.

CONTENTS

Section I. Tactical Road Marches

Troops not in contact with the enemy often travel long distances to position themselves for future operations. A successful tactical road march depends on unit discipline and the ability to execute the plan strictly in accordance with the SOP.

CRITICAL TASKS

Accomplish the following critical tasks when planning, executing, and supervising a tactical road march:

- Establish the readiness condition (REDCON).
- Issue a warning order.
- Reconnoiter the route to the start point.
- Conduct quartering party operations.
- Issue a movement order.
- Execute a tactical road march in accordance with movement order and SOP.
  - Cross and report the start point, checkpoints, and release point on time.
  - Conduct actions at halts.
  - Conduct actions on contact.
- Maintain security throughout movement and during halts.
- Operate a trail party.
- Clear the release point.

TECHNIQUES

The march discipline necessary to execute a road march with routine precision is attained only by strict adherence to SOP. Ensure the tactical road march portion of the troop SOP addresses, as a minimum, the following:

- Order of march.
- March speed.

- Distance between vehicles.
- Actions on contact (air, ground, and indirect fire).
- Actions at halts.
- Security.
- Contingency plans for vehicle breakdowns, breaks in column, and lost vehicles.
- Quartering party.
- Trail party.

The basic considerations in planning any road march are listed below.

- Time available.
- Distance of the move.
- Current situation.
- Availability and condition of routes.
- Size of the unit.
- Types, numbers, and characteristics of vehicles that must move.

The troop will most often move as part of a squadron operation, and the move will be based on the squadron order. The troop commander must, however, be familiar with planning considerations so he can plan and execute an independent troop move, if required.

March Planning Sequence. If time permits, follow this sequence when preparing for a road march:

- Prepare and issue a warning order, giving the troop's movement and any follow-on missions. Give them time to prepare for the operation.

- Analyze the situation to determine if any of the movement factors (order of march, rate of march, or interval) specified in the troop SOP must be altered to meet mission requirements. If the troop must reconnoiter two assembly areas, the two scout platoons might have to be first in the order of march.

- Conduct a map reconnaissance of the route (if assigned) or determine the best available route.

- Organize and dispatch reconnaissance and quartering parties (if occupying an assembly area).

- Prepare detailed movement plans based on mission requirements and reconnaissance information.

- Prepare and issue the march order to the orders group.

Route Reconnaissance. Reconnaissance is essential to movement planning. It reveals accurate, up-to-date information about the route. A route reconnaissance determines travel times, and identifies capacities of underpasses and bridges, locations of fords and bypasses, and critical points and obstacles. This lets the commander plan his movement and avoid areas or situations that could disrupt it.

Conduct the route reconnaissance with one of the scout platoons. Before the reconnaissance, let members know what information is needed and when and where they should submit it. Get them out as early as possible; troop plans are based on what they see and report.

If no other traffic control assets are available or if the troop is moving independent of the squadron, use the scouts from the platoon conducting the reconnaissance as traffic control teams at critical points along the route. Determine through map reconnaissance any built-up areas or key intersections that could disrupt the troop's movement, and position traffic control teams there. Based on their reconnaissance, the scouts may need to refine these positions.

If the movement is being conducted as part of the squadron, one troop will usually reconnoiter the route. The troop must, however, reconnoiter the route from its present location to the start point to determine if the route is suitable and how long it will take to reach the start point.

Quartering Party. A quartering party is used to reconnoiter and prepare a position before the troop main body arrives. The troop quartering party will often move as part of the squadron quartering party. Organize it with a scout platoon leader or the first sergeant as the party leader, with guides from each platoon, and with any additional personnel needed to clear the area.

Before the quartering party leaves, it must know the troop's route, order of march, estimated time of arrival, and any specifics on establishing the assembly area. Before the main body arrives, the quartering party reconnoiters the area, marks routes, and prepares to guide the main body as it enters the new area.

Trail Party. The troop trail party is made up of personnel and equipment (normally the troop trains) to handle emergency vehicle repair and recovery, medical aid and evacuation, and emergency refueling. The trail party moves just forward of the last maneuver platoon in the main body. The troop motor sergeant or first sergeant is in charge.

Main Body. The troop normally moves as a single march unit in column formation when conducting a tactical road march. The organization of the troop and any attached elements should be standardized during movement. Alter this organization to meet specific mission requirements as needed.

The troop commander has no prescribed place in the column. He positions himself where he can best control the movement of the troop. He will usually be well forward in the column, behind the lead platoon, to respond to contingencies while on the move. As a security measure, the troop command post should be positioned farther back in the column to disperse troop command and control.

The troop's column organization must provide adequate security against air and ground threats, while on the move and during halts. See Figure 7-1 for one way to organize a troop march column.

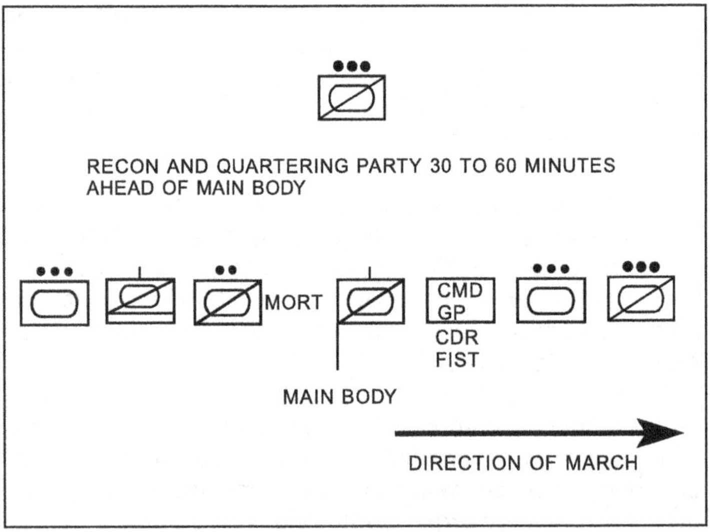

Figure 7-1. Troop column organization.

The 3d platoon, the first sergeant, and the quartering party move ahead of the troop. The troop order of march is 1st platoon, command group, 2d platoon, command post, mortars, trains, and 4th platoon. This order of march provides 360-degree security, disperses the command and control assets of the troop, and provides reconnaissance forward of the main body. Also, vehicle commanders assign sectors of observation to their crews, who search for air and ground threats (see Figure 7-2).

The troop performs the march in open or close column, depending on the situation. Use close column during limited visibility conditions. Vehicles are spaced 25 to 50 meters apart. This method takes advantage of the traffic capacity of the route, but provides little dispersion. Vehicle density is 15 to 30 vehicles per kilometer along the route of march. Use open column to provide greater dispersion and, thus, greater security. The distance between vehicles varies from 50 to 100 meters. Open column is normally used in daylight conditions. Base the troop's march speed on the slowest vehicle in the column. The troop trains will usually limit the rate of march.

Halts. Halts are used to rest personnel, provide personal comfort and relief, facilitate mess operations, refuel vehicles, maintain and inspect equipment, adjust the schedule, and allow other traffic to pass. The squadron march plan or troop SOP will specify the frequency and duration of halts, and will prioritize work to be done. For long movements, plan halts into the troop march table and ensure subordinate platoon leaders understand what actions must occur at the halt. When unscheduled halts occur, find the reason for the halt and let subordinates know how long it will last. Provide for security during halts, and establish OPs to provide early warning of enemy forces during long halts.

Vehicle commanders assign sectors of observation to their personnel so there is 360-degree observation. Each vehicle commander designates an air guard to provide air security, or specific vehicles may be designated air guard vehicles. In that case, the crews concentrate on only vehicle air observation rather than air and ground observation.

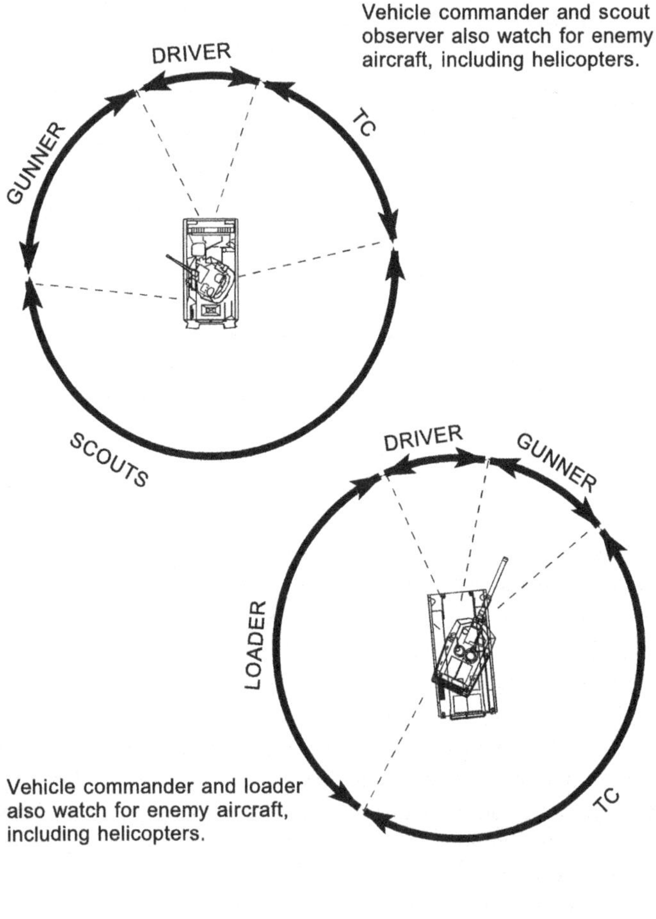

Figure 7-2. Maintain 360-degree security.

Vehicles that become disabled during movement must not obstruct traffic. The crew of the disabled vehicle must move the vehicle off the route, post guides to direct traffic, and find the problem. If the vehicle can be fixed, it rejoins the rear of the column. It does not return to its original position until the column has halted. If the vehicle cannot be readily repaired, the trail party recovers it.

Section II. Assembly Areas

An assembly area serves as a place where the troop gathers to prepare for future operations. In the assembly area, the troop prepares and issues orders, repairs and maintains vehicles, conducts resupply operations, and rests. As a minimum, assembly areas are positioned out of range of enemy light artillery.

TASKS

The following tasks are associated with assembly area operations:

- Plan the occupation of an assembly area.
- Conduct quartering party operations.
- Occupy an assembly area.
- Maintain security.
- Establish communications.
- Prepare for future combat operations.
- Depart an assembly area.

TECHNIQUES

The troop will normally be assigned a specific assembly area location. Within the area available to the troop, conduct a map reconnaissance, and if time is available, a ground reconnaissance. Look for an area that provides overhead concealment. This is extremely important if the troop will remain in the area for any length of time. Select an area that also provides—

- Cover from direct fire.
- Good drainage and a surface that will support troop vehicles.
- Good entrances and exits and an adequate internal road or trail network.
- Space for dispersion of vehicles, personnel, and equipment.

Quartering Party. The troop will often occupy an assembly area at the end of a road march. Before the quartering party leaves the troop's present location, tell them how to organize the assembly area. They must know any special requirements, such as a site for a logistics package (LOGPAC), so they can prepare the position. The quartering party must be organized to provide their own security during this operation. When the quartering party arrives at the forward assembly area, they must—

- Reconnoiter the area. If the area is not suitable, the leader of the quartering party must report immediately and provide a recommendation for another area.
- Organize the area. The leader of the quartering party selects locations for the platoons, mortars, command post, trains, and any attached elements, based on the commander's instructions. He may need to deviate from the commander's guidance to position an element in a suitable location.
- Improve and mark entrances, exits, and internal routes.

- Mark vehicle locations. Platoon representatives in the quartering party select general locations for vehicles in their platoon. Vehicle commanders and the chain of command refine these positions when they arrive.

- Perform guide duties. A platoon representative guides his platoon into position after the platoon clears the release point.

Occupation. Each platoon is guided from the release point into the assembly area by its quartering party member. Color-coded lights can be used to link up guides and lead vehicles (see FKSM 17-97-3). When the troop arrives at the assembly area, all elements move off the route without halting or slowing to keep the route of march clear. Keep this in mind when selecting routes, organizing the order of march for the road march, and allocating space in the assembly area. Once platoons have cleared the route and moved into their areas, they can adjust their positions without slowing the remainder of the troop.

Positioning troop elements in the assembly area is based on the size of the area, the terrain and avenues of approach, the length of time the troop will occupy the area, and any special requirements, such as resupply. Some considerations in allocating space for the troop are shown in Figure 7-3 and in the following list:

- If possible, space vehicles at least 100 meters apart to decrease exposure to enemy observation and fire.

- Keep vehicles in hide positions and establish OPs on terrain that provides good observation of approaches into the assembly areas for early warning of enemy movement. The vehicles can be moved up to a fighting position if necessary.

- Position platoons based on available terrain in the assembly area. Tank platoons may need more space to move vehicles in and out of position; scout platoons may occupy terrain that is more restrictive.

- Position the troop command post in the center of the assembly area. This makes establishing wire communications and issuing orders easier.
- Position troop trains where they have a good road network and space to conduct their maintenance and resupply operations. They may be used to secure the perimeter of the assembly area, but this will reduce their support ability.
- Emplace the mortars in a position from which they can support the troop with indirect fires. Overhead cover will reduce or eliminate their ability to fire.

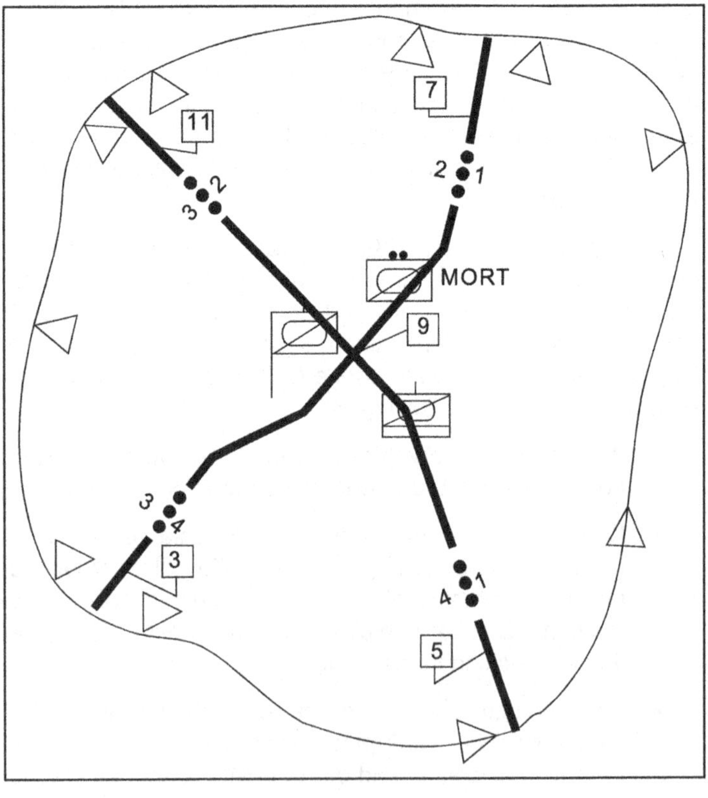

Figure 7-3. Troop assembly area.

Security. Although the assembly area is not a defensive position, the troop must be able to see and defeat enemy ground attacks. The best defense against air attacks is to remain hidden. Post guards at all entrances and exits to stop traffic that tries to enter the area. Establish OPs to observe key terrain features and likely avenues of approach for early warning of enemy approach. Each platoon must provide overlapping observation and fires within its platoon and with the platoons on its flanks. Establish a dismounted patrol plan so platoons make physical contact with their adjacent platoons. Ensure the platoons provide 360-degree coverage of the assembly area. Camouflage vehicles and equipment to prevent enemy detection from the ground and air. Place PEWS (platoon early warning system) in heavily vegetated areas or dead space to provide early warning of enemy movement in and around the position. Emplace NBC alarms upwind and about 150 meters from the troop's positions to provide early warning of an NBC attack. Assign the mortars an azimuth of fire on the most dangerous approach into the assembly area. Give the fire support officer guidance in preparing an indirect-fire plan.

Communications. Messenger and wire are the primary means of communications from the command post to each platoon and to each OP. If practicable, lay wire from the squadron to the troop. Plan on providing a messenger to the squadron command post. Use radio when no other means of communication can be used.

Preparation for Future Operations. Several tasks are routinely accomplished in an assembly area. These tasks are listed in the troop SOP (FKSM 17-97-3) under priority of tasks upon arrival in an assembly area, and include—

- Position vehicles.
- Establish local security.
- Establish OPs.
- Prepare fire plan.

- Establish wire communications.
- Maintain radio watch and man turret weapons.
- Camouflage positions.
- Prepare obstacles/mine plan.
- Select alternate and supplementary positions.
- Reconnoiter routes of withdrawal.
- Perform PMCS.
- Emplace NBC alarms and PEWS.
- Continue to improve positions.
- Conduct logistics resupply (Classes I, III, and V).
- Rest in accordance with REDCON status.

Modify this task listing to accomplish specific tasks (such as conduct rehearsals, test-fire weapons, and conduct inspections) in preparation for future operations. Ensure subordinates know how long the troop will remain in the assembly area and are told of any special requirements. Occupation of the assembly area will often be conducted and supervised by troop NCOs, while the commander and platoon leaders plan for upcoming operations.

Departure. Maintain the appropriate REDCON. Each REDCON level indicates critical tasks and time available to prepare for future operations.

- REDCON 1—be prepared to move immediately.
  - All personnel alert and ready for action.
  - Vehicles loaded and secured, and weapons manned.
  - Vehicle engines running and OPs not manned.

- REDCON 2—be prepared to move in 15 minutes.
  - All personnel alert.
  - OPs and wire pulled in.
- REDCON 3—be prepared to move in half an hour.
  - Fifty percent of each crew/squad stand down for rest, feeding, and maintenance.
  - Remaining 50 percent man vehicles, OPs, weapons, and monitor radios/phones.
- REDCON 4—be prepared to move in one hour.
  - Two men per platoon make dismounted checks of platoon area.
  - One man per vehicle monitors radios/phones and mans turret weapon.

All personnel remain at 100-percent alert until the prioritized work is complete after entering the assembly area. Initiate the appropriate REDCON when the work is finished. As the time for execution of a mission nears, increase the REDCON in accordance with guidance from squadron, assigning REDCON 1 just before the troop must move.

Section III. Relief in Place

A relief in place is an operation in which a unit in combat is replaced by another unit. Responsibilities for the combat mission and the assigned sector, battle position, or zone of the relieved unit are assumed by the relieving unit. A relief in place may be accomplished during offensive and defensive operations and may be conducted during any weather and light conditions.

The primary purpose of a relief in place is to maintain the combat effectiveness of committed units. A relief in place is conducted to replace a committed unit to give it the opportunity to reconstitute, rest, decontaminate, or perform a change in mission.

TASKS

The following tasks are associated with the relief in place:

- Plan and coordinate a relief in place.
- Establish communications.
- Establish liaison.
- Conduct reconnaissance.
- Initiate movement.
- Occupy positions.
- Maintain operations security.

TECHNIQUES

A troop will often perform a relief in place mission as part of a squadron operation to relieve or be relieved by a brigade. The troop will usually conduct a relief in place in conjunction with a battalion. The troop may relieve a battalion and operate in an economy-of-force role or be relieved by a battalion to conduct a change in mission.

To reduce confusion and maintain security, the following factors must be considered when planning a relief in place:

- The time that responsibility for the sector, battle position, or zone is to pass.
- Operations security.
- Deception plans.

- Time, method, and sequence of relief.
- Routes and critical control measures.
- Concept of subsequent missions.
- Plans for additional positions.
- Contingency plans.
- Location of obstacles and transfer of responsibility.
- Transfer of ammunition; wire lines; petroleum, oils, and lubricants (POL); and materiel to the relieving unit.

Communications. At a specified time or upon receipt of an order, the relieving unit minimizes radio traffic or begins to operate at radio-listening silence. The relieving unit command post monitors the command net of the unit to be relieved while continuing to operate on its internal command net.

Reconnaissance and Liaison. The orders group of the relieving unit moves to the command post of the unit being relieved for face-to-face coordination.

In preparation for the relief in place, the relieving unit moves to a forward assembly area under the command of the XO, if necessary.

If the situation permits, the relieving unit conducts a reconnaissance of the area when it completes coordination. The units establish a contact point behind the positions into which they will move; a guide from the relieved unit will meet them there to make any final coordination and guide them into position.

The commanders complete their plans and issue orders to their respective units after the reconnaissance and liaison are complete.

Methods. The relief in place can be conducted in several ways. Use the method appropriate to the troop's situation.

- One unit at a time. This is the slowest method, but the most secure.

- All units simultaneously. This is the quickest method, but the least secure.

- Center units first, followed by flank units.

- Flank units first, followed by relief of the center unit.

- Occupy adjacent or in-depth positions that cover the area of responsibility.

Passage of Command. The troop command post and battalion TOC collocate to facilitate command and control of the operation. The relieving unit departs its location in the prescribed sequence and lines up at the contact points with the guides from the relieved unit. The relieving unit moves into hide positions behind the relieved unit. The relieving unit moves into position after all pertinent information has been exchanged and the relieved unit moves out of position. The relieved unit moves to a rally point or assembly area to link up with the rest of the unit. When each element has completed the relief, the relieving unit assumes responsibility for the area and reports by messenger, wire, or radio to the command post. The relieved unit moves along its designated route to the new location. See Figures 7-4 and 7-5 for illustration of the relief in place.

The higher headquarters will often specify a time the relieving unit assumes responsibility for the sector, battle position, or zone. The commanders conducting the relief coordinate to ensure the relief is conducted on time. The relieving unit normally assumes responsibility when the relieved unit departs its positions. The commanders of the two units are usually collocated and responsibility passed face to face.

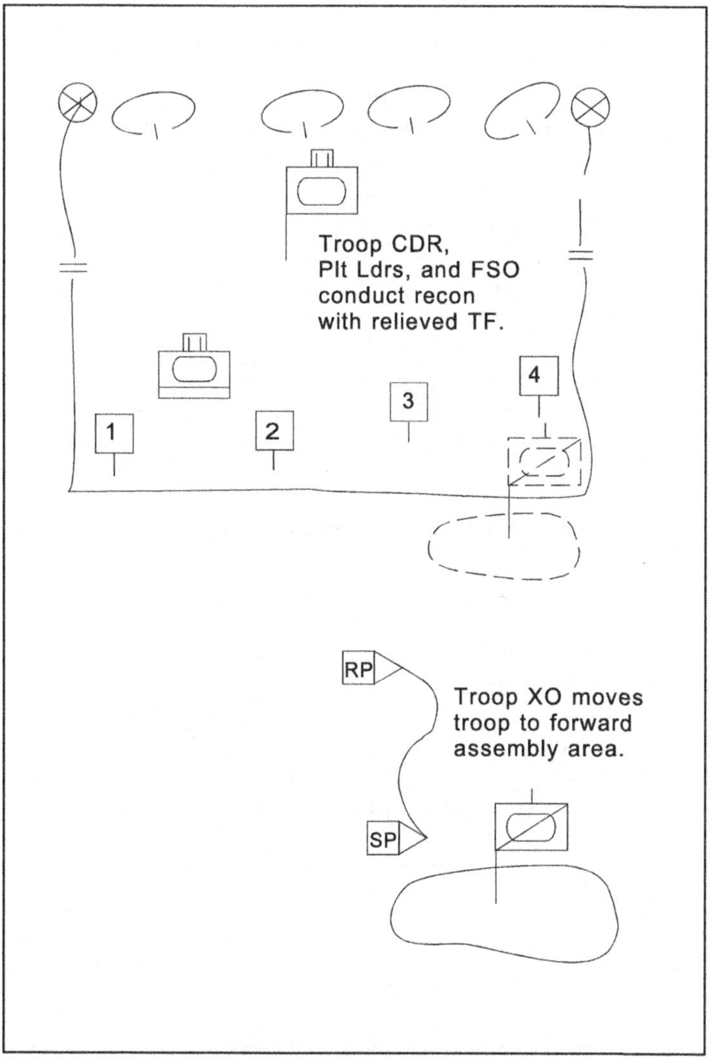

Figure 7-4. Troop prepares to conduct relief in place with task force.

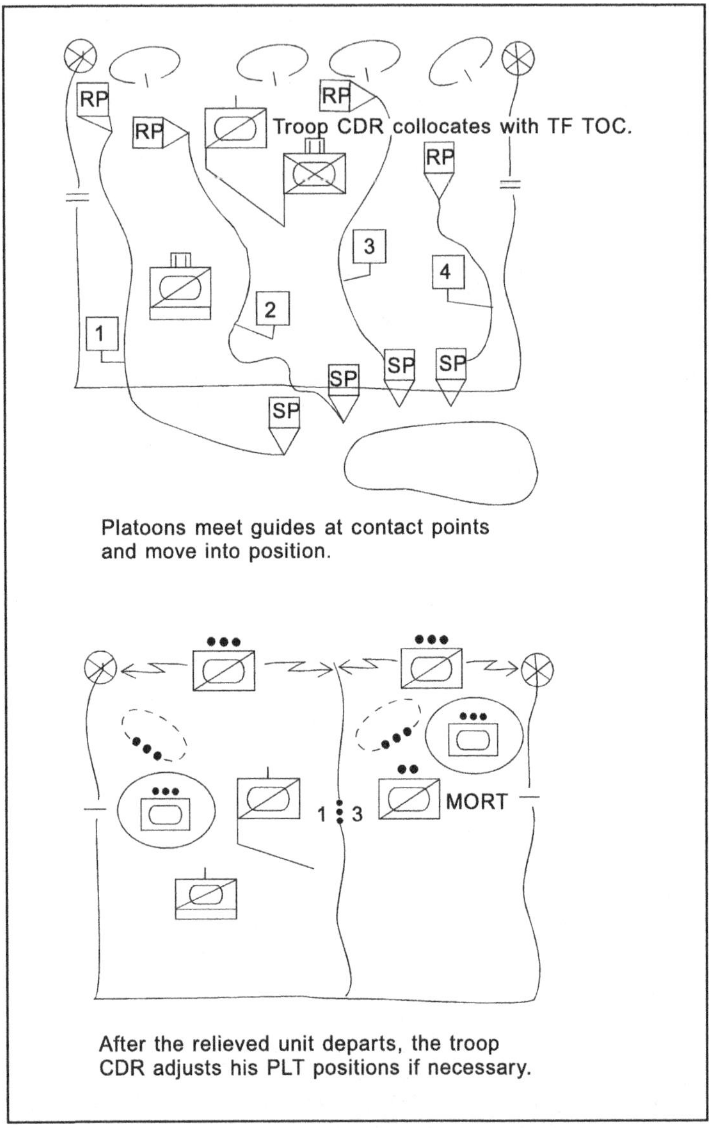

Figure 7-5. Troop completes relief in place.

Fire Support Assets. The troop fire support officer coordinates with the squadron fire support officer to ensure there is fire support throughout the operation. The fire support assets of the relieved unit remain in position throughout the relief of the maneuver forces, and are prepared to support both units. Fire support assets of the relieving unit position themselves as quickly as possible to provide additional support.

Enemy Contact. If either unit gains direct-fire contact with an enemy force, it immediately notifies the other unit and the higher headquarters directing the relief. If responsibility has not passed, the relieving unit becomes OPCON to the relieved unit. The relieving unit's mortars fire missions as directed by the relieved commander through his fire support officer. If responsibility has passed, the relieved unit becomes OPCON to the relieving unit. The collocation of the commanders and command posts facilitates rapid coordination and action in this situation.

Section IV. Battle Handover and Passage of Lines

Battle handover is an operation that transfers responsibility for fighting an enemy force from one unit to another in the close-in battle. Conducted by stationary and passing units, battle handover is designed to sustain continuity of the combined arms fight, and to prevent the enemy from getting a "free ride" anywhere on the battlefield as one force picks up the fight from another. It is also designed to preserve the fighting capabilities of both forces as they execute the operation. Battle handover is associated with almost all Army combat operations.

Passage of lines is a tactical event associated with battle handover. It is the controlled movement of one unit through the positions of another stationary unit that does not interfere with either unit's scheme of maneuver. A passage of lines is often used because the combat situation does not permit one unit to bypass another unit's positions. The passing unit must move through the positions of the stationary unit.

Passage of lines is often an integral part of the following operations:

- Deliberate attacks or counterattacks across the FLOT.
- Exploitation and pursuit.
- Route, zone, or area reconnaissance.
- Raids.
- Movement to contact.
- Defend, delay, or withdrawal.
- Defensive or offensive cover.
- Screen or guard.

Although a cavalry troop does not normally perform all of these operations, it may conduct them as part of a larger force.

CRITICAL TASKS

The three players in battle handover and passage of lines are: the *stationary unit*, the *passing unit*, and the *common commander* of both units. Each has critical tasks to perform to achieve smooth and efficient execution of the operation.

Critical Tasks of the Common Commander. The commander exercising command authority over both the stationary unit and the passing unit must accomplish three critical tasks.

- Establish where battle handover will occur by designating a phase line forward of the FEBA as the battle handover line (BHL). The line should be where combat maneuver forces of the stationary unit along the FEBA can effectively overwatch and protect the passing unit as it withdraws behind the FEBA or advances forward of the FEBA or FLOT. The distance forward of the FEBA or FLOT is limited to available fields of fire and the effective range of weapons of the stationary unit.

- Designate contact points just forward of the battle handover line at which stationary and passing units are required to conduct physical coordination (only in defensive operations).
- Ensure the passing unit is provided indirect-fire support while its artillery is displacing during battle handover and passage of lines.

The stationary unit normally recommends battle handover line and contact point locations to the common commander. It remains the responsibility of the common commander to establish the line and contact points. These control measures must be reflected graphically on an overlay and identified in the appropriate operation plan (OPLAN), OPORD, or FRAGO issued to subordinate units. The battle handover line, in effect, establishes a boundary between the ground owned and controlled by the stationary unit commander and the ground controlled by the passing unit commander. The stationary unit commander controls the ground forward of the FEBA up to the battle handover line. He can place security forces, obstacles, and direct and indirect fires into this area to support his scheme of maneuver.

While the battle handover line defines the point at which the battle handover should ideally occur, events normally preclude this from happening. The moving and stationary force commanders and their common commander should understand that the actual handover may occur in a "zone" centered on the battle handover line. Both the moving and stationary force must remain active and responsive enough to perform the battle handover anywhere in this zone, and should not dogmatically "stick to the plan."

Battle handover begins on the order of the common commander. To sustain unity of command in the passing operation, the passing unit is usually placed OPCON to the stationary unit to execute battle handover and passage of lines. OPCON by the stationary unit is limited to those actions necessary to get the passing unit through the stationary unit's area as quickly as possible. In defensive operations, battle handover from the passing unit to the stationary unit is not complete until the combat maneuver forces of the stationary unit along the FEBA have visual contact with the enemy, and the stationary unit commander indicates readiness to assume responsibility for the battle. In offensive operations, battle handover from the stationary unit to the passing unit is not complete until the combat maneuver elements of the passing unit have crossed the FEBA or FLOT, deployed for combat, and maneuvered across the battle handover line.

Passing Unit Critical Tasks. The passing unit must accomplish the following critical tasks to perform battle handover and passage of lines:

- Immediately establish communications with the stationary unit. Enter the command, operations and intelligence (OI), and fire support nets.
- Collocate a command post with the TAC CP or TOC of the stationary unit as soon as possible to enhance communications and unity of effort.
- Continuously report to the stationary unit the location, size, and composition of all enemy forces. Report the enemy's current activity. If the enemy is attacking, report his direction of movement, movement formation, and estimated rate of advance. If he is defending, report his locations, orientation, composition, fire sacks, reserves (if known), obstacle system, or flanks.
- Continuously report to the stationary unit the location, size, and activity of all subordinate elements to include combat support, combat service support, and command and control facilities.
- Given the current disposition of subordinate units, coordinate with the stationary unit to determine

contact points at which each subordinate company-size unit will physically coordinate handover and passage of lines with representatives of the stationary unit. Once contact points are determined, send a FRAGO to each subordinate unit specifying where they will physically coordinate passage with the stationary unit. Confirm recognition signals that must be displayed during passage (defensive operations).

- Once each subordinate unit acknowledges where it must physically coordinate passage, each unit will dispatch representatives to assigned contact points and coordinate passage for its unit. At the contact point, confirm recognition signals and exchange required information (defensive operations).
- Maintain visual contact with all enemy units and delay back to the battle handover line, avoiding decisive engagement (defensive operations).
- Display correct recognition signals and use correct challenge and password as specified in the CEOI during passage.
- Maintain proper weapons orientation.

Stationary Unit Critical Tasks. The stationary unit must accomplish the following critical tasks when ordered to conduct battle handover and passage of lines:

- Establish communications with the passing unit. Coordinate and direct the passing unit to contact points based on current dispositions of the subordinate units (defensive operations).
- Ensure contact points are manned and subordinate commanders have personal communications with their representatives (defensive operations).
- Ensure representatives at the contact points assign each passing unit a passage point into the area of operations and a route that extends from the passage points to the rear boundary or to an assembly area (defensive operations).

- Ensure representatives at the contact points exchange required information with the passing unit as outlined in FKSM 17-97-3.
- If security forces are employed, position them along the battle handover line to observe enemy avenues of approach. Adjust as needed for low visibility conditions (defensive operations).
- If obstacles are emplaced between the FEBA and the battle handover line, ensure routes through the obstacle system are clearly marked and physically controlled by guides, or provide an escort to the passing unit. Ensure that reserve targets on obligated routes are manned by soldiers in direct communication with their commanders.
- Ensure that all routes of withdrawal obligated to the passing unit are unobstructed and facilitate rapid movement to the release point (defensive operations).
- Ensure obligated routes of advance, attack positions, and routes to the battle handover line are unobstructed and permit rapid movement (offensive operations).

GRAPHIC CONTROL MEASURES

Figure 7-6 shows the graphic control measures that support battle handover and rearward passage of lines.

- Battle handover line. This line is established by the common commander in consultation with both commanders. The stationary commander has the major determination in locating the BHL, as his force must be able to overwatch the BHL with direct fires.
- Contact points. These are established on identifiable terrain and normally in the vicinity of the passage lanes. For rearward passage of lines, the contact points are established forward at the BHL. For forward passage of lines, the contact points are established in the stationary unit's rear area rearward of the passage lanes.

- Passage points. The passage point is that point on the passage lane where the moving unit moves through, and responsibility for the battle is passed to the stationary unit. It is usually placed where the passage lane begins.

- Passage lanes. The stationary unit establishes passage lanes to move the passing unit quickly through defending unit positions. This could include passing through gaps in friendly obstacles and moving near or through friendly engagement areas and battle positions. Lanes are restrictive; however, they should ideally be wide enough to allow the passing unit to move in a tactical formation. The passage lane begins at the passage point and ends at the rear of the stationary unit BPs. The passage is considered complete when the moving unit exits the lane.

- Routes. Routes are used to move the passing unit through the stationary unit rear area. The number of routes designated will vary based on METT-T, but as a general rule, multiple lanes/routes should be planned to facilitate rapid passage of moving units and to avoid unnecessary massing of units. The stationary unit may escort or guide the passing unit along the lane/route.

- Assembly area. An assembly area in the rear area of the stationary unit allows the passing unit to conduct hasty reorganization and emergency CSS actions. This assembly area is temporary in nature.

- Infiltration points. Units should plan infiltration points and lanes for personnel unable to complete the passage with the unit. Passing unit liaison officers may remain located with stationary unit CPs to serve as a point of contact for infiltrating personnel/equipment. Personnel who infiltrate must have some way of contacting the stationary unit before crossing into friendly territory.

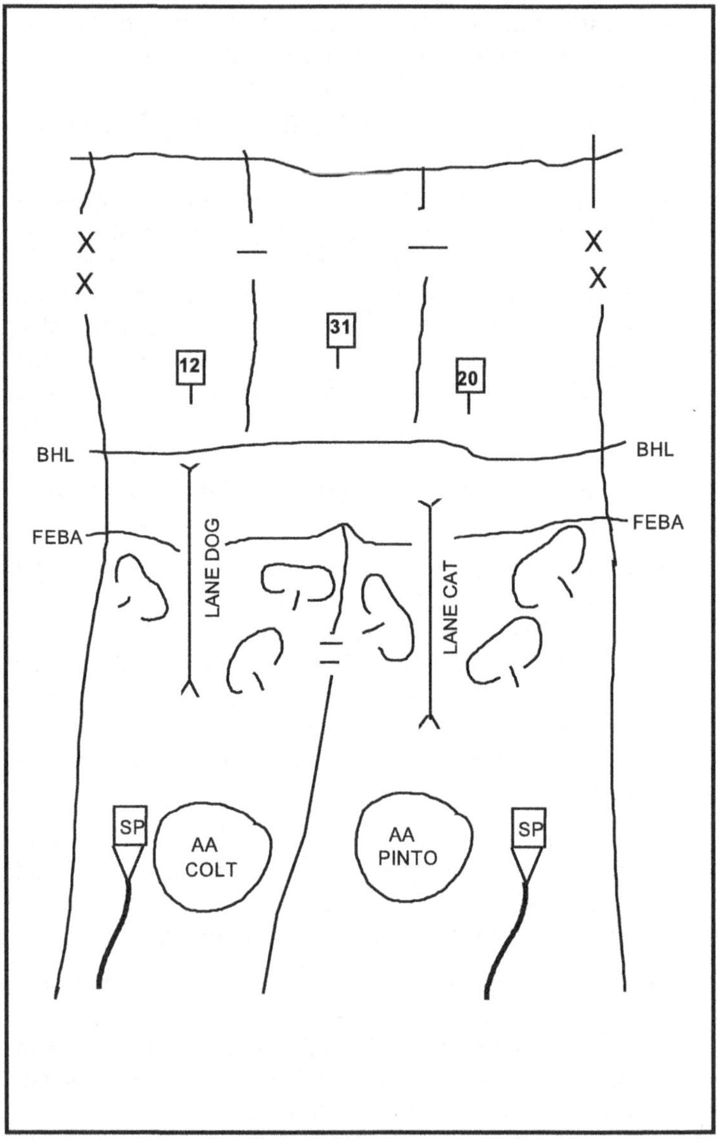

Figure 7-6. Rearward passage of lines.

TECHNIQUES

At troop level, the passage of lines will usually be performed as part of a squadron operation. The passage may be forward, such as to pass through a defending unit to conduct a counterattack, or rearward, such as when a covering force unit withdraws through units in the main battle area.

The troop is particularly vulnerable during a passage of lines. The unit may be concentrated and the fires of the stationary unit may be temporarily masked. Thorough reconnaissance and detailed coordination are critical to ensuring the operation is successful.

A troop will often perform a battle handover and passage of lines through a single battalion while the squadron conducts the operation through a brigade. Ideally, the boundaries will correspond to the battalion boundaries so that coordination for the battle handover is conducted through a single headquarters.

The troop commander has numerous considerations to ensure a successful passage of lines. During reconnaissance, he must confirm—

- The disposition of the stationary force through which the troop will pass.
- The location of contact points where both units are required to make physical contact at a predetermined time.
- The location of passage points on the battle handover line through which friendly forces will pass.
- The location of passage lanes that provide a clear route through a friendly position, and also facilitate a smooth and continuous passage. Areas selected for passage should be unoccupied or on the flanks of units in position. If possible, use multiple routes to reduce vulnerability during the operation.

- The location of an attack position (for forward passage) or assembly area. This position should provide cover and concealment and be located where the passing unit will not interfere with the stationary unit.
- The initial location for combat support and combat service support elements.

Based on the reconnaissance, the troop commander coordinates and plans for—

- Supporting fires. The stationary force supports the passing unit with direct and indirect fires up to the battle handover line. In forward passage, the stationary force supports the passing unit's move through the passage and until it crosses the battle handover line. In a rearward passage, the stationary unit supports the passing unit's move back across the battle handover line and through the passage of lines.

- Time of transfer of responsibility for control of the sector and handover of the enemy.

- Troop density. The passing troop commander should plan for multiple routes of passage to ensure rapid movement and to avoid congestion.

- Traffic control. Guides from the stationary unit pick up passing elements at each contact point and guide them through the position. The passing unit commander tells the stationary unit the type, number, and order of vehicles passing through each contact point.

- Communications. The leaders exchange CEOI information and mutually agreed upon recognition signals.

- CSS. The troop commander must coordinate the evacuation of casualties, PWs, vehicles, and resupply of fuel and ammunition. The stationary unit usually provides emergency service only. The passing unit supports itself.

- Liaison officers. The troop commander should designate a representative to perform the critical duties of a liaison officer. Commanders normally coordinate a forward passage of lines and the XO coordinates a rearward passage. Liaisons are normally located at critical points during the passage. If the commander or XO is not available, a scout platoon leader should perform liaison duties. Ensure he is thoroughly briefed on the situation and follows the checklist in the troop SOP.

Section V. Hasty Water Crossing

A hasty water crossing is the movement across an inland waterway using a crossing means at hand or readily available without significant delay once the waterway is reached. It is preplanned and conducted as a continuation of the operation underway. Although the crossing is termed hasty, detailed planning assures that fire support and crossing means are available on arrival at the water obstacle.

CRITICAL TASKS

The following critical tasks are associated with the successful conduct of a hasty water crossing:

- Plan the water crossing.
- Conduct reconnaissance.
- Establish security.
- Cross the obstacle.
- Provide continuous fire support.
- Continue the mission.

TECHNIQUES

Plan. A troop may perform a hasty water crossing independently or as part of a squadron. A hasty water crossing is performed as an extension of the ongoing operation. It gives the troop the ability to sustain the momentum of an operation, and is often associated with a movement to contact and a zone reconnaissance. The hasty water crossing is characterized by speed, surprise, minimum loss of momentum, and minimum concentration of forces.

The crossing should be preplanned to ensure fire support and crossing means are at the crossing site. The need for a water crossing should be determined in the mission analysis and added to the commander's concept of the operation. This ensures the troop's critical assets are positioned to support the water crossing. Seizing bridges intact before the enemy can destroy them is the quickest and most economical means of crossing, and is used when possible.

Reconnaissance. Scout platoons will be the first to encounter a water obstacle. When this occurs, they report it and reconnoiter it to obtain the following information, which is forwarded to the troop command post:

- Width and depth of the waterway.
- Water velocity.
- Possible entry and exit points and their conditions.
- Enemy situation on the far bank.

Scouts reconnoiter multiple crossing sites to prevent the enemy from discovering the true site and to reduce the concentration of vehicles at the site. If follow-on forces will cross at the same sites, scouts mark the entrances and exits for their use.

The troop commander, with the fire support officer, positions himself forward during the reconnaissance of the water obstacle so he can personally supervise the operation. The fire support officer will support the scouts with indirect and direct fires, if needed.

Security. A crossing can be performed with or without opposition. To avoid enemy fire while crossing, prevent him from knowing where the crossing will be.

Before the main body crosses, the scouts must secure the far bank by fording or swimming across the water obstacle and establishing positions. They should cross on both sides of the site and secure positions on the dominant terrain of the far side to provide early warning of threat activity. After securing the far side of the water obstacle, the scouts establish positions to the flanks of the crossing site to provide local security (see Figure 7-7).

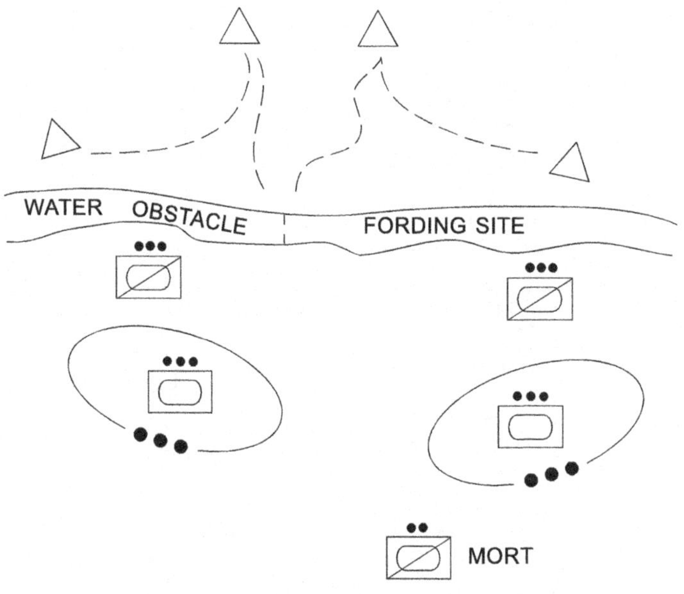

The scouts encounter a water obstacle, find fording sites, and secure the far side. Position the tank platoons and mortars to support with direct and indirect fires.

Figure 7-7. Secure the far bank.

The rest of the troop should not bunch up at the crossing site. Position them a terrain feature short of the water obstacle, so they have quick access to the crossing sites but are not exposed to enemy observation and fires.

Cross the Obstacle. Once the scouts have established positions on the far bank, the troop main body is ready to cross the obstacle. Time the movement to the crossing site so that no vehicles have to pause on the near side, but all can move directly through the crossing site and to their positions on the far side to continue the operation. Once the troop starts the crossing, complete it as quickly as possible (see Figure 7-8).

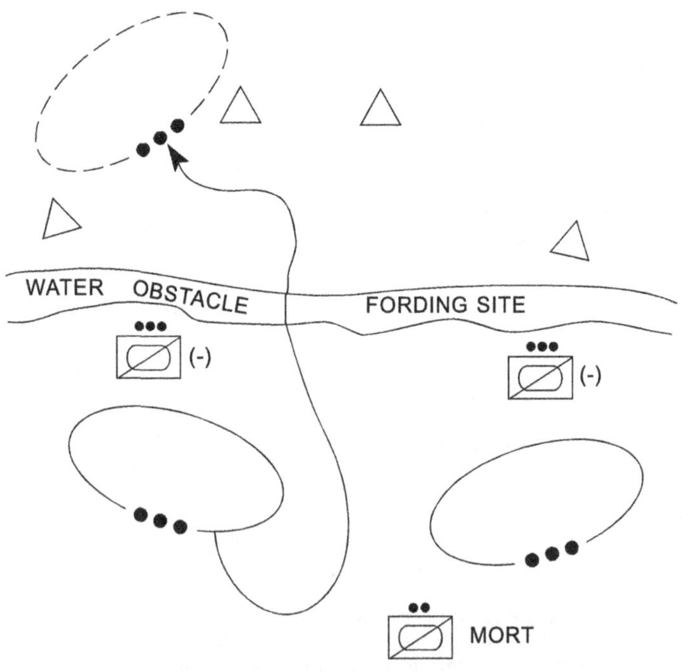

After the far side is secure, rapidly move platoons across the obstacle and continue the mission.

Figure 7-8. Cross the obstacle.

Fire Support. Fire support is used during the hasty water crossing to suppress known and suspected enemy positions at the crossing site. Smoke is employed to the front or flanks to screen the reconnaissance and the crossing. The fire support officer must time the movement of the mortars so they are in position to support the water crossing operation with continuous fire support.

Section VI. In-Stride Breach

When obstacles are encountered, an in-stride breach will maintain the tempo of offensive operations. Treat obstacles with caution and expect them to be covered by enemy observation and fires. The troop's ability to breach obstacles using available assets is important to ensure mission accomplishment. The size of the enemy force covering an obstacle with direct fire will, however, affect the troop's ability to conduct a hasty breach.

TASKS

The following combat tasks are associated with an in-stride breach:

- Detect the obstacle, reconnoiter it, and search for a bypass.
- Suppress all enemy positions with direct and indirect fires.
- Obscure enemy observation with smoke.
- Secure the near side of the obstacle.
- Breach or neutralize the obstacle.
- Move forces across the obstacle.
- Continue the mission.

TECHNIQUES

The enemy will reinforce natural obstacles with man-made ones to slow, disorganize, and canalize the troop. These obstacles may consist of—

- Minefields.
- Log obstacles, such as abatis, log cribs, stumps, and posts.
- Antitank ditches.
- Wire entanglements.

When scouts encounter an obstacle, they must follow the procedures for actions on enemy contact described in Chapter 3.

- Deploy and report.
- Develop the situation.
- Choose a course of action.
- Recommend a course of action.

As they develop the situation, the scouts should rapidly reconnoiter the obstacle to determine if bypass routes exist and whether or not they are covered by enemy fires, if breaches through the obstacle already exist, if the obstacle is part of an occupied defensive position, what the enemy strength is, and where he is located.

To conserve time and manpower, the troop should attempt to bypass all obstacles. Bypassing will not always be possible because available routes around the obstacle may lead into a fire sack. Forcing through an obstacle is the least desirable method of breaching because of the resulting loss of personnel and equipment. A hasty breach may be the only viable course of action.

Before deciding on a course of action, the troop commander should position himself to observe the situation and control the operation, if possible. Once the commander decides that a hasty breach is the only course of action, he must organize the troop into a breaching element and an overwatch element. When executing the hasty breach, the commander should avoid concentrating the troop in any one area, making it vulnerable to enemy fires.

During zone reconnaissance or movement to contact, the troop operates over a 6- to 10-kilometer-wide zone. During the hasty breach, the commander may not be able to use all the platoons in the troop. In this situation, until the breach operation is complete, give the scout platoon not involved in the breach operation a limit of advance so it does not move too far ahead of the rest of the troop. It can provide the troop with an early warning of enemy activity.

The Breaching Element. The breaching element is composed of scouts and engineers attached or in support of the troop. Their mission is to create lanes through the enemy obstacle system to allow the troop to pass.

The Overwatch Element. The overwatch element is the tank/AT platoons, scouts not in the breaching force, and any additional engineers. Its mission is to provide close, continuous overwatching fires in support of the breaching element. After the breaching element has cleared lanes through the obstacle, part of the overwatch element will assault through the lane to defeat any enemy in the area of the breach. The fire support officer coordinates the fires of the mortars and any supporting artillery to suppress enemy fires with HE and to obscure his observation with smoke.

Sequence of Events in a Hasty Breach. After encountering an obstacle and the commander chooses a hasty breach as the course of action, the sequence of events is as follows:

- Scouts.
  - Reconnoiter the near side of the obstacle to determine its front edge and lateral limits, enemy dispositions, if the obstacle is part of a defensive position, and if a partial or complete breach exists.
  - Secure flanks in support of the breaching operation.

- Overwatch element.

  – Occupy overwatch positions and provide direct and indirect suppressive fires on the enemy for elements moving to and through the obstacle.

  – Use smoke to obscure enemy observation of the obstacle (see Figure 7-9).

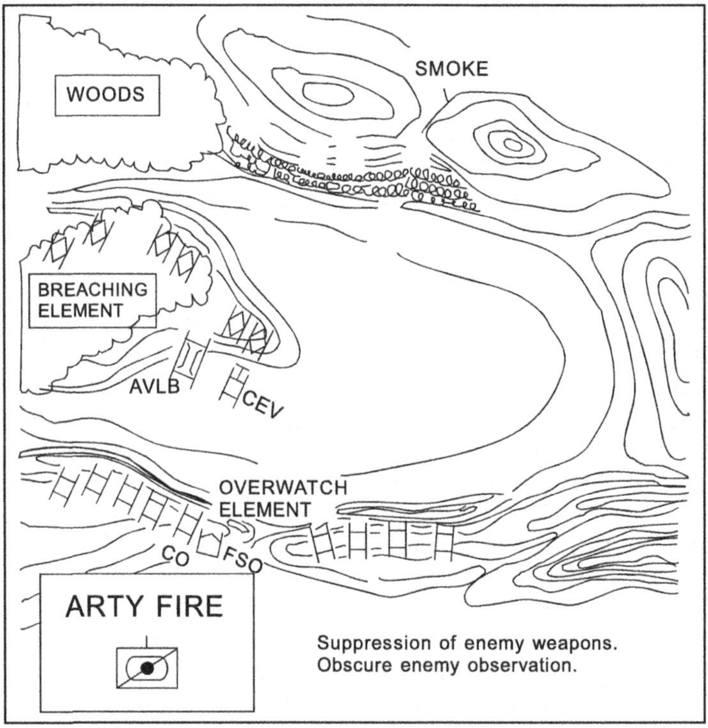

Figure 7-9. Obscure enemy observation.

- Breaching element.
  - Occupy covered and concealed positions to coordinate activities and prepare equipment, demolitions, and routes to the obstacles.
  - After enemy fire has been suppressed, breach the obstacle.
  - Secure fighting positions on the near and far sides of the obstacle (see Figure 7-10).

- Overwatch element.
  - As the breaching element breaches the obstacle, a tank platoon from the overwatch element prepares to attack through the obstacle.
  - Once the breaching element secures positions on the far side of the obstacle, the tank/AT or scout platoon attacks through the obstacle and destroys enemy elements that can fire directly on it. This assault element either continues to advance as the troop lead element or establishes hasty defensive positions while the remainder of the troop passes through the obstacle (see Figure 7-11).
- Troop. Continue the mission.

Neutralizing and Breaching Minefields. Minefields will differ in layout and composition, depending on the availability of mines and the nature of the avenue of approach. When possible, mines should be detonated in place. The objective in breaching the obstacle is to make a safe route to the far side. Multiple routes across the obstacle reduce vulnerability to enemy fires, but may be very time consuming and beyond the capabilities of the troop.

Use the following methods to neutralize and breach minefields:

- **Foot Lanes.** Establish foot lanes through the obstacle when a mounted assault breach is not at first practicable. This lane is normally two meters wide, and can later be widened into vehicle lanes. Foot lanes allow the troop to move forces through the obstacle to secure the far side.

- **The Bangalore Torpedo.** This device will clear foot lanes through mines and wire obstacles. It will clear a path three to four meters wide through wire entanglements and a narrow footpath through a minefield.

- **Vehicle Lanes.** After the first breach is made, foot lanes may be widened to one-way vehicle lanes at least eight meters wide. Vehicle lanes may also be breached separately from foot lanes. Use existing roads when possible. Vehicle-pushed rollers, if available, should be used to proof the lane cleared by line charges.

- **The M173 Rocket-Projected Line Charge.** This is an AT minefield clearing device. When the line charge explodes, it clears a vehicle lane about six meters wide.

Section VII. NBC Defensive Operations

OPERATIONS

NBC defense operations reduce casualties and damage to equipment and materiel, and minimize confusion and interruption of the troop's mission in the event of enemy NBC attacks. These operations are performed concurrently with all combat operations to preserve the fighting strength of the troop.

TASKS

The following combat tasks are associated with the conduct of troop NBC defense operations:

- Prepare for a nuclear attack.
- Respond to the initial effects of a nuclear attack.
- Respond to the residual effects of a nuclear attack.
- Cross a radiologically contaminated area.
- Conduct radiological reconnaissance.
- Perform radiological decontamination.
- Prepare for a chemical agent attack.
- Respond to a chemical agent attack.
- Cross a chemically contaminated area.
- Conduct a chemical reconnaissance.
- Perform hasty decontamination.
- Coordinate for deliberate decontamination of equipment.
- Exchange protective clothing.

ORGANIZATION

The troop's NBC defense system is composed of a trained NBC officer (usually the XO), an NBC NCO, and an enlisted alternate from troop headquarters. Troop soldiers are also designated and trained to operate all assigned NBC equipment and to assist decontamination teams.

The NBC officer supervises troop NBC defense activities and assists the commander in training NBC equipment operators and decontamination teams. The NBC NCO and his alternate directly supervise radiological monitoring, chemical detection, and decontamination operations. During combat operations, the NBC NCO is located in the command post where he—

- Receives, prepares, evaluates, and disseminates information and reports enemy and friendly NBC attacks.
- Supervises employment of detection, monitoring and survey, and decontamination teams.
- Maintains unit radiation exposure status records.
- Assists the troop commander in analyzing guidance from squadron, mission, threat, and weather as they affect NBC operations. Recommends appropriate MOPP level based on this information.

To facilitate operations in an NBC environment, designate the NBC teams from all platoons and sections by vehicle crews so vehicle commanders have responsibility for their crews and the execution of the NBC defense team mission. Each section in the troop has a chemical agent detection team and each platoon has a radiological monitoring and survey team.

Chemical Agent Detection Team. Each troop has twelve chemical agent detection teams. There are three in each scout platoon and one in each tank/AT platoon, one in the mortar section, one in the command post, and one in each combat trains and field trains. Each team has an M256 detector kit, a chemical agent alarm, detector paper, 400 meters of WD-1/TT for remoting the alarm, and enough batteries and supplies for seven days of continuous operations.

Radiological Monitoring and Survey Team. Each troop has seven teams, each equipped with an IM 174/PD dose rate instrument and an IM 93/147 dosimeter. One detection team from each platoon serves as a radiological monitoring and survey team. Also, the mortar section, maintenance section, and troop headquarters each have a radiological monitoring and survey team.

Deliberate Decontamination Team. Deliberate decontamination teams are organized by the vehicle crew. One crew from each scout and tank/AT platoon is designated as a decontamination team. This team is used to support the deliberate decontamination of the troop. One crew from each platoon should receive special training from the squadron chemical section to execute this task.

TECHNIQUES

Proficiency in NBC defense operations is attained only by strict adherence to standardized procedures as outlined in the troop SOP. The troop must be proficient in the following three fundamentals of NBC defense to survive and remain an effective fighting unit:

- Contamination avoidance (before, during, and after an attack).
- Protection (before, during, and after an attack).
- Decontamination (after an attack).

Actions Before the Attack. The best defense against a chemical or nuclear attack is to avoid being detected and targeted. To avoid becoming a lucrative target, implement and strictly enforce the following passive avoidance measures during all operations:

- **Dispersion.** Maintain dispersion between vehicles whether stationary or moving; vehicles bunched together are much easier to find. This is especially important when occupying assembly areas, conducting resupply operations, or when crossing obstacles and waterways.
- **Concealment.** Concealed positions prevent the enemy from observing vehicles and personnel from the ground or air.
- **Camouflage.** When concealment is not available or adequate, use camouflage to hide vehicles and positions.
- **Communications security (COMSEC)/signals security (SIGSEC).** The enemy can easily locate positions with direction-finding equipment if the troop does not adhere to proper communications procedures. Ensure all troop elements properly encode messages sent over unsecured nets.
- **OPSEC.** Take all possible actions to deceive the enemy about troop movement, positioning, intentions, and size. Mask movement by using folds in the terrain. Use hide positions when not moving or engaging the enemy. Enforce noise and light discipline. To deceive the enemy about the number of troop vehicles, start all vehicles simultaneously, and cover all track marks that lead into positions.

Nuclear/Chemical Vulnerability Analysis. Based on the NBC threat and the troop mission, determine the troop's vulnerability and incorporate those factors into the commander's planning.

Take the following actions to reduce the effects of a chemical or nuclear attack:

- **Improve positions.** Use available natural and man-made cover, such as folds in the earth, buildings, and dug-in positions, to protect personnel, equipment, materials, and supplies.
- **Alert personnel.** Inform soldiers to assume the appropriate MOPP level (see Table 7-1), based on the possibility of enemy attack and the MOPP analysis. Inspect unit and individual NBC equipment and prepare for operations.

Table 7-1. Mission-oriented protection posture levels.

MOPP LEVEL	OVERGARMENT	BOOTIES	MASK/HOOD**	GLOVES
0	Stored Nearby*	Stored Nearby*	Carried	Stored Nearby*
MASK ONLY	Carried	Carried	Worn	Carried
1	Worn, open or closed based on temperature	Carried	Carried	Carried
2	Same as MOPP 1	Worn	Carried	Carried
3	Same as MOPP 1	Worn	Worn, hood open or closed based on temperature	Carried
4	Worn, closed	Worn	Worn	Worn

*On combat vehicle or in fighting position.
**Includes M258A1 kit and detector paper.

- **Protect equipment and supplies.** Protect against chemical contamination inside a vehicle since it is very hard to decontaminate. Keep all hatches closed, if possible, and turn on vehicle overpressure system. Cover equipment and supplies stored outdoors or in tank bustles. Nonporous plastic sheets make good covers. Use tarpaulins if plastic is not available.

- **Position alarms.** Activate and emplace automatic alarms to provide the earliest possible warning of an attack. Position the alarms upwind and at least 150 meters away from stationary positions. Keep adjacent alarms about 300 meters apart.

- **Alert NBC personnel and teams.** Ensure NBC personnel and teams are available and prepared to perform their assigned missions.

Actions in Response to an Attack. Every soldier must be proficient at individual NBC basic skills. Masking is conducted immediately in response to these conditions:

- Chemical alarm sounds.

- Positive reading on detector paper.

- Individuals exhibiting symptoms of chemical or biological agent poisoning.

All soldiers mask automatically when an attack is possible, after the initiation of chemical/biological warfare, in response to artillery shells exploding in the area, and when aircraft drop bomblets or spray a mist or fog. In response to these conditions, all soldiers assume MOPP 4 and remain at maximum protection until the absence of contamination is confirmed or the commander has completed the analysis and determined a lower level of MOPP is appropriate.

Actions in response to a chemical or nuclear attack are standardized and included in the troop SOP.

- Initiate immediate masking and sound alarm (visual, audible, or voice).

- Continue mission.

- Assume MOPP 4 as rapidly as possible.
- Treat casualties.
- Send NBC-1 report and update it as more information becomes available.
- Identify the agent.
- Initiate decontamination operations.

These procedures must be rehearsed, and all individuals must adhere to them.

Actions After an Attack. Warn all subordinate elements immediately so they can take appropriate actions. In addition, alert squadron and all adjacent units, giving priority to the unit downwind from the attack. The troop and squadron are extremely vulnerable at the beginning of a chemical attack. Continue the troop mission while preparing for the enemy; he will take advantage of the chemical attack by conducting tactical operations.

In accordance with the troop SOP, prepare and send an NBC-1 report immediately. The personnel observing the attack send the report to the troop XO and NBC NCO in the command post. The XO forwards the report to the squadron TOC. Send the report immediately, and update it as more information is available. Treat any casualties of the attack using self-aid and buddy-aid, and report their status through the troop command post to the squadron S1. Operating in MOPP 4 will, over time, reduce the troop's effectiveness in combat operations. Soldiers become fatigued and lose mental and physical dexterity. Maintaining command and control becomes very difficult during extended MOPP 4 operations.

The troop must determine the type of agent used in the attack and the extent of contamination. Initiate tests with the M256 detector kit to identify the type of agent. The M256 kit detects and identifies field concentrations of nerve, blister, or blood agents. This test will help determine when unmasking may be safe after a chemical attack. Update the NBC-1 report with any information provided by the test. Decontaminate if required, and unmask as soon as possible.

When the troop is contaminated with persistent or semipersistent chemical agents, decontamination is necessary before reducing the MOPP level. Implement the following decontamination techniques:

- Start the skin decontamination within one minute of contamination.
- Apply decontaminating agents to equipment and vehicles to limit the spread of contamination and shorten the duration of the hazard. Use the M258A1 kit for personnel wipe down, and the M11 or M13 decontamination apparatus for vehicle/equipment spray down.

Decontaminate as far forward as possible. After leaving the contaminated area, perform hasty decontamination as soon as the situation permits. Hasty decontamination consists of exchanging MOPP gear and washing vehicles with hot, soapy water or high-pressure water to remove gross contamination. This operation is conducted at platoon or troop level, assisted by the decontamination team and other assets provided by the squadron. Select a site that is uncontaminated, near a water source, and concealed. Hasty decontamination may allow temporary relief from MOPP 4 for eating, drinking, and personal hygiene. Deliberate decontamination is required later when time and additional decontamination assets are available.

Initiate unmasking procedures as soon as possible after a chemical attack, to reduce the performance degradation that occurs while soldiers are in MOPP 4. Unmasking procedures differ based on the availability of detection equipment. When nonpersistent agents are used in the attack and an M256 detector kit is available, use the kit to determine when the agent is no longer present. Instruct two or three individuals to unmask for 5 minutes, remask, and sit in a shady area for 10 minutes. Check the individuals for chemical agent symptoms. If no symptoms appear, it is safe to unmask.

When no detector kit is available, direct two or three soldiers to perform the following procedures. Have them keep their eyes open, take a deep breath, hold it, and break the seal on their mask for 15 seconds. Reseal and clear the mask, and wait 10 minutes in a shady area. Check them for chemical agent symptoms. Have them break the seal and take three or four breaths and reseal the mask. Wait 10 minutes and check them for symptoms. Have them unmask for 5 minutes and remask for 10. Check for symptoms. If no symptoms appear, assume it is safe for all to unmask.

Following a persistent or semipersistent attack, begin unmasking procedures after hasty decontamination is complete and the M256 kit indicates no agent is present.

OPERATIONS IN A NUCLEAR ENVIRONMENT

Radiological Monitoring. This is the detection of radiation and the measurement of dose rate with radiac instruments. Monitoring finds a hazard that would otherwise go unmeasured or undetected. It is conducted by the troop radiological monitoring and survey team, which monitors—

- Total dose using the IM 93/147.
- Dose rate using the IM 174/PD.
- Food, water, and personnel contamination using the AN/PDR-27.

Conduct periodic monitoring when directed, or after nuclear war breaks out. Monitor a specific point in the area at least once an hour. Start continuous monitoring after receiving a fallout warning; when conducting a movement; when a nuclear burst is reported, seen, or heard; after detecting radiation above 1 cGy/hr during periodic monitoring; or when the squadron commander orders. Stop continuous monitoring when the squadron commander orders or, except while on the move, when the dose rate falls below 1 cGy/hr.

Based on NBC-4 monitoring reports, the troop may conduct a radiological survey to determine the degree and extent of contamination. This information is forwarded to squadron in an NBC-4 report and on DA Form 1971-R.

Exposure Control. While operating in a nuclear environment, the troop must control exposure to nuclear radiation. An operational exposure guide is a method of determining the maximum radiation dosage to which platoons can be exposed and still accomplish a mission. Determine radiation exposure by the cumulative dose or the platoon's radiation history.

Platoons record and report their total dose radiation daily, per SOP, to the troop command post where the NBC NCO compiles the data. The dose is determined by averaging the readings from both platoon dosimeters and rounding to the nearest 10. The NBC NCO totals each platoon's average dosage, and coverts it to a radiation status (see Table 7-2). He maintains a record of the radiation exposure status of each platoon and submits a daily radiation exposure status report to the squadron (see Table 7-3).

Table 7-2. Radiation dose categories.

RADIATION DOSE CATEGORIES

THE TABLE INDICATES THAT THE NUMERICAL VALUE OF cGys RECEIVED BY EACH PLATOON IS CONVERTED TO A RADIATION EXPOSURE STATUS (RES) FOR EACH PLATOON. THE CONVERSION FACTORS ARE SHOWN BELOW.

RADIATION STATUS CATEGORY	TOTAL CUMULATIVE DOSE (cGy)
RES-0	0 (NO EXPOSURE)
RES-1	GREATER THAN 0 BUT LESS THAN OR EQUAL TO 70
RES-2	GREATER THAN 70 BUT LESS THAN OR EQUAL TO 150
RES-3	GREATER THAN 150

The regimental and squadron commanders may move units on the battlefield, or relieve a particular unit that has been exposed to a high level of radiation. The squadron commander specifies the degree of exposure that units will not exceed for each operation. There are three degrees of radiation exposure risk—negligible, moderate, and emergency. Each level of risk is correlated to unit radiation status (see Table 7-4).

Table 7-4. Unit radiation status.

RADIATION STATUS (RS) CATEGORY	TOTAL PAST CUMULATIVE	EXPOSURE CRITERIA FOR A SINGLE OPERATION (cGy)
RS-0 UNITS	NO EXPOSURE	NEG RISK: 50 MOD RISK: 70 EMERG RISK: 150
RS-1 UNITS	GREATER THAN 0 BUT NOT MORE THAN 70	NEG RISK: 50-PAST DOSE MOD RISK: 70-PAST DOSE EMERG RISK: 150-PAST DOSE
RS-2 UNITS	GREATER THAN 70, NOT MORE THAN 150	ANY FURTHER EXPOSURE IS CONSIDERED EMERG RISK
RS-3 UNITS	GREATER THAN 150	ALL FURTHER EXPOSURE WILL EXCEED EMERG RISK

1. RADIATION EXPOSURE STATUS CATEGORIES ARE BASED ON PREVIOUS EXPOSURE TO RADIATION.

2. RECLASSIFICATION OF UNITS FROM ONE RADIATION EXPOSURE STATUS CATEGORY TO A LESS SERIOUS ONE IS MADE BY THE COMMANDER UPON ADVICE OF THE MEDICAL PLATOON LEADER AFTER AMPLE OBSERVATION OF ACTUAL STATE OF HEALTH OF THE EXPOSED PERSONNEL.

3. ALL EXPOSURES TO RADIATION ARE CONSIDERED TOTAL BODY AND SIMPLY ADDITIVE. NO ALLOWANCE IS MADE FOR BODY RECOVERY FROM RADIATION INJURY.

4. NEGLIGIBLE RISK WILL CAUSE NO CASUALTIES AND A LESS THAN 2.5 PERCENT INCIDENCE OF VOMITING. MODERATE RISK WILL RESULT IN A LESS THAN 5 PERCENT INCIDENCE OF VOMITING AND HEADACHES, BUT WILL CAUSE NO CASUALTIES. EMERGENCY RISK WILL RESULT IN LESS THAN 5 PERCENT CASUALTIES AND A HIGHER PERCENTAGE OF NAUSEA, VOMITING, AND HEADACHES.

Minimize the Effects of Friendly Nuclear and Chemical Attacks. A NUCWARN/CHEMWARN message is normally not sent below squadron level. Squadron alerts the troops using a code word prescribed in the OPORD. Squadron will also issue a FRAGO containing relocation instructions, the level of protection required, and the time of the friendly strike. To assure maximum reaction time, each soldier in the troop must be informed. Each crew immediately prepares its vehicles, equipment, and position for the attack as is appropriate to its location within the attack area. Crews prepare their vehicles by taking the following actions:

- Move if necessary.
- Position vehicle behind best available cover with front of vehicle toward the blast.
- Point gun away from blast.
- Lock brakes.
- Secure loose equipment in vehicle.
- Secure inside the vehicle all exterior components that could be damaged by the blast.
- Close and lock all hatches, to include ballistic shields.
- Wear helmets and protect eyes.
- Turn off and disconnect all radios but one, which is needed to retain communications with squadron.

Complete these tasks within five minutes of receiving notification. After the friendly strike, continue radiological monitoring and report if necessary. Quickly prepare to continue operations.

If the troop is contaminated with fallout, the level of contamination is monitored by the monitor and survey teams. Decontamination is accomplished by brushing or wiping the dust off the equipment. Brush off or shake out MOPP gear and wash down the equipment and vehicles to reduce radiological contamination. Use the AN/PDR-27 to determine decontamination requirements and to determine if supplies of food and water must be destroyed.

Chapter 8

Combat Support

The effective integration of combat support (CS) will spell the difference between success and failure on the battlefield. The commander must know the capabilities and limitations of CS assets, and how to employ them properly.

CONTENTS

Page

Section I. The Fire Support Team

CAPABILITIES AND LIMITATIONS

The FIST is organized, equipped, and trained to provide—

- A fire support advisor/coordinator—the FSO.
- An observation/laser designation capability.
- A communications link to all available fire support.

The FIST has secure FM communications on four radios, and digital communications on two nets through the use of digital message devices (AN/PSG-5 and AN/PSG-2). These capabilities enable the FIST to maintain communications on the following nets (see Figure 8-1):

- Troop command net.
- Troop fire support net.
- Supporting artillery fire net (digital).
- Squadron fire support net.

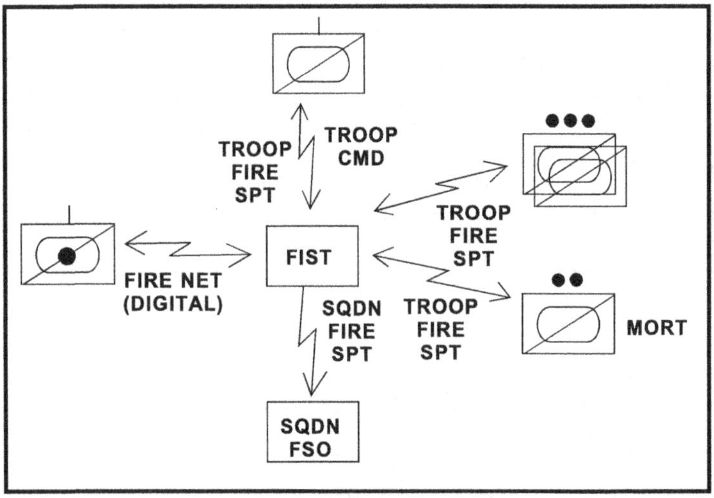

Figure 8-1. Fire support team communication nets.

The FIST is also equipped with a target laser locator designator set to designate targets for terminally guided munitions and to determine target locations accurately at ranges up to 10 kilometers (limited by line of sight). The target designator set is equipped with a thermal sight to use under limited visibility conditions.

The digital message devices allow the FIST to transmit and receive high-speed digital messages from the howitzer battery computer system, the direct support artillery battalion TACFIRE, OH-58D aircraft, and the squadron FSO's variable format message entry devices, among other digital message devices. Digital message devices can quickly transmit standard fire requests, fire adjustments, and plain text messages. The FIST can store nine fire requests on the digital message devices for rapid transmission in support of troop operations.

The FIST's abilities may be constrained by its communications with the supporting field artillery (FA). To coordinate any fire support for the troop, the FIST must be positioned where it can maintain at least a voice communications link to the supporting artillery. The FIST may not be able to see the battlefield and use the laser designator capability in all situations. When the troop is operating in a large sector, the FIST will be able to see only a small portion of the area. The scouts, therefore, must be proficient in requesting indirect-fire support. The scouts' calls for fire will be relayed through the FIST to the appropriate fire support unit.

EMPLOYMENT TECHNIQUES

Planning and Coordination. Fire support planning and coordination begin on receipt of a mission, and continue throughout its planning and execution. The troop FSO should always accompany the commander to mission briefings at squadron, so that he is fully aware of the squadron scheme of maneuver and the availability of fire support assets. He and the squadron FSO can also begin coordinating any specific requirements for fire support, such as preparations, groups, and series.

The troop commander's guidance should include the following to allow the troop FSO to plan the operation:

- What is the purpose of FA and mortar fires?
- How will FA fires support the troop's maneuver?
- What are the attack criteria? Will the commander use FA for immediate suppression, screening, or destruction?
- Are the attack and engagement criteria for mortars different from FA?
- On what and where does the commander want to mass FA and mortar fires?

Plan indirect fires on known or suspected enemy positions; at choke points; on the troop's objective or flanks; or on any other areas that would support the scheme of maneuver. These targets can be fired at a specific time, on call, or when a particular event takes place.

The FSO incorporates his approved fire plan into a target list and forwards it to the squadron FSO. The squadron FSO compiles the lists from all the troop FSOs, eliminates any duplicate targets, forwards the list to the supporting artillery, and informs the troop FSOs of any changes to their target lists.

Once he has completed his plan, the FSO will brief the commander on planned targets, scheduled fires, and fire support coordination measures that will be in effect. The commander will make the final decision on the fire support plan. Ensure it complements the scheme of maneuver and maximizes the available firepower for the troop. Once approved, the fire plan is continuously modified and updated to meet changing situations.

Silence is Consent. Once the FSO has briefed the commander and the commander approves the troop fire support plan, calls for fire are executed during the conduct of the operation by the silence is consent method. This method allows subordinate leaders within the troop to request indirect fires in support of their maneuver within the framework of the commander's guidance and the troop fire support plan. All initial requests for fires will go over the troop command net and the commander will simply acknowledge the request with "ROGER—OUT." The troop FSO will process the call for fire through the appropriate indirect-fire system based on the commander's guidance. All adjustments and corrections for the fires will be communicated on the troop fire direction net directly between the observer and the troop FSO. If the request for fire that is submitted by the subordinate leader is not within the commander's fire support guidance, the troop commander will deny fires or modify his guidance based on the situation.

Execution of the Fire Support Plan. During the operation, the FSO is responsible for—

- Keeping the commander advised of all fire support matters.

- Resolving any fire support conflicts that may arise during the planning and execution of the operation.

- Attacking targets with the most suitable fire support means available to support the scheme of maneuver.

Also, the commander may make the FSO responsible for positioning the troop mortars to best support the operation.

The FSO operates on the four radio nets mentioned previously. The scouts in the troop normally serve as the observers calling for indirect-fire support. The FIST monitors all spot reports on the troop command net and, as necessary, prepares calls for fire. The FSO will allocate fires from mortars and artillery, based on the commander's guidance. The commander may give the FSO additional guidance concerning a fire mission as it is requested (see *Silence is Consent* paragraph). Further information about the fire mission and all other calls for fire are passed on the troop fire support net. Keep the troop command net free for other traffic.

Both the FIST and the mortars monitor the troop fire support net. When scouts forward adjustments or corrections for indirect-fires, they use the troop fire support net. If the FSO forwards the mission to supporting artillery, he sends the message "LIGHTNING UP" on the troop fire support net. This tells the scouts calling for fire and the mortars that the mission is being fired by supporting artillery and that all adjustments and the "end of mission" report must be relayed through the FSO. If the mortars will fire the mission, the FSO sends the message "THUNDER DOWN" on the troop fire support net. This tells the mortar squad to fire the mission, and tells the scouts to send all adjustments and the "end of mission" report to the mortar squad (see Figures 8-2 and 8-3).

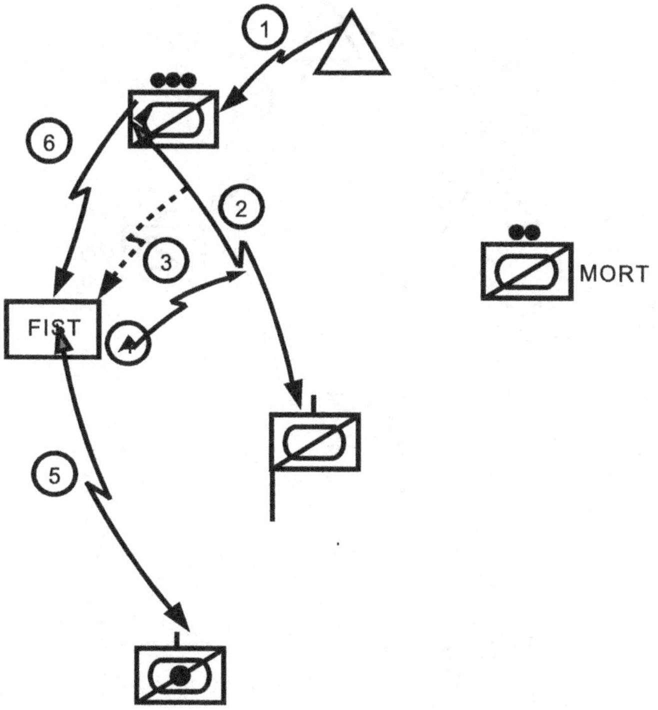

(1) SCOUTS SEND SPOT REPORT TO PLATOON LEADER (PLATOON NET).

(2) PLATOON LEADER FORWARDS THE SPOT REPORT TO THE CP (TROOP COMMAND NET) WITH REQUEST FOR FIRE.

(3) FIST MONITORS SPOT REPORT, RECEIVES PERMISSION TO FIRE THE MISSION, AND DETERMINES WHAT ARTILLERY TO USE.

(4) FIST SENDS MESSAGE "LIGHTNING UP" ON TROOP COMMAND NET. SCOUTS AND MORTARS MONITOR.

(5) FIST SENDS DIGITAL CALL FOR FIRE TO DS ARTILLERY. FIST SENDS SHOT ON TROOP COMMAND NET.

(6) SCOUTS RELAY ALL ADJUSTMENTS TO FIST ON TROOP FSN.

Figure 8-2. Fire request channels (artillery).

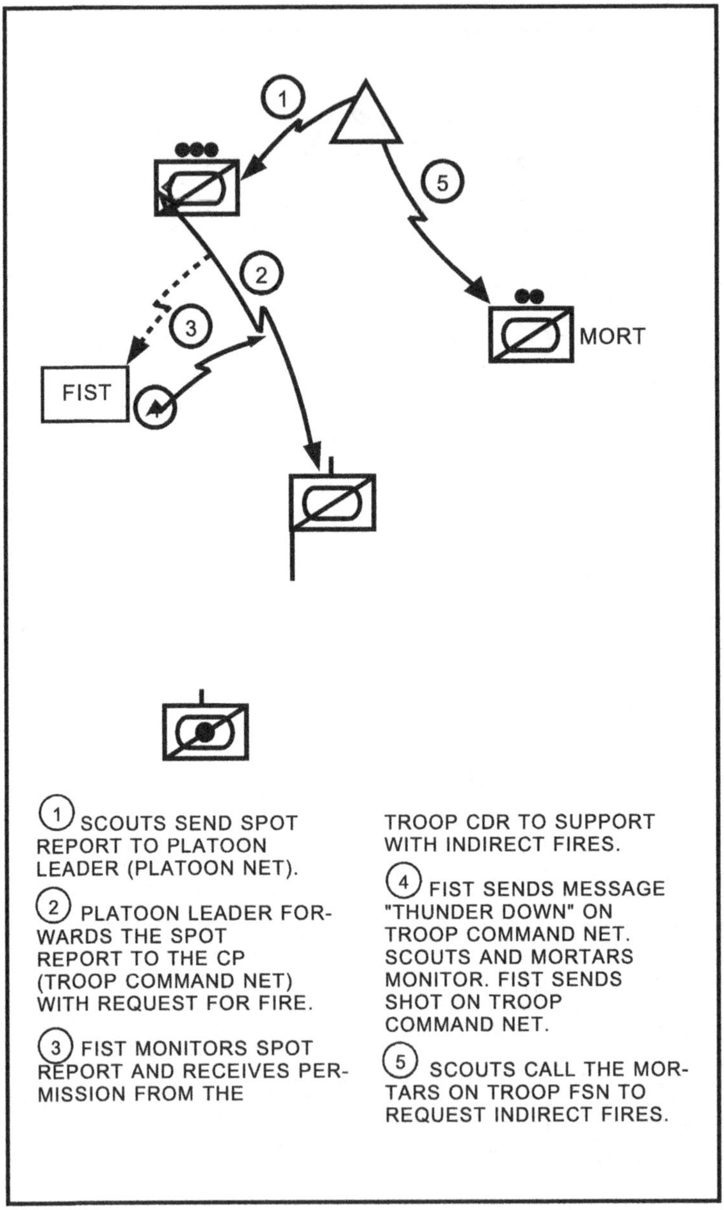

Figure 8-3. Fire request channels (mortar).

Section II. Mortar Support

CAPABILITIES AND LIMITATIONS

The mortar section provides a heavy volume of accurate, sustained indirect fires in response to the needs of the troop. They are ideal weapons for attacking targets on reverse slopes, in narrow ravines, built-up areas and in other areas that are difficult to strike with low-angle fires. Mortars are most effective for—

- **Suppression.** Suppression forces the enemy to button up or move to less advantageous positions. High explosive (HE) ammunition with a variable-time fuze or quick fuze does this best.

- **Screening.** Obscuration smoke is placed directly on or just in front of enemy positions to obscure vision. Screening smoke is placed between troop positions and the enemy to conceal the troop's movements. Smoke will degrade the effectiveness of laser range finders, so the tank platoons must adjust their gunnery techniques when engaging the enemy.

- **Illumination.** The use of illumination permits daytime engagements during some limited visibility conditions by improving the effectiveness of image intensification devices. Illumination rounds are used to mark targets for close air support and attack helicopters, and to allow observers to adjust artillery onto a target. Illumination rounds are usually adjusted so they illuminate above or behind the enemy location to silhouette him.

The height of the illumination should be adjusted to burn out about 200 meters above the ground. This gives maximum illumination time, about 90 seconds for one round, and will not degrade the effectiveness of image intensification devices. Because illumination rounds light the battlefield for both sides, it should be used when night-vision devices are not available in sufficient quantities or when ambient light levels are very low.

EMPLOYMENT TECHNIQUES

The troop mortar section is employed as a separate element during operations. It moves independently of the platoons, and provides its own security. The troop commander normally designates positions for the section sergeant and provides guidance for support forward of the scouts. The commander may find that having the FIST position the mortars is easiest and most effective. This technique allows the mortar section sergeant the ability to concentrate on providing fires and computing fire data and the commander to focus on maneuvering the troop, not one section.

The mortars can operate on two nets. If the FIST is responsible for moving the section, both nets may be on the troop fire support net. The FSO will keep them updated on the situation. However, since many spot reports over the troop net become fire missions, the mortars can eavesdrop and provide more responsive fires if they stay abreast of the situation themselves. If the commander gives the mortar section sergeant authority to select positions and to move the section, he must operate on the troop command net to stay abreast of the situation and on the troop fire support net.

Offense. During offensive operations the movement of the mortars will be based on the progress of the scout platoons. While the section is on the move, it must be prepared to provide immediate fires using direct lay, direct alignment, or hip shoot. Plan the moves of the section so it is in position to support the troop at critical times, such as during river crossings and counterattacks, and when clearing defiles.

Defense. Considerations for using mortars in the defense are similar to those for using them in the offense. Plan on prestocking ammunition at subsequent positions to reduce resupply problems during a defensive operation.

Reconnaissance and Security. During reconnaissance and security missions, the troop will often be operating over a large frontage that cannot be completely covered by the section. The commander must decide whether to position the mortars to cover the most dangerous area, or to move them to a position where they can cover a part of different areas and adjust as necessary. Knowing what other artillery assets are available will help in making that decision.

Commander's Guidance. A list of considerations the commander can use to help plan mortar section employment follows:

- Ammunition constraints by type and quantity.
- Priority of fire to a designated platoon.
- Anticipated changes in mortar employment.
- Communications constraints.
- General designation of positions.
- Movement guidance.
- Coordination requirements.
- Resupply.

Section III. Artillery Support

ORGANIZATION AND SUPPORT RELATIONSHIP

The howitzer battery organic to the regimental cavalry squadron consists of eight 155-mm self-propelled howitzers. The squadron will often be supported by an artillery battalion. The artillery battalion usually has three batteries of 155-mm self-propelled howitzers, and provides direct support to the squadron.

CAPABILITIES AND LIMITATIONS

Field artillery provides the troop its main fire support. It gives accurate fires with a wide variety of munitions. Field artillery adds a powerful dimension to troop direct-fire and maneuver capabilities.

- Artillery can—
  - Provide fire support under all weather conditions and terrain types.
  - Shift and mass fires rapidly without the requirement to displace.
  - Support the battle in depth with long-range fires.
  - Provide a variety of conventional shell/fuze combinations.
  - Provide continuous fire support by careful displacement.
  - Be as mobile as the supported unit.

- Artillery has—
  - Limited capability against moving targets.
  - Limited self-defense capability against air and ground attack.
  - Limited capability to destroy point targets without considerable ammunition expenditure. (Copperhead is the exception. It can be used to engage point targets.)
  - Vulnerability to detection by enemy target acquisition systems because of firing signatures.

Field artillery has a wide variety of munitions tailored for the engagement of different types of targets. The ammunition types include—

AMMUNITION	TARGET
High explosive	Personnel, field fortifications, vehicles
HC smoke	Obscuration and screening
White phosphorus	Obscuration
Illumination	
Cannon-launched guided projectiles (Copperhead)	Armored vehicles or high-payoff targets requiring precision target engagement

In addition, the troop must become acquainted with the characteristics of improved conventional munitions (ICM) and scatterable mines. ICMs include antipersonnel (APERS) and dual-purpose ICM. The commander must consider the danger to friendly troops in areas in which APERS is fired. Also, the high ICM dud rate makes maneuver hazardous in the area of an ICM field. Scatterable mines are area denial munitions for use against personnel and remote antiarmor mines for use against armored vehicles.

EMPLOYMENT TECHNIQUES

Artillery units normally provide direct support to the squadron, and are positioned by the artillery commander in close coordination with the squadron S3 and FSO. The troop FSO is the troop's link to the supporting artillery (see Section I, The Fire Support Team, and Chapters 1 and 2).

Section IV. Engineer Support

ORGANIZATION AND SUPPORT RELATIONSHIP

A combat engineer company is organic to the heavy and light armored cavalry regiment. Additional engineer assets are often provided from divisional and corps engineer units to support the regiment. Up to a platoon of engineers is frequently attached to or placed in direct support of a cavalry troop for combat operations. A combat engineer platoon from the heavy regimental engineer company is organized for combat as shown in Figure 8-4.

HEAVY ENGINEER CAPABILITIES AND LIMITATIONS

A combat engineer platoon is uniquely equipped and trained to conduct mobility, countermobility, and survivability operations in support of troop operations. The platoon headquarters has two M9 armored combat earthmovers, which are highly mobile, armored, and amphibious (see Figure 8-5).

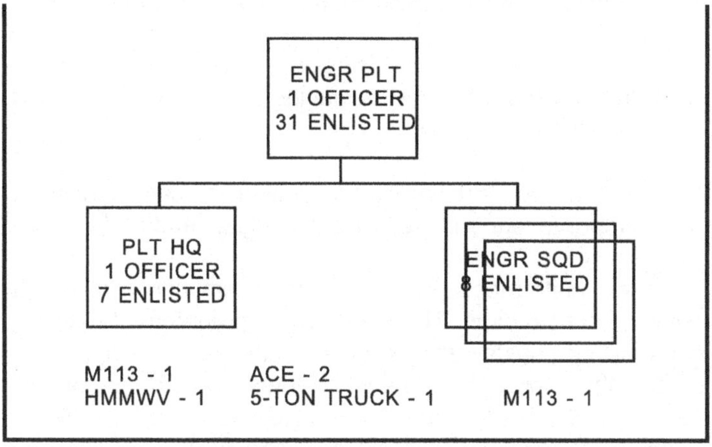

Figure 8-4. Regimental engineer company (heavy) organization.

Figure 8-5. Armored combat earthmover, M9.

In mobility operations, the engineer platoon can provide—

- Obstacle reduction. The engineers can reduce or negate the effects of obstacles to improve the troop's ability to maneuver.

- Route construction. The engineers can construct, improve, and maintain roads, bridges, and fords.

In a countermobility role, the engineers can assist with obstacle construction. The engineers can reinforce terrain by constructing obstacles to delay, canalize, disrupt, and destroy the enemy force to support the scheme of maneuver.

In survivability operations, the engineers can improve positions by constructing berms, dug-in positions, and overhead protection to reduce the effectiveness of enemy weapons.

A cavalry troop is frequently supported by additional engineer assets, such as an armored vehicle launched bridge (AVLB) and combat engineer vehicles (CEV) from the engineer company. The AVLB section of headquarters and headquarters troop of the armored cavalry squadron consists of three AVLBs that could be available during combat operations.

An AVLB is a tank chassis modified to transport, launch, and retrieve a 60-foot class-60 bridge. The bridge is capable of carrying military load class-60 track vehicles across a 17-meter gap and military load class-70 track loads across a 15-meter gap (see Figure 8-6).

Figure 8-6. Armored vehicle launched bridge.

The CEV is an M60A1 tank mounted with a hydraulically operated dozer blade, a 165-mm turret-mounted demolition gun, a retractable boom, and a winch (see Figure 8-7).

Figure 8-7. Combat engineer vehicle.

The small emplacement excavator (SEE) is a small highly mobile truck with a front-end loader and a backhoe (see Figure 8-8).

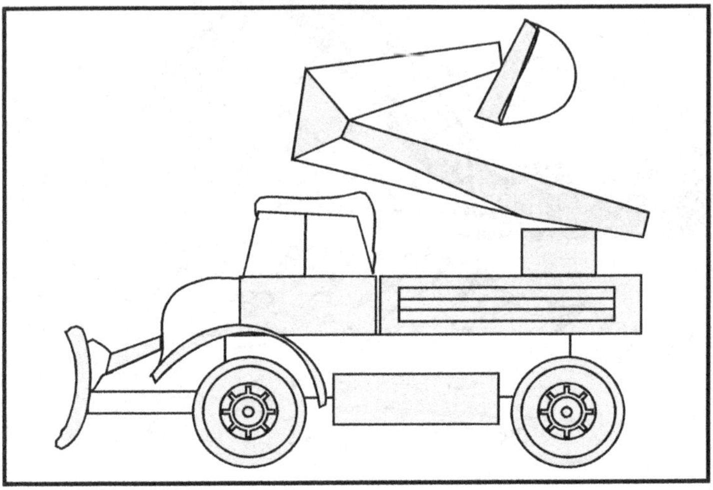

Figure 8-8. Small emplacement excavator.

LIGHT ENGINEER CAPABILITIES AND LIMITATIONS

The engineer platoon from the light regimental engineer company is equipped with five HMMWVs. The platoon can accomplish most of the same tasks that a heavy platoon can, but the time to complete most jobs is greater. The platoon can provide the troop with the following:

- In mobility operations.
  - Obstacle reduction. Engineers can reduce or negate the effects of obstacles to improve the troop's maneuver ability. To improve the platoon's ability, they may be reinforced with an M9 ACE, SEE, MICLIC, and Bangalor torpedo. The engineer company from the light regiment does not provide the regiment (troop) with any bridging capability.
  - Route reconnaissance. Engineers can conduct route reconnaissance.

- In a countermobility role, the engineers are usually left under squadron control to complete priorities of work in accordance with the squadron obstacle plan. The engineers can reinforce terrain by constructing obstacles to delay, canalize, disrupt, and destroy the enemy force to support the scheme of maneuver.

- In survivability operations, the engineers can improve positions by constructing berms, dug-in positions, and overhead protection to reduce the effectiveness of enemy weapons (see Figure 8-9).

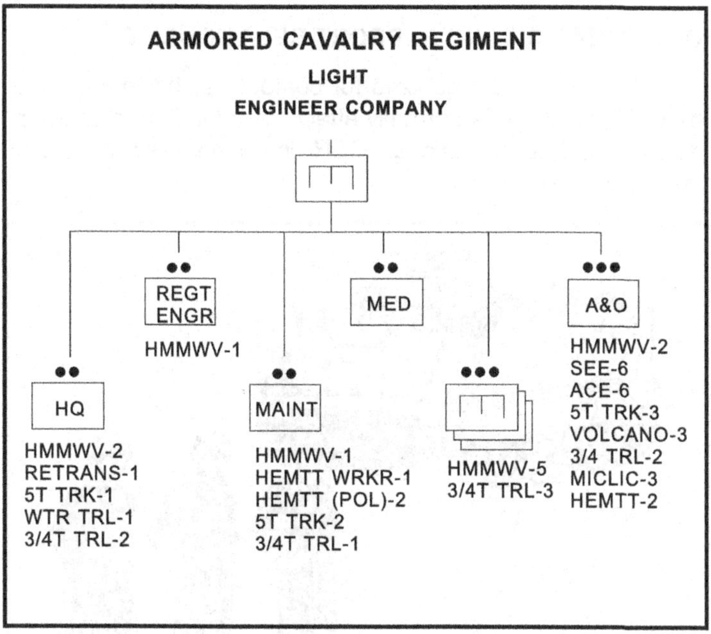

Figure 8-9. Regimental engineer company (light) organization.

EMPLOYMENT TECHNIQUES

The troop has several options in the task organization and employment of engineer assets to support its operation. Normally, the troop will keep the entire platoon under troop control and assign missions just as with a maneuver platoon. The commander can also task organize the engineer platoon, and assign squads to scout platoons to support the operation. Whatever method is chosen, the engineer platoon leader needs to be involved in the troop planning process.

Section V. Ground Surveillance Radar Support

ORGANIZATION AND SUPPORT RELATIONSHIP

A GSR team is organized for combat as shown in Figure 8-10. The team(s) is normally attached or OPCON to a troop during combat operations. GSR is found only in heavy regiments and divisions.

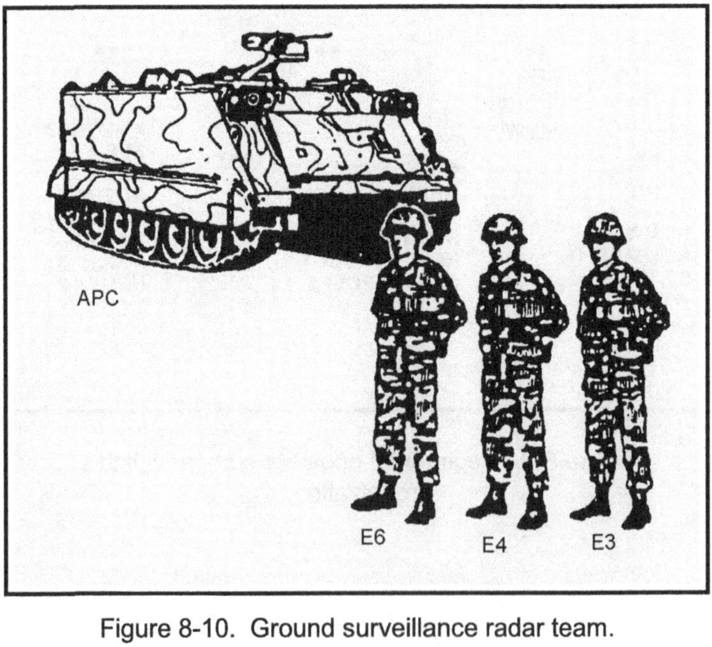

Figure 8-10. Ground surveillance radar team.

CAPABILITIES AND LIMITATIONS

GSRs provide mobile, all-weather battlefield surveillance. GSRs employed in pairs can provide observation from a given vantage point 24 hours a day, detecting targets and providing accurate range and azimuth readings.

The AN/PPS-5 has a line-of-sight range of 10,000 meters against vehicles and 6,000 meters against personnel. A GSR can detect targets through light camouflage, smoke, haze, light snow and rain, darkness, and light foliage. Heavy rain and snow seriously restrict its radar detection capability.

The GSR is generally ineffective against air targets unless the aircraft is flying close to the ground. The GSR is designed to detect targets moving in the presence of a background. It is vulnerable to enemy direction finding and jamming.

EMPLOYMENT TECHNIQUES

The troop will normally employ supporting GSR teams to augment the surveillance capabilities of its scout platoons. The commander should assign each GSR team a specific sector of surveillance and a frequency of coverage, or give the scout platoon leader guidance for employing the GSR team. The enemy can detect radar signals, so the GSR cannot conduct continuous surveillance. The tasks assigned GSR surveillance teams include—

- Searching avenues of approach or possible enemy positions on a time-schedule basis or randomly to determine location, size, composition, and nature of enemy activity.

- Monitoring point targets such as bridges, defiles, and road junctions and reporting quantity, type, and direction of enemy movement through the target.

- Extending the observation capabilities of scouts by enabling them to survey distant points and areas of special interest.
- Vectoring patrols to keep them oriented during times of limited visibility.

Position the GSRs in an area that is free of ground clutter and in one that affords them long-range observation and a wide field of view. Normally, they should be assigned a general area by the troop commander or scout platoon leader and the GSR team leader will select the specific location. The team should be prepared for rapid displacement, and have several alternate positions selected and reconnoitered to avoid enemy suppressive fires.

Offensive Operations. Employ GSRs in the offense to augment reconnaissance and security efforts. Fast-moving operations may preclude the continuous, effective use of GSRs.

In a movement to contact or a zone reconnaissance, employ GSRs on high-speed avenues of approach to the front or flanks of the troop to provide information on enemy movement. A GSR team can provide long-range observation while the scouts conduct a detailed reconnaissance of the local area. Employ GSR teams on the flanks of the troop's movement in pairs so they can bound forward. In this manner, the GSR can provide continuous early warning and keep pace with the operation.

Defensive Operations. In defensive operations, employ GSR teams to augment scouts. Position them to provide long-range observation on expected avenues of approach or to maintain surveillance of flanks.

Position the GSR teams behind the screen line if the terrain affords them long-range observation. Use GSR to assist scouts in maintaining contact with the enemy as the scouts deploy to subsequent screen lines. This also reduces their vulnerability to enemy fires.

Section VI. Air Defense

SUPPORT RELATIONSHIP

An air defense artillery (ADA) battery is organic to the regiment. For combat operations, these assets are normally attached or direct support to the squadron. The squadron commander may provide a slice of ADA to the troop.

Air defense assets are scarce, so the troop cannot plan on dedicated air defense protection. It is the troop's responsibility to protect itself from enemy air attack. The commander must take measures to avoid enemy air attack, to limit the damage if attacked and, if necessary, to fight back.

AIR DEFENSE EMPLOYMENT

Include air defense planning in all tactical operations. Air defense includes passive and active air defense and the employment of ADA assets.

Passive Air Defense

Passive air defense measures are the troop's first line of defense against enemy air attack. They include all measures, other than active defense, taken to minimize the effects of hostile air action. There are two types of passive air defense—attack avoidance and damage limiting measures.

Attack Avoidance. If the enemy pilot cannot find the troop, he cannot attack it. Use concealment, camouflage, obscuration, and deception to hide from the enemy.

Select positions that provide good concealment. When adequate concealment is not available, camouflage the vehicles to blend with their surroundings. Cover all track marks leading into the troop's position. Cover all shiny objects that could reflect light and attract attention. Take all actions necessary to reduce the enemy's ability to find the troop.

Damage Limiting Measures. Damage limiting measures are taken to reduce the effects of enemy air attack. Disperse vehicles while occupying static positions such as assembly areas or when preparing to cross a water obstacle or a breached obstacle. Dispersion reduces the effects of munitions. While on the move, air guards must be alert for enemy air attacks. When an enemy air attack is identified, quickly disperse, go to a concealed position, if possible, and then stop moving. A stationary vehicle is more difficult to see than a moving vehicle, and dispersion will reduce the effect of the air attack.

Use natural or man-made cover to reduce the effects of enemy air attack. Use folds in the earth, depressions, buildings, and sandbagged positions as damage limiting cover. These methods to reduce the effects of air attack must be used scrupulously by troop trains and command posts to reduce their vulnerability on the battlefield.

Active Air Defense

Although passive measures are the first line of defense against air attack, the troop must be prepared to fight back if necessary. The troop gains two things by fighting back. It may kill or drive off the attacker, and the morale and spirit of the troops improve by fighting back.

The decision to fight an air threat is based on the situation and the capabilities of the troop's weapon systems. Generally the troop may defend itself against direct attack, but unless attacked, it does not engage aircraft unless directed to do so by the squadron commander.

Stinger/Avenger Employment

Stingers/Avengers are used most effectively in point defense. However, mobility during tactical operations requires that Stinger crews move with the troop during maneuvering to provide the best possible air defense.

The troop commander must determine the air defense priorities for each phase of the operation and brief the Stinger/Avenger section chief accordingly. In order for the section chief to design an effective troop air defense plan for the commander, the commander must furnish the section chief the mission, objective, routes of march, intended scheme of maneuver, battle formations, and intended response to hostile air activity.

Unit air defense falls under two major categories: air defense of a fixed asset (assembly area or defensive position), and air defense of a mobile asset (convoy or maneuvering troop).

Fixed Asset. The air defense section leader places his crews from 2 to 4 kilometers apart near the troop's position, which provides overlapping and mutually supporting fires for troop defense. He provides command and control to his section, and air defense early warning to the troop unit and his deployed crews. The commander must provide logistical support for the section, indirect protective fires for their defense, and security within the troop defensive perimeter during hours of darkness.

Mobile Asset. While the unit moves, either in a tactical road march or during movement to contact, the air defense crews must move with each element to provide uninterrupted air defense protection.

Convoy operations require the air defense section leader to weight his crews near the front and rear of the troop's column(s), and to spread the remainder of his crews evenly throughout the column(s). While traveling, he provides command and control to his crews and air defense early warning to the troop and his crews.

Section VII. Army Aviation Support

SUPPORT RELATIONSHIP

Army aviation units are seldom OPCON to the troop; however, the troop frequently conducts joint operations with air cavalry or attack helicopter units within the troop area of operations. Air and ground units work together to make an effective team in conducting cavalry operations.

AVIATION TROOP COMMANDER

Air operations are controlled and coordinated by one man, the aviation troop commander. He coordinates directly with the ground troop commander, on the command net, when air operations extend into the troop's area of operations. When time is available, coordinate face-to-face to prepare for an operation. Let him know—

- Situation.
  - Enemy. Ground and ADA units by type and location.
  - Friendly. Locations and FLOT to include supporting artillery.
- Mission.
- Execution.
  - Scheme of maneuver.
  - Fire support.

The aviation troop commander tells the ground troop commander—

- His capabilities.
  - Number and type of aircraft.
  - Armament.
  - Time available.

- Execution. Scheme of maneuver to include approach direction into area.

- Command and signal.
  - Aviation troop commander, frequency, and call sign.
  - Succession of command and designated platoon leader in the troop commander's absence.

The aviation troop commander reports all spot reports and fire requests through the troop command post or FIST while operating in the ground troop's area of operations.

Reconnaissance. The capabilities of air and ground scouts complement each other. Air scouts can quickly reconnoiter a large area while ground scouts are conducting a detailed terrain reconnaissance. Employ the air scouts forward to provide information about areas where the ground scouts need to concentrate their efforts, such as at possible fording sites or route restrictions. The ground scouts can focus their efforts where the air scouts have directed them, instead of conducting a complete and time-consuming reconnaissance. The air scouts can also provide early warning of enemy movement to the front or flanks of the troop.

Security. During screen missions, employ the air assets to the front or flanks of the troop to observe areas between OPs and those areas that are difficult to observe from the ground. The air scouts can place effective indirect fires on the enemy and in maintaining contact with the enemy when the ground scouts move to subsequent screen lines. In offensive and defensive operations, Army aviation will normally work for the squadron to complement the squadron scheme of maneuver.

Section VIII. Close Air Support

CAS sorties are usually allocated to the regiment by the corps commander to meet preplanned requests submitted by the regimental fire support element. CAS for the troop is obtained by submitting an immediate CAS request.

CAS is requested by submitting an immediate request to the squadron air liaison officer. When CAS is on station, the tactical air control party may talk directly to the troop commander to determine enemy strengths, descriptions, and locations, as well as friendly unit locations. A forward air controller (FAC) normally directs the air strike. In the absence of a FAC, the troop FSO is trained to direct air strikes.

The ground troop may be required to mark the target for CAS aircraft. Use the mortar section for this. Fire white phosphorus or illumination rounds set for a ground burst. The FIST can designate targets for aircraft equipped with laser-spot trackers, such as Pave Penny or laser-guided bombs.

CAS capabilities are listed below.

- High speed and long range.
- Versatile weapon and ammunition mix.
- Accurate delivery.
- Excellent air-ground communications in A-10 and A-16 aircraft.
- Can locate and strike moving targets.

CAS limitations include—

- Limited resources.
- Delivery restrictions caused by limited visibility and weather or the proximity of friendly forces.
- Flight restrictions imposed by enemy air defense.
- Delayed response and short time on station.

Chapter 9

Combat Service Support

CONTENTS

Mission accomplishment depends on the troop's ability to obtain, transport, and distribute fighting resources such as fuel, ammunition, replacement personnel, and rations. Dependable ways to treat and evacuate wounded soldiers are crucial to good morale. Evacuation and repair of combat equipment sustain troop combat power and readiness to fight.

Section I. Organization

Troop trains are organized as combat trains and field trains. The combat trains are made up of immediate battlefield service support. The rest of the troop CSS elements are collocated with squadron CSS assets in the squadron field trains and at the unit maintenance collection point (UMCP). The troop CSS team consists of the troop XO, first sergeant, supply sergeant, communications sergeant, maintenance sergeant, and senior troop aidman.

There is no dedicated service support radio net at troop level. All logistical reports and initial requests for logistical support are conducted on the troop command net. Additional coordination is conducted on the platoon radio nets. Routine reports are sent before and after combat operations, and are delivered by messenger when possible. All service support coordination with squadron is conducted on the squadron administration and logistics (A/L) net. The first sergeant operates on this net. The command post does not routinely operate on the A/L net, but uses it when necessary to forward reports and to conduct logistical coordination.

COMBAT TRAINS

The troop combat trains provide CSS for the troop during combat operations. They are organized for combat as shown in Figures 9-1 and 9-2.

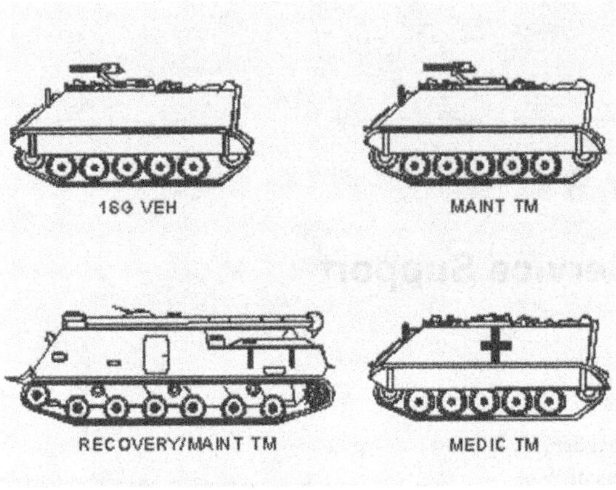

Figure 9-1. Heavy troop combat trains.

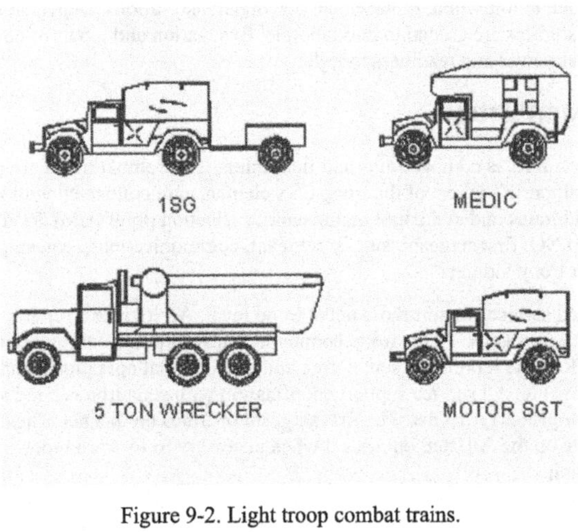

Figure 9-2. Light troop combat trains.

The first sergeant, assisted by the maintenance sergeant, directs movement and employment of the combat trains. He receives logistical reports directly from the platoon sergeants or through the troop command post, coordinates logistical matters with the XO, and assists the XO in ensuring the troop is logistically ready for battle. During operations, the first sergeant directs aidman teams and recovery teams forward as needed, leads the troop LOGPAC forward from the logistics release point (LRP) to resupply the troop and then leads it back to the LRP, and organizes the LOGPAC to conduct troop resupply operations. He also directs the medical evacuation of soldiers who are wounded in action. Additionally, he directs the evacuation of soldiers killed in action, enemy prisoners of war, and disabled vehicles.

The troop combat trains can be consolidated and moved under the control of the first sergeant, or they can be split and moved. If the combat trains are split, the motor sergeant and first sergeant will divide the medics, mechanics, special tools, and high-use PLL items into equal portions and move parallel and behind troop combat elements to provide responsive medical evacuation and battle damage assessment across the troop zone. Recovery assets will continue to move center of zone to facilitate recovery. This technique may be used if the troop zone is wide and split operations are necessary to provide proper support.

Position the combat trains where they can support the troop, but will not be exposed to enemy direct fires. This location is usually from 2 to 5 kilometers to the rear.

Offensive and Reconnaissance Operations.During such operations as a movement to contact or a zone reconnaissance, position the trains in the center of the troop zone and about 2 kilometers, or one terrain feature, behind the tank platoons. Use the established road network, if possible, to move the trains. Bound the trains forward to successive concealed positions, based on the movement of the troop, to keep them in position to support the troop.

Defensive and Security Operations.During defensive and security operations, position the combat trains about 3 kilometers behind the tank platoon positions in the center of the troop sector. If possible, hide the trains in a small built-up area, which provides cover and concealment and a road network to facilitate the movement of the trains. The trains must be ready to move to support the troop or displace to a new location. The XO and first sergeant plan subsequent locations, and as the situation requires, the first sergeant or maintenance sergeant moves the trains to the next location.

Security.Trains must provide their own local security. They do not have the manpower or equipment to clear large areas, so they should move into areas that have already been cleared by scouts. The trains' best defense is to avoid detection. The trains should disperse their vehicles and camouflage the positions by using natural concealment like a wooded area or a small built-up area. A built-up area is best, because there are a limited number of approaches and the thermal signature of the trains is concealed. Enforce strict noise and light discipline. Hide vehicles in barns or garages, if available. The buildings in a built-up area and the trees in a wooded area provide some protection against the effects of indirect fires and air-delivered munitions. Dispersion also limits the damaging effects of these weapons.

Establish OPs around the trains to provide early warning of enemy movement toward the position. Put them where they can cover major avenues of approach. An OP has at least two soldiers, a crew-served weapon (if available), a map, binoculars, a night observation device, and a radio or field phone. The OP must immediately report, by radio or phone, enemy ground and air attacks so the trains can take appropriate actions.

Passive defense is the trains' best defense against air attack. The steps taken to locate them properly, disperse vehicles, camouflage positions, and use the cover and concealment of built-up or wooded areas help prevent the enemy from detecting and attacking, and limit the effects of enemy munitions if attacked.

Air attacks cannot always be avoided. If attacked, take active air defense measures. The first sergeant or senior soldier present must coordinate the small arms fires of the trains for them to be effective against an air threat. Refer to Chapter 8, Section VI, Air Defense, for a complete discussion of passive and active air defense.

FIELD TRAINS

Troop CSS assets not in the combat trains are collocated with the squadron field trains or the UMCP, and sustain the troop's fighting capability by moving rations, ammunition, fuel, repair parts, and replacements from the squadron rear area to the troop on the battlefield. Troop field trains are controlled primarily by the supply sergeant, and consist of the personnel and equipment shown in Figures 9-3 and 9-4. Troop maintenance assets in the field trains may be consolidated under the control of the squadron maintenance officer.

The troop supply sergeant is the troop representative in the squadron field trains. The supply sergeant is responsible for the following tasks:

- Maintaining records of troop clothing and equipment.
- Processing or forwarding requests for all Class I, II, III, IV, V, VI, and VII. He also assists the maintenance section in ordering Class IX repair parts and the medics in resupplying Class VIII (medical).
- Assembling and leading all vehicles in the troop LOGPAC from the field trains to the LRP. If the squadron is conducting resupply operations, the supply sergeant may move under the control and direction of one of the following squadron personnel: squadron support platoon leader or platoon sergeant, S4, HHT commander, or XO.
- Assisting the first sergeant in organizing the LOGPAC site for troop resupply and in evacuating soldiers who are killed in action, enemy prisoners of war, and disabled vehicles; and in medically evacuating wounded and injured soldiers.
- Delivering mail, when available, via the troop LOGPAC.

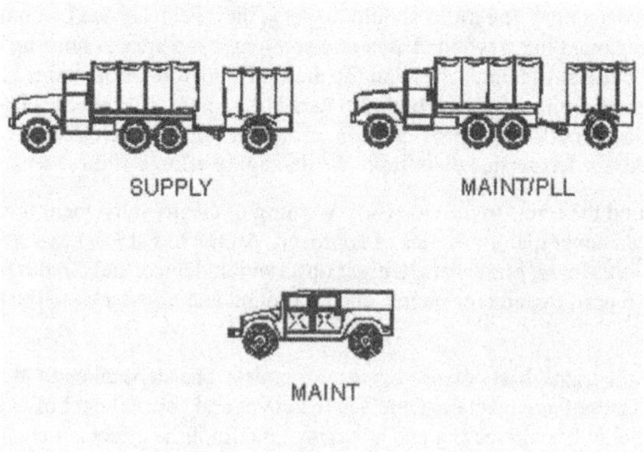

Figure 9-3. Heavy troop field trains.

Figure 9-4. Light troop field trains.

At the squadron field trains, the supply sergeant coordinates with the S4 and support platoon leader to ensure the LOGPAC meets the troop's CSS requirements.

The supply sergeant will coordinate with the troop PLL clerk for parts and requisitions and with the squadron personnel and administration center (PAC) section for mail, promotions, awards, Standardization Installation/Division Personnel System (SIDPERS), and other personnel actions. He will coordinate with squadron maintenance and communication for repaired vehicles and equipment and with the squadron food service sergeant for rations. The supply sergeant must frequently check with the HHT command post in the field trains for any additional troop requests reported on the A/L net.

The squadron field trains are responsible for their own local security. The supply sergeant will be tasked to assist the HHT commander in maintaining security of the field trains. He must execute and supervise security operations in accordance with the HHT commander's plan. During the movement through nonsecure areas, the supply sergeant should ensure that all crew-served and individual weapons are manned and ready. He should also ensure that all members of the troop field trains are briefed on immediate action drills for enemy contact or vehicle breakdowns.

If available, the troop should provide the supply sergeant with a radio to enhance communications between the supply sergeant and the first sergeant.

SQUADRON COMBAT SERVICE SUPPORT

In addition to organic troop CSS, one or two medical/evacuation teams are attached to the troop for combat operations from the squadron medical platoon. During resupply operations, the troop normally receives two fuel and two ammunition carriers from the transportation section of the support platoon. These carriers are not usually attached, but are provided to the troop as needed or as part of a standardized LOGPAC. The LOGPAC is discussed in Section II, Logistics.

To ensure responsive CSS for the squadron, the S4 and squadron maintenance officer organize the squadron support into echeloned trains and establish a main supply route (MSR), a UMCP, and an LRP. These trains are made up of combat trains (which provide immediate recovery, maintenance, medical, and emergency resupply support), and field trains (which provide the remainder of the squadron service support and limited direct-support maintenance). The UMCP is established to provide maintenance support for combat operations. It is positioned near or collocated with the squadron combat trains, and is the closest point to which damaged or failed equipment and systems are recovered. The UMCP is the focal point of the squadron maintenance effort. The MSR links troop combat trains to the squadron combat trains or UMCP and the field trains. LRPs are established along the MSR to facilitate the handover of LOGPACs and equipment between troops and squadron. An example of a squadron support plan is shown in Figure 9-5.

Section II. Logistics

SUPPLY OPERATIONS

The supply sergeant is responsible for getting supplies and delivering them to the troop. He delivers small items, but is largely dependent on support platoon assets to deliver bulky or high expenditure items. The commander establishes priorities for delivery, but the demands of combat normally dictate Classes I, III, V, VIII, and IX as most critical.

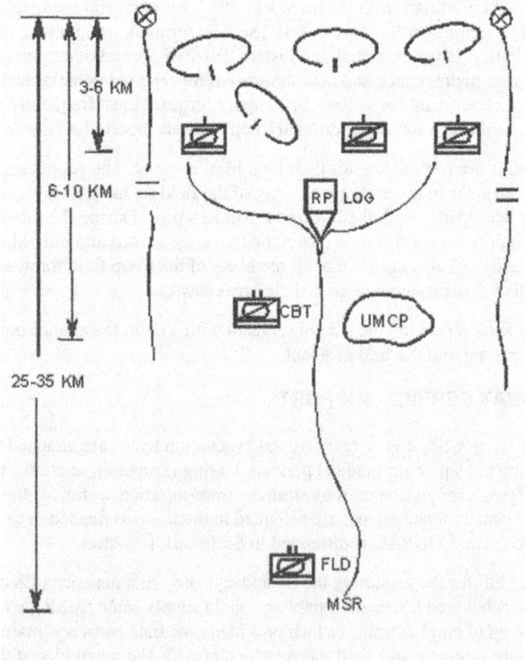

Figure 9-5. Squadron combat service support plan.

Class I (Rations).Meals ready to eat (MRE), C rations, are stocked on board each troop vehicle in a basic load prescribed by SOP (three-day supply). Class I is delivered daily by the supply sergeant as part of the

LOGPAC from the squadron field trains. Hot meals (B or tray-pack rations) are served when possible, but the normal ration cycle is C-C-C during combat operations. This requires resupply of at least 34 cases of MRE per day per troop. Water is a critical item, and must be resupplied daily. Soldiers may require up to 5 gallons of water per soldier per day, or more than 600 gallons for the entire troop. The troop can transport only 400 gallons of water on the water trailer. Soldiers should top off their water cans when possible. The supply sergeant should carry 5-gallon cans of water on his truck to supplement the needs of the troop. When he resupplies the troop, he exchanges the full cans for empty ones.

Class II (Supplies and Equipment).Class II items are requisitioned through the S4 and delivered by the supply sergeant as needed. The supply sergeant also maintains the following supplies, as a minimum, in his troop supply:

- One set of NBC overgarments and filter per soldier.
- One case of BA-30s.
- One mile of WD-1 wire.
- Additional batteries as necessary for night-vision devices and other equipment.
- Five sets of TA-50, to include personal clothing of various sizes, stored in duffel bags.
- Fifteen body bags.
- Map cases, grease pencils, permanent markers, etc.

Class III (POL).Class III is delivered by support platoon assets both as part of the troop LOGPAC and on an as-needed basis. Platoon sergeants report their Class III status to the command post daily, when status is requested, or when any Class III item is below a certain percentage, usually 59 percent.

Each troop vehicle carries a small basic load of packaged products as prescribed by SOP, and the fuel carriers normally have only small amounts of packaged products on board. Requests for large quantities of packaged products must be requested through the S4.

Class IV (Construction Materials). Class IV supplies are requested through the CTCP (combat trains command post) and delivered by the supply sergeant with the LOGPAC. Class IV supplies needed at troop level may include concertina wire, sandbags, and lumber.

Class V (Ammunition).The troop deploys with a basic load of Class V as prescribed by the unit for each type of vehicle. Ammunition status is reported by platoon sergeants to the command post daily, upon completion of enemy contact, or as needed. Normal resupply of Class V is delivered by the support platoon vehicles with the LOGPAC. Emergency resupply is requested through the first sergeant to the S4, and is delivered from the combat trains if it is available. Emergency resupply of Class V will normally consist of ammunition for major weapon systems, such as tank main guns, TOW missiles, and 25-mm guns in the heavy troop; and TOW missiles and caliber .50, 40-mm grenades, 7.62-mm, and individual weapons ammunition in the light troop.

Class VI (Personal Demand Items).Class VI items (such as soap, toothpaste, and cigarettes) are requested through the S4 by the first sergeant. They are usually picked up by the supply sergeant in the field trains and delivered as part of the LOGPAC.

Class VII (Major End Items).Class VII items, such as vehicles and night-vision devices, are automatically requested based on equipment shortages and battle losses. Items are delivered to the S4, who notifies the troop of the availability of the equipment. The equipment is delivered with a LOGPAC or brought forward immediately to an LRP where the first sergeant meets it and delivers it to the troop.

Class VIII (Medical Supplies).Class VIII is provided by the squadron medical platoon. The troop senior medic requests supplies through the squadron aid station. During intense combat, supplies may be pushed forward. They are then delivered to the combat trains where the troop medic picks them up from the squadron aid station, or are brought forward on ambulances.

Class IX (Repair Parts).The troop PLL clerk, who is in the field trains or the UMCP, works under the control of the squadron maintenance officer and the squadron maintenance technician. The PLL clerk requisitions, receives, and maintains records for all Class IX items. All troop PLLs are consolidated under the squadron maintenance officer during combat operations, and one or more PLL trucks remain in the UMCP at all times to provide a supply of Class IX for forward area support. When the troop maintenance sergeant needs a repair part, the request is sent through the squadron maintenance officer. The squadron maintenance officer determines if the part is available through the PLL, and fills the request if possible. If the part is not in the PLL, he then requests it through the squadron maintenance team in the field trains. They have the troop PLL clerk requisition the part for immediate use or replenish the part used from the PLL. The troop PLL clerk submits the requisition through squadron maintenance to the supporting maintenance unit. The part is delivered through squadron maintenance to the troop when the requisition is filled. The PLL clerk updates the records to reflect the requisition.

Maps.Maps are requested through the troop command post to the S4. As maps are available, the supply sergeant picks them up in the field trains and delivers them to the troop as part of a LOGPAC.

RESUPPLY OPERATIONS

Resupply of combat resources is accomplished using standardized procedures to rearm, refuel, and refit the troop as fast as possible to sustain its combat potential. There are two types of resupply operations—routine and supplementary. Methods of resupply are tailgate issue and service station.

Routine Resupply

Routine resupply operations include daily resupply of Classes I, III, V, and IX, and mail and other items needed by the troop. Routine resupply takes place when the troop is not in heavy contact, and may be conducted in an assembly area or behind troop positions when the troop is deployed in sector or zone. Class III is required more than once a day when the vehicles are in continuous operation.

Routine resupply is conducted using the LOGPAC from the field trains and the troop combat trains. The LOGPAC is organized in the field trains. Its composition is based on the troop's needs as reported to the S4, on requisitions, and on the availability of supplies. The troop supply sergeant reports to the S1 and S4 sections, squadron mess section, and squadron maintenance section to ensure all available supplies are picked up and loaded on trucks for troop resupply. The S4, support platoon leader, and troop supply sergeants assemble the LOGPACs for each troop by adding the troop supply trucks to the troop slice of Class III and V trucks from the support platoon. The support platoon leader leads the LOGPACs to the LRP, where the first sergeants meet them. Each first sergeant leads his LOGPAC to the troop resupply site. When it arrives, members of the troop combat trains guide the LOGPAC vehicles into position. Once the LOGPAC is established, the first sergeant reports to the command post that he is set. The troop commander or the XO coordinates resupply operations, and ensures all platoons and sections are resupplied. Resupply is conducted by one of two methods—tailgate issue or service station.

Tailgate Issue Method.The tailgate method is used in static positions such as assembly areas. Class III and V supply vehicles and other bulk-issue vehicles move from vehicle to vehicle to conduct resupply. The rest of the service support vehicles are centrally located in the troop area. Little or no movement is required by the combat vehicles. Personnel move to a centralized location to receive supplies, Class I, and mail. This method provides 360-degree security throughout the resupply operation; however, it is very time-consuming and requires an adequate road network for the wheeled supply vehicles to reach each vehicle. An example of this method is shown in Figure 9-6.

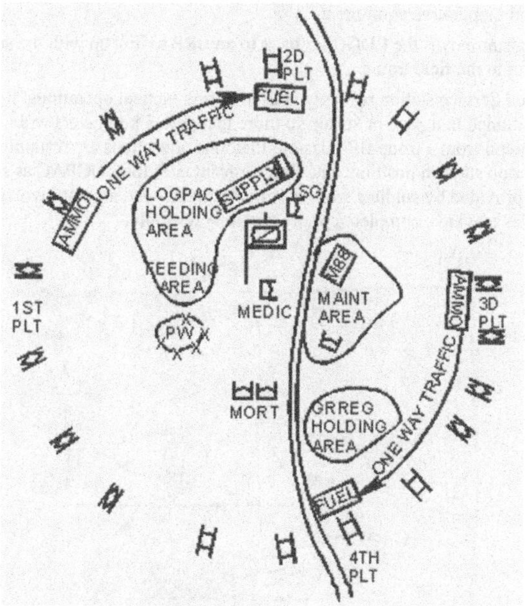

Figure 9-6. Tailgate issue method in a troop assembly area.

The following takes place during tailgate resupply:

- Combat vehicles remain in place. POL and ammunition trucks travel in a clockwise direction around the assembly area to each vehicle position, in turn, to conduct resupply.
- Crewmen rotate through the feeding area and pick up supplies, water, and mail.
- The first sergeant and platoon sergeants arrange for pickup of those killed in action (KIA) and their personal effects. The KIA are brought to a holding area near the medical aid station, but kept out of view.
- Ambulances pick up, treat, and evacuate seriously wounded soldiers. Other wounded soldiers are

carried to the ambulance or walk to it for emergency medical treatment.

- Prisoners are kept together and guarded. As soon as possible, they are moved to the squadron trains on a returning supply vehicle.
- The troop armorer, the radio repairman, and the organizational mechanics repair known problems and spot-check other vehicles.
- Vehicles needing maintenance are brought to the maintenance area.
- The first sergeant and platoon sergeants closely monitor the resupply operation.
- Empty LOGPAC vehicles are moved to a holding area, where they are loaded with KIA, prisoners of war (PW) and inoperative equipment.
- The supply sergeant moves the LOGPAC back to an LRP to link up with the support platoon leader and return to the field trains.

Service Station Method.Service station resupply is used during tactical operations. It is most effective when the troop is positioned in a zone or sector no more than 3 to 5 kilometers wide, such as is found in a defend in sector or defend from a troop BP mission. Platoons or sections are resupplied at the LOGPAC while the rest of the troop stays in position. The first sergeant sites the LOGPAC as shown in Figure 9-7. LOGPAC security is provided by soldiers from the combat trains who are not involved in the resupply, and by platoon vehicles that have completed or are awaiting resupply.

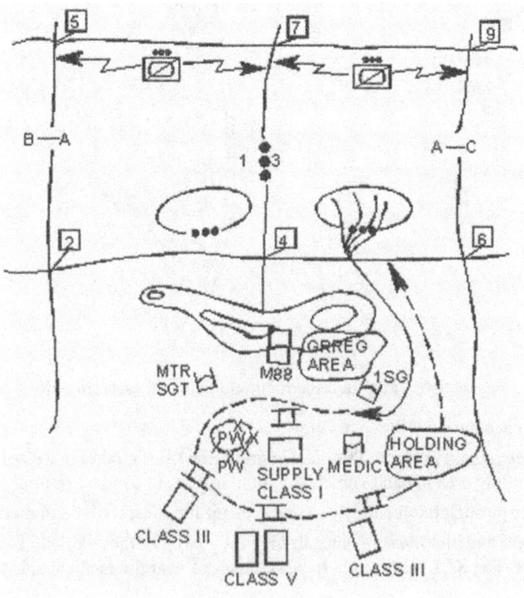

Figure 9-7. Service station method.

The following takes place during service station resupply:

- Vehicles of the first platoon or section enter the LOGPAC at a designated location and follow one-way traffic flow.
- The vehicle carrying the KIA in body bags moves to the holding area, out of view of the troop, and leaves the remains and personal effects.
- Soldiers requiring medical attention are delivered to medics and treated or prepared for evacuation.
- The platoon sergeant supervises the operation and coordinates face-to-face with the first sergeant for any special requirements.
- The platoon leader dismounts his vehicle and uses the first sergeant's vehicle and radio to contact the troop commander for orders and situation reports.
- Crews requiring unit-level maintenance remain in the maintenance holding area.
- Vehicles rotate through stations.
- Unit-level mechanics, the armorer, and the radio repairman repair known problems and spot-check other vehicles.
- Crews rotate to the supply truck to pick up mail, supplies, and Class I.
- The platoon leader and platoon sergeant conduct precombat inspection.
- When the platoon or section has completed resupply, it moves to its designated position.
- The rest of the platoons rotate individually through the LOGPAC for resupply.

At the end of the resupply operation, the troop LOGPAC returns to the LRP where it links up with the support platoon leader and returns to the field trains. In the field trains, the supply sergeant returns the fuel and ammunition carriers to the support platoon, returns ration-serving equipment to the mess team, delivers the KIA to the graves registration (GRREG) collection point, requests any additional supplies from the squadron supply sergeant, and returns to his position in the field trains.

Supplementary Resupply

In the heavy troop, daily routine resupply will not sustain the combat power of the M1/M3 fleet when operating without a break for 8 to 10 hours. The tanks will consume so much fuel that they will have to supplement routine resupply of Class III one or two times daily. The troop may also need to resupply Class V. Supplementary resupply can be conducted in a couple of ways depending on the mission and if the troop is in contact with the enemy. Regardless of the technique used, the commander must push the supplies forward to his platoons; the tempo of combat operations cannot be disrupted in order to resupply. Conduct emergency resupply when in heavy contact with the enemy, and prestock resupply when not. Once the supplies are brought forward, either the tailgate issue or service station methods are used.

Supplementary resupply must be planned when continuous operations are expected. Resupply must be requested through the S4 several hours in advance to ensure the service support assets are prepared to support the operation. The troop must quickly resupply once fuel trucks are available, and then return them to the field trains. There are few fuel carriers and drivers available in the squadron and these valuable assets cannot be wasted.

Platoon Prestock.When operating over wide frontages, such as during screen, zone reconnaissance, or movement to contact operations, break the supplies down into platoon packages and push the supplies forward to each platoon position. Position the vehicles behind the platoon and resupply individual vehicles or sections at a time using the service station method. If the situation allows, the resupply vehicles can move through the platoon position and resupply the platoon in place using the tailgate issue method.

Controlling platoon prestock is difficult because resupply vehicles are moving to several locations rather than remaining under the centralized control of the first sergeant. The platoon leader must also provide security for their own prestock (see Figure 9-8).

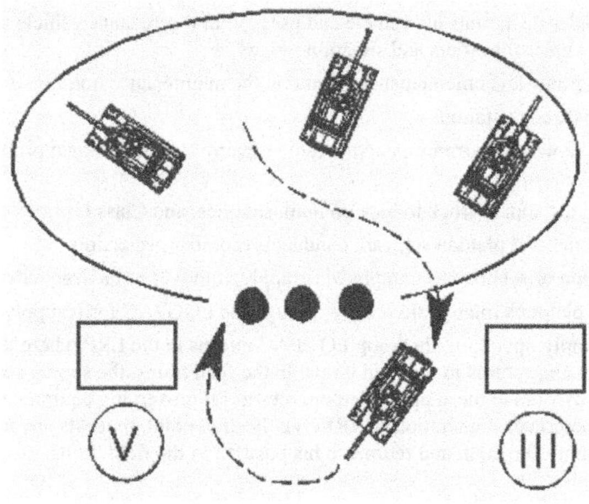

Figure 9-8. Platoon prestock.

Troop Prestock.Troop prestock is used to supplement routine resupply of Classes III and V. Troop prestock is used when the troop is operating in a narrow zone or sector. The resupplies are pushed forward to the troop position using either the service station or tailgate method. Each platoon rotates through the resupply site, using the service station method, while the rest of the troop remains in position; or each vehicle is resupplied in position using the tailgate method (see Figure 9-9).

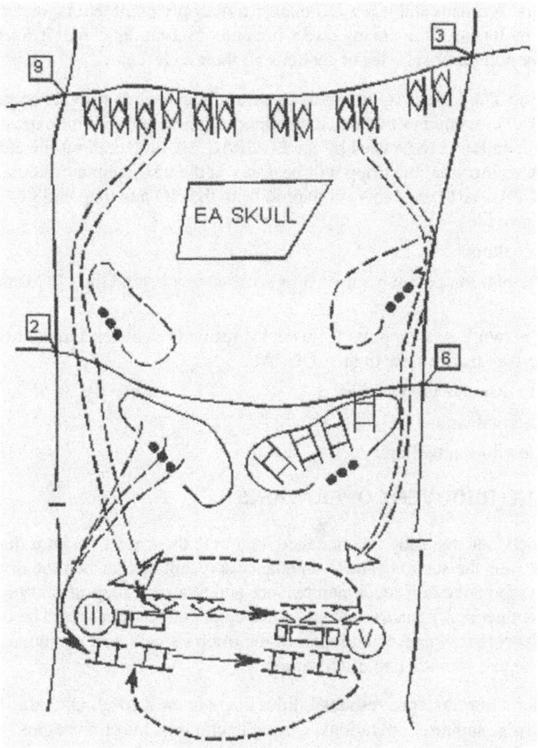

Figure 9-9. Troop prestock.

Variations of these methods can be used to meet the troop's situation. The troop must, however, rehearse these techniques to ensure it can quickly and efficiently conduct resupply operations and continue the mission.

Emergency Resupply

Emergency resupply normally involves only fuel and ammunition and is conducted while in contact with the enemy. The resupply begins at section and platoon level by redistributing ammunition between vehicles to cross-level loads. The platoon sergeant reports his need for emergency resupply to the first sergeant, who relays the request to the S4. The squadron combat trains maintain a small load of Class III and V for these situations. The S4 or support platoon leader coordinates a linkup between the squadron combat trains and the troop first sergeant. The first sergeant meets the resupply trucks and moves back to the troop area. The first sergeant and troop XO choose a resupply point that is just behind the troop position and masked by terrain from enemy direct fire and observation. If fuel is needed, the fuel truck is moved to the resupply point and vehicles or sections go there to refuel.

Resupply Site Selection.The LOGPAC site must be carefully chosen to provide responsive support for the troop, to support the movement of wheeled resupply vehicles, and to limit exposure to enemy fires. The XO selects the general area to be used by the LOGPAC. He knows about the current tactical situation, and can determine what the troop will be doing in the next couple of hours. The first sergeant selects the exact LOGPAC site based on information from the XO and map and ground reconnaissance. He selects a site that provides—

- Cover and concealment.
- Proximity to the platoon positions, from 3 to 8 kilometers behind the FLOT and center of the sector.
- A road or trail network that supports the wheeled resupply vehicles and the heaviest troop vehicles and allows one-way traffic flow to the LOGPAC.
- Enough room to disperse the vehicles.
- Reduction of thermal signature.
- Level enough to allow refueling.

MAINTENANCE AND RECOVERY OPERATIONS

The first sergeant supervises the troop maintenance section in the troop combat trains and the maintenance sergeant runs the section. The maintenance sergeant operates on the troop command net and stands by to provide responsive unit-level maintenance and recovery support for combat vehicles and radio equipment. When needed, squadron maintenance augments troop assets. The troop armorer is responsible for unit-level maintenance and repair of the troop's small arms weapons. Position him in the combat trains where he can provide battlefield support.

Maintenance Operations.Maintenance responsibilities begin at crew level. Operator-level PMCS must be performed before, during, and after operations. During operations, the crew begins the maintenance process when it identifies the fault. Once a problem develops, the vehicle commander determines support requirements, to include self-recovery, assistance from another vehicle, or assistance from troop or squadron maintenance, and takes the following actions:

- Reports the situation to the platoon sergeant, providing condition(s), location, and circumstances.
- Attempts to self-recover, if mired.
- Uses another vehicle to move to a secure location if exposed to enemy observation and fire. Uses smoke to screen the area between the mired vehicle and the enemy.

When repairs are beyond the capability of the crew, the platoon sergeant uses the troop command net to notify the first sergeant of the situation and to request assistance. The crew must maintain radio contact (if the radio is operational) on the platoon net and maintain local security. The maintenance sergeant monitors the radio traffic. He switches his radio to the platoon frequency to coordinate maintenance support and dispatches the appropriate assets to the "down" vehicle. The maintenance team checks whether the problem can be corrected in place within the time criteria (30 minutes in defensive situations and 2 hours in offensive situations). They move the vehicle to a more secure location, if necessary, and fix the vehicle. If repairs will take too long, the maintenance sergeant uses the squadron A/L net to notify the squadron maintenance officer who will coordinate for a handover of the equipment from troop to squadron at the LRP or along the MSR. The squadron maintenance officer needs the following information:

- Identification of the troop.
- Equipment involved.
- Location of equipment.
- Nature of damage.
- Pertinent information about the tactical situation, and any other necessary information.

Based on the information above, the squadron maintenance officer will designate a time and place for equipment handover.

The troop does not have the personnel, equipment, or time to create a holding area in the troop combat trains. If repairs will take more than the allowable time, evacuate the equipment to squadron.

Recovery Operations.When a vehicle cannot be repaired within the allowable time or is damaged beyond repair, the maintenance section must recover it and pull it to the MSR where it is turned over to squadron. Squadron maintenance will pull it to the UMCP or the field trains.

If a vehicle is catastrophic, or not able to be recovered based on the enemy situation, the commander may authorize the destruction of the vehicle. Before using thermite grenades to destroy a vehicle, remove as many valuable items as time permits, such as—

- Classified equipment and documents.
- Communications equipment.
- Weapons.
- Ammunition.

FIELD SERVICES

GRREG is performed by the field services platoon from corps, located in the regimental support area. The initial collection, identification, safeguarding of personal effects, and evacuation of the dead is the troop's responsibility.

When remains are discovered, be careful to preserve all items that may be used for identification. If metal identification tags (dog tags) are on the remains, do not remove them. Secure all personal effects in a bag or poncho and tie it to the remains. Place each casualty in a body bag, poncho, or shelter half and evacuate with the first available means of transportation, such as LOGPAC vehicles or disabled vehicles, to the squadron field trains. The supply sergeant receives all remains, and is responsible for turning them over to the GRREG collection point.

If the tactical and logistical situation makes evacuation impossible, emergency on-site burial is performed. On-site burial requires the permission of the squadron commander. If an on-site burial is performed, do the following:

- Complete two copies of DD Forms 551 and 1077 for each body. These forms are available through the squadron PAC. The first sergeant and supply sergeant should carry them.

- If there are several remains, engineer support may be needed to dig trenches.
- Dig a trench 6-1/2 feet wide and 3-1/2 feet deep. The length is determined by the number of remains.
- Remove one identification tag from each body, and string the tags on a wire in the order in which the remains are buried.
- Place the remains in the grave shoulder to shoulder.
- Bury all personal effects with the remains.
- Mark the ends of each row with a stake. Tag each stake to identify the marker as a grave. Indicate the length of the grave.
- Prepare an overlay that shows the location of the grave site.
- Cover the grave with earth removed from the trench.
- Forward the completed forms, overlay, and identification tags to squadron.

If the remains are contaminated, the grave site must be clearly marked and separated from noncontaminated grave sites. This must also be indicated on the grave-site overlay.

BATH AND LAUNDRY SERVICES

Bath and laundry services are provided by supply and service units from the corps support command. When available, these services are coordinated through the S4.

Section III. Personnel Service Support

PERSONNEL AND ADMINISTRATIVE

All troop administrative and personnel actions are handled by the squadron PAC in the field trains. The troop chain of command is responsible for ensuring that soldiers receive passes, leaves, promotions, awards, mail, legal assistance, financial services, and other personnel and welfare services on a fair and prompt basis. The first sergeant interfaces with PAC daily through the supply sergeant, at the administrative/logistics operations center with the S1/S4, or at the LRP.

Casualties and missing personnel are reported on DA Forms 1155 and 1156. These forms are initiated by the individual who witnessed the incident and are forwarded to the first sergeant. This action is the basis for notification of the next of kin and awarding benefits such as Serviceman's Group Life Insurance (SGLI), so accuracy is a must.

The commander should continuously manage the troop personnel situation to ensure trained personnel are manning key positions, crew shortages are filled with available personnel, and replacement personnel are trained and incorporated into crews. During combat, platoon leaders and platoon sergeants must balance their crews after casualties are evacuated. They must fill key positions on vehicles with the most qualified soldier. If necessary the commander may direct cross-leveling of personnel within the troop.

The troop submits daily personnel reports to the squadron as prescribed by SOP. The personnel reports are important because they allow the squadron commander to make tactical decisions for the employment of the troop. The reports also allow the S1, with the squadron commander's guidance, to properly distribute personnel replacements. Replacement personnel are received by the supply sergeant in the field trains. The supply sergeant in-processes the soldier into the troop, ensures he has the proper equipment, then delivers him to the first sergeant during the troop LOGPAC.

combat health support

First Aid.First aid is the responsibility of all soldiers; they use first aid, self-aid, buddy-aid, and combat lifesavers. All must be trained to take action after a soldier is wounded to keep him breathing, stop the bleeding, prevent shock, and dress the wound until medical personnel are available to treat the soldier. Training must include treatment of NBC casualties and crew evacuation drills to get soldiers out of a vehicle without creating further injury. Selected soldiers from each platoon will be trained in advanced medical skills in the combat lifesaver program.

Treatment.The first sergeant should position troop medical aid and evacuation teams on the battlefield where they can be most responsive. They will usually operate under the control and direction of the first sergeant in the troop combat trains. If the troop zone is pushing 10 kilometers wide, the troop may operate split combat trains with the first sergeant controlling medical evacuation on one side of the zone and the motor sergeant controlling the other half. The medics must know the locations of and routes to each platoon, the troop combat trains, the squadron combat trains, and each collection point.

Combat Stress Control.The psychological effects of combat on soldiers influence their ability to execute their missions. Individual self-control and self-discipline in the face of danger are maintained through unit discipline and firm leadership. The commander and his subordinate leaders are key to the mental toughness of soldiers. Commanders should be visible to soldiers and share their hardships with them. They should talk to their soldiers individually and as a troop to keep them informed. The commander's tone of voice on the radio must indicate that he has the situation under control. He should transmit the successful accomplishments of the troop and tell the soldiers they are doing a good job. Keep details about any casualties in the troop off the radio and ensure that soldiers who are killed in action are covered up and kept out of view of the troop. The chaplain can be a great asset in maintaining troop morale. Do not neglect the mental fitness of soldiers, because uncontrolled fear is contagious and can quickly lead to chaos.

Evacuation.To receive medical assistance, a platoon leader or platoon sergeant calls the first sergeant on the troop command net. The medic team monitors the call and goes to the platoon frequency to make any other necessary coordination. The medic team moves to the reported location to treat the soldier(s). The medics transport as many wounded soldiers as possible in their armored ambulance. Do not transport the dead with the wounded; evacuate them separately. If the medics determine the soldiers are seriously wounded, they either transport them directly to the squadron combat trains, or coordinate to hand them over to squadron at a collection point. Soldiers with slight wounds should not be evacuated unless it is necessary. They should be treated and returned to duty. Secure the seriously wounded soldiers on their litters to prevent any further injury during the evacuation.

As a general rule, do not use combat vehicles to evacuate the wounded; this reduces combat strength. Use combat vehicles only if absolutely necessary, and no other transportation is available.

Weapons and military equipment (except NBC protective equipment) of personnel being evacuated are secured with the platoon sergeant, first sergeant, or supply sergeant. Ensure the soldier keeps his protective mask and overgarments. When the situation permits, all equipment that was secured by the first sergeant or platoon sergeant is turned over to the supply sergeant.

disease and nonbattle injuries

More soldiers are lost in combat to illness and disease than to combat-related wounds. Maintaining the health and fighting fitness of the troop is a leadership responsibility. Personal hygiene, field sanitation, and rest must be incorporated into all troop operations. If these are ignored, the combat effectiveness of the troop will decrease rapidly.

UNIT MINISTRY TEAM

The unit ministry team (UMT), consisting of an appropriate number of chaplains and chaplain assistants, is assigned to provide unit, area, and denominational coverage to all troops in the cavalry squadron. UMT provides the following eight subfunctions of worship opportunities:

- Administration of sacraments, rites, and ordinances.
- Pastoral care and counseling.
- Development and management of the UMT.
- Management of material resources.
- Advising the commander on matters of religion, morale, and morals as affected by religion.
- Ministry in support of soldiers suffering from battle fatigue.
- Development of programs that enhance the total well-being of the soldier.
- Development of activities to enhance unit cohesion.

PERSONAL HYGIENE

The personal hygiene of the troop's soldiers is a leader's responsibility. Rules of hygiene must be observed to ward off disease and improve troop morale. Ensure soldiers wash and change their socks and underwear daily, if possible. All soldiers should shower as time and resources are available. The senior aidman plays a big part in personal hygiene by educating soldiers and by checking for signs of trench foot or immersion foot and frostbite.

Disease can quickly spread through the troop and incapacitate its soldiers. Watch for symptoms and ensure the soldiers' immunizations are current.

REST

Rest is extremely important for both commanders and their soldiers. When the troop is at REDCON 3 or 4, troop leadership should ensure a sleep plan is planned and executed.

FIELD SANITATION

Field sanitation is vital to the prevention of the spread of disease. The medics must assist in this effort by checking troop water supplies to ensure they are potable, and by ensuring mess utensils and equipment are properly cleaned. In static situations, such as when in assembly areas, soldiers will use slit trenches or latrines; at other times soldiers will use cat holes. Slit trenches and cat holes must be covered up after use to prevent the spread of disease.

Section IV. Prisoners of War

CAPTURE AND HANDLING

Maps, military documents, letters, and diaries obtained on the battlefield, as well as PWs, are valuable sources of combat intelligence. Scout platoons usually obtain information while searching the battlefield in the conduct of reconnaissance operations or after contact with the enemy. They should rapidly report this information to the troop commander for evacuation instructions. The scouts may be ordered to take the documents to the troop command post or to rendezvous with the first sergeant at a specific location to turn over the documents. Proper handling and evacuation of captured documents and equipment are important. The materials must be immediately turned over to trained intelligence personnel. In most cases, captured documents lose their value over time.

EVACUATION TECHNIQUES

In most situations, returning supply vehicles, aircraft, or troop headquarters vehicles can be used to evacuate PWs from the troop combat trains to the squadron collection point. The troop is responsible for guarding PWs until they are turned over to the S1. Wounded PWs are treated through normal medical channels, but are kept separate from US soldiers.

When support vehicles are not readily available and the troop must continue its operation, secure the PWs in a holding area such as a basement or compound. Notify the first sergeant and the S1 of the PW location and continue the mission. The first sergeant will pick up the PWs and complete their evacuation to the collection point.

Appendix A

Fratricide Prevention

Fratricide is as old as warfare itself, a complex problem that defies simple solutions. Fratricide can be broadly defined as the employment of friendly weapons and munitions, with the intent to kill the enemy or destroy his equipment or facilities, that results in unforeseen and unintentional death or injury to friendly personnel. This appendix focuses on actions leaders can take with current resources to reduce the risk of fratricide.

CONTENTS

Section I. Magnitude of the Problem

The modern battlefield is more lethal than any in history. The tempo of operations is rapid, and the nonlinear nature of the battlefield creates command and control challenges for all unit leaders.

The accuracy and lethality of modern weapons make it possible to engage and destroy targets at these extended acquisition ranges. At the same time, however, the ability of US forces to acquire targets using thermal imagery and other sophisticated sighting systems exceeds our ability to accurately identify these targets as friend or foe. As a result, friendly elements can be engaged unintentionally and destroyed in a matter of seconds.

Added to this is the problem of battlefield obscuration, which becomes a critical consideration whenever thermal sights are the primary source of target identification. Rain, dust, fog, smoke, and snow degrade identification capability by reducing the intensity and clarity of thermal images.

On the battlefield, positive visual identification cannot be the sole engagement criteria at ranges beyond 1,000 meters. Situational awareness is the key; it must be maintained throughout an operation.

Section II. Risk Identification and Preventive Measures

Reduction of fratricide risk begins during the planning phase of an operation and continues throughout preparation and execution. Risk identification must be conducted at all levels during each phase, and the results clearly communicated up and down the chain of command so risk assessment can begin. This section covers considerations that influence risk identification; it also focuses on measures the platoon leader can implement both to make the identification process more effective and to help prevent friendly fire incidents from occurring. Section III of this appendix covers the risk assessment process; Section IV lists additional fratricide reduction measures and guidelines.

Planning Phase

A plan that is thoroughly developed and understood helps to minimize fratricide risk. The following considerations help indicate the potential for fratricide in a given operation:

- The clarity of the enemy situation.
- The clarity of the friendly situation.
- The clarity of the commander's intent.
- The complexity of the operation.
- The planning time available at each level.

Graphics are a basic tool that commanders at all levels use to clarify their intent, add precision to their concept, and communicate their plan to subordinates. As such, graphics can be a very useful tool in reducing the risk of fratricide. Commanders at all levels must understand the definitions and purpose of operational graphics and the techniques of their employment.

Note. See FM 101-5-1 for the definitions of each type of graphic control measure.

Preparation Phase

The following factors may cause fratricide risks to become evident during rehearsals:

- Number and type of rehearsals.
- Training and proficiency levels of units and individuals.
- The habitual relationships between units conducting the operation.
- The physical readiness (endurance) of the troops conducting the operation.

Backbriefs and rehearsals are primary tools in identifying and reducing fratricide risk during the preparation phase. The following are some considerations for their use:

- Backbriefs ensure subordinates understand the commander's intent. They often highlight areas of confusion, complexity, or planning errors.
- The type of rehearsal conducted determines what types of risks are identified.
- Rehearsals should extend to all levels of command and involve all key players.

Execution Phase

During execution, in-stride risk assessment and reaction are necessary to overcome unforeseen fratricide risk situations. The following are factors to consider when assessing fratricide risks:

- Intervisibility between adjacent units.
- Amount of battlefield obscuration.
- Ability or inability to positively identify targets.
- Similarities and differences in equipment, vehicles, and uniforms among friendly and enemy forces.
- Vehicle density on the battlefield.
- The tempo of the battle.

Maintaining situational awareness at all levels and at all times is another key to fratricide reduction as an operation progresses. Units must develop and employ effective techniques and SOPs to aid leaders and crewmen in this process. These techniques include—

- Monitoring on the next higher net.
- Radio cross-talk between units.
- Accurate position reporting and navigation.
- Training and use/exchange of liaison officers (LO).

Section III. Risk Assessment

Risk assessment must be conducted whenever fratricide risk factors are identified. It must take place at all levels during each phase of operations. As with risk identification, the results of the assessment must be passed on to all levels of the chain of command so that fratricide reduction measures can be developed and implemented. Refer to Section IV for specific reduction measures.

Figure A-1 is a worksheet for evaluating fratricide risk in the context of mission requirements. The worksheet lists six mission-accomplishment factors that affect the risk of fratricide, along with related considerations for each factor. Assess the potential risk in each area as low, medium, or high, and assign a point value to each (one point for low risk, two for medium risk, three for high risk). Add the point values for the overall fratricide assessment score. Use the resulting score only as a guide, however. The final assessment must be based both on observable risk factors like those on the worksheet and on your "feel" for the intangible factors affecting the operation. Note that descriptive terms are listed only in the low- and high-risk columns of the worksheet. The assessment of each factor will determine whether the risk matches one of these extremes or lies somewhere between them as a medium risk.

FACTORS	LOW (1)	MEDIUM (2)	HIGH (3)
1. UNDERSTAND PLAN			
• Commander's Intent	Clear		Foggy
• Complexity	Simple		Complex
• Enemy Situation	Known		Unknown
• Friendly Situation	Clear		Unclear
• ROE	Clear		Unclear

2. ENVIRONMENT ● Intervisibility ● Obscuration ● Battle tempo ● Positive target ID	Favorable Clear Slow 100%		Unfavorable Obscured Fast 0%
3. CONTROL MEASURES ● Command relationships ● Audio ● Visual ● Graphic ● SOPs ● LOs ● Location/Navigation	Organic Loud/Clear Well Seen Standard Standard Proficient Sure		Joint/Combined Jammed Obscured Not understood Not used Untrained Unsure
4. EQUIPMENT (Compared to US) ● Friendly ● Enemy	Similar Different		Different Similar
5. TRAINING ● Individual proficiency ● Unit proficiency ● Rehearsal ● Habitual relationship ● Endurance	MOS Qual Trained Multiple Yes Alert		Untrained Untrained None No Fatigued
6. PLANNING TIME (1/3 - 2/3 Rule) ● Higher HQ ● Own HQ ● Lower HQ	Adequate Adequate Adequate		Inadequate Inadequate Inadequate
OVERALL FRATRICIDE ASSESSMENT	LOW 26-46%*	MEDIUM 42-62%*	HIGH 58-78%*
● Commander may use numbers as the situation dictates. ● Numbers alone may not give accurate fratricide risk.			

Figure A-1. Fratricide risk assessment worksheet.

Section IV. Fratricide Reduction Measures

The following measures are provided as a guide to actions that can reduce fratricide risk. They are not directive in nature, nor are they intended to restrict initiative. Apply them as appropriate based on the specific situation and METT-T factors.

- Identify and assess potential fratricide risks in the estimate of the situation. Express these risks in the OPORD or FRAGO.
- Maintain situational awareness, focusing on such areas as current intelligence; unit locations and dispositions; denial areas (minefields/FASCAM); contaminated areas, such as ICM and NBC; SITREPs; and METT-T factors.
- Ensure positive target identification. Review vehicle/weapon ID cards. Know at what ranges and under what conditions positive identification of friendly vehicles/weapons is possible.
- Establish a command climate that stresses fratricide prevention. Enforce fratricide prevention measures, emphasize the use of doctrinally sound tactics, techniques, and procedures. Ensure constant supervision in the execution of orders and the performance of all tasks and missions to standard.
- Recognize the signs of battlefield stress. Maintain unit cohesion by taking quick, effective action to alleviate it.
- Conduct individual, leader, and collective (unit) training covering fratricide awareness, target identification and recognition, and fire discipline.
- Develop a simple, decisive plan.
- Give complete and concise mission orders.
- Use SOPs that are consistent with doctrine to simplify mission orders. Periodically review and change SOPs as needed.
- Strive for maximum planning time for you and your subordinates.
- Use common language/vocabulary and doctrinally correct standard terminology and control measures, such as fire support coordination line, zone of engagement, and restrictive fire lines.
- Ensure thorough coordination is conducted.
- Plan for and establish effective communications.
- Plan for collocation of command posts whenever it is appropriate to the mission, such as during a passage of lines.
- Designate and employ LOs as appropriate.
- Ensure rules of engagement are clear.
- Include fratricide risk as a key factor in terrain analysis (OCOKA).
- Conduct rehearsals whenever the situation allows time to do so.
- Be in the right place at the right time. Use position location/navigation devices (GPS and POSNAV); know your location and the locations of adjacent units (left, right, leading, and follow-on); and synchronize tactical movement.
- Include discussion of fratricide incidents in after-action reports.

Section V. Fratricide Risk Considerations (OPORD Format)

This section, which parallels the five-paragraph OPORD, contains key factors and considerations in fratricide reduction. This is not a change to the OPORD format; rather, it should be used during OPORD development to ensure fratricide reduction measures are included in the order. It is not a strict guide. The factors and considerations are listed where they would likely appear in the OPORD, but they may warrant evaluation during preparation of other paragraphs.

1. Situation.

a. Enemy forces.

(1) Are there similarities among enemy and friendly equipment and uniforms that could lead to fratricide?
(2) What languages do enemy forces speak? Could these contribute to fratricide risk?

(3) What are the enemy's deception capabilities and his past record of deception activities?

(4) Do you know the locations of enemy forces?

b. Friendly forces.

(1) Among the allied forces, are there differences (or similarities with enemy forces) in language, uniform, and equipment that could increase fratricide risk during combined operations?
(2) Could differences in equipment and uniforms among US armed forces increase fratricide risk during joint operations?

(3) What differences in equipment and uniforms can be stressed to help prevent fratricide?

(4) What is the friendly deception plan?

(5) What are the locations of your unit and adjacent units (left, right, leading, follow-on)?
(6) What are the locations of neutrals and noncombatants?

c. Own forces.

(1) What is the status of training activities? What are the levels of individual, crew, and unit proficiency?

(2) Will fatigue be a factor for friendly forces during the operation? Has an effective sleep plan been developed?

(3) Are friendly forces acclimatized to the area of operations?

(4) What is the age (new, old, or mix) and condition of equipment in friendly units? What is the status of new equipment training?

(5) What are the expected MOPP requirements for the operation?

d. Attachments and detachments.

(1) Do attached elements understand pertinent information regarding enemy and friendly forces?
(2) Are detached elements supplied this pertinent information by their gaining units?

e. Weather.

(1) What are the expected visibility conditions (light data and precipitation) for the operation?
(2) What effect will heat and cold have on soldiers, weapons, and equipment?

f. Terrain.

(1) Do you know the topography and vegetation (such as urban, mountainous, hilly, rolling, flat, desert, swamp/marsh, prairie/steppe, jungle, dense forest, open woods) of the expected area of operations?
(2) Have you evaluated the terrain using the factors of OCOKA?

2. Mission. Is the mission, as well as all associated tasks and purposes, clearly understood?

3. Execution.

a. Task organization.

(1) Has the unit worked under this task organization before?
(2) Are SOPs compatible with the task organization (especially with attached units)?

(3) Are special markings or signals (for example, cats' eyes, chemlites, or panels) needed for positive identification of uniforms and equipment?

(4) What special weapons and/or equipment will be used? Do they look or sound like enemy weapons and/or equipment?

b. Concept of the operation.

(1) Maneuver. Are main and supporting efforts identified to ensure awareness of fratricide risks and prevention measures?

(2) Fires (direct and indirect).

(a) Are priorities of fires identified?

(b) Have target lists been developed?
(c) Has the fire execution matrix/overlay been developed?

(d) Have locations of denial areas (minefields, FASCAM) and contaminated areas (ICM, NBC) been identified?

(e) Are the locations of all supporting fire targets identified in the OPORD/OPLAN overlays?

(f) Are aviation and CAS targets clearly identified? Have signals been established to positively identify these targets for the aircraft? Have airspace coordination areas been developed? Have enemy air defense systems been suppressed?

(g) Has the direct-fire plan been developed and synchronized with the fire support plan?

(h) Have final protective fires been designated?

(i) Have you identified and verified sector limits?

(j) Have executors for each target been assigned and do they understand when and where to shoot? Do the shooters have "eyes on" the target?

(k) Are the observers surveyed in or are they using a map spot? Target location errors can cause big problems.

(l) Do all leaders and executors understand where the fire support coordination measures are and when they go into effect? Rehearsal is the key.

(m) Can the fire support officer hear what targets are being called on the maneuver nets?

(n) Have all targets been rehearsed with the executors and the field artillery battalion?
(o) Does the reinforcing or general support reinforcing field artillery have all the proper graphics and understand where they fit in? Did they attend the rehearsal?

(p) Have restrictions on specific munitions been established and does everyone know where they are planned and emplaced?

(3) Engineer tasks.

(a) Are friendly minefields, including FASCAM and ICM dud-contaminated areas, known?

(b) Are obstacles identified, along with the approximate time needed for reduction/breaching of each?

(4) Tasks to each subordinate unit. Are friendly forces identified, as appropriate, for each subordinate maneuver element?

(5) Tasks to CS/CSS units. Have locations of friendly forces been reported to CS/CSS units?

(6) Coordinating instructions.

(a) Will rehearsals be conducted? Are they necessary? Are direct and indirect fires included?

(b) Is a backbrief necessary?

(c) Are appropriate control measures clearly explained and illustrated in the OPORD and overlays? Have they been disseminated to everyone who has a need to know? What is the plan for using these control measures to synchronize the battle and prevent fratricide?

(d) Have target/vehicle identification drills been practiced?

(e) Do subordinate units know the immediate action, drill, or signal for "CEASE FIRE" or "I AM FRIENDLY" if they come under unknown or friendly fire? Is there a backup action?

(f) Is guidance in handling dud munitions, such as ICM and CBUs, included?

4. Service Support.

a. Are train locations and identification markings known by everyone?

b. Do medical and maintenance personnel know the routes between train units?

5. Command and Signal.

a. Command.

(1) What are the locations of the commander and key staff members?
(2) What is the chain of command and the succession of command?

b. Signal.

(1) Do instructions include backup code words and visual signals for all special and emergency events?

(2) Do instructions cover how to identify friendly forces to aircraft?

(3) Are SOI distributed to all units with a need to know, such as higher, lower, adjacent, leading, and follow-on elements?

Appendix B

Standing Operating Procedures

The intent of this appendix is to provide an outline as an aid for developing a squadron tactical SOP.

I. GENERAL

A. Purpose: This tactical SOP prescribes standard procedures for use during all combat situations and provides a comprehensive reference for conducting operations in a field environment.

B. Conformity: All assigned, attached and OPCON personnel will read and comply with the provisions of this tactical SOP.

II. BATTLE COMMAND

A. Command.

1. Organization.

a. Succession of command.

b. Cues for assuming command.

c. Operation of the command post.

(1) Shifts.

(2) Displacement/set-up/tear-down.

2. Troop-leading procedures.

a. Estimate input (checklist).

b. Precombat inspection (checklist).

c. Backbriefs.

d. Rehearsals.

e. Combat orders.

(1) Formats.

(2) Preparation.

(3) Reproduction.

(4) Dissemination.

3. Coordination with adjacent units (checklist).

4. Liaison with main body elements (checklist).

B. Control.

1. Combat graphics and symbols.

2. Control measures designation (numbering system).

3. Terrain index reference system (TIRS).

4. Operational terms.

5. Vehicle identification marking system.

6. Unit recognition signals.

7. Communication.

a. Net diagrams.

b. Fixed call signs.

c. Brevity codes/cue words.

d. Antijamming actions.

e. Alternate means.

8. Reports.

a. Battle.

b. Intelligence.

c. Logistical.

III. MANEUVER

A. Readiness condition (REDCON).

B. Quartering party.

C. Road marches.

D. Assembly areas.

E. Squadron organization for combat.

F. Formations.

G. Battle plays.

1. Actions on contact.

2. Fix and bypass.

3. Close assault.

4. In-stride breach.

5. Bridges/defile.

6. Formation changes.

7. Passage of lines.

8. Hasty defense.

9. Consolidate on the objective.

10. Ambush.

11. Other plays.

H. Reconnaissance operations.

1. Zone reconnaissance.

a. Graphics.

b. Critical tasks.

2. Area reconnaissance.

a. Graphics.

b. Critical tasks.

3. Route reconnaissance.

a. Graphics.

b. Critical tasks.

I. Security operations.

1. Screen.

a. Graphics.

b. Critical tasks.

2. Area security.

a. Route.

b. Convoy.

J. Offensive operations.

1. Movement to contact.

2. Hasty attack.

3. Offensive graphics.

K. Defensive operations (defensive graphics).

L. Limited visibility operations (checklist).

M. Break in action (checklist).

1. Redistribution of ammunition, personnel, and equipment.

2. Evacuation of casualties and PWs.

3. Redistribution of ammunition under fire.

N. Relief in place.

1. Relief in place graphics.

2. Critical tasks.

O. Air-ground team.

1. Command relationship.

2. Formations/techniques of movement.

IV. MOBILITY AND SURVIVABILITY

A. Mobility.

1. Standard tasks.

2. Standard priorities.

B. Countermobility.

1. Standard tasks.

2. Standard priorities.

3. Engineer target turnover (checklist).

C. Survivability.

1. Standard tasks.

2. Standard priorities.

3. Fighting position construction.

D. NBC defense.

1. NBC team organization and equipment.

2. Unit NBC equipment.

3. Defense against nuclear attack.

4. Defense against chemical/biological attack.

5. Decontamination.

6. NBC reconnaissance.

V. FIRE SUPPORT

A. Fire support request sequence.

B. Fire support planning and execution matrix format.

C. Graphics.

VI. AIR DEFENSE

A. Air defense warning and cue words.

B. Weapons control status and cue words.

VII. INTELLIGENCE/ELECTRONIC WARFARE

A. Intelligence.

1. Standard tasks.

2. Standard priorities.

B. Electronic warfare.

1. Standard tasks.

2. Standard priorities.

3. Countermeasures.

VIII. LOGISTICS

A. Resupply procedures.

1. LOGPAC procedures.

2. Battle loss actions.

B. Combat Health Support.

1. Emergency medical treatment.

2. Medical evacuation.

3. Field hygiene and sanitation.

4. Combat stress control.

5. Routine sick call.

C. Maintenance support.

1. Battle damage assessment and repair.

2. Exchange criteria.

3. Cannibalization criteria.

4. Destruction criteria.

5. Maintenance repair time guidelines.

D. Personnel.

1. Replacements.

2. Accountability.

3. Personnel actions.

E. Combat trains.

1. Layout.

2. Operations.

IX. SAFETY

Appendix C

Operations Order

Operations order format

A sample OPORD is outlined below.

1. Situation.

a. Enemy.

(1) Weather (note effects - include light data).

(a) Light data.
(b) Weather forecast for operation.
(c) Effects of weather and light data on operations.

- Trafficability.
- Visibility.
- Effect on lasers/thermals.
- Effect on air operations.

(2) Terrain.

(a) Obstacles, hills, valleys, road types and conditions, streams, rivers, bridges, towns.
(b) Avenues of approach.

- Size unit supported.
- Start and end point.
- Objective.

(c) Key terrain.

(d) Observation.
(e) Cover and concealment.
(f) Engagement areas.
(g) Effect of terrain on the operation.

(3) Enemy forces.

(a) Identification of enemy forces.

(b) Activity of enemy forces.
(c) Location of enemy units.
(d) Disposition of enemy forces.
(e) Strength of enemy forces.
(f) Composition of enemy forces, to include type of equipment.
(g) Other enemy information critical to the upcoming operation, to include the following:

- Chemical and nuclear capabilities.
- Air defense artillery (ADA).
- Aviation, including helicopters.
- Electronic warfare.

(h) Enemy courses of action.
(i) Probable enemy course of action.

b. Friendly.

(1) Mission of higher headquarters (squadron) and commander's intent.
(2) Mission of adjacent units (left, right, front, rear).
(3) Mission of other organic units in higher headquarters.
(4) Mission of reserves in higher headquarters.
(5) Mission of supporting units who are in direct support/reinforcing (DS/R) to higher headquarters (field artillery, engineer, ADA).
(6) Which element from higher headquarters has priority of fires.
(7) Reinforcing units with a reinforcing/general support (R/GS) role to supporting units.
(8) Close air support (CAS) and number of sorties allocated to higher headquarters.

c. Attachments and detachments to the troop.

2. Mission. The *who, what, when, where,* and *why* for troop/company. State the essential task or tasks to be accomplished by the entire unit, to include on-order missions. Clearly define the troop's objective.

3. Execution.

a. Concept of the Operation. This paragraph explains the commander's intent by stating the *purpose*, *method*, and *endstate* of the operation. The *purpose* tells the "why" of the operation. The *method* tells how the commander visualizes achieving success with respect to the troop as a whole and the utilization of any combat multipliers, in general terms. The *endstate* tells what the final disposition of forces will be and how the endstate will facilitate future operations. For example--

We will conduct the reconnaissance to determine the location, disposition, and composition of enemy main defensive belt in order to facilitate movement through zone, and penetration of the main defensive belt by 52d ID (Mech), allowing 10th Corps to continue offensive operations to the north. We will conduct a forward passage of lines, and perform zone reconnaissance in a troop vee with the engineer platoon conducting a detailed route reconnaissance of Route Cherry behind the scout platoons. Initially movement will be rapid and aggressive until we make contact with the enemy main defensive belt, then transition to a more deliberate, detailed reconnaissance. Indirect fires will be used to provide immediate suppression and assist in the development of the situation by scout platoons. Success is determining the disposition, composition, and location of the enemy main defensive belt then determining the axis and point for the 52d ID (Mech) penetration of the enemy's defense. Our endstate is the troop screening along the enemy main defensive belt ready to pass elements of the 52d ID (Mech) forward for their attack.

(1) Scheme of Maneuver. How major units (maneuver and combat support) will be employed; including movement techniques, bypass criteria (if any), and engagement/disengagement criteria. See the following example.

C Troop moves at 130430 from TAA Viper along Route Dog in troop column. Order of march on Route Dog is Red, Blue, Mortars, FIST, Engineers, Black 6, White, Green, TOC, and combat trains.

C Troop crosses LD at 0500 with Red on left (south), Blue on right (north), Mortars, Engineers, White, Green, TOC, and combat trains moving center of zone (troop vee).

Scout platoons reconnoiter in zone abreast, establishing contact at designated troop internal and flank contact points. Upon enemy contact, develop the situation at platoon level, initially with indirect fire, and report. Destroy squad- and section-sized light-armored and soft-skinned vehicles, within capability. Anything beyond capability, be prepared to assist hasty attack by White and/or Green. Bypass only enemy dismount elements squad-sized or smaller. Bypass all built-up areas larger than 1 kilometer square not on Route Cherry. Upon contact with enemy main defensive belt, conduct dismounted patrols to ascertain enemy disposition, composition, orientation, and location of obstacles.

Mortars cross LD and establish MFP 10 following Blue. Maneuver from MFP 11 to MFP 12 through MFP 25 in sequence. Maintain ability to range at least one-third maximum range forward of lead scouts. Key movement on scouts reporting crossing of phase lines throughout zone.

Engineers cross LD following the FIST and conduct a detailed route reconnaissance of Route Cherry from SP to RP. Maneuver on Route Cherry approximately 500 meters rearward of scouts. Upon identification of enemy's main defensive belt, be prepared to assist scouts with obstacle reconnaissance. White and Green cross LD in platoon columns, White leading. Deploy into platoon wedges forward of LD, terrain permitting. Maneuver in center of troop zone, Green following White by 500 meters. White and Green maneuver approximately one phase line behind Red and Blue through hide positions T-44 through T-52 in sequence. Key movement on scouts reporting crossing of phase lines throughout zone. Scouts will assist in positioning tanks for conduct of hasty attacks by fire or by fire and maneuver upon approval of the troop commander.
TOC and combat trains cross LD in column, following Green. TOC maneuvers center troop zone, establishing positions H-44 through H-48 in sequence. Remains approximately two phase lines rearward of lead scouts. Trains move center troop zone in column, establishing positions T-27 through T-32 in sequence. Remain approximately two phase lines rearward of lead scouts. Key movement on scouts reporting crossing of phase lines throughout zone.

(2) Fires.

(a) Purpose for field artillery and mortar fires. (How will indirect fires support the scheme of maneuver?)

(b) Preparation starting time, duration and description of any fires landing in the area of operations.
(c) Allocation of final protective fires (FPF).
(d) Which element will have priority of fires.
(e) Number of priority targets allocated and who will control them.
(f) Special fires, restrictions, allocation/use of smoke, illumination, or CAS.

(g) Description of scheduled fires (offensive).
(h) Reference to fire support or target annexes.
(3) Obstacles, mines, fortifications.

(a) Priority of engineer effort.

(b) Priority of engineer mission.
(c) Obstacle overlay.
(d) Obstacle list.
(e) Logistical constraints.
(f) On-order missions.
b. Specific tasks to subordinate units. List specific missions in "battle sequence" for each subordinate unit, including attached units. Include movement techniques, flank coordination requirements, other details, and any "be-prepared" missions.

c. Coordinating instructions.
(1) Movement instructions.

(2) Time schedule of events.

- Rehearsal.
- Backbrief times.
- Precombat inspection time.
- First movement time.

(3) Passage of lines.

- Contact points.
- Passage points.
- Lanes and identification.

(4) Mission-oriented protection posture (MOPP) level.

(5) Operational exposure guide (OEG).
(6) Actions on contact.
(7) Actions at danger areas.
(8) Priority intelligence requirements (PIR) and intelligence requirements (IR).
(9) Effective time of attachment or detachment.
(10) Rally points.
(11) Rules of engagement (ROE).
(12) Priority targets for direct-fire weapons.
(13) Uniform and equipment, to include weapon and ammunition.
(14) Any changes regarding battlecarry and battlesight ranges.
(15) Air defense posture and weapons control status.
(16) Any instructions not provided in concept of the operation or specific instructions.

4. Service Support.
a. General.

(1) Organization of trains (squadron and troop).

- Combat.
- Field.

(2) Location of trains (initial and subsequent).

(3) Movement of trains (movement instructions).

b. Materiel and services.
(1) Supply.

- Class I: Time and type.
- Class III: Time, location, and method.
- Class V: Time, amount, and type.
- LOGPAC instructions (include squadron logistics release point [LRP]).

(2) Transportation. Location of main supply route.

(3) Services.

- Mortuary affairs.
- Clothing exchange and bath.
- Maintenance.

c. Medical evacuation and treatment.
(1) Location of squadron aid station.

(2) Displacement of squadron aid station.
(3) Location of regimental clearing station.
(4) Aero-medical evacuation information.
(5) Location of ambulance exchange points.
(6) Handling of contaminated wounded personnel.

d. Personnel.
(1) Prisoner of war (PW) handling and disposition instructions.

(2) PW guard instructions.
(3) Location of PW collection point.
(4) Instructions for interaction with local civil populace (ROE).
(5) Number of expected replacements.
(6) Cross-leveling procedures.
e. Miscellaneous.

5. Command and Signal.

a. Command.

(1) Location of troop commander/troop TOC/squadron command post.
(2) Succession of command.

b. Signal.

(1) SOI index and edition in effect.

- Key frequencies.
- Key call signs.
- Current item number identifier.

(2) KY-57 fill and change over.

(3) Current sign and countersign.
(4) Code words.
(5) Action if jamming or "hot mike" occurs.
(6) Pyrotechnics use.
(7) Periods and conditions of listening silence.

Appendix D

Operations Other Than War

Operations other than war are military activities during peacetime and conflict that do not necessarily involve armed clashes between two organized forces. A cavalry troop will conduct reconnaissance, security, offensive, defensive, and other tactical operations in support of operations other than war. Some examples of operations other than war that a cavalry troop may be involved in are as follows:

PEACETIME CONFLICT
Counterdrug Strikes and raids

Disaster relief Peace enforcement

Civil support Support to insurgency

Peace-building Anti-terrorism

Nation assistance Peacekeeping

Noncombatant evacuation

operations (NEO)

The primary operational constraints during operations other than war are defined by the rules of engagement (ROE). These rules range from the use of force for self-defense only to total impartiality when applying force. The rules of engagement for peacetime will be more restrictive than those for conflict. The troop commander should consider the following in relation to the rules of engagement:

- The degree of force used must only be sufficient to achieve the task at hand and prevent, as far as possible, loss of human life or serious injury.
- Leaders at all levels must ensure that the ROE do not limit the soldiers' ability to defend themselves.
- Leaders must never tape over magazines to keep soldiers from chambering rounds.
- Leaders must ensure all soldiers understand what conditions must be met before chambering rounds.
- The ROE must be realistic, simple, and easy to understand. Develop a single card that clearly outlines the ROE and issue it to all soldiers for reference, keeping in mind that the card in itself is not the answer. Soldiers must know and understand the ROE.
- During peacekeeping, forces have no mandate to prevent violations of peace agreements by the active use of force. (Observe and report only.) To maintain the peace, units may be positioned

between belligerents. Troop commanders must realize that soldiers are being placed at risk. The protection of the troop must be emphasized.

- Peace enforcement missions allow the active use of force. The peacetime ROE resemble the ROE for hostilities (wartime).
- The formulation of ROE should consider the cultural differences of multinational forces.
- Soldiers should be trained in the ROE, using tactical vignettes or simulated events. Train soldiers to avoid unnecessary collateral damage to property.

Actions on contact and the troop's ability to develop the situation will be affected by the rules of engagement for the theater in which the troop is operating. The troop commander should rehearse battle drills based on those rules.

In preparing for operations other than war, a troop may require specialized training. They could benefit from training in the following areas:

- Nature of peacekeeping.
- Regional orientation/culture of belligerents.
- Negotiating skills.
- Mine/booby trap/unexploded ordnance training.
- Checkpoint operations.
- Investigating and reporting.
- Information collection.
- Patrolling.
- Media inter-relationships.
- Establishing a buffer zone.
- Supervising a truce or cease-fire.
- Contributing to maintenance of law and order.
- Demilitarizing cities or geographical areas.
- Border surveillance/security.

Glossary

AAR	after-action review
A&O	assault and obstacle
ACE	armored combat earthmover
ACR	armored cavalry regiment
ACT	air cavalry troop
ADA	air defense artillery
A/L	administrative/logistics
ALO	air liaison officer
ALOC	administrative/logistics operations center
AO	area of operations
APERS	antipersonnel (ammunition)
AT	antitank
APC	armored personnel carrier
arty	artillery
AVLB	armored vehicle launched bridge
BHL	battle handover line
BP	battle position
BUA	built-up area
C2	command and control
cal	caliber
CAS	close air support
cbt	combat
CBU	cluster bomb unit
CEOI	communications-electronics operation instructions
CEV	combat engineer vehicle
CDR	commander
CFV	cavalry fighting vehicle
cGy/hr	centigray(s) per hour
CHEMWARN	chemical warning (message sent in the event of a friendly strike)
COA	course of action
COMSEC	communications security
COT	counter obstacle team
CP	command post
CS	combat support
CSS	combat service support

CTCP	combat trains command post
DA	Department of the Army
DS	direct support
DS/R	direct support/reinforcing
EA	engagement area
engr	engineer
EPW	enemy prisoner of war
evac	evacuate; evacuation
EW	electronic warfare
1SG	first sergeant
FA	field artillery
FASCAM	family of scatterable mines
FAC	forward air controller
FEBA	forward edge of the battle area
FIST	fire support team
FLOT	forward line of own troops
FM	frequency modulation (radio); field manual
FPF	final protective fires
FRAGO	fragmentary order
FS	fire support
FSCL	fire support coordination line
FSE	fire support element
FSO	fire support officer
FSN	fire support net
GLD	ground laser designator
GPS	global positioning system
GRREG	graves registration
GSR	ground surveillance radar
HC	heavy concentration (smoke)
HE	high explosive
HEMTT	heavy expanded mobile tactical truck
HHT	headquarters and headquarters troop
HMMWV	high mobility multipurpose wheeled vehicle
HQ	headquarters
hvy	heavy
ICM	improved conventional munitions

ID	identification
IPB	intelligence preparation of the battlefield
IR	intelligence requirements
KIA	killed in action
km	kilometers
LACR	light armored cavalry regiment
LC	line of contact
LD	line of departure
ldr	leader
LID	light infantry division
LO	liaison officer
LOA	limit of advance
LOGPAC	logistics package
LRP	logistics release point
lt	light
maint	maintenance
MBA	main battle area
MCOO	modified combined obstacle overlay
mech	mechanized
METT-T	mission, enemy, terrain, troops and time available
MFP	mortar firing point
MICLIC	mine clearing line charge
MOPP	mission-oriented protection posture
mort	mortar
MRB	motorized rifle battalion
MRC	motorized rifle company
MRE	meals ready to eat
MRP	motorized rifle platoon
MSR	main supply route
MTOE	modified table(s) of organization and equipment

NAI	named area(s) of interest
NBC	nuclear, biological, chemical
NCO	noncommissioned officer
NCOIC	noncommissioned officer in charge
NEO	noncombatant evacuation operations
NET	new equipment training
NPL	no penetration line
NUCWARN	nuclear warning (message sent in the event of a friendly strike)
obj	objective
OCOKA	observation and fields of fire, cover and concealment, obstacles, key terrain, avenues of approach (considerations in evaluating terrain as part of METT-T)
OEG	operational exposure guidance
OI	operations and intelligence
OIC	officer in charge
OP	observation post
OPCON	operational control
OPLAN	operation plan
OPORD	operations order
OPSEC	operations security
PAC	personnel and administration center
PD	point of departure
PEWS	platoon early warning system
PIR	priority intelligence requirements
PL	phase line
PLL	prescribed load list
plt	platoon
PMCS	preventive maintenance checks and services
POL	petroleum, oils, and lubricants
POSNAV	position navigation
PSG	platoon sergeant
PW	prisoner of war

recon	reconnaissance; reconnoiter
REDCON	readiness condition
RFA	restricted fire area
RFL	restricted fire line
R/GS	reinforcing/general support
ROE	rules of engagement
ROM	refuel on the move
RP	release point
rte	route
S1	adjutant (US Army)
S2	intelligence officer (US Army)
S3	operations and training officer (US Army)
S4	supply officer (US Army)
SCO	squadron commander
sec ldr	section leader
SEE	small emplacement excavator
SGLI	Serviceman's Group Life Insurance
sgt	sergeant
SIDPERS	Standardization Installation/Division Personnel System
SIGSEC	signal security
SITREP	situation report
SOI	signal operation instructions
SOP	standing operating procedure
SP	start point
sqd ldr	squad leader
TAA	tactical assembly area
TAC CP	tactical command post
TACFIRE	tactical fire direction system
TC	tank commander
TIRS	terrain index reference system
TOC	tactical operations center
TOE	table(s) of organization and equipment
TOW	tube-launched, optically-tracked, wire-guided (missile)
TRADOC	Training and Doctrine Command
TRP/trp	target reference point; troop
TTP	tactics, techniques, and procedures
UMCP	unit maintenance collection point

UMT	unit ministry team
US	United States
veh	vehicle
WIA	wounded in action
wrkr	wrecker
XO	executive officer

References

SOURCES USED

These are the sources quoted or paraphrased in this publication.

Army Publications

FM 17-15	Tank Platoon. 7 October 1987.
FM 17-95	Cavalry Operations. 19 September 1991.
FM 17-95-10	Armored Cavalry Regiment and Squadron. 22 September 1993.
FM 17-98	Scout Platoon. 9 September 1994.
FM 71-1	Tank and Mechanized Infantry Company. 22 November 1988.

Command Publications

Command publications cannot be obtained through Armywide resupply channels. Determine availability by contacting the address below.

Commander
USAARMC
ATTN: ATZK-IMO-RS
Fort Knox, KY 40121-5000

FKSM 17-97-3	Cavalry Troop Common SOP. 15 July 1994.

READINGS RECOMMENDED

These readings contain relevant supplemental information.

Army Publications

ARTEP 17-57-10-MTP	Mission Training Plan for the Scout Platoon. 27 December 1988.
ARTEP 17-237-10-MTP	Mission Training Plan for the Tank Platoon. 3 October 1988.
ARTEP 17-487-30-MTP	Mission Training Plan for the Regimental Armored Cavalry Troop. 3 September 1991.

DA Form 1155	Witness Statement on Individual. June 1966.
DA Form 1156	Casualty Feeder Report. June 1966.
DA Form 1971-R	Radiological Data Sheet Monitoring and Point Technique. September 1986.
DA Form 2028	Recommended Changes to Publications and Blank Forms. February 1974.
DD Form 551	Record of Interment. August 1984.
DD Form 1077	Collecting Point Register of Deceased Personnel. July 1984.
FM 1-114	Tactics, Techniques, and Procedures for the Regimental Aviation Squadron. 20 February 1991.
FM 1-116	Tactics, Techniques, and Procedures for the Air Cavalry/Reconnaissance Troop. 20 February 1991.
FM 3-3	Chemical and Biological Contamination Avoidance. 16 November 1992.
FM 3-4	NBC Protection. 20 May 1992.
FM 3-5	NBC Decontamination. 23 July 1992.
FM 3-50	Smoke Operations. 4 December 1990.
FM 3-100	NBC Defense, Chemical Warfare, Smoke, and Flame Operations.

	23 May 1991.
FM 3-101	Chemical Staffs and Units. 22 April 1987.
FM 5-36	Route Reconnaissance and Classification. 10 May 1985.
FM 5-100	Engineer Combat Operations. 22 November 1988.
FM 5-250	Explosives and Demolitions. 15 June 1992.
FM 6-20	Fire Support in the AirLand Battle. 17 May 1988.
FM 7-91	Tactical Employment of Antiarmor Platoons, Companies, and Battalions. 30 September 1987.
FM 7-98	Operations in a Low-Intensity Conflict. 19 October 1992.
FM 8-55	Planning for Health Service Support. 15 February 1985.
FM 10-63-1	Graves Registration Handbook. 17 July 1986.
FM 17-12-1-1	Tank Gunnery (Abrams), Volume I. 19 March 1993.
FM 17-12-1-2	Tank Gunnery (Abrams), Volume II. 19 March 1993.
FM 17-12-8	Light Cavalry Gunnery. 28 October 1994.
FM 17-18	Light Armor Operations. 8 March 1994.
FM 20-32	Mine/Countermine Operations. 30 September 1992.
FM 21-10	Field Hygiene and Sanitation. 22 November 1988.
FM 21-10-1	Unit Field Sanitation Team. 11 October 1989.
FM 21-11	First Aid for Soldiers. 27 October 1988.
FM 22-100	Military Leadership. 31 July 1990.

FM 23-1	Bradley Fighting Vehicle Gunnery. 1 March 1991.
FM 34-3	Intelligence Analysis. 15 March 1990.
FM 44-1	US Army Air Defense Artillery Employment. 9 May 1983.
FM 44-3	Air Defense Artillery Employment: Chaparral/Vulcan/Stinger. 15 June 1984.
FM 44-8	Small Unit Self-Defense Against Air Attack. 30 December 1981.
FM 44-18-1	Stinger Team Operations. 31 December 1984.
FM 71-2	The Tank and Mechanized Infantry Battalion Task Force. 27 September 1988.
FM 100-2-1	Soviet Army Operations and Tactics. 16 July 1984.
FM 100-2-3	The Soviet Army: Troops, Organization, and Equipment. 6 June 1991.
FM 100-5	Operations. 14 June 1993.
FM 101-5-1	Operational Terms and Symbols. 21 October 1985.
TC 6-40A	Field Artillery Automated Cannon Gunnery. 21 April 1989.

Command Publications

Contact the address shown on page References-1 for these command publications.

FKSM 17-15-3	Tank Platoon SOP. February 1989.

FKSM 17-98-3	Scout Platoon SOP. November 1994.
FKSM 17-98-4	Scout Platoon Leader's Notebook.

Standardization Agreements (STANAGs)

To requisition these documents, complete DD Form 1425 and send it to—

Naval Publications and Forms Center
5801 Tabor Avenue
Philadelphia, PA 19120

STANAG 2002	Warning Signs for the Marking of Contaminated or Dangerous Land Areas, Complete Equipment, Supplies and Stores. 2 December 1989.
STANAG 2014	Operation Orders, Warning Orders and Administrative/Logistics Orders. 14 August 1991.
STANAG 2029	Method of Describing Ground Locations, Areas and Boundaries. 10 June 1991.
STANAG 2036	Land Mine Laying, Marking, Recording and Reporting Procedures. 29 April 1988.
STANAG 2044	Procedures for Dealing with Prisoners of War (PW). 12 July 1988.
STANAG 2047	Emergency Alarms of Hazard or Attack (NBC and Air Attack Only). 10 October 1990.
STANAG 2082	Relief of Combat Troops. 24 July 1991.
STANAG 2083	Commanders' Guide on Nuclear Radiation Exposure of Groups.

	10 September 1990.
STANAG 2084	Handling and Reporting of Captured Enemy Equipment and Documents. 26 June 1986.
STANAG 2103	Reporting Nuclear Detonations, Biological and Chemical Attacks, and Predicting and Warning of Associated Hazards and Hazard Areas. 10 June 1991.
STANAG 2104	Friendly Nuclear Strike Warning. 12 December 1989.
STANAG 2112	NBC Reconnaissance. 5 August 1986.
STANAG 2113	Denial of a Unit's Military Equipment and Supplies to an Enemy. 11 July 1991.
STANAG 2129	Identification of Land Forces on the Battlefield. 16 May 1989.
STANAG 2889	Marking of Hazardous Areas and Routes Through Them. 8 February 1990.

FIELD MANUAL
No. 3-22.68

HEADQUARTERS
DEPARTMENT OF THE ARMY
Washington, DC, 31 January 2003

CREW-SERVED MACHINE GUNS, 5.56-mm AND 7.62-mm

CONTENTS

DISTRIBUTION RESTRICTION: Approved for public release; distribution is unlimited.

*This publication supersedes FM 23-67, 29 February 1984 and FM 23-14, 26 January 1994.

PREFACE

This manual provides technical information, training techniques, and guidance on the crew-served machine guns, 5.56-mm and 7.62-mm (M249/M60/M240B). The purpose of this manual is, to provide a **one-source document for all three weapons.** This prohibits having several sources to rely on. Unit leaders, trainers, and the designated gunners will find this information invaluable in their efforts to successfully integrate these automatic weapons into their combat operations.

Trainers must ensure that safety procedures are observed at all times. Leaders, trainers, and soldiers must remember that safety is everyone's responsibility. All training should be conducted as though the weapon is fully loaded. At no time while using this manual does speed or accuracy override the safety procedures.

Unless this publication states otherwise, masculine nouns and pronouns do not refer exclusively to men.

The proponent of this publication is the United States Army Infantry School. Send comments and recommendations on DA Form 2028 directly to Commandant, U.S. Army Infantry School, ATTN: ATSH-IN-S3, Fort Benning, GA 31905-5596, or send email to doctrine@benning.army.mil.

CHAPTER 1

M249 MACHINE GUN

The 5.56-mm M249 machine gun supports the soldier in both the offense and defense. The M249 provides a medium volume of close and continuous fire the soldier needs to accomplish the mission. With it, units can engage the enemy along with the capability of individual weapons with controlled and accurate fire. The medium-range, close defensive, and final protective fires delivered by the M249 MG form an integral part of a unit's defensive fires. Although the M249 MG is described here as a machine gun, it also plays the role of the automatic rifleman. This FM or chapter supersedes FM 23-14, which describes the M249 MG in the automatic rifle role. This chapter also describes the weapon and the types of ammunition in detail and provides a table of general data.

Section I. DESCRIPTION AND COMPONENTS

This section describes the M249 machine gun and its components and purposes. It also discusses the different types of ammunition that is fired from the M249 machine gun. This section describes how to install the blank firing adapter for the M249 machine gun and how to take care of the machine gun while using the blank firing adapter.

1-1. DESCRIPTION

The M249 machine gun is a gas-operated, air-cooled, belt or magazine-fed, automatic weapon that fires from the open-bolt position (Figure 1-1). It has a maximum rate of fire of 850 rounds per minute. Primarily, ammunition is fed into the weapon from a 200-round ammunition box containing a disintegrating metallic split-link belt. As an *emergency* means of feeding, the M249 machine gun can use a 20- or 30-round M16 rifle magazine but increases the chance of stoppages. This gun can be fired from the shoulder, hip, or underarm position; from the bipod-steadied position; or from the tripod-mounted position. See Table 1-1, on page 1-2, for general data.

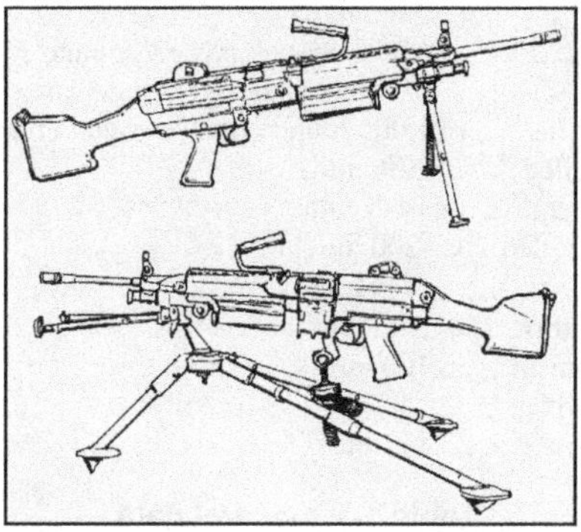

Figure 1-1. M249 machine gun, bipod and tripod mounted.

Ammunition	5.56-mm ball and tracer (4:1 mix) ammunition is packaged in 200-round drums, each weighing 6.92 pounds; other types of ammunition available are ball, tracer, blank, and dummy.
Tracer burnout	900 meters (+)
Length of M249	40.87 inches
Weight of M249	16.41 pounds
Weight of tripod mount M122 with traversing and elevating mechanism and pintle	16 pounds
Maximum range	3,600 meters
Maximum effective range	1,000 meters with the tripod and T&E
Area:	
Tripod	1,000 meters
Bipod	800 meters
Point:	
Tripod	800 meters
Bipod	600 meters
Suppression	1,000 meters
Maximum extent of grazing fire obtainable over uniformly sloping terrain	600 meters
Height of M249 on tripod mount M122A1	16 inches
Rates of Fire:	
Sustained	100 rounds per minute Fired in 6- to 9-round bursts with 4 to 5 seconds between bursts (change barrel every 10 minutes)
Rapid	200 rounds per minute Fired in 6- to 9-round bursts 2 to 3 seconds between bursts (change barrel every 2 minutes)
Cyclic	650 to 850 rounds per minute Continuous burst (change barrel every minute)
Basic load, ammunition	1,000 rounds (in 200-round drums)
Elevation, tripod controlled	+200 mils
Elevation, tripod free	+445 mils
Depression, tripod controlled	-200 mils
Depression, tripod free	-445 mils
Traverse, controlled by traversing and elevating mechanism	100 mils
Normal sector of fire (with tripod)	875 mils

Table 1-1. General data.

1-2. COMPONENTS

The components of the M249 machine gun and their purposes are described in Table 1-2 and shown Figure 1-2. The item numbers in Table 1-2 correspond to the callout numbers in Figure 1-2. The sights and safety button are shown in Figures 1-3 and 1-4. (See Table 1-1 for general data.)

COMPONENTS		PURPOSES
(1)	Barrel assembly	Houses cartridges for firing, directs projectile, and supports the gas regulator.
(2)	Heat shield assembly	Provides protection for the gunner's hand from a hot barrel.
(3)	Rear sight assembly	Adjusts for both windage and elevation.
(4)	Cover and feed mechanism assembly	Feeds linked belt ammunition, and positions and holds cartridges in position for stripping, feeding, and chambering.
(5)	Feed tray assembly	Positions belted ammunition for firing.
(6)	Cocking handle assembly	Pulls the moving parts rearward. Moves in a guide rail fixed to the right side of the receiver.
(7)	Buttstock and buffer assembly	Contains a folding buttplate. Serves as a shoulder support for aiming and firing M249. Contains a folding shoulder rest and a hydraulic buffer to absorb the recoil.
(8)	Bolt assembly	Provides feeding, stripping, chambering, firing, and extraction, using the projectile gases for power.
(9)	Slide assembly	Houses firing pin and roller assembly.
(10)	Return rod and transfer mechanism assembly	Absorbs recoil for bolt and operating rod assembly at the end of recoil movement.
(11)	Receiver assembly	Serves as a support for all major components and houses action of weapon. Through a series of cam ways, controls functioning of weapon.
(12)	Trigger mechanism	Controls the firing of the weapon. Provides storage area for lubricant in grip portion.
(13)	Handguard assembly	Provides thermal insulation to protect the gunner's hand from heat or extreme cold and houses the cleaning equipment.
(14)	Sling and snap hook assembly	Provides a means of carrying the weapon.
(15)	Bipod	Supports the M249 machine gun in the prone position. The telescopic legs can be individually adjusted to three different lengths.

Table 1-2. Components and purposes.

COMPONENTS		PURPOSES
(16)	Gas cylinder assembly	Locks bipod in place and provides passageway for gases.
(17)	Piston assembly	Holds the bolt and slide assemblies and houses the return spring.
(18)	Return spring	Returns bolt, slide, and piston assemblies to locked position during counterrecoil cycle.
(19)	Tripod (M122) (Not shown)	The tripod and T&E mechanism, with pintle, provides a stable mount and permits a higher degree of accuracy and control.
(20)	M145 straight telescope (Not shown)	Machine gun optic (MGO) provides target acquisition and identification at greater ranges.

Table 1-2. Components and purposes (continued).

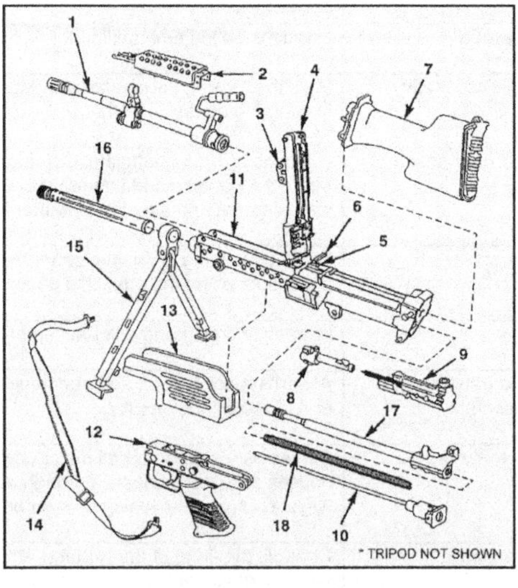

Figure 1-2. M249 machine gun components.

a. **Sights.** The M249 machine gun has a hooded and semifixed front sight (Figure 1-3). The rear sight assembly mounts on the top of the cover and feed mechanism assembly. The elevation knob drum has range settings from 300 meters to 1,000 meters. Range changes are made on the M249 machine gun sight by rotating the elevation knob to the desired range setting. Rotation of the rear sight aperture (peep sight) is used for fine changes in elevation or range adjustments, such as during zeroing. Each click of the peep sight. One click moves the sight 180 degrees, or one-half turn. This equals a one-half-mil change in elevation, which is .5 cm at 10 meters. The sight adjusts for windage by rotating the windage knob. Each click of windage adjustment also equals a one-half-mil change, which is .5 cm at 10 meters. There is also a windage sliding scale marked with index lines for centering the rear sight aperture.

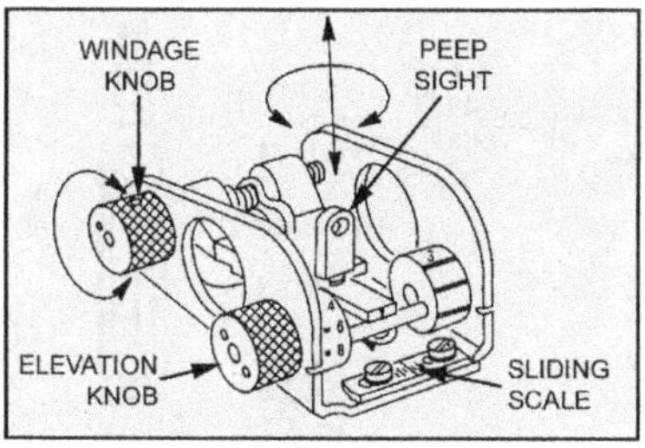

Figure 1-3. Sights.

b. **Safety**. The safety (Figure 1-4) is in the trigger housing. The safety is pushed from left to right (red ring not visible) to render the weapon safe, and the bolt cannot be released to go forward. The safety is pushed from right to left (red ring visible) to render the weapon ready to fire. The cocking handle on the right side of the weapon is used to pull the bolt to the rear.

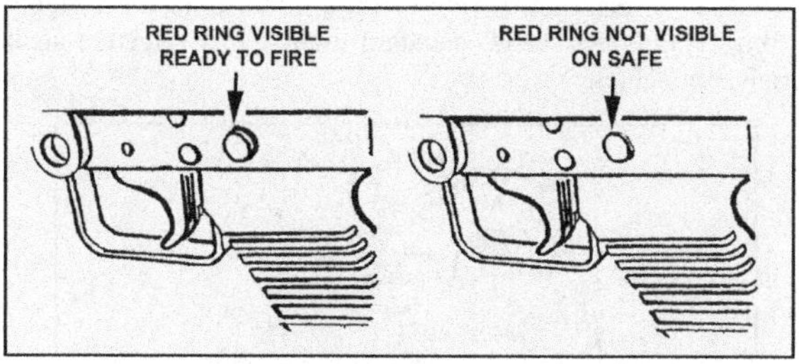

Figure 1-4. M249 machine gun safety.

1-3. AMMUNITION

The M249 machine gun uses several different types of 5.56-mm standard military ammunition. Soldiers should use only authorized ammunition that is manufactured to US and NATO specifications. The 5.56-mm NATO cartridge is identified by its appearance, the painted projectile tips, the stamped manufacturer's initials and year of manufacture on the base of the cartridge case, and the markings on the packing containers. When removed from the original packing container, the cartridge can be identified by its physical characteristics. The M193 and M196 cartridge for the M16 can be fired with the M249, but accuracy is degraded; therefore, it should only be used in emergency situations when M855 or M856 ammunition is not available.

a. **Type and Characteristics**. The specific types of ammunition (Figure 1-5) and its characteristics are as follows.

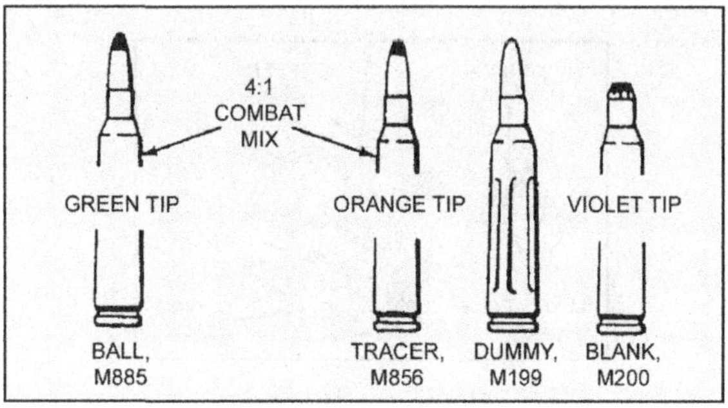

Figure 1-5. Cartridges for the M249.

(1) *Cartridge, 5.56-mm ball M855 (A059).* The M855 cartridge has a gilding, metal-jacketed, lead alloy core bullet with a steel penetrator. The primer and case are waterproof. This ammunition is linked by a disintegrating metallic split-linked belt so that the ammunition can feed from the ammunition box (Figure 1-6). In an emergency, the M855 round can also be fired from the M16A2, A3, or A4 when loaded in a 20- or 30-round magazine. It is identified by a green tip, has a projectile weight of 62 grains, and is 2.3 cm long. This is the NATO standard round. It is effective against personnel and light materials, not vehicles.

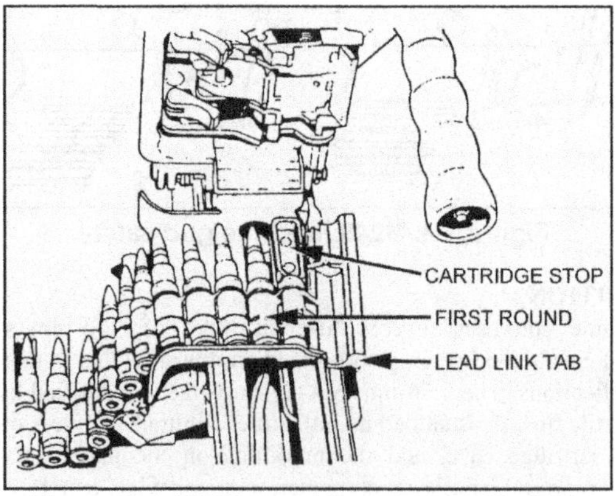

Figure 1-6. M855 cartridges in metallic belt.

(2) *Cartridge, 5.56-mm tracer, M856 (A064).* This cartridge has a projectile weight of 63.7 grains and lacks a steel penetrator. It is identified by an orange tip. The tracer is used for adjustments after observation, incendiary effects, and signaling. When tracer rounds

are fired, they are mixed with ball ammunition in a ratio of four ball rounds to one tracer round. The DODAC for ball and tracer mix is A064.

(3) *Cartridge, 5.56-mm dummy M199 (A060).* This cartridge can be identified by the six grooves along the side of the case beginning about one-half inch from its head. It contains no propellant or primer. The primer well is open to prevent damage to the firing pin. The dummy round is used during mechanical training, dry-fire exercises, and function checks.

NOTE: The 5.56-mm NATO cartridge may be identified by its appearance, the painting of projectile tips, the stamping of the manufacturer's initials and year of manufacture on the base of the cartridge case, and the markings on the packing containers. When removed from the original packing container, the cartridge can be identified by its physical characteristics. The M193 and M196 cartridge for the M16 can be fired with the M249, but accuracy is degraded; therefore, it should only be used in emergency situations when M855 or M856 ammunition is not available.

(4) *Cartridge, 5.56-mm blank M200 (M2 link, A075*). The blank cartridge has no projectile. The case mouth is closed with a seven-petal rosette crimp and has a violet tip. The original M200 blank cartridge had a white tip. Field use of this cartridge resulted in residue buildup, which caused malfunctions. Only the violet-tipped M200 cartridge should be used. The blank round is used during training when simulated live fire is desired. The M249 blank-firing attachment (NSN 1005-21-912-8997) must be used to fire this ammunition. (See paragraph 1-4.)

DANGER

DO NOT FIRE BLANK AMMUNITION AT ANY PERSON WITHIN 20 FEET BECAUSE FRAGMENTS OF A CLOSURE WAD OR PARTICLES OF UNBURNED PROPELLANT CAN CAUSE INJURY OR DEATH.

b. **Storage**. Ammunition is stored under cover. If ammunition is in the open, it must be kept at least 6 inches above the ground and covered with a double thickness of tarpaulin. The cover must be placed so that it protects the ammunition yet allows ventilation. Trenches are dug to divert water from flowing under the ammunition.

c. **Care, Handling, and Preservation**. Ammunition should not be removed from the airtight containers until ready for use. Ammunition removed from the airtight containers, particularly in damp climates, may corrode.

(l) Ammunition must be protected from mud, dirt, and moisture. If it gets wet or dirty, the ammunition must be wiped off before use. Lightly corroded cartridges are wiped off as soon as the corrosion is discovered. Heavily corroded, dented, or loose projectiles should not be fired.

(2) Ammunition must be protected from the direct rays of the sun. Excessive pressure from the heat may cause premature detonation.

(3) Oil should never be used on ammunition. Oil collects dust and other abrasives that may possibly damage the operating parts of the weapon.

d. **Packaging**. The ammunition can contains two plastic ammunition drums. Each drum contains 200 rounds and weighs 6.92 pounds. Dummy ammunition (M199) is packed in boxes of 20 rounds each.

1-4. BLANK FIRING ATTACHMENT

The M249 BFA is the only BFA (NSN 1005-21-912-8997) authorized for use with the M249 MG.

a. **Installation**. The BFA is attached to the M249 machine gun by using the three steps shown in (Figure 1-7).

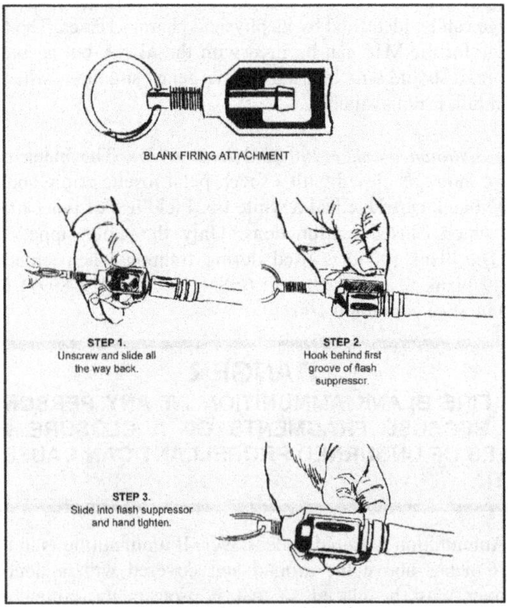

Figure 1-7. M249 blank firing attachment.

b. **Care of the M249 While Using the BFA**. A buildup of carbon inside the weapon causes friction between the moving parts. Carbon deposits build rapidly when blanks are fired. When these deposits become excessive, stoppages occur. Therefore, keeping the weapon—especially the gas system and chamber—clean, during blank firing is very important. To get the best performance with the BFA, the automatic rifleman performs the following:

(1) Inspects the weapon for damaged parts, excessive wear, cleanliness, and proper lubrication before firing.

(2) When feasible, test fires the weapon using ball ammunition before attaching the BFA.

(3) Adjusts the BFA to fit the weapon.

(4) Applies immediate action when stoppages occur.

(5) Cleans the gas system after firing 500 rounds.

(6) Cleans and lubricates the entire weapon after firing 1,000 rounds.

Section II. MAINTENANCE

Proper maintenance contributes to weapon effectiveness as well as to unit readiness. This section discusses the maintenance aspects of the M249 machine gun to include inspection; cleaning and lubrication; maintenance before, during, and after firing, and during NBC conditions. Associated tasks essential to maintenance (clearing, general assembly and disassembly, and function checks) are provided in detail.

1-5. CLEARING PROCEDURES

The first step in maintenance is to clear the weapon (Figure 1-8). This applies in all situations, not just after firing. The gunner must always assume the M249 machine gun is loaded. To clear the M249, the gunner performs the following procedures:

a. Moves the safety to the fire "F" position by pushing it to the left until the red ring is visible.

b. With his right hand, palm up, pulls the cocking handle to the rear, locking the bolt in place.

c. While holding the resistance on the cocking handle, moves the safety to the SAFE position by pushing it to the right until the red ring is not visible. (The weapon cannot be placed on safe unless the bolt is locked to the rear.)

d. Returns and locks the cocking handle in the forward position.

DANGER

WHEN OPENING THE FEED COVER ON A HOT GUN, MAKE SURE THE WEAPON IS ON THE GROUND AWAY FROM YOUR FACE. WITH THE WEAPON ON YOUR SHOULDER, POSSIBLE DEATH OR INJURY COULD OCCUR IF A ROUND GOES OFF WHEN THE COVER IS RAISED.

e. Raises the cover and feed mechanism assembly, and conducts the *five-point safety check* for brass, links, or ammunition:

(1) Checks the feed pawl assembly under the feed cover.

(2) Checks the feed tray assembly.

(3) Lifts the feed tray assembly and inspects the chamber.

(4) Checks the space between the bolt assembly and the chamber.

(5) Inserts two fingers of his left hand in the magazine well to extract any ammunition or brass.

f. Closes the cover and feed mechanism assembly and moves the safety to the "F" position. With his right hand, palm up, returns the cocking handle to the rear position. Presses the trigger and at the same time eases the bolt forward by manually riding the cocking handle forward.

CAUTION

The cocking handle must be manually returned to the forward and locked position each time the bolt is manually pulled to the rear.

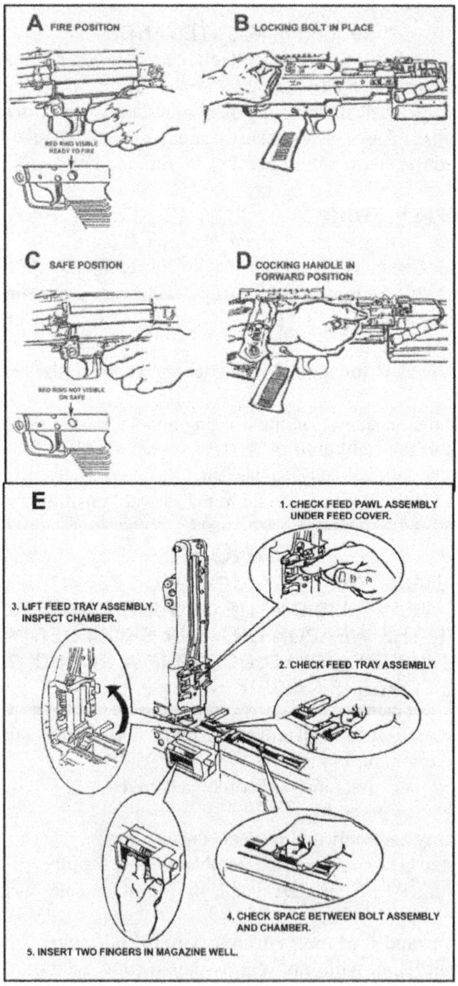

Figure 1-8. Clearing procedures.

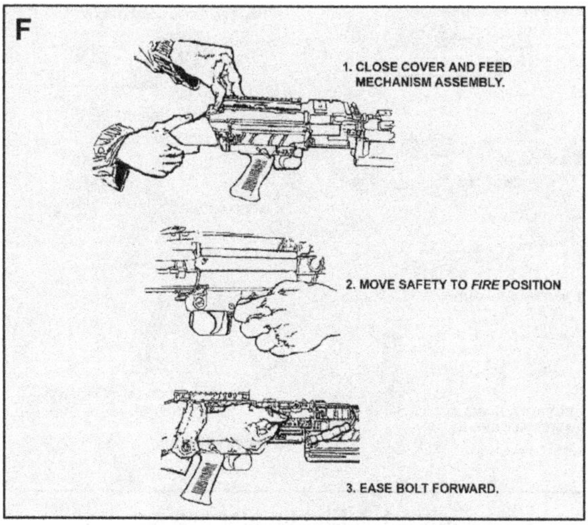

Figure 1-8. Clearing procedures (continued).

1-6. GENERAL DISASSEMBLY

General disassembly is removing and replacing the eight major groups (Figure 1-9, page 1-12). The unit armorer performs detailed disassembly. Disassembly beyond what is explained in this manual is prohibited, except by ordnance personnel. During general disassembly, each part is placed on a clean flat surface such as a table or mat. This aids in assembly in reverse order and avoids the loss of parts. Before disassembly, the bipod legs must be released from under the receiver and placed into the bipod mode position.

DANGER

BE SURE THE BOLT IS IN THE FORWARD POSITION BEFORE DISASSEMBLY. THE GUIDE ROD CAN CAUSE DEATH OR INJURY IF THE GUIDE SPRING IS RETRACTED WITH THE BOLT PULLED TO THE REAR.

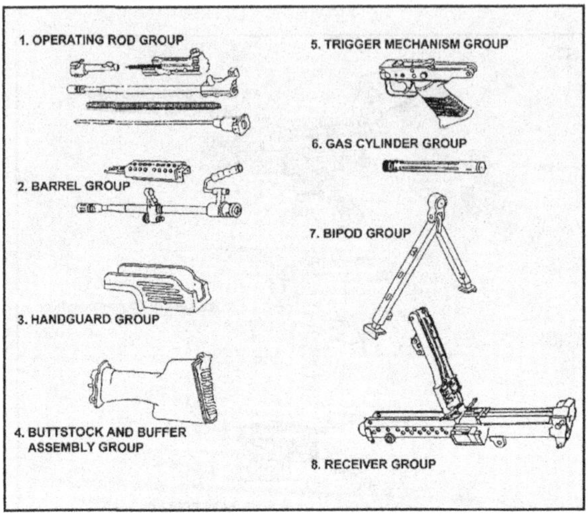

Figure 1-9. Eight major groups.

a. **Removing the Operating Rod Group.** The operating rod group (Figure 1-10), operating rod spring, slide assembly, piston assembly, and bolt assembly consists of the spring guide rod.

(1) To remove the operating rod, pull the upper retaining pin at the rear of the receiver to the left. Allow the buttstock to pivot downward and place it on a surface to support the weapon for disassembly.

(2) To release the operating rod assembly from the positioning grooves inside the receiver, hold the weapon with one hand on the buttstock assembly. Use the thumb of the other hand to push in and upward on the rear of the operating rod assembly.

(3) Pull the operating rod and spring from the receiver group and separate the parts.

(4) Hold the buttstock assembly with your left hand to stabilize the weapon. With your right hand, pull the cocking handle to the rear to lock the bolt. Return the cocking handle to the forward position. Place a finger on the face of the bolt and push until your finger makes contact with the bridge at the end of the receiver. This leaves the piston, slide, and bolt assemblies exposed.

(5) Hold the slide assembly while pulling the moving parts out the rear of the receiver. This leaves the piston, slide, and bolt assemblies exposed.

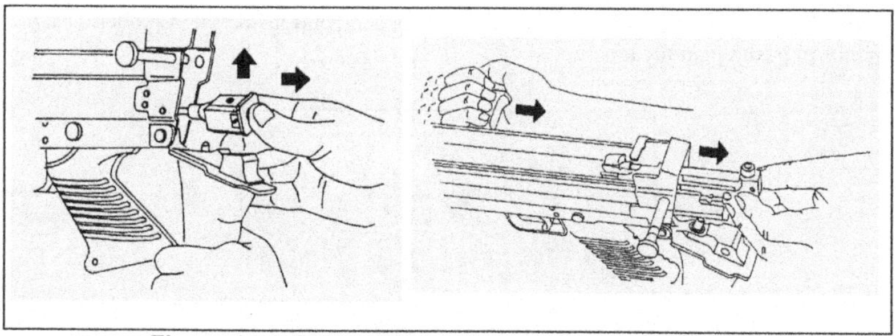

Figure 1-10. Removal of the operating rod group.

(6) To separate the operating rod group (Figure 1-11), hold the piston assembly in one hand, place your other hand on the bolt assembly, and rotate the bolt to disengage it bolt from the slide assembly. Remove the firing pin spring from the firing pin, but be careful not to break the spring. If the spring sticks, rotate it clockwise to free it. The weapon will function without the spring, but this weakens the firing pin action. To separate the slide assembly from the piston, press the retaining pin at the rear of the slide assembly to the left and lift the slide assembly.

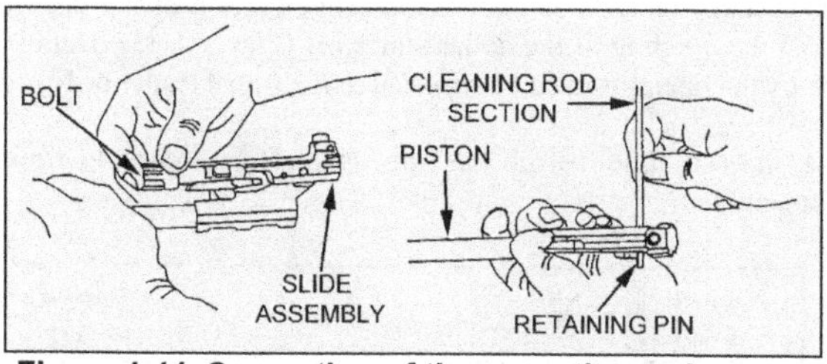

Figure 1-11. Separation of the operating rod group.

b. **Removing the Barrel Group.** The barrel group consists of barrel, heat shield, flash suppressor, front sight, gas regulator, and gas regulator collar. The following steps correspond to the callouts in Figure 1-12, page 1-14.

CAUTION

Barrels must not be interchanged with those from other M249s unless the headspace has been certified for that weapon by direct support personnel.

(1) To remove the barrel from the receiver, close the cover and feed mechanism assembly. Depress the barrel locking lever with your left hand, then lift the carrying handle using your right hand and push the barrel forward. To remove the heat shield, place the barrel with the muzzle end on a hard, flat surface, with the heat shield facing

away from your body. Place the index fingers of each hand inside the chamber. Use your thumbs to push up on the top clip.

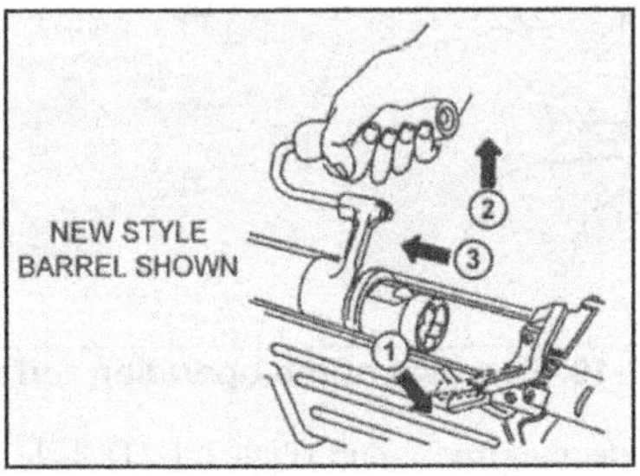

Figure 1-12. Removal of the barrel.

(2) Raise the feed cover.

(3) To remove the gas regulator and collar, rotate the gas collar pin out of the notch. Place the tip of the scraper with the concave side facing the pin of the collar inside the notch. (Be careful not to use too much pressure, so as not to break the tip of the scraper.) Rotate the collar counterclockwise over the concave portion of the tip, which is on the scraper, and past the notch until the collar slides off (Figure 1-13). Deficiencies that are not correctable by the operator must be reported to the squad leader or NCOIC.

NOTE: The newest style barrel has an internal gas system, which cannot be disassembled.

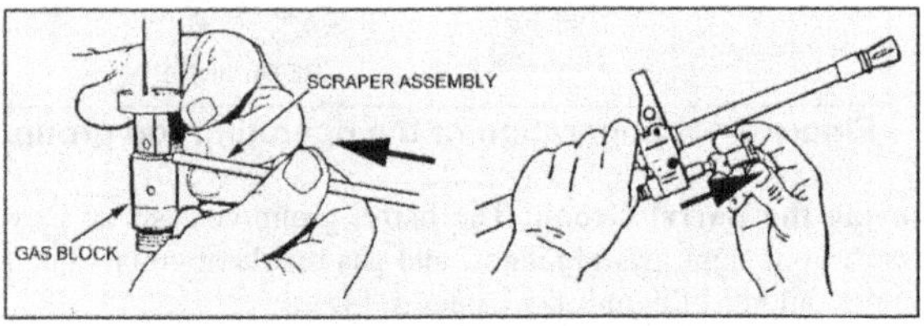

Figure 1-13. Removal of the collar.

(5) To remove the gas regulator (Figure 1-14), separate it from the gas block.

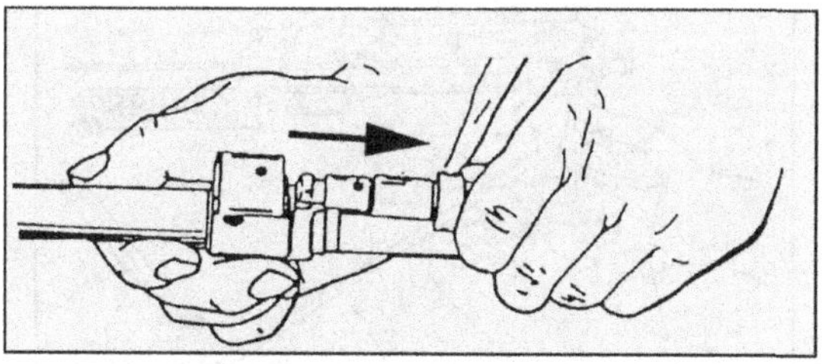

Figure 1-14. Removal of the gas regulator.

c. **Removing the Handguard Group.** The handguard group (Figure 1-15) consists of the handguard, handguard retaining pin, and cleaning equipment retaining clip. Push the handguard retaining pin to the left using a cartridge or the spring guide rod; then pull the handguard down.

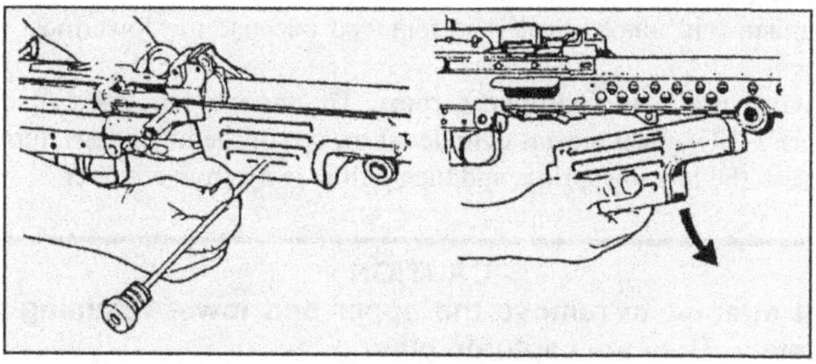

Figure 1-15. Removal of the handguard.

CAUTION

Do not attempt to remove the handguard retaining pin completely. It is a captured pin.

d. **Removing the Buttstock and Buffer Assembly Group**. To remove the buttstock and buffer assembly (Figure 1-16, page 1-15), use a cartridge or the spring guide rod to push the lowermost retaining pin on the rear of the receiver to the left. It is a captured pin; it is not removable. Remove the buttstock and shoulder assembly by pulling them rearward, while supporting the trigger mechanism.

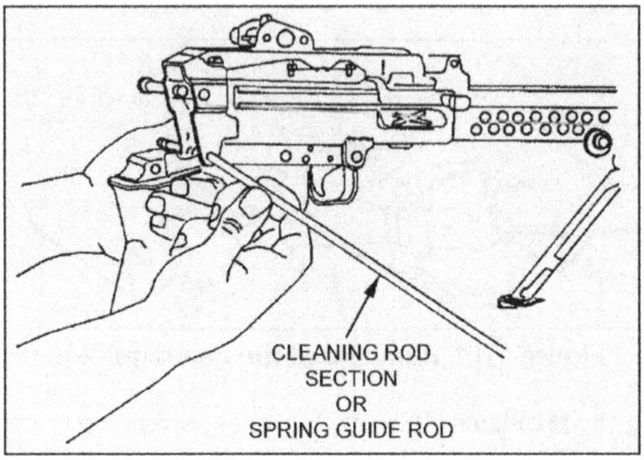

Figure 1-16. Removal of the buttstock and buffer assembly.

e. **Removing the Trigger Mechanism Group**. After the release of the support, the trigger mechanism will automatically be removed because the lowermost retaining pin holds it on.

f. **Removing the Gas Cylinder Group**. To remove the gas cylinder from the receiver (Figure 1-17), grasp the gas cylinder at the top of the bipod legs, turn it to the left or right to release the locking spring, and then pull it away from receiver.

CAUTION
Do not attempt to remove the upper and lower retaining pins completely. They are captured pins.

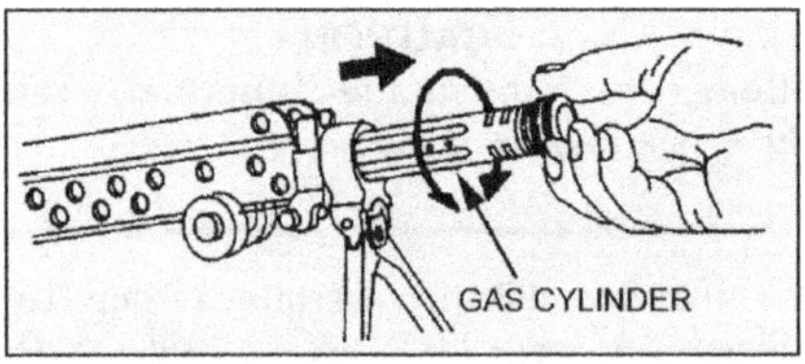

Figure 1-17. Removal of the gas cylinder group.

g. **Removing the Bipod Group**. Once the gas cylinder group is removed, remove the bipod group (Figure 1-18) by pulling it away from the receiver.

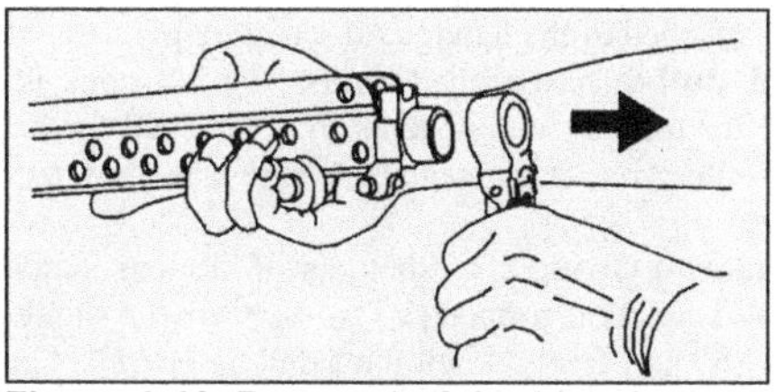

Figure 1-18. Removal of the bipod group.

h. **Removing the Receiver Group**. Once the bipod group is removed, the part remaining is the receiver group, and disassembly is complete.

1-7. INSPECTION

Inspection begins with the weapon disassembled in its major groups. Shiny surfaces do not mean the parts are unserviceable. The parts of the weapon and related equipment are inspected. Any broken or missing parts are repaired or replaced IAW TM 9-1005-201-10. The gunner performs preventive maintenance checks and services (PMCS) every 90 days. If the weapon has not been used in 90 days, PMCS is performed as stated in the operator's manual. If rust is seen on the weapon, perform PMCS immediately:

a. **Operating Rod Group**. The operating rod should not be bent, broken, or cracked. The buffer spring should not have breaks. Lug pins should protrude equally on both sides of the buffer spacer. The operating rod spring should not have kinks or separated strands or broken strands. It can have a maximum of one break on any one strand.

(1) Check the bolt assembly for visible damage. The cartridge extractor should not be cracked or chipped.

(2) Check the slide assembly for visible damage. Check the feed roller for spring tension when compressed and to ensure that the pivot slide is locked onto the slide assembly.

(3) Check the firing pin for straightness and cracks. Ensure the tip is completely rounded.

(4) Ensure the firing pin spring is not crushed or bent. Ensure the beveled end is not stretched.

(5) Check the sear notch on the piston assembly for signs of excessive wear or burring. Slight rotation of the piston on its housing is normal and is not cause for rejection.

b. **Barrel Group**. The flash suppressor should not be cracked, and it should be fastened securely. The front sight post and front sight base must not be bent, cracked, or broken. Weapons already zeroed should not be adjusted. The heat shield assembly is inspected for damage, cracks, or broken retaining clamps. The gas regulator and collar are checked for cracks or burrs. The barrel is checked for bulges, cracks, bends, obstructions, or pits in the chamber or bore. The gas plug is checked for obstructions, cracks, and bulges. The carrying handle is checked to ensure it is not cracked, broken, or missing; that it can be folded under spring pressure to the right and left; and that it remains locked in an upright position.

c. **Handguard Group**. The handguard should not be cracked or broken. The retaining clip must be attached to the handguard retaining pin.

d. **Buttstock and Buffer Assembly Group**. The buttstock is checked for cracks, bends, or breaks; and for missing components. It is checked for linkage and tension on the buffer rod. The shoulder rest is checked to ensure it is not bent or broken and that it locks in both positions.

e. **Trigger Mechanism Group**. The shoulder of the sear should not show excessive wear. The safety should function properly. That is, the sear should move only slightly when the safety is on "S" and freely when the safety is on "F". The sear pin should not protrude from the trigger mechanism, because, if it does protrude, the trigger mechanism will not go back in place.

f. **Gas Cylinder Group**. The gas cylinder should not be cracked, bent, or broken.

g. **Bipod Group**. The bipod group should not be cracked, bent, or broken. The bipod legs should extend and collapse easily.

h. **Receiver Group**. The cover latch should work properly. All parts inside the cover assembly should move under spring tension. All spotwelds are checked for cracks. The cover assembly should remain open without support. The belt-holding pawl must be under spring tension. The receiver should not be bent or cracked. The cocking handle should slide freely within its guide and lock in its forward position. The windage and elevation knobs on the rear sight should be movable and legible. The windage scale screws should not be worn or burred.

1-8. CLEANING, LUBRICATION, AND PREVENTIVE MAINTENANCE

The M249 machine gun should be cleaned immediately after firing. It should be disassembled into its major groups before cleaning. After it has been cleaned and wiped dry, a thin coat of CLP is applied by rubbing with a cloth. This lubricates and preserves the exposed metal parts during all normal temperature ranges. When not in use, the M249 should be inspected weekly and cleaned and lubricated when necessary.

a. **Cleaning**. All metal components and surfaces that have been exposed to powder fouling should be cleaned using CLP on a bore-cleaning patch. The same procedure is used to clean the receiver.

CAUTION

When using CLP, no other type cleaner can be used. Never mix CLP with RBC or LSA.

(1) Clear and disassemble the weapon.

(2) Clean the bore and chamber using CLP and fresh swabs.

(3) Clean the gas regulator with the special tool (scraper). Remove all carbon dust. *Do not use CLP on the collar, gas block, or body.*

(a) Clean the gas-vent hole (Figure 1-19).

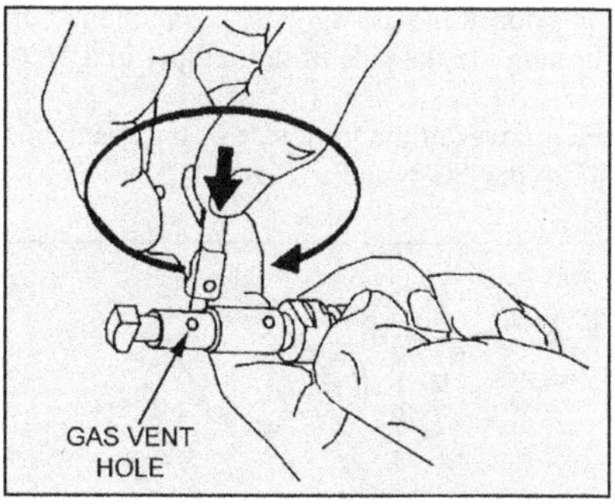

Figure 1-19. Cleaning of the gas vent hole.

(b) Clean the central hole with the appropriate part of the scraper by turning it clockwise and pushing it inward toward the bottom of the housing (Figure 1-20).

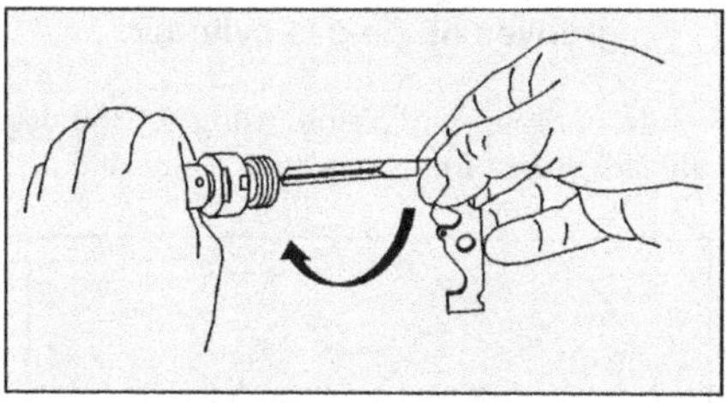

Figure 1-20. Cleaning of the central hole.

(c) Use the protruding tips of the scraper to clean the two grooves of the body (Figure 1-21).

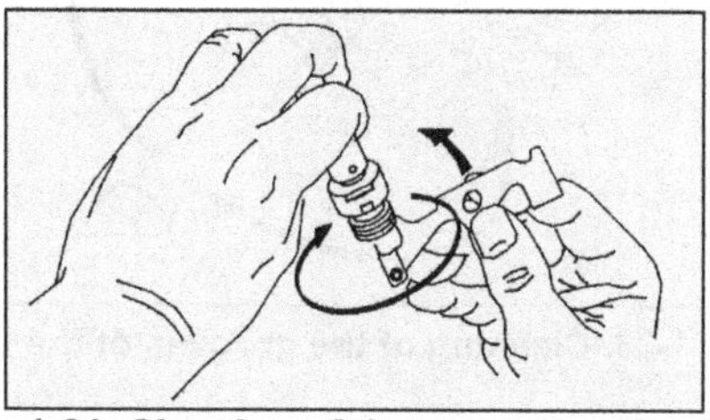

Figure 1-21. Cleaning of the grooves of the body.

(4) Clean the gas cylinder and piston with the special tool (scraper). Do not use CLP on the gas cylinder or piston.

(a) Clean the front interior of the gas cylinder (repositioned in receiver with bipod in place) by inserting and turning the flat side of the scraper in a 360-degree circular motion (Figure 1-22).

(b) Clean the internal grooves of the front side of the gas cylinder the same as, except insert the scraper farther into the gas cylinder (Figure 1-22).

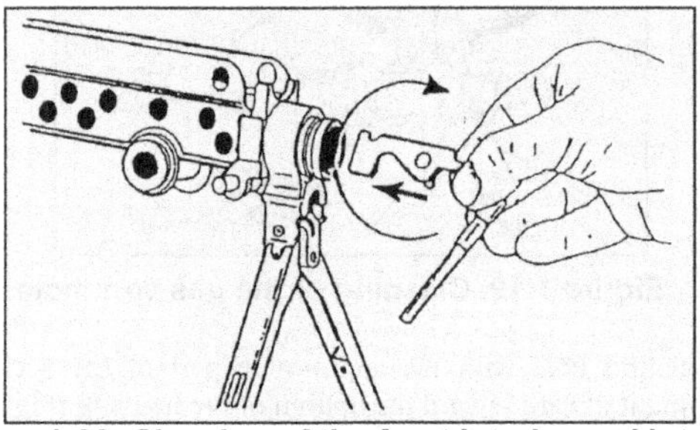

Figure 1-22. Cleaning of the front interior and internal grooves of the gas cylinder.

(c) Clean the three grooves of the piston using a 360-degree circular motion (Figure 1-23). Remove all carbon dust from the piston, inside and out.

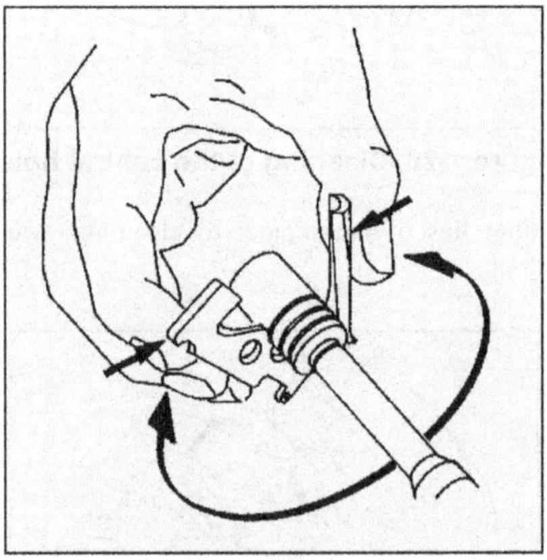

Figure 1-23. Cleaning of the grooves of the piston.

(d) Clean the hole in the front of the piston by inserting and turning the flat side of the scraper in a 360-degree circular motion (Figure 1-24).

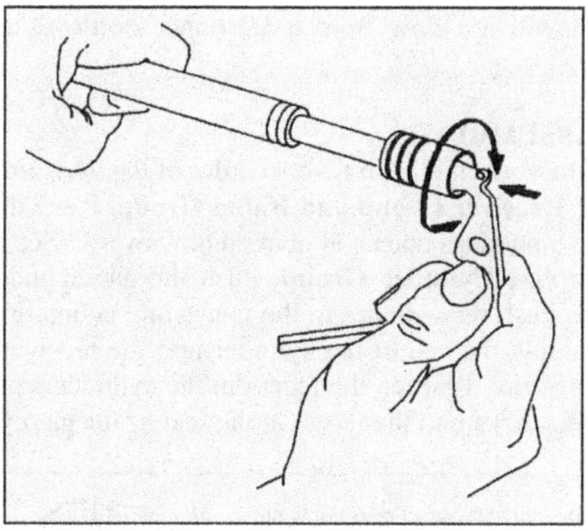

Figure 1-24. Cleaning of the hole in the front of the piston.

(5) Clean carbon and dirt from all other parts of the weapon.

(6) A cloth saturated in CLP is used on exterior surfaces to prevent corrosion.

b. **Lubrication**. After the M249 machine gun is cleaned and wiped dry, a thin coat of CLP is applied by rubbing it on with a cloth. This lubricates and preserves the exposed metal parts during all normal temperature ranges. The moving parts are also lubricated with CLP. After lubricating, the components are rubbed by hand to spread the CLP.

(1) *Operating rod group*. Use CLP on the operating rod and spring, the slide assembly, the feed roller, and the bolt-locking lug.

(2) *Barrel group*. Use CLP on the cam surfaces of the bolt-locking lugs, the heat shield, and along the outer surfaces of the barrel clamp.

(3) *Receiver group*. Use CLP on all moving parts on the cover assembly and the receiver rails.

c. **Preventive Maintenance**. Weapons that are seldom fired or stored for prolonged periods should have a light film of CLP applied to the interior of the gas plug, the gas regulator, and the piston immediately after cleaning or inspecting. Preventive maintenance is performed every 90 days, unless inspection reveals more frequent servicing is necessary. The use of the lubricant does not eliminate the need for cleaning and inspecting to ensure that corrosion has not formed. The gas regulator, gas plug, and piston must be clean and free of oil and lubricants before using the weapon. If it is not clean and oil free, stoppages will occur. CLP is the only lubricant to use on the M249 machine gun. The following procedures apply to cleaning and lubricating the M249 machine gun during unusual conditions:

(1) Extremely hot—use CLP, grade 2.

(2) Damp or salty air—use CLP, grade 2. Clean and apply frequently.

(3) Sandy or dusty areas—use CLP, grade 2. Clean and apply frequently. Remove excess with a rag after each application.

(4) Below -18 degrees Celsius (0 degrees Fahrenheit)—use CLP, grade 2, generously. Lubricate heavily enough so that the lubricant can be spread with finger. Although CLP provides required lubrication at temperatures between 0 degrees Fahrenheit and -35

degrees Fahrenheit, it will not flow from a 1/2-ounce bottle at temperatures below 0 degrees Fahrenheit.

1-9. GENERAL ASSEMBLY

The M249 machine gun is assembled in reverse order of the disassembly.

a. **Replacing the Receiver Group and Bipod Group**. Place the bipod group on the receiver group with the bipod legs open and pointed downward. (See Figure 1-18.)

b. **Replacing the Gas Cylinder Group**. Push the gas cylinder through the bipod yoke into the receiver. Push the cylinder to the rear while countering the pressure of the locking spring and guiding the end of the cylinder into the receiver with the other hand applying downward pressure. Position the recess in the cylinder near the spring. Turn the cylinder until the spring clicks into the recess at the rear of the gas cylinder (Figure 1-25).

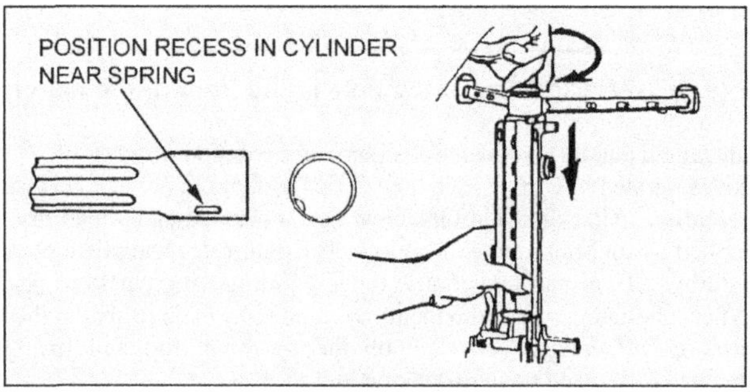

Figure 1-25. Replacement of gas cylinder group.

c. **Replacing the Trigger Mechanism Group**. Align the trigger mechanism (Figure 1-26) with the slot on the bottom of the receiver. Hold the trigger mechanism in position to accomplish the next step.

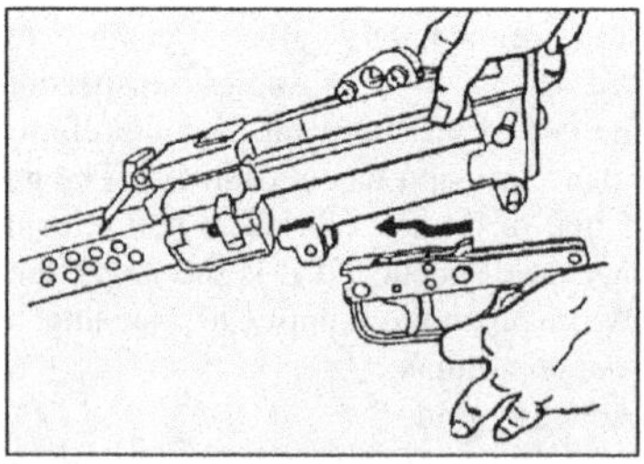

Figure 1-26. Replacement of the trigger mechanism group.

d. **Replacing the Buttstock and Buffer Assembly Group**. Align the lower hole in the buttstock and buffer assembly with the rear hole in the trigger mechanism; then push the lower retaining pin to the right (Figure 1-27).

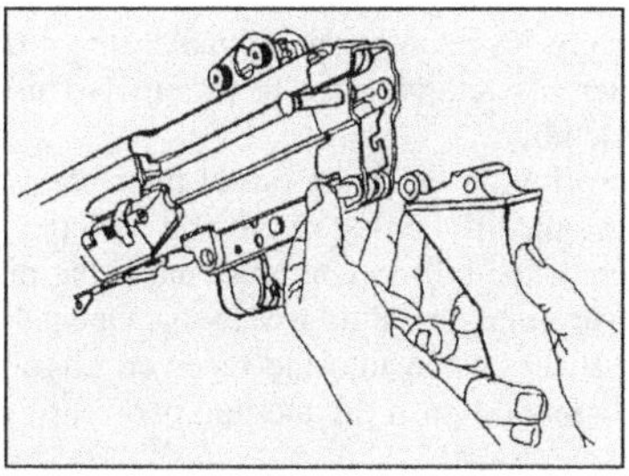

Figure 1-27. Replacement of the buttstock and buffer assembly group.

e. **Replacing the Handguard Group**. To replace the handguard (Figure 1-15), place it on the receiver from the bottom and push it to the rear until it stops. Using the guide rod, push the handguard retaining pin to the right, which locks the handguard into position. Push the handguard down to make sure it is locked.

f. **Replacing the Barrel Group**. Insert the gas regulator into the gas block and align the notch on the gas regulator with the notch of the gas block. With the gas regulator already installed and supported on a firm surface, place the gas regulator collar onto the protruding end of the body and align the spring with the stud. Push the gas regulator collar downward firmly and rotate it until it slips into place. Then, press it in and rotate it to lock it in place. Depress the barrel locking lever to the rear with your left hand, while holding the carrying handle with your right hand. Pull the barrel rearward and push downward; align the gas regulator with the gas cylinder and lock it by releasing the barrel locking lever. Check the barrel to ensure it is locked into the receiver by pulling or lifting on the carrying handle. Replace the heat shield by placing the hook end of the heat shield under the front sight post and press down until the clamps lock on the barrel. (Figure 1-28).

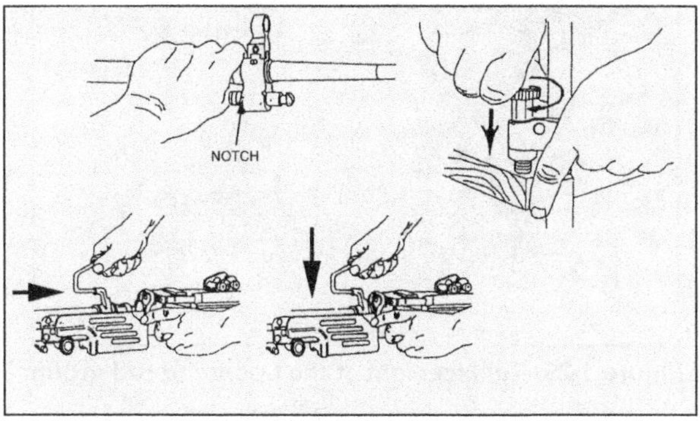

Figure 1-28. Replacement of the barrel group.

g. **Replacing the Operating Rod Group**. Hold the piston in one hand with the face of the piston facing outward and the sear notches downward. With the other hand, place the slide assembly onto the rear of the piston with the firing pin toward the front of the piston. (Check the slide assembly retaining pin to make sure it is out.) (Figure 1-29).

(1) Push the slide assembly retaining pin to the right. This locks together the piston assembly and the slide assembly.

(2) Put the firing pin spring on the firing pin of the slide assembly. Place the bolt on the slide assembly, aligning the driving lug of the bolt with the slot of the slide assembly. Apply pressure to the face of the bolt to compress the firing pin spring. Then, rotate the bolt to hook the driving lug into the slide assembly. Open the cover assembly on the receiver. Insert the face of the piston into the receiver, aligning the bolt lugs onto the receiver rails. Pull the trigger and push the moving parts forward until the bolt is seated into the chamber.

(3) Place the operating rod tip into the operating rod spring. Then, insert the free end of the operating rod and spring into the rear of the piston. Depress the rear of the operating rod assembly until the two lugs on the buffer are positioned in the receiver grooves.

(4) Pivot the buttstock upward into position and push the upper retaining pin to the right, locking the buttstock to the receiver.

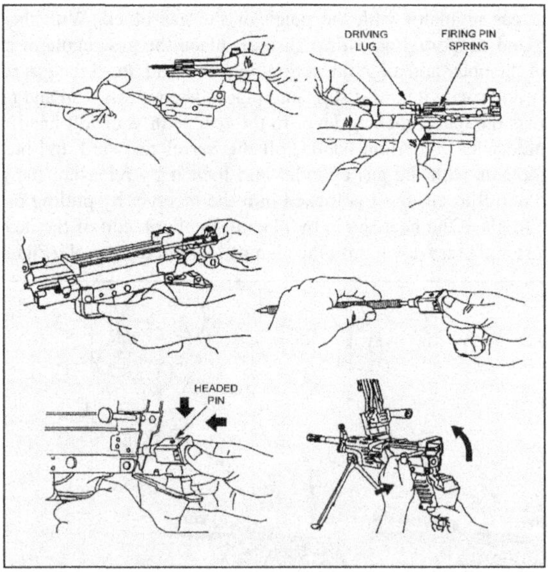

Figure 1-29. Replacement of the operating rod group.

1-10. FUNCTION CHECK

A function check must be performed to ensure that the M249 machine gun has been assembled correctly. The procedures, in order, are as follows:

a. Grasp the cocking handle with the right hand, palm up, and pull the bolt to the rear, locking it in place.

b. While continuing to hold the resistance on the cocking handle, use the left hand to move the safety to the SAFE position.

c. Push the cocking handle forward into the forward lock position.

d. Pull the trigger (The weapon should not fire).

e. Grasp the cocking handle with the right hand, palm up, and pull and hold it to the rear.

f. Move the safety to the FIRE position.

g. While continuing to hold resistance on the cocking handle, use the left hand to pull the trigger and ease the bolt forward to prevent it from slamming into the chamber area and damaging the face of the bolt.

h. If the weapon fails the function check, check for missing parts or repeat the reassembly procedures. Before disassembling the weapon, make sure it is positioned where the guide rod and spring cannot cause bodily harm if the bolt is locked to the rear. The cover and feed mechanism assembly can be closed with the bolt in either the forward or the rearward position.

CAUTION

Ease the bolt forward to prevent damage to the hardened surfaces on the bolt, barrel, and so forth.

1-11. MAINTENANCE PROCEDURES

There are certain actions that must be taken before, during, and after firing to properly maintain the M249 machine gun.

a. Before firing—

(1) Wipe the bore dry.

(2) Inspect the weapon as outlined in the operator's TM.

(3) Lubricate the weapon.

b. During firing—

(1) Inspect the weapon periodically to ensure that it remains lubricated.

(2) When malfunctions or stoppages occur, follow the procedures in Section IV.

c. After firing—

(1) Immediately clear and clean the weapon.

(2) Every 90 days during inactivity, clean and lubricate the weapon, unless inspection reveals more frequent servicing is necessary.

1-12. MAINTENANCE DURING NUCLEAR, BIOLOGICAL, CHEMICAL CONDITIONS

If the M249 machine gun is contaminated by chemical, biological, or radiological agents, the appropriate action must be taken to reduce exposure and penetration.

a. **Chemical.** Use towelettes from the M258A1 kit to wipe off the weapon. If these are not available, wash the weapon with hot, soapy water, and rinse.

b. **Biological.** Use towelettes or hot, soapy water and rinse the weapon as above.

c. **Radiological**. Brush or wipe the weapon, or wash with water, and rinse. For more details, see FM 3-5.

Section III. OPERATION AND FUNCTION

This section discusses the operation and function of the M249 machine gun. They include loading, firing, unloading, cycle of functioning, adjusting the sight, and using the bipod.

1-13. OPERATION

The M249 machine gun operations are loading, firing, unloading, and using belted ammunition or, in an emergency, a 20- or 30-round M16 magazine. The firing operation works on gas pressure created as a fired round passes through the barrel. The M249 is loaded, fired, unloaded, and cleared from the open-bolt position. The safety must be in the FIRE position before the bolt can be pulled to the rear. Before using belted ammunition, it must be checked to ensure it is properly linked with the double link or the link tab at the open end of the box. It must be free of dirt and corrosion. When using a magazine of ammunition, it must be loaded into the magazine well and be free of dirt and corrosion.

1-14. LOADING

To load the M249, the weapon must be cleared as described. (With the feed cover raised, the gunner makes sure his face is not exposed to the open chamber area while loading.) (Figure 1-30).

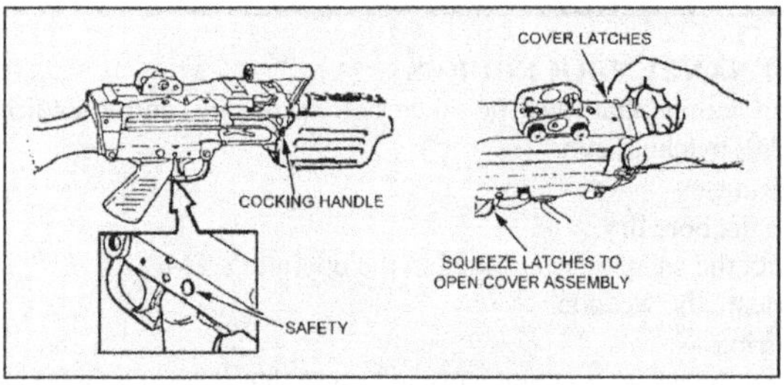

Figure 1-30. Loading.

a. **Belt**. When loading belted ammunition (Figure 1-31), always cant the weapon to the right. Make sure the open side of the links is facing down, and place the lead link tab or first round of the belt in the tray groove against the cartridge stop. The rounds should be placed flat across the feed tray. With your left hand, count five to six rounds down to hold ammunition in place on the feed tray, while at the same time closing the feed cover with your right hand. When closing the feed cover, always place your hand in front of the rear sight to prevent accidentally changing the sight adjustment.

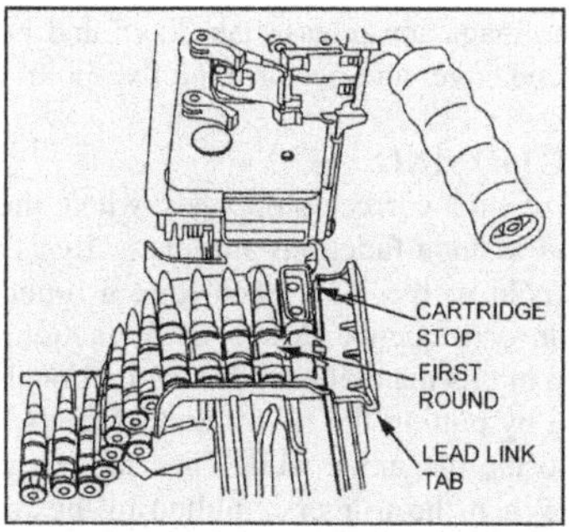

Figure 1-31. Belt-fed ammunition.

NOTE: Use the 20- or 30-round magazine for emergency use only when linked ammunition is not available.

b. **Magazine.** Load the 20- or 30-round magazine by inserting it into the magazine well on the left side of the receiver. Push the magazine firmly into the well until it seats and the release tab clicks into the recess on the magazine (Figure 1-32).

Figure 1-32. Loading of a magazine.

1-15. UNLOADING

To unload the weapon, pull the bolt and, if it is not already locked lock it in the rear position, it is locked there at this time. The safety is placed on "S". Depending on whether belt-fed or magazine-fed ammunition is used, the following procedures are used:

> **WARNING**
>
> **With a hot gun, before you raise the feed cover, move the weapon away from your face so that you are not exposed to the open chamber.**

a. **Belt.** Raise the feed cover and remove any ammunition or links from the feed tray. Perform the five-point safety check.

b. **Magazine**. Push the magazine release tab down and pull the magazine from the magazine well. Raise the feed cover and perform the five-point safety check.

1-16. CYCLE OF FUNCTIONING

The gunners can recognize and correct stoppages when they know how the M249 machine gun functions. The weapon functions automatically as long as ammunition is fed into it, and the trigger is held to the rear. Each time a round is fired the parts of the weapon function in a cycle or sequence. Many of the actions occur at the same time. These actions are separated in this manual only for instructional purposes.

a. The cycle is started by putting the first round of the belt in the tray groove or by inserting the magazine into the magazine well. Then the trigger is pulled, releasing the sear from the sear notch. When the trigger is pulled to the rear, the rear of the sear is lowered and disengaged from the sear notch. This procedure allows the piston and bolt to be driven forward by the expansion of the operating rod spring. The cycle stops when the trigger is released and the sear again engages the sear notch on the piston.

b. The sequence of functioning is as follows:

(1) *Feeding*. As the bolt starts its forward movement, the feed lever is forced to the right, causing the feed-pawl assembly to turn in the opposite direction. This forces the feed-pawl assembly over the next round in the belt, and the feed-pawl assembly is ready to place the next round into the tray groove when the rearward action occurs again. As the bolt moves to the rear after firing, the feed roller forces the feed lever to the left. The feed lever is forced to turn, moving the feed pawl to the right. This places a round in the tray groove.

(2) *Chambering*. As the bolt travels forward, the upper stripping (belt-fed or magazine-fed) lug engages the rim of the round. The pressure of the front and rear cartridge guides holds the round so that a positive contact is made with the upper stripping lug of the bolt. The front cartridge guide prevents forward movement of the link as the round is stripped from the belt. The upper locking lug carries the round forward. The chambering ramp causes the nose of the round to be cammed downward into the chamber. When the round is fully seated in the chamber, the extractor snaps over the rim of the round, and the ejector on the rail inside the receiver is depressed.

(3) *Locking*. As the round is chambered, the bolt enters the barrel socket. The upper and lower locking lugs contact the bolt camming surfaces inside the barrel and start the bolt turning clockwise. The action of the bolt into the slide assembly, as the piston continues forward, turns the bolt to complete its 90-degree (one-quarter turn) clockwise rotation. Locking is now complete.

(4) *Firing*. After the bolt is fully forward and locked, the piston continues to go forward independently of the bolt for a short distance. The piston assembly carries the firing pin through the face of the bolt. The firing pin strikes the primer of the round, and the primer fires the round.

(5) *Unlocking*. After the round is fired and the bullet passes the gas port, part of the expanding gases go into one block (new style) or into the gas regulator through the gas plug. The rapidly expanding gases enter into the gas cylinder from the gas regulator, forcing the piston to the rear. As the piston continues to the rear, the slide assembly's simultaneous movement to the rear causes the bolt to begin its counterclockwise rotation. The upper and lower locking lugs of the bolt contact the bolt camming surfaces inside the

barrel socket and, as the bolt continues toward the rear, it completes a one-quarter turn counterclockwise. The rotation and movement to the rear unlocks the bolt from the barrel socket.

(6) *Extracting.* Extracting begins during the unlocking cycle. The rotation of the bolt loosens the cartridge case in the chamber. As the piston and bolt move to the rear, the extractor pulls the cartridge case from the chamber.

(7) *Ejecting.* As the cartridge case is pulled from the chamber, the bolt passes by the ejector. This procedure causes the ejector clip to expand, forcing the ejector to push the expended cartridge. The extractor grips the right side of the cartridge and causes it to spin from the weapon as it reaches the ejection port. The empty belt links are forced out of the link ejection port as the rearward movement of the bolt causes the next round to be positioned in the tray groove.

(8) *Cocking.* The piston assembly acts against the firing pin, pulling the firing pin from the primer of the spent cartridge case. The action of the piston assembly, continuing to the rear with the firing pin, releases the compression of the firing pin spring. As long as the trigger is held to the rear, the M249 will continue to complete the eight steps of functioning automatically. When the trigger is released and the sear again engages the sear notch, the cycle of functioning is stopped and the weapon is cocked. To prevent undue wear to the sear and sear notch, the automatic rifleman must hold the trigger firmly to the rear during firing.

1-17. SIGHTS

This paragraph provides information on how to set the sights for elevation and windage for the M249 machine gun. It also includes information on how to make corrections if the initial setting on the windage knob or peep sight is not accurate. On a 10-meter target, each paster is 1 cm. Therefore, two clicks on the windage knob in either direction moves the strike of the round left or right 1 cm, and two turns on the peep sight moves the strike of the round up or down 1 cm. For example, if the shot group was 2 cm above and 1 cm to the right of the paster, sight corrections are made first by correcting the windage. In this case, the windage knob is rotated two clicks toward the buttstock (clockwise). The elevation knob is rotated four turns toward the buttstock (clockwise) to lower the strike of the round. (See Figure 1-33, below and Table 1-3, on page 1-31.)

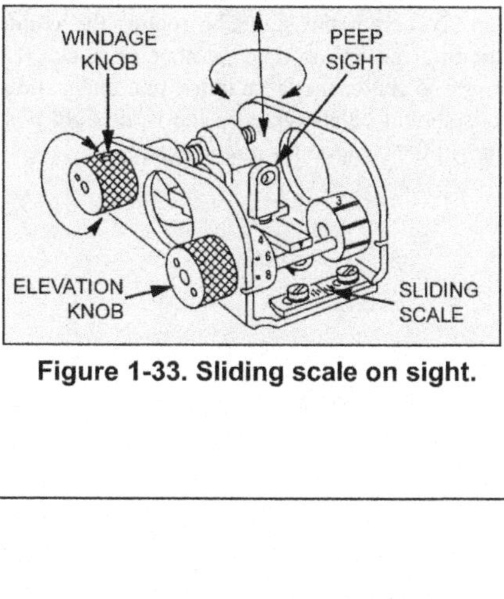

Figure 1-33. Sliding scale on sight.

a. **Elevation.** Adjustments for elevation (range) require the automatic rifleman to turn the elevation knob (closest to the buttstock) on the rear sight to the desired range setting. Range settings are graduated increments from 300 to 1,000 meters. Even-numbered settings are on the left side of the scale wheel and are numbered 4, 6, 8, 10, which represent 400, 600, 800, and 1,000 meters, respectively. Odd-numbered settings are on the right side of the scale wheel and marked with the number 3 and three index lines, which represent 300, 500, 700, and 900 meters, respectively. Rotation of the elevation knob toward the muzzle (counterclockwise) increases the range, while rotation toward the buttstock (clockwise) decreases the range. Fine adjustments, like zeroing, are made by adjusting the peep sight. Each 180-degree turn equals a half-mil change in elevation, which equals a half-cm change in impact at a range of 10 meters. Clockwise (to the right) rotations decrease elevation, while counterclockwise (to the left) rotations increase elevation. The peep sight can be turned nine 180-degree turns from top to bottom. To make the peep sight easier to grasp, the elevation knob is turned to its highest point (1,000 meters). The appropriate adjustment is made for the peep sight, and then the sight is returned to the desired range. Whenever readjusting the range, the point of aim is never changed. The point of aim is the center base of the target.

b. **Windage.** Adjustments for windage are made by traversing the rear sight right and left along the sliding scale. The sliding scale is marked or graduated with index lines. Each index line is equal to a half-mil change in direction or a half-cm change of impact at 10 meters. Rotation of the windage knob (closest to the muzzle end) toward the muzzle (counterclockwise) moves the rear sight aperture right, which moves the strike of the rounds right. Rotation toward the buttstock (clockwise) moves the aperture left, which moves the strike of the rounds left.

c. **10-Meter Zero, Setting of the Sights (Mechanical Zero).** The gunner indexes or places the elevation knob on a range of 700 meters. He centers the rear peep sight by rotating it clockwise (right) as far as it will go, then rotating counterclockwise (left) five clicks or half-turns. He rotates the windage knob toward the muzzle until the peep sight is *completely* to the right, then rotates the windage knob toward the buttstock twelve clicks to the left. This places the peep sight in the approximate center of the sight. Each sight may vary as to how many clicks are needed. To check the sight, the gunner starts with the sight all the way to the right and, while counting the clicks, rotates the windage knob until it stops on the left side. He divides the clicks by two. If the click is an uneven number, he rounds it up. To center the sight, he rotates the windage knob toward the center (right) while counting the appropriate number of clicks. He adjusts the sliding scale at the rear of the sight to center the large index line under the zeroed windage mark on the sight. Two threads should be showing on the front sight post. If more or less are showing, the gunner turns in the weapon for maintenance.

100 meters—one click moves the strike 5 cm (2 inches).
200 meters—one click moves the strike 10 cm (4 inches).
300 meters—one click moves the strike 15 cm (6 inches).
400 meters—one click moves the strike 20 cm (8 inches).
500 meters—one click moves the strike 25 cm (10 inches).
600 meters—one click moves the strike 30 cm (12 inches).
700 meters—one click moves the strike 35 cm (14 inches).
800 meters—one click moves the strike 40 cm (16 inches).
900 meters—one click moves the strike 45 cm (18 inches).

Table 1-3. Windage and elevation (peep sight) correction chart.

NOTE: The primary and spare barrels are zeroed by making adjustments on the front sight.

1-18. M122 TRIPOD

The M122 tripod provides a stable mount for the M249, and it permits a higher degree of accuracy and control. The tripod is recommended for all marksmanship training and defensive employment.

a. **Mounting the M122 Tripod**. The tripod assembly provides a stable and relatively lightweight base that is far superior to the bipod. The tripod may be extended and collapsed without difficulty. It consists of a tripod head, one front leg and two rear legs, and traversing bar. The traversing bar connects the two rear legs. The traversing bar is hinged on one side with a sleeve and sleeve latch on the other. This procedure allows the tripod to collapse to a closed position for carrying or storage, or to lock in an open, extended position for use. The traversing bar also supports the T&E mechanism. Engraved on the bar is a scale, which measures direction in mils. It is graduated in 5-mil increments. It is numbered every 100 mils to 425 mils right of center, and it is numbered every 100 mils to 450 mils left of center.

(1) The T&E mechanism provides controlled manipulation and the ability to engage predetermined targets.

(a) The traversing portion of the mechanism consists of the traversing handwheel and traversing slide-lock lever. As the traversing handwheel is turned, the muzzle of the weapon will turn to the left or right, depending on the direction it is turned. Each click of the traversing handwheel indicates a 1 mil change in direction of the muzzle: 1 click equals 1 mil. There is a total of 100 mils traverse (50 mils right and 50 mils left of center).

(b) The elevating portion of the mechanism consists only of the elevating handwheel. The elevating handwheel has a mil-click device built into it (1 click equals 1 mil). Engraved into the handwheel is a scale divided into 5-mil divisions and 1 mil subdivisions, for a total of 50 mils increments. There are 200 mils above and 200 mils below the zero mark ,for a total of 400 mils in elevation change. Elevation readings are taken in two parts. First, the major reading is taken from the elevation screw plate. The second, minor reading is from the handwheel. The two readings are separated by a slash ("/") when they are recorded.

(c) The traversing slide-lock lever allows rapid lateral adjustments along the traversing bar. Direction readings are taken from the scale on the traversing bar, using the left side of the traversing slide as an index. The direction of the reading comes from the position of the muzzle, not the position of the slide.

(2) To set up the tripod, unfold the front leg and spread the rear legs until the leg lock engages. Insert the pintle assembly and rotate the pintle lock-release cam to lock. Ensure that the locking lever of the pintle is facing forward toward the front leg (Figure 1-34).

(3) Attach the traversing and elevating mechanism (which requires a special adapter). Ensure that the adapter pin is to the right and the opening between it is to the rear. Center the elevating and traversing handwheels. To do this, he rotates the elevation handwheel until about 1-1/2 inches (two fingers) are visible on the upper elevating screw; he rotates the traversing slide until about two fingers are visible on the lower elevating screw. He rotates the traversing handwheel towards his body as far as it will go, then turns it away two complete revolutions. He checks the traversing handwheel scale to ensure the "0" on the scale is aligned with the "0" index line before and after the two revolutions. The T&E is now roughly centered. At night, he positions the traversing mechanism by turning the traversing handwheel toward his body as far as it will go, and then turning it away 50 clicks (two revolutions) (Figure 1-34).

(4) With the T&E roughly centered, he lowers the traversing slide on to the traversing bar with the locking lever to the rear, and the traversing handwheel to the left, and secures it by turning the locking lever clockwise (Figure 1-34).

(5) The weapon attaches to the M122 tripod. First, he extends the bipod legs forward. Then, he engages the mounting pins (Located between the front of the handguard and the bipod legs) of the M249 into the pintle of the tripod by squeezing the locking lever of the pintle. He lowers the rear of the weapon so that the hole above the trigger guard can be engaged with the locking pin of the T&E adapter. He aligns the hole with the pin of the adapter and pushes the pin from right to left to secure the M249 to the M122 tripod (Figure 1-34).

(6) After the M249 is attached and secured to the tripod, the gunner must attach a special ammunition adapter to the M249. He inserts the adapter into the magazine well, as if inserting a magazine. This procedure allows the gunner to use the 200-round drum of ammunition (Figure 1-34).

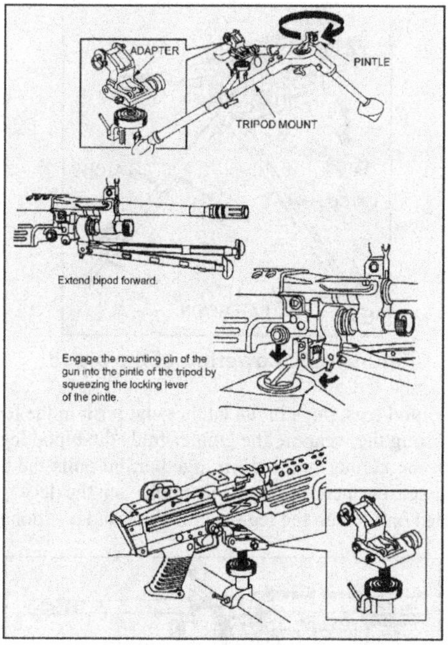

Figure 1-34. Tripod mount.

b. **Dismounting The M249 From The M122 Tripod**. The gunner dismounts the M249 from the tripod by first removing the traversing and elevation mechanism from the weapon. He pulls the locking pin of the adapter to release the T&E from the trigger guard. He grasps the carrying handle with his left hand and squeezes the pintle-locking lever with his right hand. He lifts the weapon from the pintle assembly and the tripod.

1-19. BIPOD OPERATIONS

The bipod group is used to fire from the prone position. The shoulder rest on the buttstock provides support for the gunner when fired in the bipod mode. The gas cylinder group holds the bipod group in place. Once the gas cylinder is removed, the bipod group can also be removed from the receiver.

a. To lower the bipod legs, hold the legs together and pull down and away from the handguard. Release the legs so that they lock in the vertical position. To extend the bipod legs, grasp the foot of each leg and pull down (Figure 1-35, page 1-34).

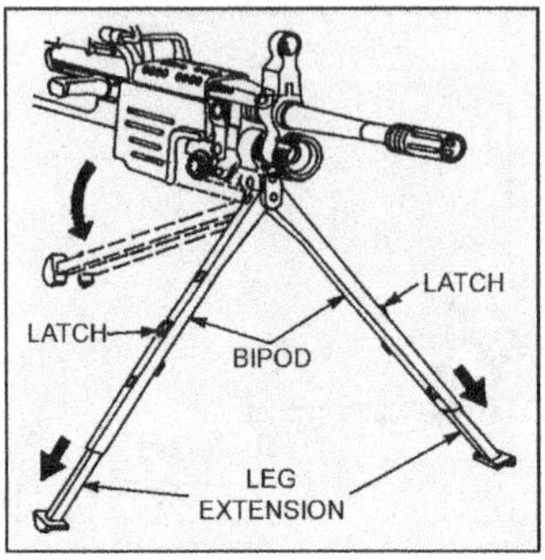

Figure 1-35. Lowering of the bipod.

b. To retract the bipod legs, push in the latches and push in the legs.

c. Before transporting the weapon, the gunner folds the bipod legs. To place the legs in the closed position, the gunner holds them together; he pulls the them back under the handguard; he then releases them so that the hooks on the legs grip the handguard. The bipod can be folded only when the legs are in the closed position (Figure 1-36).

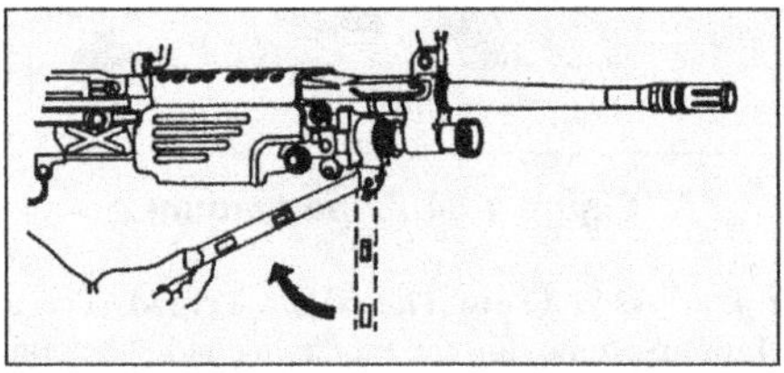

Figure 1-36. Folding of bipod under the handguard.

1-20. VEHICULAR MOUNT

The standard vehicular mount for the M249 machine gun is the M6 pedestal mount used on the (HMMWV). One component of the pedestal mount is the M197 machine gun mount (travel lock). This mount also adapts to other vehicles (Figure 1-37).

a. To mount the weapon, the gunner ensures that the release lever of the pintle is facing forward. To extend the bipod legs forward, he places the front mounting pins of the M249 into the pedestal by squeezing the locking lever of the pintle. He ensures that the M60 machine gun adapter assembly is pivoted away from the M249 fork (clevis). He lowers the rear of the weapon so that the locking pin of the machine gun mount can engage the hole above the trigger guard. The gunner engages this part of the weapon into the fork of the mount and pushes in the locking pin.

b. To dismount the weapon, the gunner pulls the locking pin of the mount. He raises the rear of the weapon slightly and squeezes the locking lever of the pintle. Once the front mounting pins are released, the gunner lifts the weapon from the mount.

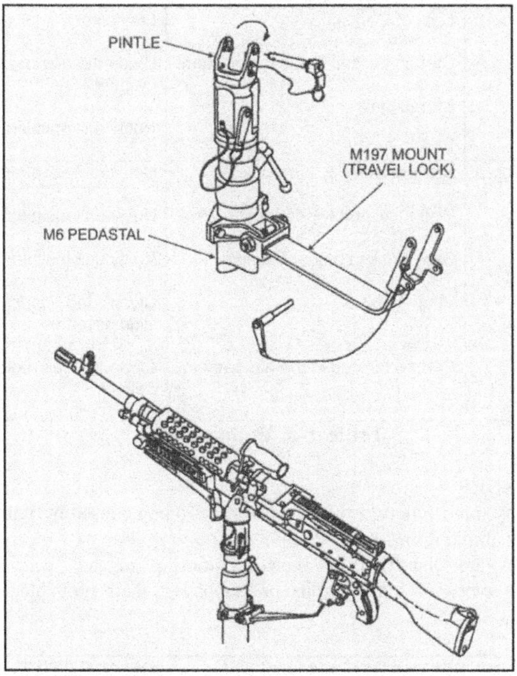

Figure 1-37. Vehicular mount.

1-21. TRIPOD POSITIONING

The M122 tripod provides a stable mount for the M249, and it permits a high degree of accuracy and control. The gunner unfolds the front leg and positions it toward the target and spreads the rear legs until the leg lock engages.

Section IV. PERFORMANCE PROBLEMS AND DESTRUCTION

This section identifies some of the problems that cause the M249 to perform improperly. It also explains how to identify unserviceable parts and how to destroy the M249 when authorized to do so.

1-22. MALFUNCTIONS

A malfunction occurs when a mechanical failure causes the M249 to fire improperly. Defective ammunition or improper operation by the automatic rifleman is not considered a malfunction. If cleaning and or lubricating the weapon does not fix the problem, then it is turned in to the unit armorer. Table 1-4 shows malfunctions, their probable causes, and the corrective actions.

MALFUNCTION	PROBABLE CAUSES	CORRECTIVE ACTIONS
Sluggish operation.	Lack of lubricant.	Lubricate.
	Carbon buildup in the gas system.	Clean the gas regulator, piston, and cylinder.
	Burred parts.	Notify unit maintenance.
Failure to cock or a runaway weapon.	Broken, worn, or burred sear.	Notify unit maintenance.
	Piston assembly sear notch worn.	Notify unit maintenance.
	Sear stuck in trigger housing.	Notify unit maintenance.
	Short recoil.	Clean and lubricate the bolt and slide assembly.
	Carbon buildup in the gas system.	Clean the gas regulator, piston, and cylinder.

Table 1-4. Malfunctions.

1-23. STOPPAGES

A stoppage is any interruption in the cycle of functioning caused by faulty action of the weapon or faulty ammunition.

a. Stoppages are classified by their relationship to the cycle of functioning. Table 1-5 shows types of interruptions or stoppages, their probable causes, and the corrective actions.

STOPPAGE	PROBABLE CAUSE	CORRECTIVE ACTION
Failure to feed.	Insufficient lubrication.	Lubricate as required.
	Defective ammunition link.	Remove and replace ammunition.
	Obstruction in receiver.	Remove obstruction.
	Insufficient gas pressure.	Clean the gas regulator, piston, and cylinder.
	Unlatched cover.	Latch cover.
	Long or short rounds.	Align rounds in the link belt.
	Inverted link belt.	Reinstall link belt with the open end of the link facing down.
	Damaged, weak, or worn operating parts.	Notify the unit maintenance.

Table 1-5. Stoppages.

STOPPAGE	PROBABLE CAUSE	CORRECTIVE ACTION
Failure to fire.	Safety on.	Push safety to the left, exposing red ring.
	Link belt improperly loaded.	Remove and reinstall link belt properly.
	Defective ammunition.	Eject round.
	Faulty ammunition.	Replace ammunition.
	Broken or damaged firing pin.	Notify unit maintenance.
	Broken or weak driving spring.	Notify unit maintenance.
Failure to extract.	Dirty chamber or bolt and slide assembly.	Clean chamber or bolt slide assembly. If problem continues, notify unit maintenance.
	Carbon buildup in the gas system.	Clean the gas regulator, cylinder, and piston.
	Damaged extractor or spring.	Notify unit maintenance.
Failure to chamber.	Dirty ammunition.	Clean the ammunition.
	Carbon buildup in the gas cylinder.	Clean the gas cylinder.
	Carbon buildup in the receiver.	Clean the receiver.
	Damaged round.	Remove the round and recock the weapon.
	Damaged or weak driving spring.	Notify unit maintenance.
	Dirty chamber.	Clean the chamber.
	Damaged gas regulator.	Notify unit maintenance.
Failure to eject.	Short recoil.	Clean and lubricate the eject bolt and slide assembly. If problem still exists, notify unit maintenance.
	Damaged ejector or spring.	Notify unit maintenance.
	Carbon buildup in the gas system.	Clean the gas regulator, piston, and cylinder.

Table 1-5. Stoppages (continued).

DANGER

IF NOTHING IS EJECTED AND THE WEAPON IS HOT (200 OR MORE ROUNDS FIRED IN LESS THAN 2 MINUTES), DO NOT OPEN THE COVER. PUSH THE SAFETY TO THE RIGHT (RED RING NOT VISIBLE), WHICH PLACES THE WEAPON ON SAFE. KEEP THE WEAPON POINTED DOWNRANGE FOR 15 MINUTES, THEN CLEAR THE WEAPON. BE CAREFUL CLEARING THE WEAPON WHEN THE BARREL IS HOT; A ROUND MAY FIRE (COOK OFF) DUE TO THE BARREL'S HEAT INSTEAD OF THE FIRING MECHANISM. DURING TRAINING OR ON A FIRING RANGE, AFTER THE WEAPON HAS FIRED 200 ROUNDS, THE BARREL IS CONSIDERED "HOT."

- **DURING COMBAT, WAIT 5 SECONDS (BECAUSE OF THE POSSIBILITY OF A *HANGFIRE*) BEFORE APPLYING IMMEDIATE ACTION OR REMEDIAL ACTION.**
- **DURING TRAINING, WAIT 15 MINUTES BEFORE CLEARING A HOT WEAPON AND APPLYING IMMEDIATE OR REMEDIAL ACTION.**

b. If a round is in the chamber (this means any part of the round, ranging from the tip of the bullet to the rim) when applying immediate or remedial action on a cold or hot gun. The gunner removes the ammunition from the feed tray only, then closes the cover and attempts to fire. If the weapon fires, he reloads and continues firing. If it does not fire, he clears the weapon (removes the round using a clearing rod, with the cover closed). He does not use anything other than a cleaning rod. Then, he inspects weapon and ammunition.

1-24. IMMEDIATE ACTION

Immediate action is action taken to *reduce a stoppage without looking for the cause.* Immediate action should be taken in the event of a misfire or a cook off. A misfire is the failure of a chambered round to fire. Such failure can be due to an ammunition defect or faulty firing mechanism. A cook off is the firing of a round due to the heat of a hot barrel and not to the firing mechanism. Cook offs can be avoided by applying immediate action within 10 seconds of a failure to fire. The gunner keeps the M249 on his shoulder while performing immediate action procedures. If the M249 stops firing, he takes the following immediate actions. An effective memory aid is POPP, which stands for Pull, Observe, Push, and Press:

a. Pull and lock the cocking handle to the rear while observing the ejection port to see if a cartridge case, belt link, or round is ejected. Ensure that the bolt remains to the rear to prevent double feeding if a round or cartridge case is not ejected.

b. If a cartridge case, belt link, or round is ejected, push the cocking handle to its forward position, take aim on the target, and press the trigger. If the weapon does not fire, take remedial action. If a cartridge case, belt link, or round is not ejected, take remedial action.

1-25. REMEDIAL ACTION

Remedial action is any action taken to determine the cause of a stoppage and to restore the weapon to an operational condition. This action is taken only after immediate action does not remedy the problem.

a. **Cold Weapon Procedures.** When a stoppage occurs with a cold weapon, and if immediate action has failed, use the following procedures:

(1) While the weapon is on your shoulder, grasp the cocking handle with the right hand, palm up; pull the cocking handle to the rear, locking the bolt. While holding the resistance on the cocking handle, move the safety to SAFE and return the cocking handle.

(2) Place the weapon on the ground or away from your face. Open the feed cover and perform the five-point safety check. Reload and continue to fire.

(3) If the weapon does not fire, clear the weapon and inspect it and the ammunition.

b. **Hot Weapon Procedures**. If the stoppage occurs with a hot weapon (200 or more rounds in less than 2 minutes, or as noted previously for training), move the safety to SAFE, wait 5 seconds (during training, let the weapon cool for 15 minutes), and use the same procedures as outlined for cold weapon procedures.

c. **Jammed Cocking Handle**. If a stoppage occurs, and if the cocking handle cannot be pulled to the rear by hand (the bolt may be fully forward and locked or only partially forward), the gunner takes the following steps:

(1) Tries once again to pull the cocking handle *by hand.*

CAUTION

Do not try to force the cocking handle to the rear with your foot or a heavy object. This could damage the weapon.

(2) If the weapon is hot enough to cause a cook off, moves all Soldiers a safe distance from the weapon and keeps them away for 15 minutes.

(3) After the gun has cooled, opens the cover and disassembles the gun. Ensures rearward pressure is kept on the cocking handle until the buffer is removed. (The assistant gunner helps the gunner do this.)

(4) Removes the round or fired cartridge. Uses cleaning rod or ruptured cartridge extractor if necessary.

(a) In a training situation, after completing the remedial action procedures does not fire the gun until an ordnance specialist has conducted an inspection.

(b) In a combat situation, after the stoppage has been corrected, the gunner changes the barrel and tries to fire. If the weapon fails to function properly, he sends it to the unit armorer.

1-26. DESTRUCTION PROCEDURES

Destruction of any military weapon is only authorized as a last resort to prevent enemy capture or use. This paragraph discusses the field-expedient means of this destruction; it does not replace published policies. In combat situations, the commander has the authority to destroy weapons, but he must report this destruction through channels.

a. Disassemble the weapon as completely as time permits. Use the barrel to destroy the bolt, operating rod group, bipod, sights (rear and front), and receiver.

b. Bury the disassembled weapon or dump the parts into a stream, a sump, or a latrine.

c. Burn the weapon by placing an incendiary grenade on the receiver group over the bolt (with the cover feed mechanism assembly resting on the grenade) and detonating the grenade.

CHAPTER 2
M60 MACHINE GUN

The 7.62-mm M60 machine gun supports the rifleman in offense and defense. It provides the heavy volume of close and continuous fire the rifleman needs to accomplish his mission. The M60 is used to engage targets beyond the range of individual weapons, with controlled and accurate fire. The long-range, close defensive, and final protective fires delivered by the M60 form an integral part of a unit's defensive fires. This chapter describes the weapon and the types of ammunition it uses in detail and provides a table of general data.

Section I. DESCRIPTION AND COMPONENTS

This section describes the M60 machine gun and its components and purposes, ammunition, and blank firing adapters, and how to install them. The primary use of the M60 is to support the rifleman in both offense and defense. It provides the heavy volume of close and continuous fire he needs to accomplish the mission. It can engage targets beyond the capability of individual weapons with controlled, accurate bursts.

2-1. DESCRIPTION

The M60 is a gas-operated, air-cooled, belt-fed, automatic machine gun that fires from the open-bolt position (Figure 2-1). It has a maximum rate of fire of 550 rounds per minute. Ammunition is fed into the weapon from a 100-round bandoleer containing a disintegrating metallic split-link belt. It can be fired from the shoulder, hip, or underarm position; from the bipod-steadied position; or from the tripod-mounted position. (See Table 2-1, on page 2-2, for general data.)

DANGER

DO NOT INTERCHANGE THE BARREL ASSEMBLY OR THE BOLT ASSEMBLY FROM ONE WEAPON TO ANOTHER. IF YOU DO SO, IT MAY RESULT IN DEATH OR INJURY.

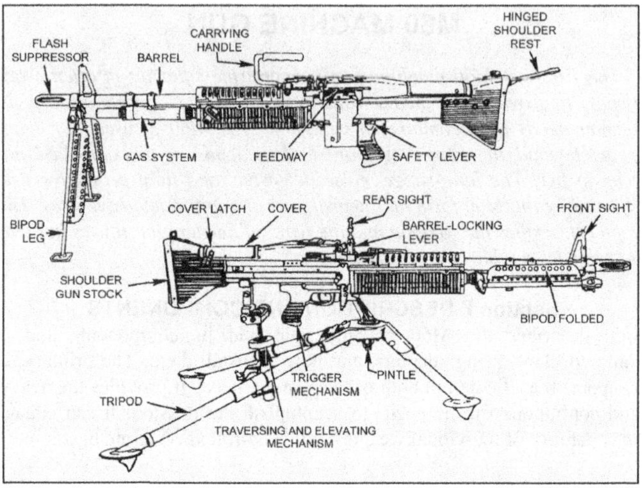

Figure 2-1. M60 machine gun, bipod- and tripod-mounted.

Ammunition	7.62-mm ball, tracer, armor-piercing, blank, dummy. Armor-piercing is not authorized for training.
Tracer burnout	900 meters
Length of the M60	43.5 inches
Weight of the M60	23 pounds
Weight of tripod mount M122 with traversing and elevating mechanism and pintle	16 pounds
Maximum range	3,725 meters
Maximum effective range	1,100 meters with tripod and T&E
Area:	
Tripod	1,100 meters
Bipod	800 meters
Point:	
Tripod	800 meters
Bipod	600 meters
Maximum extent of grazing fire obtainable over level or uniformly sloping terrain	600 meters
Height of the M60 on tripod mount M122	16.5 inches

Table 2-1. General data.

Rates of fire:	
Sustained	100 rounds per minute. 6- to 9-round bursts. 4 to 5 seconds between bursts. Barrel changed every 10 minutes.
Rapid	200 rounds per minute. 10- to 13-round bursts. 2 to 3 seconds between bursts. Barrel changed every 2 minutes.
Cyclic	550 rounds per minute. Fire continuous burst. Barrel changed every minute.
Basic load of ammunition for a three-man crew	600 to 900 rounds.
Elevation, tripod controlled	+200 mils
Elevation, tripod free	+445 mils
Depression, tripod controlled	-200 mils
Depression, tripod free	-445 mils
Traverse, controlled by traversing and elevating mechanism	100 mils
Normal sector of fire (with tripod)	875 mils

Table 2-1. General data (continued).

2-2. COMPONENTS

The major components of the M60 and their purposes are shown in Table 2-2 and Figures 2-2 and 2-3. The sights and safety lever are discussed in paragraphs a and b and shown in Figures 2-4 and 2-5.

COMPONENTS	PURPOSES
(1) Stock assembly	Provides a shoulder support for aiming and firing the M60 machine gun. The stock assembly has a shoulder rest.
(2) Buffer assembly	Absorbs recoil from the bolt and operating rod assembly at the end of the recoil movement.
(3) Bolt assembly	Feeds, chambers, fires, and extracts, using the propellant gases and recoil spring for power. It houses the firing pin.
(4) Operating rod assembly	Transfers power from propelling gases to the bolt and slide assemblies.

Table 2-2. Components and purposes.

COMPONENTS	PURPOSES
(5) Cover, hanger, and feed assembly	Feeds linked-belt ammunition. Holds the cartridges in position for stripping, feeding, and chambering.
(6) Barrel assembly	Houses cartridges for firing, directs projectiles, supports fixed front sight, and provides passageway for operating gases. The bipod supports the M60 in the prone position. The telescopic legs can be adjusted to three different lengths.
(7) Trigger assembly	Controls the firing of the weapon.
(8) Forearm assembly	Provides thermal insulation to protect the gunner's hands from heat. Has a slotted top to allow air to circulate around the barrel for cooling purposes.
(9) Receiver assembly	Supports all assembly major components and houses the action of the weapon.
(10) Rear sight assembly	Adjustable for both windage and elevation allowing the gunner to make changes rapidly.
(11) Cocking handle assembly	Pulls the moving parts rearward. Moves in a guide rail fixed to the right side of the receiver.
(12) Tripod (M122) assembly	Together with the tripod, T&E mechanism, and pintle, provides a stable mount and permits a higher degree of accuracy and control.

Table 2-2. Components and purposes (continued).

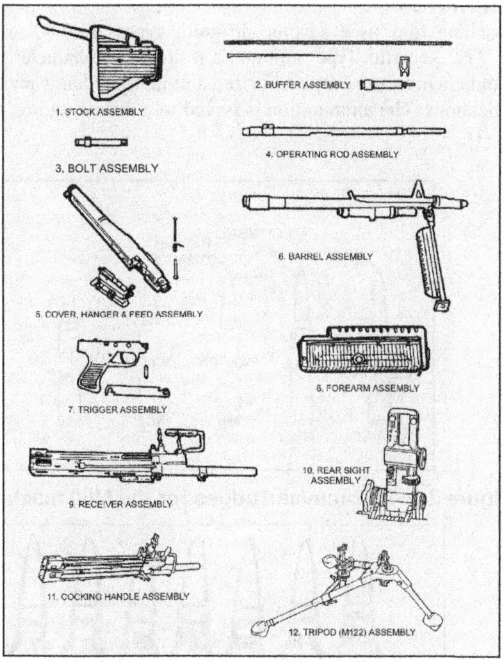

Figure 2-2. M60 components and tripod.

a. **Sights.** The front sight is attached to the barrel. The rear sight is mounted on a spring-type dovetail base, and it can be folded down when the gun is moved (See Figure 2-1).

b. **Safety Lever.** The safety lever is on the left side of the trigger group (Figure 2-1). It has an "S" and an "F" position. On the "S" position, the bolt cannot be pulled to the rear or released to go forward.

WARNING

Manually return the cocking handle to its forward position each time the bolt is pulled to the rear to prevent damage to the cocking and injury to the gunner.

2-3. AMMUNITION

The M60 machine gun uses several different types of 7.62-mm standard military ammunition. The specific type ammunition and its characteristics are shown in Figure 2-3. Soldiers must use only authorized ammunition that is manufactured to US and NATO specifications. The ammunition is issued in a disintegrating, metallic, split-linked belt (Figure 2-4).

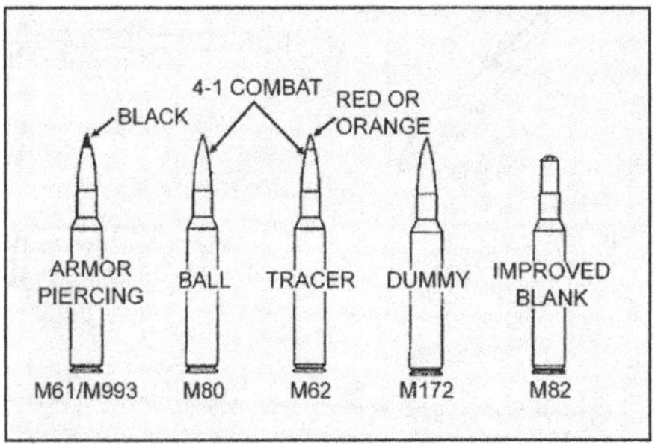

Figure 2-3. 7.62-mm cartridges for the M60 machine gun.

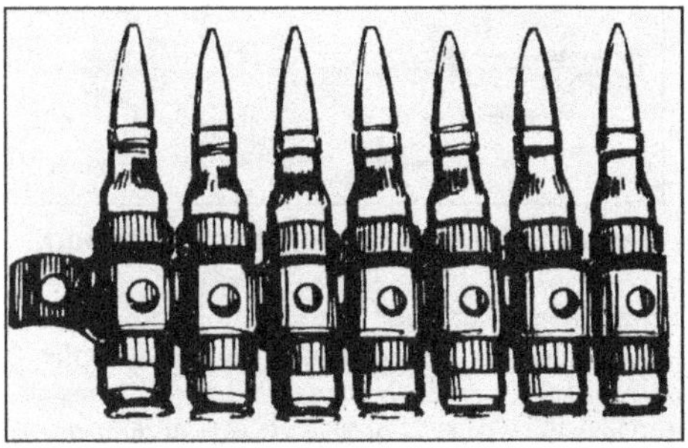

Figure 2-4. 7.62-mm cartridges in metallic belt.

a. **Classification**. The M60 machine gun ammunition is classified as follows:

(1) *Cartridge, 7.62-mm Ball M80.* This cartridge is used against light materials and personnel, and for range training.

(2) *Cartridge, 7.62-mm Armor-Piercing M61.* This cartridge is used against lightly armored targets.

(3) *Cartridge, 7.62-mm Tracer M62.* This cartridge is used for observation of fire, incendiary effects, signaling, and for training. When tracer rounds are fired, they are mixed with ball ammunition for a mix of four ball rounds to one tracer round.

(4) *Cartridge, 7.62-mm Dummy M63.* This cartridge is used during mechanical training.

(5) *Cartridge, 7.62-mm Blank M82.* This cartridge is used during training when simulated live fire is desired. A blank firing attachment is used to fire this ammunition.

b. **Storage**. The ammunition is stored under cover. If ammunition is in the open, it must be kept at least 6 inches above the ground and covered with a double thickness of tarpaulin. The cover must be placed so that it protects the ammunition yet allows ventilation. Trenches are dug to divert water from flowing under the ammunition.

c. **Care, Handling, and Preservation.** Ammunition should not be removed from the airtight containers until ready for use. Ammunition removed from the airtight containers, especially in damp climates, may corrode.

(l) Ammunition must be protected from mud, dirt, and moisture. If ammunition gets wet or dirty, it must be wiped off before use. Lightly corroded cartridges are wiped off as soon as the corrosion is discovered. Heavily corroded, dented, or loose projectiles should not be fired.

(2) Ammunition must be protected from the direct rays of the sun. Excessive pressure from the heat may cause premature detonation.

(3) Oil should never be used on ammunition. Oil collects dust and other abrasives that may possibly damage the operating parts of the weapon.

d. **Packaging.** The ammunition box contains two cartons. Each carton has a bandoleer for carrying purposes. Each carton contains 100 rounds and weighs about 7 pounds. Ammunition in the bandoleers may be linked together, attached to the hanger assembly, and fired from the container; or, the bandoleers may be removed for firing.

2-4. BLANK FIRING ATTACHMENTS, M13 AND M13A1

The M13 BFA is used on the M60 machine gun when blank cartridges are fired to simulate live firing during training where live firing is not practical. An M13A1 BFA is also available for the M60 machine gun. The difference between the M13 and the M13A1 is that, on the M13A1, the restrictor tube can be tightened (using a flat tip screwdriver) for a more secure fit, to prevent gas leakage.

DANGER

DO NOT FIRE BLANK AMMUNITION AT ANY PERSON WITHIN 20 FEET. FRAGMENTS OF A CLOSURE WAD OR PARTICLES OF UNBURNED PROPELLANT CAN CAUSE DEATH OR INJURY.

a. **Installation of the M13**. The BFA must be adjusted to fit the machine gun barrel. The orifice tube fits inside the flash suppressor, flush against the muzzle and flush with the forward end of the flash suppressor. The BFA is clamped tightly to the front sight (Figure 2-5, page 2-8). When properly adjusted, it will fit snugly against the muzzle, thus preventing the escape of gas during firing. When the BFA is manufactured, the distance the orifice tube screws into the restrictor bushing is fixed and fitted by *staking* the restrictor bushing. This fixed distance, set in the restrictor bushing, does not provide the correct adjustment for every machine gun, because the distance from the muzzle to the forward end of the flash suppressor varies from weapon to weapon. In some instances, the *stake* mark must be broken, the orifice tube adjusted to fit the barrel, and "restaked."

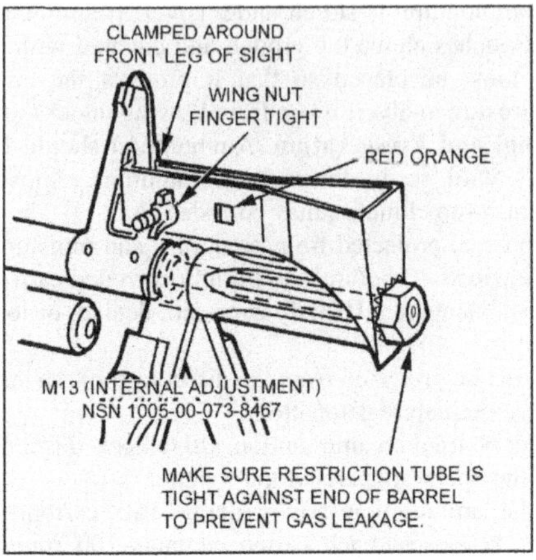

Figure 2-5. M13 blank firing attachment.

b. **Installation of the M13A1**. The M13A1 is installed by (Figure 2-6) loosening the lock nut, turning the restrictor tube out a few turns, and loosening the wing nut a few turns. The restrictor tube is inserted in the flash suppressor as far as possible and is clamped around the front sight. The wing nut is tightened finger tight. The restrictor tube is screwed in until it seats tightly against the muzzle end of the barrel to prevent gas leakage, and then the lock nut is tightened.

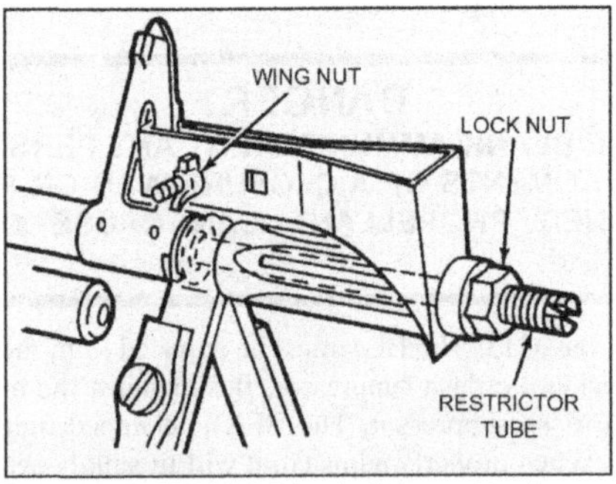

Figure 2-6. M13A1 blank firing attachment.

c. **Care of the M60 While Using the BFA**. A buildup of carbon inside the weapon causes friction between the moving parts. Carbon deposits build rapidly when blanks are fired. Stoppages occur when these deposits become excessive. Therefore, the weapon must be kept clean, especially the gas system and chamber, during blank firing. To obtain the best performance when using the BFA, the gunner performs the following:

(1) Inspects the weapon before firing for damaged parts, excessive wear, cleanliness, and proper lubrication.

(2) Test fires the weapon using ball ammunition when feasible before attaching the BFA.

(3) Adjusts the BFA to fit the weapon.

(4) Applies immediate action when stoppages occur.

(5) Cleans the weapon including barrel assembly, gas cylinder, gas piston, gas port, chamber bore, and BFA.

(6) Cleans and lubricates the entire weapon after 1,000 rounds.

DANGER

NEVER LOAD ANY AMMUNITION OTHER THAN BLANKS WHEN THE BFA IS IN PLACE. NEVER FIRE THE BFA-FITTED MACHINE GUN AT PERSONNEL WHO ARE WITHIN 20 FEET OF THE WEAPON. IT COULD CAUSE DEATH OR INJURY.

Section II. MAINTENANCE

Proper maintenance contributes to weapon effectiveness as well as unit readiness. Maintenance aspects of the M60 include inspection; cleaning and lubrication; and maintenance before, during, and after firing, and during NBC conditions. Associated tasks essential to maintenance (clearing, general assembly and disassembly, and function checks) are discussed in detail.

2-5. CLEARING PROCEDURES

The first step in maintenance is to clear the weapon. This applies in all situations, not just after firing. The gunner must always assume the M60 is loaded. To clear the M60, the gunner performs the following procedures. (Subparagraph letters are keyed to the lettered callouts in Figure 2-7, page 2-10.)

a. Moves the safety lever to the "F" position.

b. With his right hand, palm up, pulls the cocking handle in the rear. Moves the safety lever to the "S" position. Returns and locks the cocking handle in the forward position.

c. Raises the cover and conducts the *four-point safety check* for brass, links, or ammunition.

(1) Checks the feed pawl assembly under the cover.

(2) Checks the feed tray.

(3) Lifts the feed tray and hanger assembly and inspects the chamber.

(4) Checks the space between the face of the bolt and the chamber.

d. Closes the cover and moves the safety lever to the "F" position. Pulls the cocking handle to the rear position. Pulls the trigger and at the same time eases the bolt forward by manually riding the cocking handle forward.

e. Places the safety lever on "S" and raises the cover. (If not disassembling the gun, keep the cover down.)

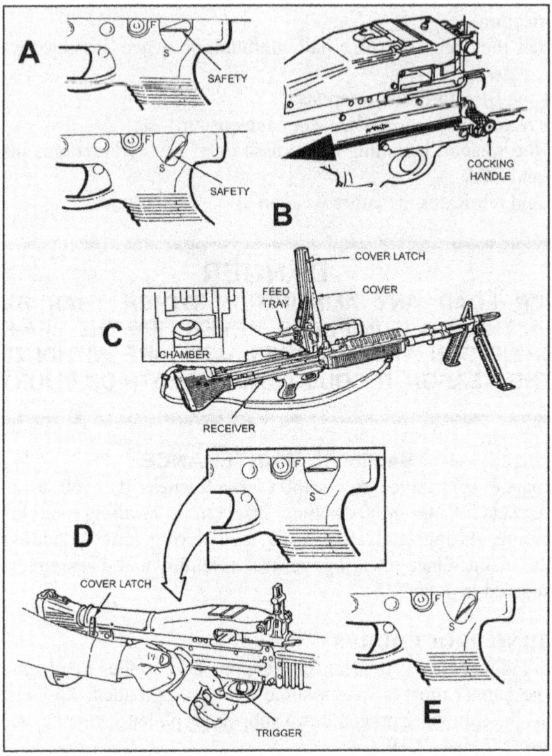

Figure 2-7. Clearing procedures.

> **CAUTION**
> **Manually return the cocking handle to the forward and locked position each time the bolt is manually pulled to the rear, or it could cause damage to the weapon.**

2-6. GENERAL DISASSEMBLY

The gunner performs general disassembly, which is removing and replacing the eight major groups (Figure 2-8). (The unit armorer performs the detailed disassembly. Disassembly beyond this point is prohibited except by ordnance personnel.) During general disassembly, the gunner clears the weapon and places each part on a clean flat surface such as a table or mat. This aids in assembly in reverse order and avoids the loss of parts.

DANGER

BE SURE THE BOLT IS IN THE FORWARD POSITION BEFORE DISASSEMBLY. THE SPRING GUIDE CAN CAUSE DEATH OR INJURY IF THE OPERATING ROD SPRING IS RETRACTED WITH THE BOLT PULLED TO THE REAR.

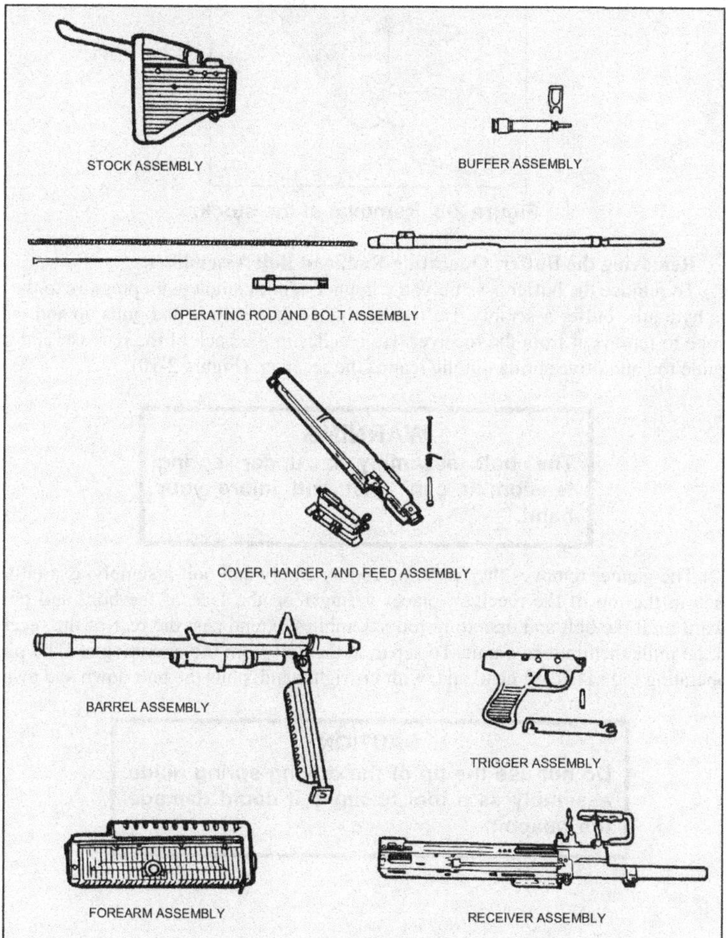

Figure 2-8. Eight major assemblies.

a. **Removing the Stock Assembly**. First, the gunner makes sure the bolt is in the forward position. To remove the stock, he raises the shoulder rest, inserts a cleaning rod into the hole to release the latch. He pulls the shoulder stock from the receiver, turns the latch lever, and opens the cover (Figure 2-9).

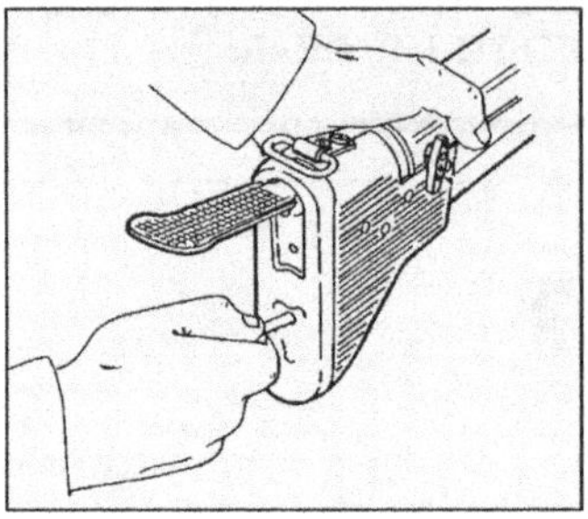

Figure 2-9. Removal of the stock.

b. **Removing the Buffer, Operating Rod, and Bolt Assemblies.**

(1) To remove the buffer assembly, the gunner applies slight palm pressure to the rear of the hydraulic buffer assembly. He reaches inside the receiver and pulls up and out on the yoke to remove it from the receiver. He reaches in the back of the receiver and pulls the guide rod and drive spring out the rear of the receiver. (Figure 2-10).

WARNING

The bolt assembly is under spring tension; it can twist and injure your hand.

(2) The gunner removes the operating rod assembly and bolt assembly as a unit. He reaches in the top of the receiver, places a finger on the face of the bolt, and pushes rearward until the bolt and operating rod assemblies extend past the rear of the receiver. Then, he pulls them out as a unit. To separate the bolt from the operating rod, he places the operating rod in his left hand and, with his right hand, pulls the bolt down and away.

CAUTION

Do not use the tip of the driving spring guide assembly as a tool because it could damage the weapon.

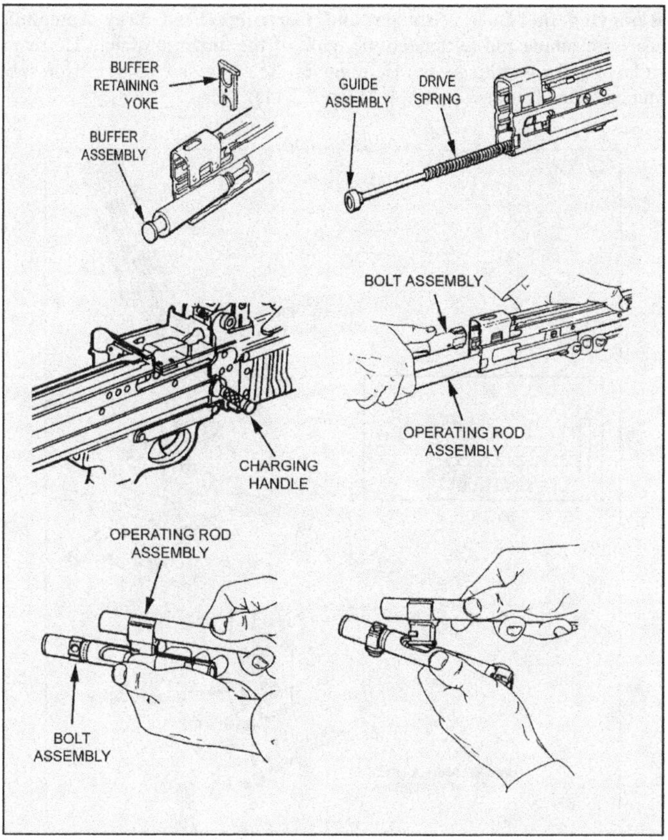

Figure 2-10. Removal of the buffer, operating rod, and bolt assemblies.

c. **Removing the Cover, Hanger and Cartridge Feed Tray Assemblies.** The gunner uses a cleaning rod to unlatch the hook of the hinge pin latch. He removes the hinge pin latch and cover hinge pin. He removes the cover assembly, torsion spring, and hanger and cartridge feed tray assembly (Figure 2-11).

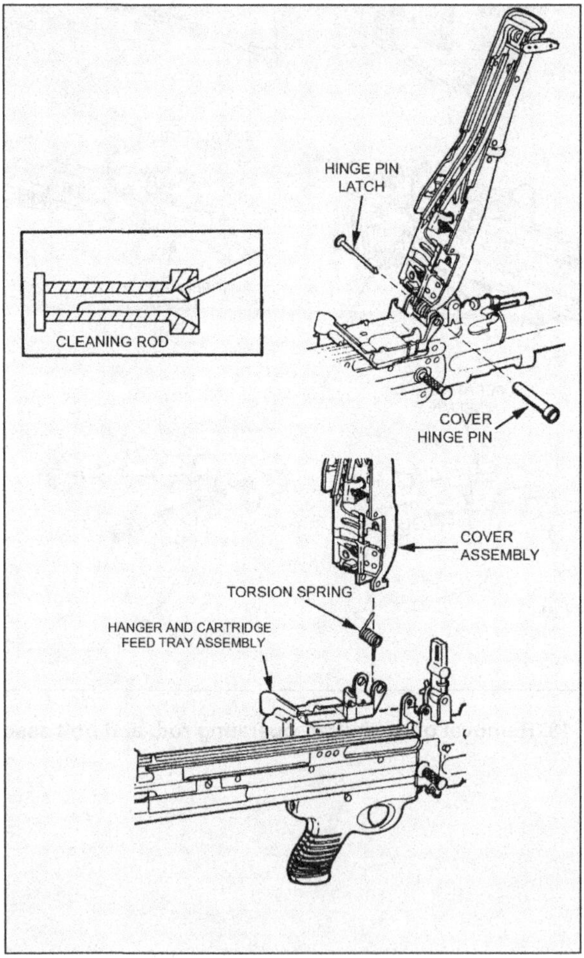

Figure 2-11. Removal of the cover, hanger and cartridge feed tray assemblies.

d. **Removing the Barrel Assembly**. The gunner pushes in the spring detent, raises the barrel lock, and removes the barrel assembly (Figure 2-12).

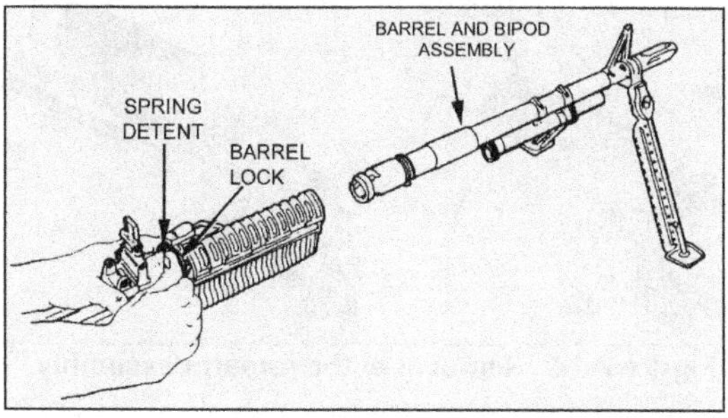

Figure 2-12. Removal of the barrel assembly.

e. **Removing the Trigger Mechanism Grip Assembly**. The gunner pushes in and removes the flat leaf spring. He pushes out the *front pin* and slides the trigger mechanism grip assembly slightly forward, then pulls it out to remove it (Figure 2-13).

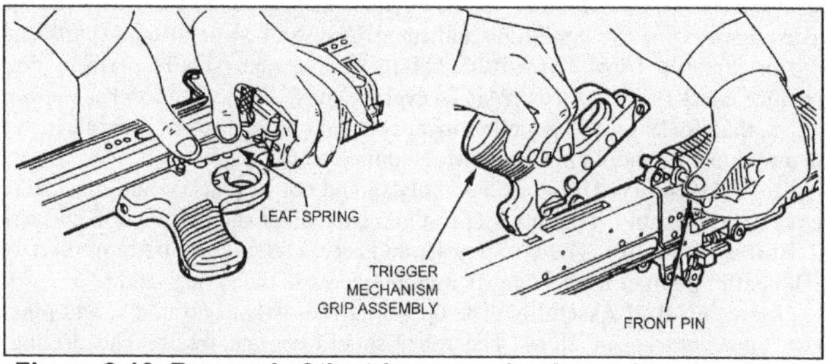

Figure 2-13. Removal of the trigger mechanism grip assembly.

f. **Removing the Forearm Assembly**. The gunner inserts a cleaning rod or the reamer portion of a combination wrench through the round opening in the forearm assembly, and then he pushes down on the spring. He lifts and gently slides the forearm assembly from the receiver (Figure 2-14, page 2-16).

CAUTION

Be careful not to damage the internal ribs of the forearm assembly. It can cause damage to the weapon.

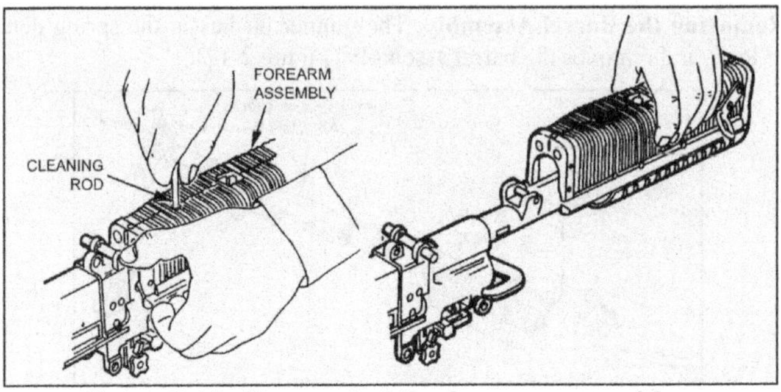

Figure 2-14. Removal of the forearm assembly.

g. **Removing the Receiver Assembly**. Once the forearm assembly is removed, the part remaining is the receiver assembly, and general disassembly is complete.

2-7. INSPECTION

Inspection begins with the weapon disassembled in its major groups. Shiny surfaces do not mean the parts are unserviceable. The gunner inspects each area of the weapon and related equipment for the conditions indicated. Any broken or missing parts should be repaired or replaced IAW TM 9-1005-224-10. The gunner should perform Preventive maintenance checks and services (PMCS) every 90 days. If the weapon has not been used in 90 days, the PMCS in the operator's manual should be performed regardless. If he sees rust on a weapon, he should perform PMCS immediately.

a. **Stock Assembly**. The stock assembly should not be cracked and must fit securely on the receiver assembly. The shoulder rest and latch lever should function correctly.

b. **Buffer Assembly**. The buffer yoke and recess should not be burred, cracked, or bent. The buffer plunger must fit easily into the recess in the spring guide.

c. **Operating Rod Assembly**. The operating rod, yoke, sear notch, and pins should not have burrs, cracks, or chips. The roller should operate freely. The driving spring should have tension and should not have kinks, breaks, or wear. The guide assembly stop should be tight. If it is loose, the gunner notifies unit maintenance.

d. **Bolt Assembly**. The bolt assembly is checked for burrs or cracks, especially in the locking lug area.

(1) The roller should operate freely and not be cracked.

(2) The spring should not be kinked.

(3) The threads on the bolt plug assembly and in the breech bolt should not be damaged.

e. **Cover Assembly**. The cover assembly is checked for burrs, cracks, looseness, or missing parts. The spring action of the front and rear cartridge guides, feed pawls (beneath the cartridge guides), and feed cam assembly are checked. The feed cam assembly should be secure and operate freely.

f. **Hanger and Cartridge Feed Tray Assembly**. The hanger and cartridge feed tray assembly is checked for burrs, cracks, and missing or loose parts. The feed tray is checked to make sure it fits on the receiver and that the cartridge-retaining pawl works properly. The hanger is checked for obstructions.

g. **Barrel Assembly**. The barrel assembly is checked for burrs, cracks, and wear, especially in the barrel socket area. The sight and flash suppressor are checked to make sure they are tight. The bipod should work properly, the legs should be straight, and connections should be tight. The bipod plungers should operate smoothly.

h. **Trigger Mechanism Grip Assembly**. The sear is checked for chips, cracks, or signs of wear. The sear plunger and spring are checked for wear.

i. **Forearm Assembly**. The forearm assembly is checked for damage.

j. **Receiver Assembly**. The receiver rails are checked for burrs and wear. The cocking handle should move freely.

k. **M122 Mount**. The T&E mechanism should not bind. The numbers on the scales and dials must be legible.

(1) Distinct clicks must be heard when the handwheels are turned. Index lines should be calibrated with the indicator pointer.

(2) The pintle should fit snugly in the pintle bushing, and the pintle lock should hold the pintle securely.

(3) The sleeve latch should function properly, and the traversing bar should be tight when the tripod legs are extended and latched.

l. **Carrying Case**. Maintenance tools and equipment should be complete and serviceable. The case should be serviceable. Frequent washing of the case should be avoided. Such washing may destroy the waterproofing and shrink the case.

2-8. CLEANING, LUBRICATION, AND PREVENTIVE MAINTENANCE

The M60 machine gun should be cleaned immediately after firing. The gunner disassembles the M60 into its major groups for cleaning. All metal components and surfaces that have been exposed to powder fouling should be cleaned using CLP on a bore-cleaning patch. The CLP is used on the bristles of the receiver brush to clean the receiver. After the M60 is cleaned and wiped dry, a thin coat of CLP is rubbed on with a cloth. This lubricates and preserves the exposed metal parts during all normal temperature ranges.

CAUTION

Do not get CLP in the gas cylinder when cleaning the barrel. Turn the barrel upside down so that the gas cylinder is above the barrel during cleaning.

a. The gas cylinder components are removed and cleaned only when inspection shows that the piston will not move within the cylinder when the barrel is tilted end-for-end. Unit maintenance personnel must supervise disassembly of the gas system. The receiver brush and swab-holding section of the cleaning rod may be used to clean the interior of the gas cylinder. When CLP is used, the gas cylinder and gas piston must be wiped dry before assembly. After assembly, the piston is checked for free movement. The unit armorer replaces safety wire. The gunner cleans the rest of the weapon as follows:

(1) Cleans the bore using CLP and a bore brush attached to a cleaning rod. Does not reverse the direction of the bore brush while in the bore.

(a) Runs the brush through the bore several times until most of the powder fouling and other foreign matter have been removed.

(b) Swabs the bore several times using a cleaning rod and a swab wet with CLP.

(c) Swabs the bore several times using a cleaning rod and dry swab.

(2) Cleans the chamber using CLP and a chamber brush attached to a cleaning rod.

(a) Runs the brush through the chamber several times until most of the powder fouling and other foreign matter have been removed.

(b) Swabs the chamber several times using a cleaning rod and a swab wet with CLP.

(c) Swabs the chamber several times using a cleaning rod and dry swab.

(3) Cleans the receiver using a receiver brush and CLP.

(a) Brushes the receiver until most of the powder fouling and other foreign matter have been removed.

(b) Swabs the receiver several times using a cleaning rod section and a swab wet with CLP.

(c) Swabs the receiver several times using a cleaning rod section and dry swab.

(4) Wipes all the parts of the weapon except those that are rubber coated, using a rag wet with CLP.

(5) Dries completely all parts cleaned with CLP.

(6) Lubricates the following moving parts with CLP as instructed:

(a) Barrel Assembly. Lubricates on the camming surfaces of the bolt-locking lugs.

(b) Operating Rod Assembly. Lubricates on the rollers and those surfaces immediately below the yoke that ride within the receiver rails.

(c) Cover assembly. Lubricates inside the feed cam assembly.

(d) Bolt assembly. Lubricates on the bolt locking lugs and cam actuator roller, and in the camming recess (for the operating rod).

b. After lubricating, the components are cycled by hand to spread the CLP. Weapons fired infrequently or stored for prolonged periods should have a light film of CLP applied to the interior of the gas cylinder and the gas piston immediately after cleaning or inspection. Preventive maintenance is performed every 90 days, unless inspection reveals more frequent servicing is necessary. The use of the lubricant does not eliminate the requirement for cleaning and inspecting to ensure that corrosion has not formed. Before the weapon is used, the gas system and components must be clean and free of oil and lubricants.

c. All exposed surfaces of the M122 tripod, pintle assembly, and T&E mechanism are wiped down with a clean rag. For stubborn areas with hard-to-remove dirt, a steel brush or bore brush is used to loosen the particles. Then a clean rag is used to wipe them down and CLP is used to lubricate them.

d. The following procedures apply to cleaning and lubricating the M60 during unusual conditions:

(1) Below 0 degrees Fahrenheit—use lubricating oil, arctic weather (LAW). Oil lightly to avoid freeze-up.

(2) Extreme heat—use a light coat of CLP.

(3) Damp or salty air—use CLP. Clean and apply frequently.

(4) Sandy or dusty areas—use CLP. Clean and apply frequently. Wipe with a rag after each application to remove excess.

2-9. GENERAL ASSEMBLY

After cleaning, lubricating, and inspecting the weapon, the gunner reassembles it in the reverse order of disassembly.

CAUTION

Be careful not to damage the internal ribs when installing the forearm assembly.

a. **Replacing Receiver Assembly and Forearm Assembly**. To replace the forearm assembly, the gunner slides it onto the receiver. He presses in on the bottom of forearm assembly to latch. He shakes the forearm assembly up and down to ensure it is seated.

b. **Replacing the Trigger Mechanism Grip Assembly**. To replace the trigger mechanism grip assembly, the gunner positions it on the bottom of the receiver, aligns it with the T-slot, and installs the front pin from the left side. He slides the slotted end of the flat leaf spring on the front pin. (The hooked end of the flat leaf spring should bend outward.) He pushes down and slides the hooked end of the flat leaf spring onto the grooved pin (Figure 2-15).

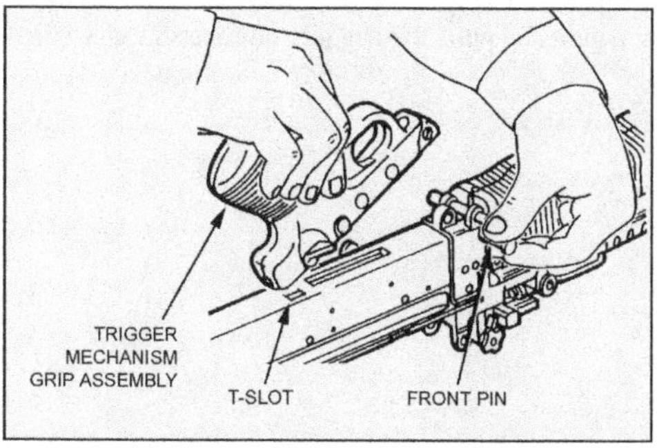

Figure 2-15. Replacing the trigger mechanism grip assembly.

c. **Replacing the Barrel Assembly**. To replace the barrel assembly, the gunner ensures that the barrel lock is up (unlocked) before inserting the barrel in the forearm assembly. He pushes the barrel lock down to lock it. Then, he lifts it up by the bipod and shakes it to ensure it is seated.

d. **Replacing the Cover, Hanger and Cartridge Feed Tray Assembly**. The gunner positions the hanger and cartridge feed tray assembly on the receiver. He installs the torsion spring and cover assembly and makes sure the ends of the torsion spring stick in the holes of the cover and receiver. (The hinge pin is inserted from right to left.) The gunner applies slight pressure to line up the torsion spring and inserts the cover hinge pin. He installs the hinge pin latch through the hinge pin until it locks. He keeps the cover assembly open until the machine gun is assembled.

e. **Replacing the Bolt Assembly, Operating Rod Assembly, and Buffer Assembly**. The gunner places the yoke of the operating rod assembly against the spool of the firing pin in the slot of the bolt assembly (Figure 2-16). Then, he pushes the spool forward. He seats the yoke between the spools of the firing pin and lets the yoke slide back in the bolt assembly.

(1) Slides the operating rod assembly with the bolt assembly into the weapon. Turns the bolt assembly so that the stripping lugs line up with the upper rails. Pushes in until the stripping lugs engage the rails.

(2) Turns the roller straight up and pushes the bolt assembly and operating rod assembly as a unit into the weapon.

(3) Pulls and holds the trigger while pushing the bolt assembly into the receiver until it locks in place.

(4) Installs the drive spring and guide assembly in the rear of the receiver.

(5) Inserts the buffer assembly against the end of the guide assembly and pushes until the groove on the buffer assembly lines up with the slot for buffer retaining yoke. Inserts the buffer retaining yoke.

f. **Replacing the Stock**. The gunner positions the stock on the rear of the receiver and pushes until it snaps in place. He pulls the cocking handle to the rear to lock the bolt assembly to the rear, and then he places the safety on "S." He closes the cover. While holding the charging handle, he pulls the trigger and eases the charging handle forward.

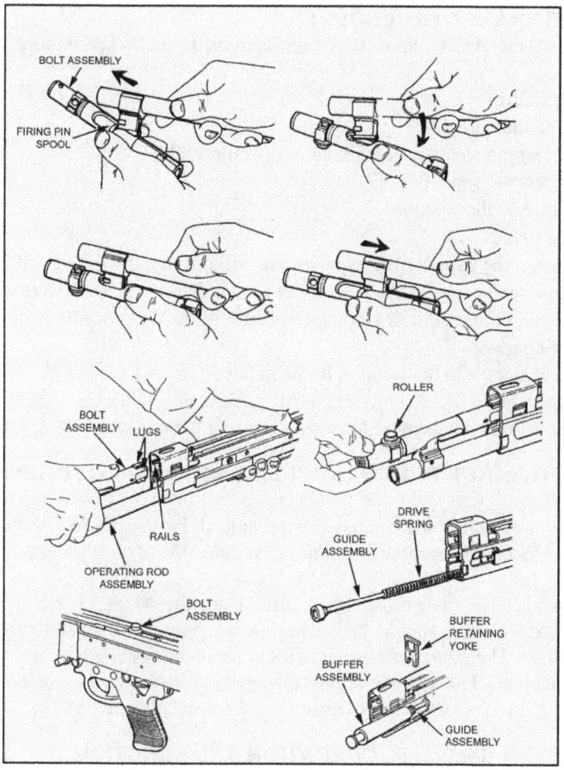

Figure 2-16. Replacing the bolt assembly, operating rod assembly, and buffer assembly.

2-10. FUNCTION CHECK

The gunner must perform a function check to ensure that the M60 is correctly assembled. The procedures, in order, are:

- Open the cover and pull the cocking handle to the rear.
- Place the safety on "S" position.
- Return the cocking handle to the forward position.
- Close the feed tray cover.
- Place the safety on "F" position.
- Grasp the cocking handle with the right hand, pull the trigger with the left hand, and ease the bolt forward.

2-11. MAINTENANCE PROCEDURES

Maintenance of the M60 requires certain actions to be taken before, during, and after firing.

a. **Before firing—**
- Wipe the bore dry.
- Inspect the weapon as outlined in operator's TM.
- Inspect the spare barrel.
- Lubricate the weapon.

b. **During firing—**
- Change the barrels. Barrel changing will prolong the life of both barrels.
- Periodically inspect the weapon to ensure that it is properly lubricated.
- When malfunctions or stoppages occur, follow the procedures in Section IV.

c. **After firing—**
- Clear and clean the weapon immediately.
- Every 90 days during inactivity, clean and lubricate the weapon unless inspection reveals more frequent servicing is necessary (TM 9-1005-224-10).

2-12. MAINTENANCE DURING NUCLEAR, BIOLOGICAL, CHEMICAL CONDITIONS

If the M60 machine gun is contaminated by chemical, biological, or radiological agents, the gunner takes the appropriate action to reduce the exposure and minimize the penetration.

a. **Chemical**. The gunner uses towelettes from the M258A1 kit to wipe off the weapon. If these are not available, he washes the weapon with soap and water.

b. **Biological**. The gunner uses towelettes or soap and water as above.

c. **Radiological**. The gunner wipes the weapon with warm soapy water if it is available. If not, he uses towelettes or rags. (For more details, see FM 3-5.)

Section III. OPERATION AND FUNCTION

This section discusses the operation of the M60 machine gun. This includes loading, firing, unloading, cycle of functioning, adjustment of the sight, using both the bipod and tripod, and use of the M60 on a vehicular mount.

2-13. OPERATION

The M60 machine gun is loaded, fired, unloaded, and cleared from the open-bolt position. The safety lever must be in the "F" position before the bolt can be pulled to the rear. Before use belted ammunition must be linked with the double link at the open end of the bandoleer. It must be free of dirt and corrosion.

2-14. LOADING

The gunner makes sure the weapon is cleared as previously described. He places the safety lever on "F". With his palm up, he pulls the cocking handle to the rear, then pushes it forward until it locks. With the bolt held to the rear, he places the safety lever on "S" and manually returns the cocking handle to the forward position. He turns the latch lever and opens the cover. He raises the cartridge feed tray and places the bandoleer on the bandoleer hanger. He places the first round of the belt in the feed groove with the double link first and the open side of the links down. He ensures that the round remains in the feed groove and holds the belt up (about six rounds from the loading end) while closing the cover (Figure 2-17).

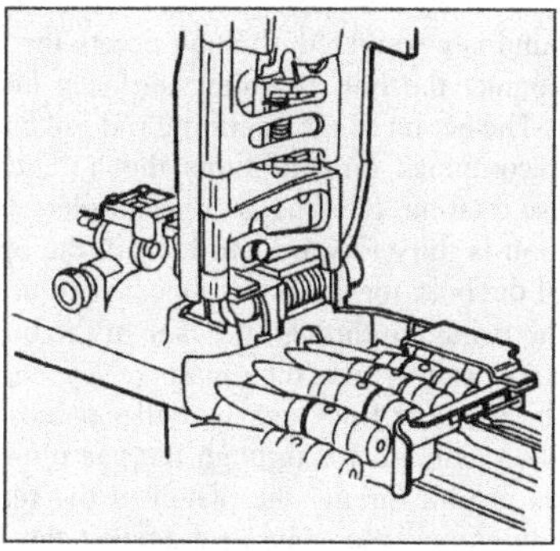

Figure 2-17. Loading.

2-15. UNLOADING

The gunner unloads the M60 by pulling and locking the bolt to the rear position. He places the safety lever on "S", raises the cover and tray, and removes any ammunition or links from the tray. He performs the *four-point safety check* (see Section II).

2-16. CYCLE OF FUNCTIONING

The M60 gunners can recognize and correct stoppages when they know how the weapon functions. The weapon functions automatically as long as ammunition is fed into it and the trigger is held to the rear. Each time a round is fired, the parts of the weapon function in a cycle or sequence. Many of the actions occur at the same time. These actions are separated in this manual only for instructional purposes.

a. The first round of the belt is placed in the tray groove to start the cycle. Then, the trigger is pulled, releasing the sear from the sear notch. The rear of the sear is lowered and disengaged from the sear notch when the trigger is pulled to the rear. This procedure allows the operating rod and bolt to be driven forward by the expansion of the operating rod spring. The cycle stops when the trigger is released and the sear again engages the sear notch on the operating rod.

b. The sequence of functioning is as follows:

(1) *Feeding.* As the bolt starts its forward movement, the feed cam is forced to the right causing the feed cam lever to turn in the opposite direction and forcing the belt feed pawl over the next round in the bolt. The next round is then ready to be placed into the tray groove when the rearward action occurs again. As the bolt moves to the rear after firing, the cam roller forces the feed cam to the left. The feed cam lever is forced to turn, which moves the feed pawl to the right and places a round in the tray groove.

(2) *Chambering.* As the bolt travels forward, the upper locking lug engages the rim of the round. The pressure of the front and rear cartridge guides holds the round so that positive contact is made with the upper locking lug of the bolt. The front cartridge guide prevents forward movement of the link as the round is stripped from the belt. The upper

locking lug carries the round forward. The chambering ramp causes the nose of the round to be cammed downward into the chamber. When the round is fully seated in the chamber, the extractor snaps over the rim of the round, and the ejector on the rail inside the receiver is depressed.

(3) *Locking.* As the round is chambered, the bolt enters the barrel socket. The upper and lower locking lugs contact the bolt camming surfaces inside the barrel and start turning the bolt clockwise. The action of the operating rod yoke against the bolt camming slot as the operating rod continues forward turns the bolt, completing its 90-degree (one-quarter turn) clockwise rotation. Locking is now complete.

(4) *Firing.* After the bolt is fully forward and locked, the operating rod continues to go forward, independent of the bolt, for a short distance. The yoke, engaged between the firing pin spools, carries the firing pin through the face of the bolt. The firing pin strikes the primer of the round and the primer fires the round.

(5) *Unlocking.* After the round is fired and the bullet passes the gas port, part of the expanding gases go into the gas regulator through the gas plug. The rapidly expanding gases enter the hollow gas piston, forcing the piston to the rear. As the operating rod continues to the rear, the operating rod yoke acts against the bolt camming slot. This causes the bolt to begin its counterclockwise rotation. The upper and lower locking lugs of the bolt contact the bolt camming surfaces inside the barrel socket and, as the bolt continues toward the rear, it completes a one-quarter turn counterclockwise. The rotation and movement to the rear unlocks the bolt from the barrel socket. Unlocking begins as the yoke of the operating rod contacts the curve of the bolt camming slot and ends as the bolt clears the end of the barrel socket.

(6) *Extracting.* Extracting begins during the unlocking cycle. The rotation of the bolt loosens the cartridge case in the chamber. As the piston and bolt move to the rear, the extractor pulls the cartridge case from the chamber.

(7) *Ejecting.* As the cartridge case is pulled from the chamber, the bolt passes by the ejector. This procedure causes the ejector clip to expand, forcing the ejector to push the expended cartridge. The extractor grips the right side of the cartridge and causes it to spin from the weapon as it reaches the ejection port. The empty belt links are forced out the link ejection port as the rearward movement of the bolt causes the next round to be positioned in the tray groove.

(8) *Cocking.* As the expanding gases force the gas piston to the rear, the operating rod first moves independently of the bolt. The yoke of the operating rod acts against the rear firing pin spool, pulling the firing pin from the primer of the spent cartridge case. The action of the piston assembly, continuing to the rear with the firing pin, releases the compression of the firing pin spring. As long as the trigger is held to the rear, the M60 continues to complete the eight steps of functioning automatically. When the trigger is released and the sear engages the sear notch, the cycle of functioning is stopped and the weapon is cocked. To prevent undue wear to the sear and sear notch, the gunner must hold the trigger firmly to the rear during firing.

2-17. SIGHTS

The range scale on the rear sight is marked for each 100 meters from 300 to 1,100 meters. It can be adjusted for zeroing. Range changes are made by using either the scale retaining/adjusting screw or elevation knob. The scale retaining/adjusting screw is used to make major adjustments in elevation. The elevation knob is used to make minor adjustments, such as during zeroing (Figure 2-18, page 2-26). Four clicks on the elevation knob are equal to a 1-cm change in elevation or point of aim 1 cm. From the rear of the weapon, the elevation knob is turned clockwise to raise the rear sight and lower the strike of the round. It is turned counterclockwise to lower the sight and raise the strike of the round. The rear sight is adjustable for windage, 5 mils right or left of the zero index line. The windage knob is on the left side of the rear sight. One click on the windage knob equals a 1-cm change in deflection or point of aim 1 cm. The windage knob is turned toward the muzzle of the weapon to move the sight and the strike of the round to the right. It is turned toward the rear to move the sight and the strike of the round to the left.

a. **Elevation**. Before making elevation adjustments, the range knob must be at its highest setting. If the center of the shot group is above or below the aiming point, the gunner rotates the elevation knob by moving the rear sight slide in the direction of the desired change. One 180-degree turn in either direction moves the strike of the round 1/4 cm at 10 meters, or four clicks equals 1-cm change.

b. **Windage**. If the center of the group is to the left or right of the black aiming paster, the gunner must correct for windage. To correct windage, he rotates the windage knob to move in the direction of the desired change. For example, he rotates the windage knob toward the muzzle (counterclockwise) to move the strike of the round to the right; he rotates the windage knob toward the buttstock [clockwise] to move the strike of the round to the left. One click in either direction moves the strike of the round 1/2 cm at 10 meters.

c. **10-Meter Zeroing, Set the Sights (Mechanical Zero)**. The gunner indexes or places the range scale on a range of 500 meters. He aligns the windage by placing zero windage on the index line. He assumes a prone position and sights on the target.

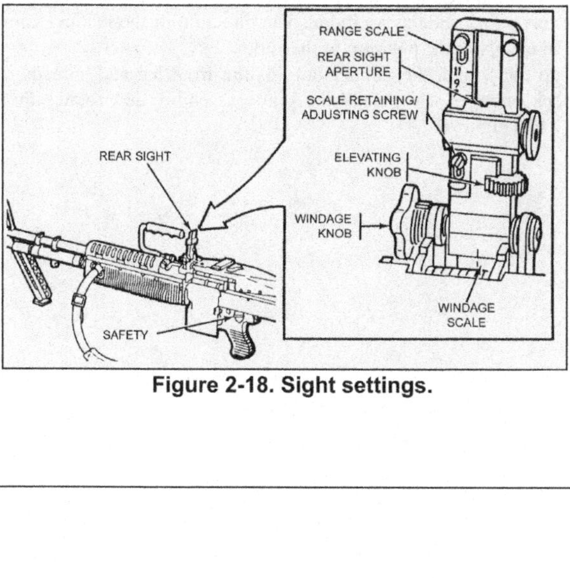

Figure 2-18. Sight settings.

2-18. M122 TRIPOD

The M122 tripod provides a stable mount for the M60, and it permits a high degree of accuracy and control. The tripod is recommended for marksmanship training and defensive employment.

a. **Mounting M60 on the Tripod**. The tripod assembly provides a stable and relatively lightweight base that is far superior to the bipod. The tripod may be extended and collapsed without difficulty. It consists of a tripod head, one front leg and two rear legs, and traversing bar. The traversing bar connects the two rear legs. It is hinged on one side with a sleeve and sleeve latch on the other, which allows the tripod to collapse to a closed position for carrying or storage or to lock in an open extended position for use. The traversing bar also supports the T&E mechanism. Engraved on the bar is a scale that measures direction in mils. It is graduated in 5-mil increments. It is numbered every 100 to 425 mils right of center and numbered every 100 to 450 mils left of center.

(1) The T&E mechanism is to provide controlled manipulation and the ability to engage predetermined targets.

(a) The traversing portion of the mechanism consists of the traversing handwheel and traversing slide-lock lever. As the traversing handwheel is turned, the muzzle of the weapon turns to the left or right, depending on the direction it is turned. Each click of the traversing handwheel indicates a 1-mil change in direction of the muzzle: 1 click equals 1 mil. The total of 100 mils traverse includes 50 mils right and 50 mils left of center.

(b) The elevating portion of the mechanism consists only of the elevating handwheel. The elevating handwheel has a mil-click device built into it (1 click = 1 mil). Engraved into the handwheel is a scale divided into 5-mil divisions and 1-mil subdivisions, for a total of 50-mil increments. There are 200 mils above and 200 mils below the zero mark, for a total of 400 mils in elevation change. Elevation readings are taken in two parts. First, the major reading is taken from the elevation screw plate. The second, minor reading is taken from the handwheel. The two readings are separated by a slash (/) when they are recorded.

(c) The traversing slide-lock lever allows rapid lateral adjustments along the traversing bar. Direction readings are taken from the scale on the traversing bar, using the left side of the traversing slide as an index. The direction of the reading comes from the position of the muzzle, not the position of the slide.

(2) To set up the tripod, the gunner unfolds the front leg and spreads the rear legs until the leg lock engages. He inserts the pintle assembly and rotates the pintle lock release cam to lock (Figure 2-19).

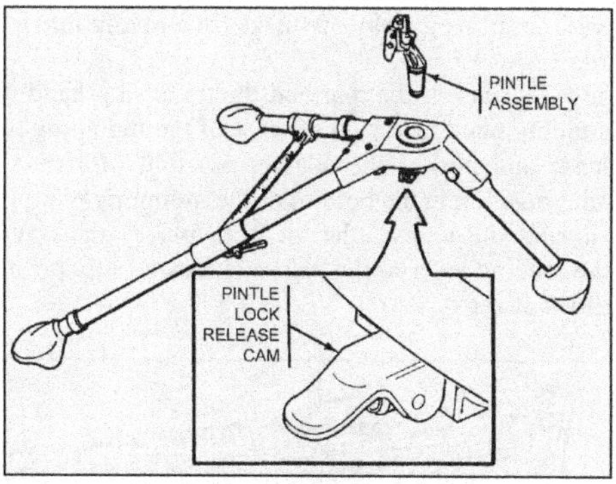

Figure 2-19. M122 tripod.

(3) To mount the M60, the gunner places the ends of the M60's front mounting pin on top of the pintle assembly. He presses the bottom of the latch to open the pintle assembly. The ends of the M60 mounting pin should lock in place on the pintle assembly (Figure 2-20).

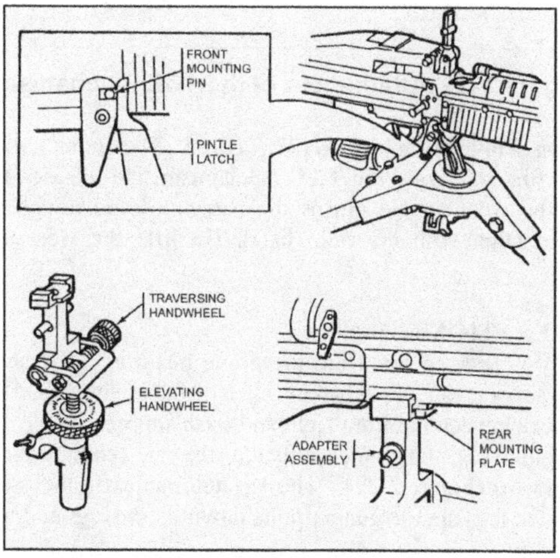

Figure 2-20. Mounting of the M60 on the tripod.

(4) To attach the T&E mechanism to the weapon mounted on the tripod (Figure 2-21), the gunner centers the elevating and traversing handwheels. To do this, he turns the traversing handwheel toward his body as far as it will go, then turns it away two complete revolutions. He checks the traversing handwheel scale to ensure the "0" on the scale is aligned with the "0" index line before and after the two revolutions. At night, the

gunner positions the traversing mechanism by turning the traversing handwheel toward his body as far as it will go, turns it away 50 clicks (two revolutions), and then allows the weapon to go forward.

(a) With the slide lock lever to the rear and the traversing handwheel to the left, the gunner places the mounting plate recess on the rear of the mounting plate. He pulls down the locating pin release and pushes the adapter assembly forward. (The locating pin automatically locks into position in the bottom of the mounting plate.)

(b) The gunner lowers the rear of the weapon, places the traversing slide on the traversing bar with the locking lever to the rear, and locks it into position. He indexes the left edge of the slide lock at zero.

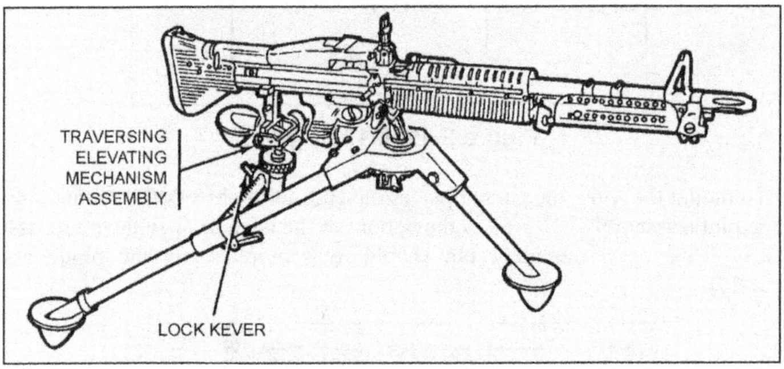

Figure 2-21. Attachment of the T&E mechanism.

b. **Dismounting M60 From the M122 Tripod**. The gunner dismounts the M60 from the tripod by first removing the T&E mechanism. He releases the lock lever and raises the rear of the weapon. He grasps the carrying handle with his left hand and depresses the pintle latch with his right hand. He lifts the weapon from the pintle assembly.

2-19. BIPOD OPERATIONS

The bipod assembly is used to fire from the prone position. The shoulder rest on the buttstock provides support for the weapon when fired in the bipod mode. The bipod group is held in place between the front sight and flash suppressor.

a. To lower a bipod leg, the gunner pulls it to the rear (compressing the lock spring) and rotates it downward (Figure 2-22). The leg automatically locks when in the down position. To return the legs up, the gunner pulls down on the legs and rotates upward.

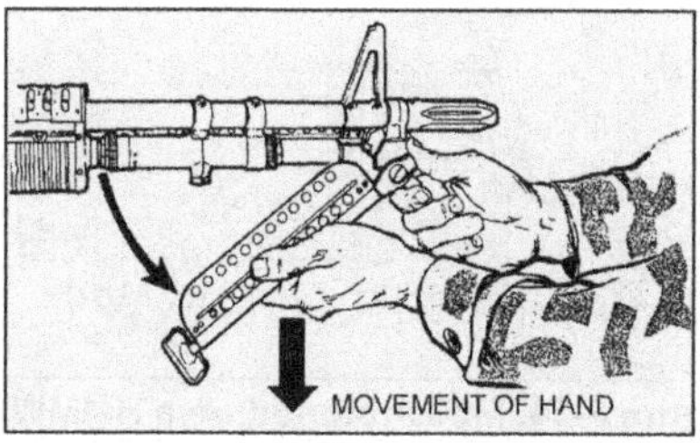

Figure 2-22. Lowering of the bipod.

b. To extend a bipod leg, the gunner pulls down on the foot. The bipod leg plunger engages a notch in the bipod leg extension and holds it in the desired position. To shorten the bipod leg, the gunner depresses the bipod leg plunger and pushes up on the bipod foot (Figure 2-23).

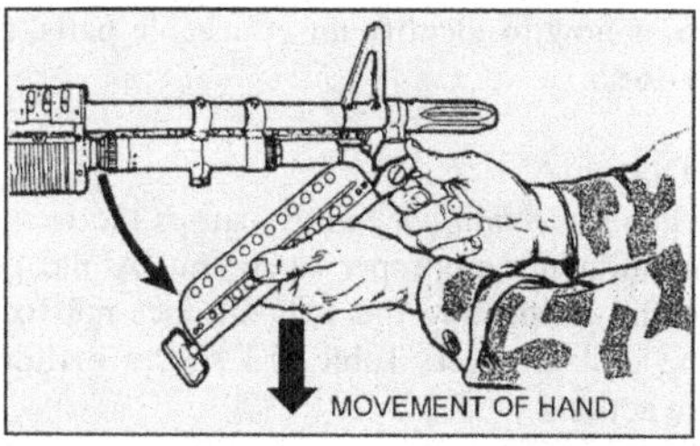

Figure 2-23. Adjustment of the bipod leg extension.

2-20. VEHICULAR MOUNT

The standard vehicular mount for the M60 machine gun is the M4 pedestal mount used on the HMMWV. One component of the pedestal mount, the M142 machine-gun mount, (which serves as a cradle for the weapon), is also adaptable to other vehicles (Figure 2-24).

a. To mount the weapon, the gunner locks the platform in the horizontal position by inserting the travel-lock pin into the travel lock. He places the front mounting pin (in the forearm assembly) into the front mounting lug. He lowers the receiver so that the rear locating pin snaps into the platform latch.

b. To dismount the weapon, the gunner ensures that the travel lock is engaged (holding the platform in a horizontal position). He grasps the carrying handle with one hand and depresses the platform latch with the other. He raises the rear of the weapon slightly and lifts it from the mount.

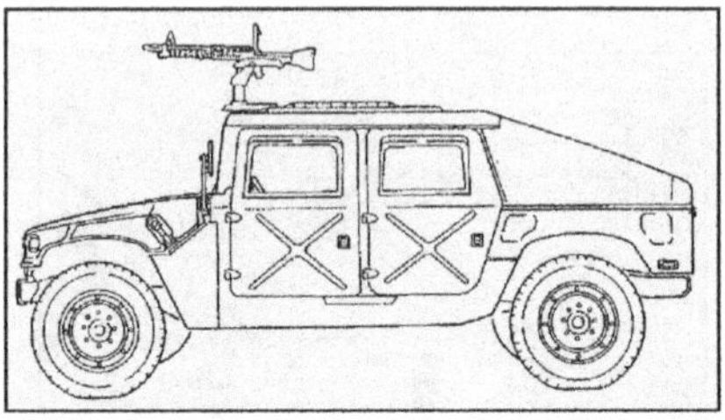

Figure 2-24. M60 mounted on a HMMWV.

2-21. TRIPOD POSITIONING

The M122 tripod provides a stable mount for the M60, and it permits a high degree of accuracy and control. The gunner unfolds the front leg, positions it toward the target, then spreads the rear legs until the leg lock engages.

Section IV. PERFORMANCE PROBLEMS AND DESTRUCTION

This section identifies some of the problems that cause the weapon to perform improperly. It also explains how to identify unserviceable parts, and how to destroy the M60 when authorized to do so.

2-22. MALFUNCTIONS

A malfunction occurs when a *mechanical failure* causes the *weapon to fire improperly*. Neither defective ammunition nor improper operation by the gunner is considered a malfunction. If cleaning and lubricating the weapon does not fix the problem, then the gunner turns it in to the unit armorer. Table 2-3 shows malfunctions, their probable causes, and the corrective actions.

MALFUNCTION	PROBABLE CAUSE	CORRECTIVE ACTION
Sluggish operation.	Carbon buildup in the gas system.	Clean gas regulator, piston, and cylinder.
	Lack of lubricant.	Lubricate.
	Burred parts.	Notify unit maintenance.
Uncontrolled fire (runaway weapon).	Worn or broken sear.	Send to DS maintenance.
	Worn or broken operating rod sear notch.	Send to DS maintenance.
	Sear installed backwards.	Install correctly.
		Install correctly. **NOTE**: Always install the rod yoke between the two firing pin spools.
	Carbon buildup in gas port.	Clean gas port.

Table 2-3. Malfunctions.

2-23. STOPPAGES

A stoppage is any *interruption* in the cycle of functioning caused by *faulty action* of the weapon or *faulty ammunition*. Stoppages are classified by their relationship to the cycle of functioning. Table 2-4 shows types of interruptions or stoppages, their probable causes, and the corrective actions.

STOPPAGE	PROBABLE CAUSE	CORRECTIVE ACTION
Failure to chamber.	Ruptured cartridge case.	Remove cartridge IAW TM 9-1005-224-10.
	Carbon buildup in gas cylinder.	Remove carbon.
	Carbon buildup in receiver.	Remove carbon.
	Damaged round.	Remove round and reload gun.
	Dirty chamber.	Clear barrel and clean and lubricate as required.
	Weak or short operating rod.	Replace.
Failure to lock.	Weak or short operating rod spring.	Replace.
	Foreign matter in chamber of receiver.	Clean and lubricate as required.
Failure to fire.	Faulty ammunition.	Replace.
	Broken or damaged firing pin or firing pin spring.	Replace.
	Defective trigger.	Send to DS maintenance.
	Broken or deformed sear plunger or spring.	Send to DS maintenance.
	Failure to lock.	See "Failure to Lock."
Failure to extract.	Gas piston installed backwards.	Install properly IAW TM 9-1005-224-10.
	Broken extractor spring.	Replace.
	Chipped or broken extractor.	Replace.
	Defective extractor plunger.	Replace.
	Short recoil.	Clean gas port and operating rod tube, and lubricate as required. Replace operating rod spring.

Table 2-4. Stoppages.

STOPPAGE	PROBABLE CAUSE	CORRECTIVE ACTION
Failure to cock.	Broken sear.	Send to DS maintenance.
	Worn operating rod sear notch.	Send to DS maintenance.
	Broken, defective, or missing sear plunger or spring.	Send to DS maintenance.
	Short recoil.	Clean gas port and operating rod tube, and lubricate as required. Replace operating rod spring.

Table 2-4. Stoppages (continued).

DANGERS

1. **IF NOTHING IS EJECTED AND THE WEAPON IS HOT (200 OR MORE ROUNDS FIRED IN LESS THAN 2 MINUTES), DO NOT OPEN THE COVER. MOVE THE SAFETY TO "S", WHICH PLACES THE WEAPON ON "S". KEEP THE WEAPON POINTED DOWNRANGE AND KEEP AWAY FROM THE WEAPON FOR 15 MINUTES, THEN CLEAR THE WEAPON.**
2. **BE CAREFUL IN CLEARING THE WEAPON WHEN THE BARREL IS HOT, A ROUND MAY FIRE (COOK OFF) DUE TO THE BARREL'S HEAT INSTEAD OF DUE TO THE FIRING MECHANISM. DURING TRAINING OR ON A FIRING RANGE, AFTER THE WEAPON HAS FIRED 200 ROUNDS, ITS BARREL IS CONSIDERED A HOT BARREL.**
3. **DURING COMBAT, WAIT 5 SECONDS, BECAUSE OF THE POSSIBILITY OF A "HANGFIRE," BEFORE APPLYING IMMEDIATE AND REMEDIAL ACTION. DURING TRAINING, WAIT 15 MINUTES BEFORE CLEARING A HOT WEAPON AND APPLYING IMMEDIATE OR REMEDIAL ACTION.**

NOTE: When applying immediate or remedial action on a cold or hot gun, if any part of the round (ranging from the tip of the bullet to the rim) is in the chamber, the gunner removes the ammunition from the feed tray only, then closes the cover and attempts to fire. If the weapon fires, he reloads and continues firing. If it does not fire, he clears the weapon (removes the round using a clearing rod with the cover closed, not using anything other than a clearing rod), then he inspects the weapon and ammunition.

2-24. IMMEDIATE ACTION

Immediate action is action taken to *reduce a stoppage without looking for the cause.* Immediate action should be taken in the event of either a misfire or a cookoff. A *misfire* is the failure of a chambered round to fire. Such failure can be due to an ammunition defect or faulty firing mechanism. A *cookoff* is the firing of a round caused by the heat of a hot barrel and not by the firing mechanism. A cookoff can be avoided by applying immediate action within 10 seconds after a failure to fire. If the M60 stops firing, the gunner performs the following immediate actions. An effective memory aid is POPP, which stands for Pull, Observe, Push, and Press:

a. Pulls and locks the cocking handle to the rear while observing the ejection port to see if a cartridge case, belt link, or round is ejected. Ensures that the bolt remains to the rear to prevent double feeding if a round or cartridge case is not ejected.

b. If a cartridge case, belt link, or a round is ejected, returns the cocking handle to the forward position, aims on the target, and presses the trigger. If the weapon still does not fire, takes remedial action. If a cartridge case, belt link, or round is not ejected, takes remedial action.

2-25. REMEDIAL ACTION

Remedial action is any action taken to determine the cause of a stoppage and to restore the weapon to an operational condition. This action is taken only after immediate action fails to remedy the problem.

a. **Cold Weapon Procedures**. When a stoppage occurs with a cold weapon and immediate action has failed, the gunner uses the following procedures:

(1) Pulls the cocking handle to the rear, locking the bolt. Moves the safety to “S” and returns the cocking handle.

(2) Places the weapon on the ground or away from his face, opens the cover, and performs the four-point safety check. Reloads and continues to fire.

(3) If the weapon does not fire, clears the weapon and inspects it and the ammunition.

b. **Hot Weapon Procedures**. If the stoppage occurs with a hot weapon (200 rounds or more fired in 2 minutes or noted as previously for training), the gunner moves the safety to “S”, waits 5 seconds (during training, lets the weapon cool for 15 minutes), then uses the same procedures as outlined for cold weapon procedures.

c. **Jammed Cocking Handle**. If a stoppage occurs and the cocking handle cannot be pulled to the rear by hand (the bolt may be fully forward and locked or only partially forward), the gunner takes the following steps:

(1) Tries once again to pull the cocking handle *by hand.*

> **CAUTION**
> **Do not try to force the cocking handle to the rear with your foot or a heavy object. This could damage the weapon.**

(2) If the weapon is hot enough to cause a cookoff, moves all soldiers a safe distance from the weapon and keeps them away for 15 minutes.

(3) After the gun has cooled, opens the cover and disassembles the gun. Ensures rearward pressure is kept on the cocking handle until the buffer is removed. (The assistant gunner helps the gunner do this.)

(4) Removes the round or fired cartridge. Uses cleaning rod or ruptured cartridge extractor if necessary.

(a) In a training situation, after completing the remedial action procedures, the gunner does not fire the weapon until it has been inspected by an ordnance specialist.

(b) In a combat situation, after the stoppage has been corrected, the gunner changes the barrel and tries to fire. If the weapon fails to function properly, the gunner sends it to the unit armorer.

2-26. DESTRUCTION PROCEDURES

Destruction of any military weapon is only authorized as a last resort to prevent enemy capture or use. This paragraph discusses the field-expedient means of this destruction; it does not replace published policies. In combat situations, the commander has the authority to destroy weapons, but he must report this destruction through channels.

a. Disassemble the weapon as completely as time permits. Use the barrel or tripod mount to destroy the bolt, buffer, and operating rod group, barrels, rear and front sights, and mounts.

b. Bury the disassembled weapon or dump the parts into a stream, a sump, or a latrine.

c. Burn the weapon by placing an incendiary grenade on the receiver group over the bolt (with the cover resting on the grenade) and detonating the grenade.

d. Smash the T&E mechanism and pintle assembly. Bend the tripod legs.

CHAPTER 3

M240B MACHINE GUN

The M240B machine gun supports the rifleman in both offensive and defensive operations. The M240B provides the heavy volume of close and continuous fire needed to accomplish the mission. The M240B is used to engage targets beyond the range of individual weapons, with controlled and accurate fire. The long-range, close defensive, and final protective fires delivered by the M240B form an integral part of a unit's defensive fires. This chapter describes the weapon and the types of ammunition in detail and provides a table of general data.

Section I. DESCRIPTION AND COMPONENTS

This section describes the M240B machine gun and its components and purposes. It also discusses the different types of ammunition that are fired from the M240B machine gun. This section also discusses the ammunition adapter and the blank firing adapter for the M240B machine gun.

3-1. DESCRIPTION

The M240B is a general-purpose machine gun. (Figure 3-1) It can be mounted on a bipod, tripod, aircraft, or vehicle. The M240B is a belt-fed, air-cooled, gas-operated, fully automatic machine gun that fires from the open bolt position. Ammunition is fed into the weapon from a 100-round bandoleer containing a disintegrating metallic split-link belt. The gas from firing one round provides the energy for firing the next round. Thus, the gun functions automatically as long as it is supplied with ammunition and the trigger is held to the rear. As the gun is fired, the belt links separate and are ejected from the side. Empty cases are ejected from the bottom of the gun. A spare barrel is issued with each M240B, and barrels can be changed quickly as the weapon has a fixed head space. However, barrels from different weapons should not be interchanged. The bore of the barrel is chromium plated, reducing barrel wear to a minimum.

DANGER

DO NOT INTERCHANGE THE BARREL ASSEMBLY OR THE BOLT ASSEMBLY FROM ONE WEAPON TO ANOTHER. IF YOU DO SO, IT MAY RESULT IN DEATH OR INJURY.

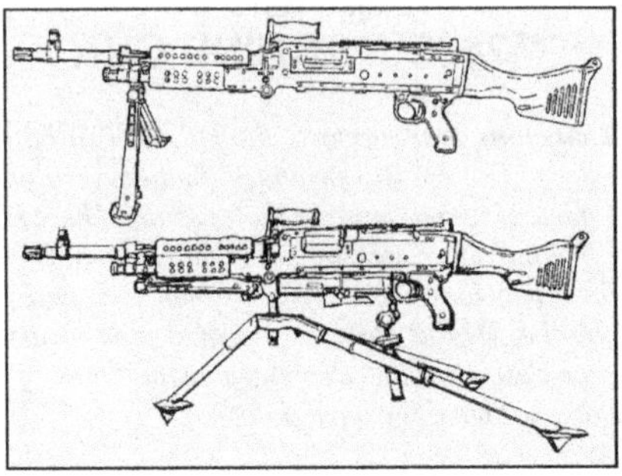

Figure 3-1. M240B machine gun, bipod and tripod mounted.

Ammunition	7.62-mm ball, tracer, armor-piercing, blank, dummy. Armor-piercing round is not authorized for training.
Tracer burnout	900 meters
Length of the M240B	49 inches
Weight of the M240B	27.6 pounds
Weight of tripod-mount M122A1 tripod with/flex-mount, complete	20 pounds
Maximum range	3,725 meters
Maximum effective range	1,100 meters with tripod and T&E
Area:	
Tripod	1,100 meters
Bipod	800 meters
Point:	
Tripod	800 meters
Bipod	600 meters
Suppression	1,800 meters
Maximum extent of grazing fire obtainable over level or uniformly sloping terrain	600 meters
Height of the M240B on the tripod mount M122A1	17.5 inches
Rates of fire:	
Sustained	100 rounds per minute fired in 6- to 9-round bursts and 4 to 5 seconds between bursts (barrel change every 10 minutes).
Rapid	200 rounds per minute fired in 10- to 13-round bursts and 2 to 3 seconds between bursts (barrel change every 2 minutes).

Table 3-1. General data.

Cyclic	650 to 950 rounds per minute in continuous bursts (barrel change every minute).
Basic load of ammunition (three-man crew)	900 to 1200 rounds.
Elevation, tripod controlled	+247 mils
Elevation, tripod free	+300 mils
Depression, tripod controlled	-200 mils
Traverse, controlled by T&E mechanism	100 mils
Normal sector of fire (with tripod)	875 mils
Free gun	6,400 mils

Table 3-1. General data (continued).

3-2. COMPONENTS

The components of the M240B machine gun and their purpose are shown in Table 3-2 and Figure 3-2, page 3-4.

	COMPONENTS	PURPOSES
(1)	Barrel assembly.	Consists of the barrel, flash suppressor, carrying handle, heat shield, front sight assembly, and gas-regulator plug. Houses the cartridge for firing and directs the projectile.
(2)	Heat shield assembly.	Protects the gunner's hand from a hot barrel.
(3)	Buttstock/buffer assembly and buffer and spade grip assembly/buttstock and buffer assembly.	Composite buttstock: buffer housing that contains spring washers to absorb recoil from bolt and operating rod assembly at the end of recoil movement.
(4)	Receiver assembly.	Consists of receiver, handguard, bipod, and rear sight assembly. Serves as support for all major components. Houses action of the weapon and, through a series of cam ways, controls function of weapon.
(5)	Handguard assembly (not shown).	Provides thermal insulation to protect the gunner's hands from heat or extreme cold.
(6)	Cocking handle assembly.	Pulls the moving parts rearward. Moves in a guide rail fixed to the right side of the receiver.
(7)	Trigger housing assembly.	Controls the firing of the machine gun. Contains trigger and safety components
(8)	Sling and snap hook assembly.	Provides a means of carrying the weapon.
(9)	Bipod.	Supports M240B machine gun in prone position.

Table 3-2. Components and purposes.

COMPONENTS		PURPOSES
(10)	Drive spring rod assembly.	Provides energy for returning bolt and operating rod assembly to firing position.
(11)	Bolt/operating rod assembly.	Provides feeding stripping, chambering, firing, extracting, and ejecting of cartridges using propellant gases for power.
(12)	Cover assembly.	Feeds linked belt, positions and holds cartridges in position for stripping, feeding, and chambering. Top rail configuration allows mounting of optical and electronic sights.
(13)	Feed tray.	Serves as guide for positioning cartridge to assist in chambering. Has a slotted top to allow air to circulate around barrel for cooling purposes.
(14)	Tripod assembly (not shown).	The tripod T&E mount assembly is flexible, provides a stable mount, absorbs recoil, and improves accuracy.
(15)	Ejection port.	Provides guide for ejection of spent cartridges.

Table 3-2. Components and purposes (continued).

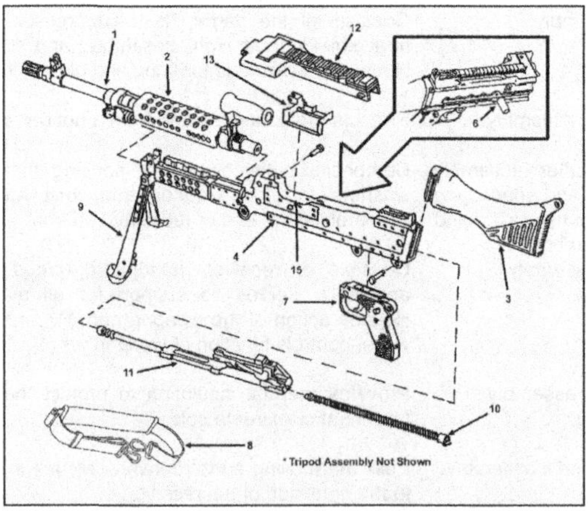

Figure 3-2. Major components of the M240B.

a. **Sights.** The front sight is attached to the barrel and can be adjusted for elevation and windage to allow the gunner to zero his weapon. Since the sight on the barrel is adjusted to zero the machine gun, both barrels must be zeroed before combat and training. The rear sight is attached to the rear of the receiver and is marked for each 100 meters of range, from 200 to 800 meters on the upper surface of the leaf, and on the reverse side of the leaf from 800 to 1,800 meters. (Figure 3-1)

b. **Safety Mechanism**. The safety mechanism is located on the pistol grip just behind the trigger well. When the safety is pushed to the right, the letter "S" is visible indicating the weapon is on safe. When pushed to the left, the letter "F" is visible on the safety indicating the weapon is on fire. The safety can only be engaged when the bolt is in the rear position. On the "S" position, the bolt cannot be released to go forward (Figure 3-1).

3-3. AMMUNITION

The M240B machine guns use several different types of 7.62-mm standard military ammunition. The specific type ammunition and its characteristics are as shown in (Figure 3-3). Soldiers use only authorized ammunition that is manufactured to US and NATO specifications. The ammunition is issued in a disintegrating, metallic, split-linked belt (Figure 3-4).

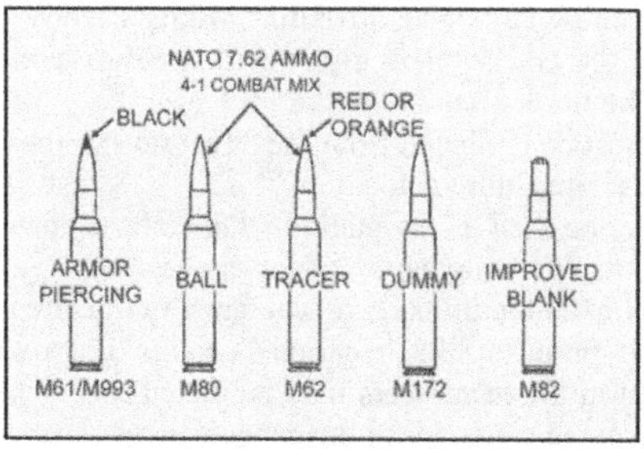

Figure 3-3. 7.62-mm cartridges for the M240B machine gun.

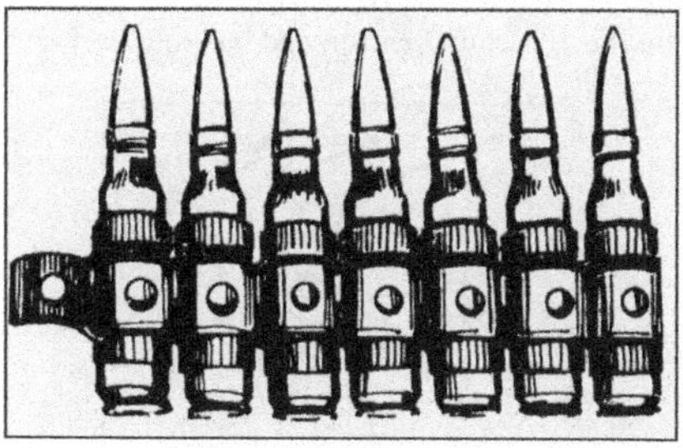

Figure 3-4. 7.62-mm cartridge in metallic belt.

a. **Classification.** The M240B machine guns ammunition are classified as follows:

(1) *Cartridge, 7.62-mm Ball M80*—for use against light materials and personnel, and for range training.

(2) *Cartridge, 7.62-mm Armor-Piercing M61*—for use against lightly armored targets.

(3) *Cartridge, 7.62-mm Tracer M62*—for observation of fire, incendiary effects, signaling, and for training. When tracer rounds are fired, they are mixed with ball ammunition in a ratio of four ball rounds to one tracer round.

(4) *Cartridge, 7.62-mm Dummy M63*—for use during mechanical training.

(5) *Cartridge, 7.62-mm Blank M82*—for use during training when simulated live fire is desired. A BFA should be used to fire this ammunition.

b. **Storage**. Ammunition is stored under cover. If ammunition is in the open, it must be kept at least 6 inches above the ground and covered with a double thickness of tarpaulin. The cover must be placed so that it protects the ammunition yet allows ventilation. Trenches are dug to divert water from flowing under the ammunition.

c. **Care, Handling, and Preservation**. Ammunition should not be removed from the airtight containers until ready for use. Ammunition removed from the airtight containers, particularly in damp climates, may corrode.

(1) Ammunition must be protected from mud, dirt, and moisture. If it gets wet or dirty, the ammunition must be wiped off before using. Lightly corroded cartridges are wiped off as soon as the corrosion is discovered. Heavily corroded, dented, or loose projectiles should not be fired.

(2) Ammunition must be protected from the direct rays of the sun. Excessive pressure from the heat may cause premature detonation.

(3) Oil should never be used on ammunition. Oil collects dust and other abrasives that may possibly damage the operating parts of the weapon.

d. **Packaging.** The ammunition box contains two cartons. Each carton has a bandoleer for carrying purposes. Each carton contains 100 rounds and weighs about 7 pounds. Ammunition in the bandoleers may be linked together, attached to the hanger assembly, and fired from the container or the bandoleers may be removed for firing.

3-4. AMMUNITION ADAPTER

The ammunition adapter is used on the M240B machine gun when firing. This adapter allows the gunner to use the 100-round carton and bandoleer. (Figure 3-5)

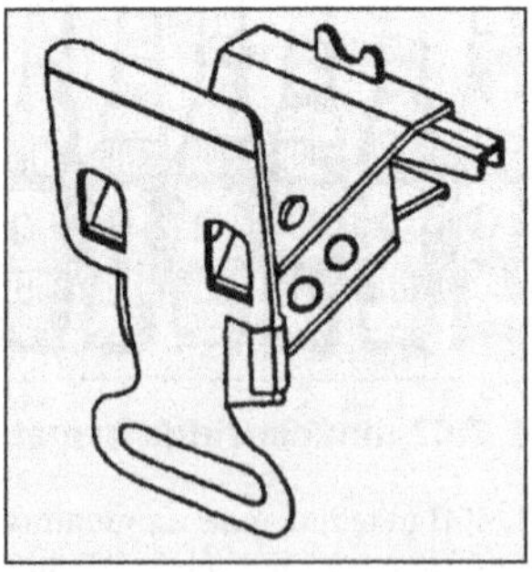

Figure 3-5. Ammunition adapter.

a. **Attaching the Ammunition Adapter**. The ammunition adapter is fitted to the left side and under the feed tray of the receiver. When looking at the left side of the receiver, you will see a slot and a button under the feed tray (Figure 3-5). The gunner first attaches the bandoleer holder to the base of the adapter by inserting the tapered end (green plastic) of the holder into the adapter. Then the gunner opens the cover assembly, and raises the feed tray. He inserts the curved lip of the adapter assembly into the slot located in the rail on the left of the receiver, below the feed tray, depressing the lever on the adapter assembly, and pushing the assembly towards the receiver, until it is against the receiver. Releasing the lever to allow the adapter assembly to secure itself onto the button on the receiver (Figure 3-5).

b. **Care of the Ammunition Adapter**. Over a prolonged period, the moving parts, to include plastic, will start to wear out and break.

(1) Inspect the adapter for damaged parts, excessive wear, and cleanliness when every the weapon is taken out of the arms room.

(2) When feasible, test-fit the adapter.

(3) After using the adapter, inspect to ensure it is still operational.

3-5. BLANK FIRING ATTACHMENT

The BFA is used on the M240B machine gun when blank cartridges are fired to simulate live firing during training where live firing is not practical (Figure 3-6).

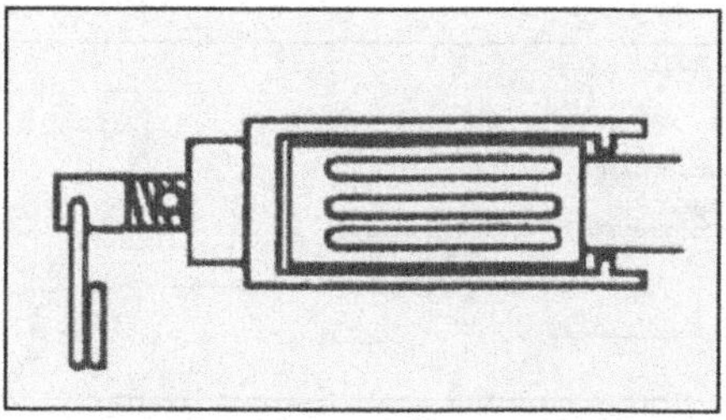

Figure 3-6. Blank firing attachment.

DANGER

DO NOT FIRE BLANK AMMUNITION AT ANY PERSON WITHIN 20 FEET BECAUSE FRAGMENTS OF A CLOSURE WAD OR PARTICLES OF UNBURNED PROPELLANT CAN CAUSE DEATH.

a. **Install the BFA on the M240B Machine Gun**. The BFA is used on the M240B machine gun when blank cartridges are fired to support MILES force-on-force operations to simulate live-fire exercises. The BFA fits any M240B barrel. The tube fits inside the flash suppressor with the remaining portion fitting over the outside of the flash suppressor, flush against the gun muzzle and flush with the forward end of the flash suppressor. The BFA is secured by using the following procedures:

(1) ***Attach the BFA*** (Figure 3-7). Unscrew the shaft (1) until it slides all the way to the rear. Install the chamber device (2) over the flash suppressor (3). Slide the shaft (1)

into the throat of the flash suppressor. Engage the threads on shaft into the body of the chamber device (2); turn clockwise until it is hand tight.

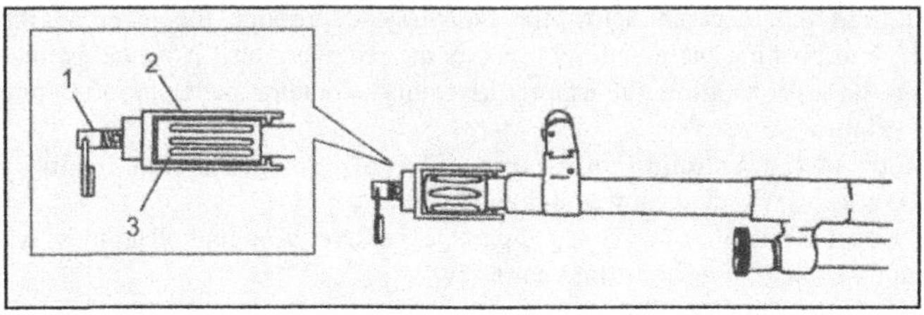

Figure 3-7. Attach the blank firing attachment.

(2) ***Remove the BFA*** (Figure 3-8). Hold the barrel and rotate the chamber of the body (2) about 180 degrees counterclockwise to break any carbon sealed between the shaft (1) and the suppressor (3). Unscrew the shaft (1) until the threads disengage. Remove the chamber device from the suppressor (3).

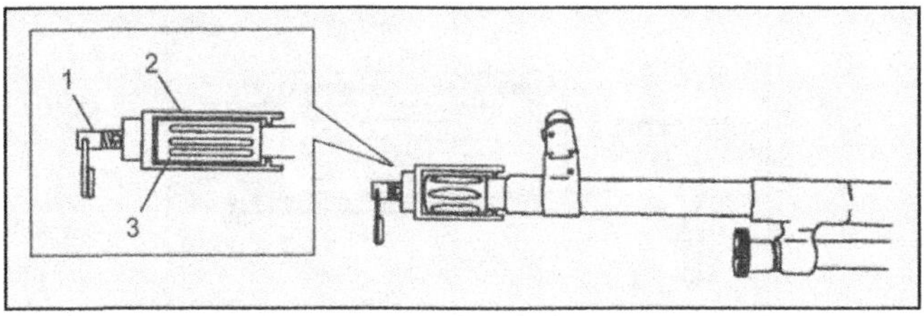

Figure 3-8. Remove the blank firing attachment.

b. **Care of the M240B While Using the BFA.** A buildup of carbon inside the weapon causes friction between the movi0Ong parts. Carbon deposits build up rapidly when blanks are fired. When these deposits become excessive, stoppages occur. Therefore, the weapon must be kept clean, especially the gas system and chamber, during blank firing. To get the best performance with the BFA, the gunner performs the following:

(1) Inspects the weapon for damaged parts, excessive wear, cleanliness, and proper lubrication before firing.

(2) When feasible, test fires the weapon using ball ammunition before attaching the BFA.

(3) Adjusts the BFA to fit the weapon.

(4) Applies immediate action when stoppages occur.

(5) Cleans the weapon including barrel assembly, gas cylinder, gas piston, gas port, chamber bore, and BFA.

(6) Cleans and lubricates the entire weapon after firing 400 blank rounds.

DANGER
1. NEVER LOAD ANY AMMUNITION OTHER THAN BLANKS WHEN THE BLANK FIRING ATTACHMENT IS IN PLACE.
2. NEVER FIRE THE BFA-FITTED MACHINE GUN AT PERSONNEL WHO ARE WITHIN 20 FEET OF THE WEAPON. DEATH OR INJURY COULD OCCUR.

Section II. MAINTENANCE

Proper maintenance contributes to weapon effectiveness as well as unit readiness. Maintenance aspects of the M240B include inspection; cleaning and lubrication; and maintenance before, during, and after firing, and during NBC conditions. Associated tasks essential to maintenance (clearing, general disassembly and assembly, and function checks) are provided in detail.

3-6. CLEARING PROCEDURES

The first step in maintenance is to clear the M240B (Figure 3-9). This applies in all situations, not just after firing. The gunner must always assume the M240B is loaded. To clear the M240B, the gunner performs the following procedures:

a. Move the safety to the fire "F" position.

b. With his right hand, (palm up) pulls the cocking handle to the rear, ensuring the bolt is locked to the rear (bipod mode).

c. Return the cocking handle to its forward position.

d. Place the safety on safe "S."

e. Raise the cover assembly and conduct the four-point safety check for brass, links, or ammunition.

(1) Check the feed pawl assembly under the cover.

(2) Check the feed tray.

(3) Lift the feed tray and inspects the chamber.

(4) Check the space between the face of the bolt and chamber to include the space under the bolt and operating rod assembly.

f. Close the feed tray and cover assembly and place the safety to the fire "F" position. Pull cocking handle to the rear, and pull the trigger while manually riding the bolt forward. Close the ejection port cover.

CAUTION
MANUALLY RETURN THE COCKING HANDLE TO THE FORWARD AND LOCKED POSITION EACH TIME THE BOLT IS MANUALLY PULLED TO THE REAR.

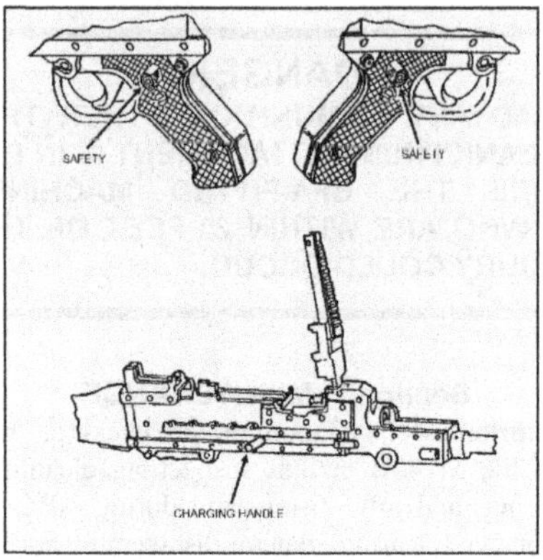

Figure 3-9. Clearing procedures.

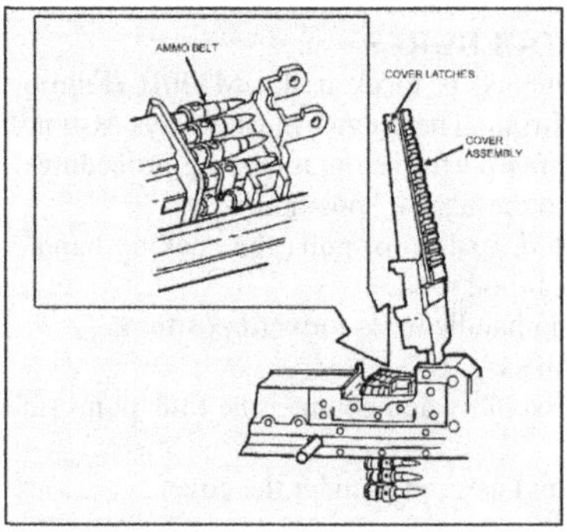

Figure 3-9. Clearing procedures (continued).

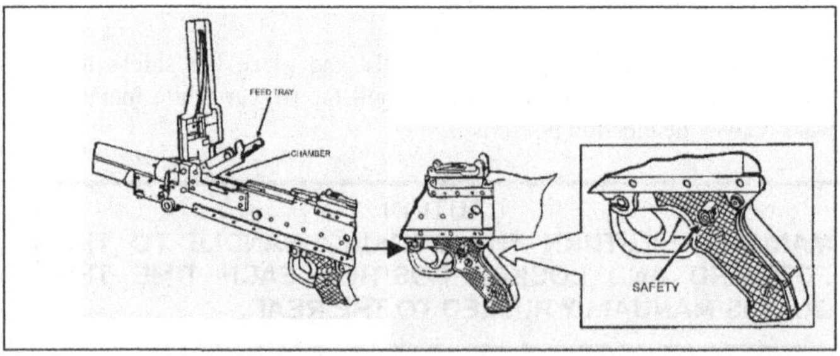

Figure 3-9. Clearing procedures (continued).

3-7. GENERAL DISASSEMBLY

The gunner performs general disassembly, which is removing and replacing the eight major groups (Figure 3-10). (The unit armorer performs detailed disassembly. Disassembly beyond what is explained in this manual is prohibited, except by ordnance personnel.) During general disassembly, the gunner clears the weapon. He ensures the bolt is forward before disassembly, and he places each part on a clean, flat surface such as a table or mat. This aids in assembly in reverse order and avoids the loss of parts.

DANGER

BE SURE THE BOLT IS IN THE FORWARD POSITION BEFORE DISASSEMBLY. THE SPRING GUIDE CAN CAUSE DEATH OR INJURY IF THE OPERATING ROD SPRING IS RETRACTED WITH THE BOLT PULLED TO THE REAR.

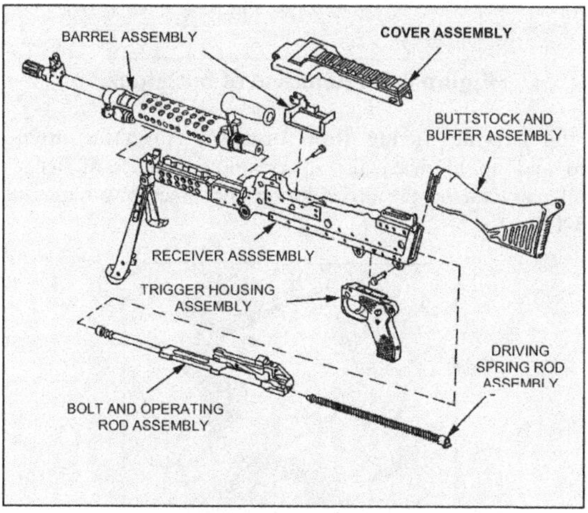

Figure 3-10. Eight major assemblies.

a. **Remove the Buttstock and Buffer Assembly**. Depress the back plate latch located on the underside of the butt stock where it joins the receiver. Slide the butt stock upward (straight up) and remove it from the receiver (Figure 3-11, page 3-12).

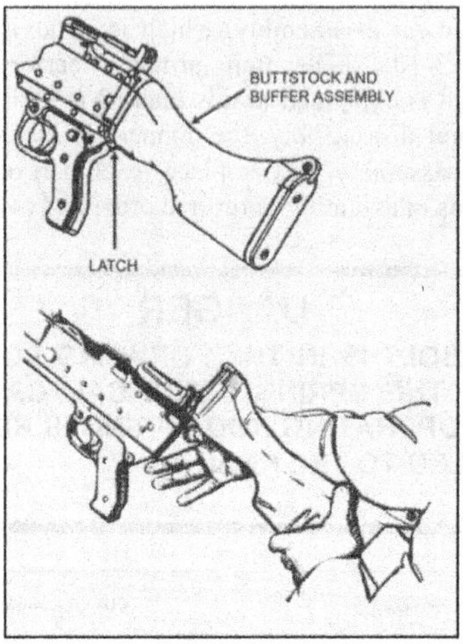

Figure 3-11. Removal of buttstock.

b. **Remove the Driving Spring Rod Assembly**. Push the driving spring rod assembly forward and up to disengage its retaining stud from inside the receiver (Figure 3-12). Pull rearward on the drive spring rod assembly, removing it from the receiver (Figure 3-13).

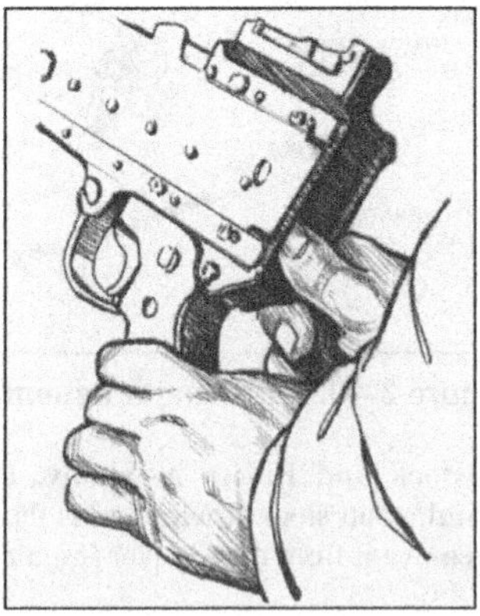

Figure 3-12. Removal of driving spring rod assembly.

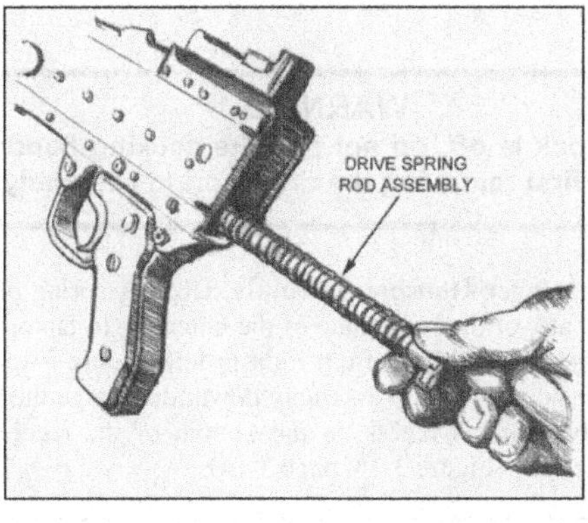

Figure 3-13. Driving spring rod assembly.

WARNING

To avoid injury, keep face away from rear of receiver. Hold rod assembly securely as it is under tension.

c. **Remove the Bolt and Operating Rod Assembly**. Pull the cocking handle to the rear to start the rearward movement of the bolt and operating rod assembly inside of the receiver. With the index finger, reach inside the top of the receiver and push rearward on the face of the bolt until the bolt and operating rod assembly are exposed at the rear of the receiver. Grasp the bolt and operating rod and remove them from the rear of the receiver. Return the cocking handle to the forward position (Figure. 3-14).

NOTE: Pulling the trigger maybe necessary to lower the sear and allow the bolt to release.

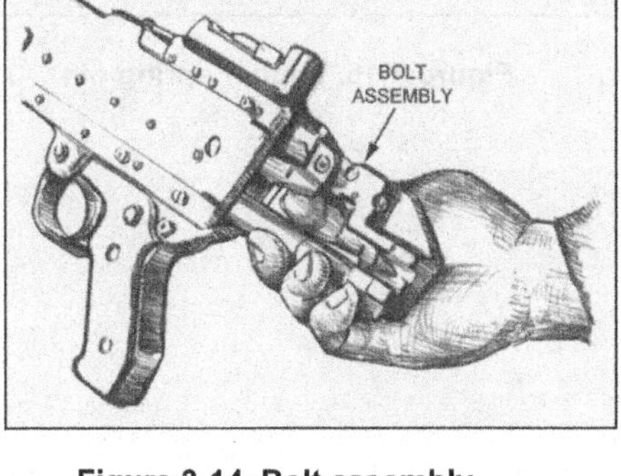

Figure 3-14. Bolt assembly.

WARNING
When buttstock is off, do not pull the cocking handle to the rear without first removing the drive spring assembly.

d. **Remove the Trigger Housing Assembly**. Depress spring pin and remove. You may need to use the back of the back plate of the buttstock to tap on the spring pin, then remove pin with fingers. All pins go from right to left (Figure 3-15, page 3-14). Rotate the rear of the trigger-housing group assembly down, disengage the holding notch at the front of the assembly from its recess on the bottom of the receiver, and remove the assembly from the receiver (Figure 3-16, page 3-14).

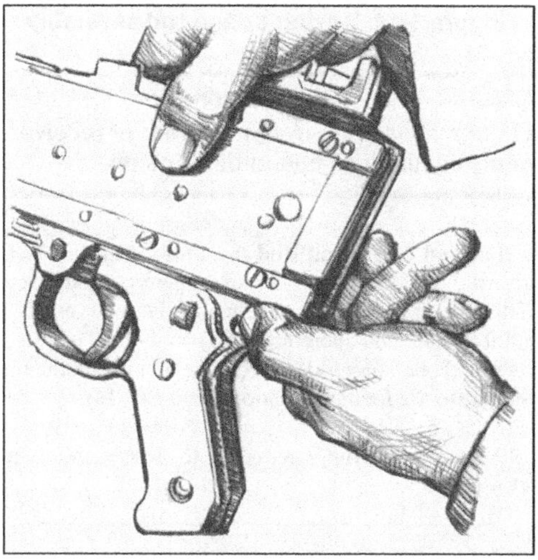

Figure 3-15. Trigger spring pin.

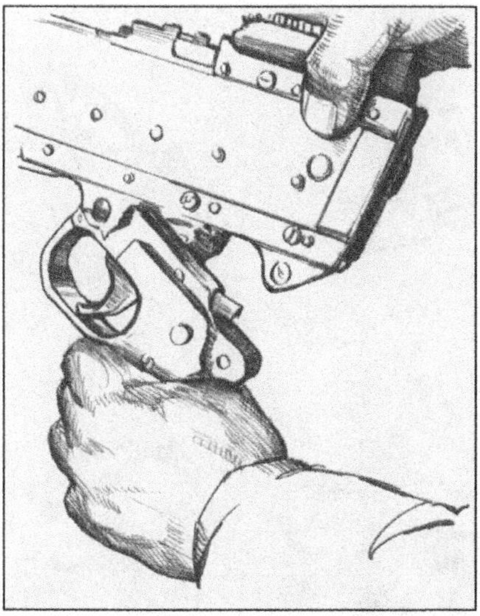

Figure 3-16. Removal of trigger housing.

e. **Remove the Cover Assembly**. (A) Close cover. Depress spring pin and remove. You may need to use the back of the back plate of the buttstock to tap on the spring pin. (B) Then remove pin with fingers. All pins go from right to left. (C) Depress cover latches, lift upwards and remove cover assembly. (D) Remove feed tray (Figure 3-17, page 3-16).

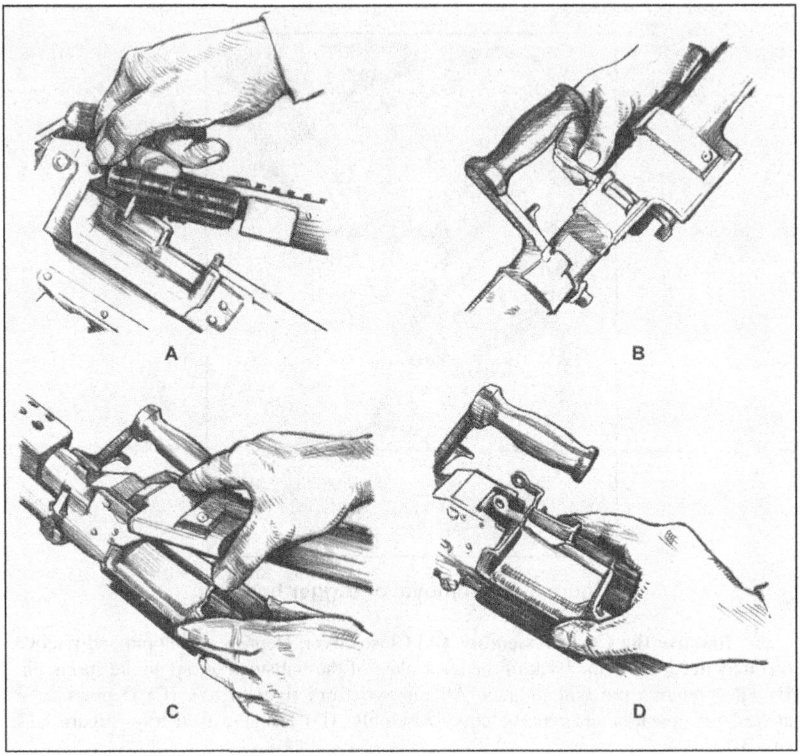

Figure 3-17. Removal of cover, feed tray, and spring pin.

f. **Remove the Barrel Assembly**. (A) Make sure that the barrel-carrying handle is to the right side. Depress the barrel-locking latch located on the left side of the receiver where the barrel joins the receiver and hold. (B) Grasp the barrel carrying handle and rotate the carrying handle to the upright position (without pulling up on the barrel release). (C) Then push forward and pull up, separating the barrel from the receiver (Figure 3-18).

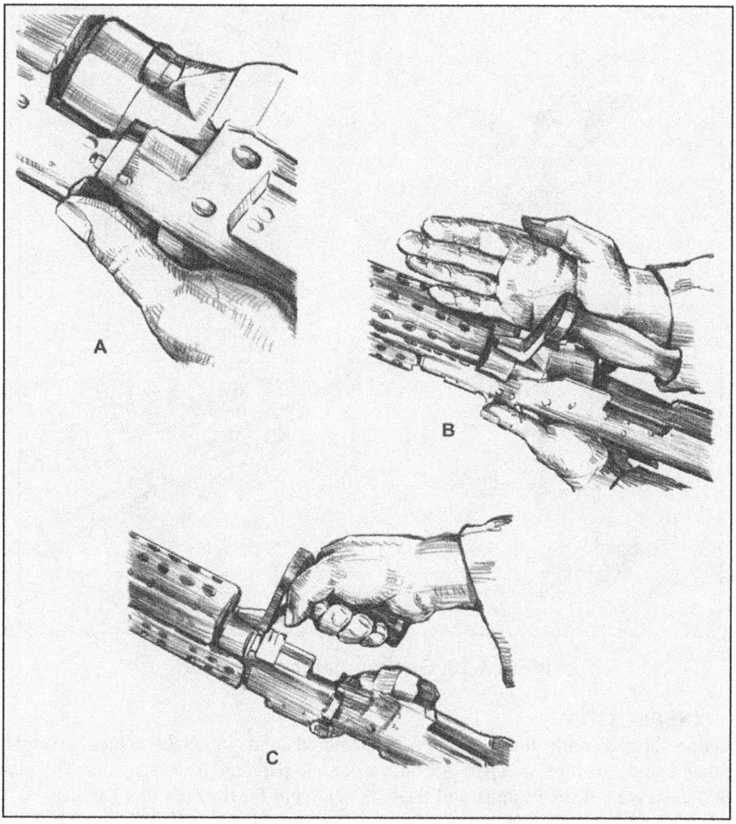

Figure 3-18. Removal of barrel.

g. **Disassemble the Barrel Assembly.** (A) Hold the barrel at the point where the gas system attaches to it. (B) Grasp and rotate the gas collar clockwise until it releases from the gas plug. Remove the collar from the gas plug. (C) Slide the gas regulator plug from front to rear, removing it from the gas hole bushing. (D) Remove heat shield. Lift the rear of heat shield assembly off the barrel, then pry one of the front metal tabs out of hole on gas hole bushing, rotate the heat shield towards the other metal tab, and remove heat shield from the barrel. This completes the general disassembly (Figure 3-19, page 3-18).

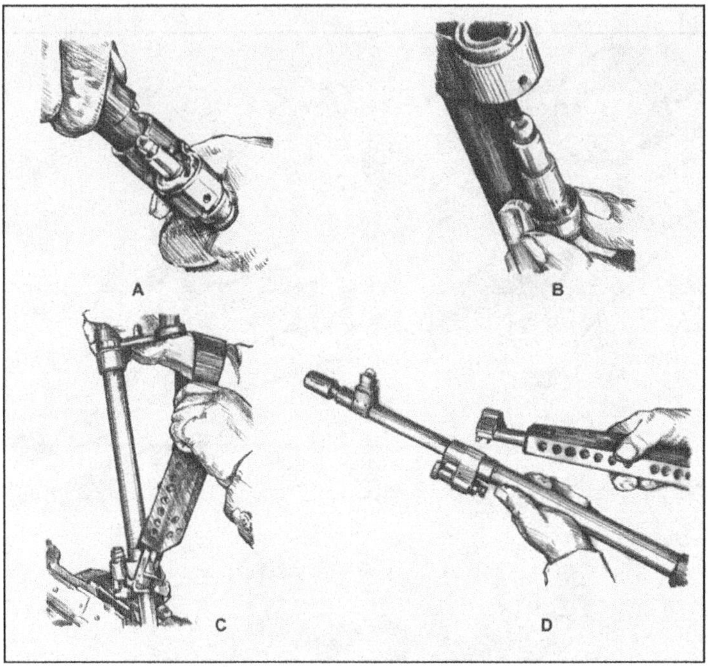

Figure 3-19. Gas regulator and collar.

3-8. INSPECTION

Inspection begins with the weapon disassembled into its eight major assemblies. Note that a shiny surface on a part does not mean the parts are unserviceable. The gunner inspects each area of the weapon and related equipment for the conditions indicated. Any broken or missing parts should be repaired or replaced IAW TM 9-1005-313-10. The gunner should perform PMCS every 90 days. If the weapon has not been used in 90 days, the PMCS in the operator's manual should be performed regardless. If you see rust on a weapon, the PMCS should be done immediately. Inspect all of the components for broken or missing parts. Inspect parts for cracks, dents, burrs excessive wear, rust, or corrosion. Inspect external surfaces for adequate finish.

a. **Barrel Assembly.** Check barrel for bulges, bends, burrs, obstructions and obstructions or pits in the chamber or bore. Disassemble, inspect, and clean the gas collar and plug. Ensure the flash suppressor is fastened securely. Inspect the front sight for damage or looseness. Inspect carrying handle assembly for bent, broken, or missing parts. Assure the heat shield is present, on the barrel assembly, and is not bent or broken, and does not have any missing parts.

NOTE: Some heat distortion or charring may be observed on the outer nonmetallic portion of the heat shield and is not cause for replacement. Do not apply lubricants to composite or rubber components.

b. **Buttstock and Buffer Assembly**. Check for burrs and rough edges on mating grooves and flanges. Check to be sure the back plate latch locks the buffer assembly securely to the receiver assembly when installed. Make sure the buffer plug sticks out through the back plate and is flush or higher than the protrusion below it. Make sure there is no rattling sound when the buffer is shaken and that the plug cannot rotated by finger pressure. Inspect the butt stock for cracks. Check to make sure the back plate locks the butt stock securely to the receiver assembly when installed.

c. **Driving Spring Rod Assembly**. Check the spring for broken strands. Ensure the rod assembly is not bent.

d. **Bolt And Operating Rod Assembly.** Inspect entire area of the bolt and operating rod assembly for missing parts, broken or cracked areas, burrs, bends, or pits on the surface. Looking at the bolt, you can see if the firing pin is broken. The extractor should not move. The operating rod piston should have a slight movement from left to right (about 1/8-inch turn). When the bolt and operating rod are pulled to the rear, the piston should move freely without binding.

NOTE: Always turn both barrels in with the weapon if damage is found on the bolt assembly.

e. **Trigger Mechanism/Housing Assembly**. Inspect the tripping lever and sear for burrs on edges. Push the tripping lever back to raise the sear, put the safety on "S," and pull the trigger. The sear should not drop down far enough to lock in the downward position. Place the safety on "F," and pull the trigger. The sear should drop down and lock in the downward position. Check the sear spring, ensuring the leg of the spring is behind the trigger pin and not between the trigger and the pin. Check grip assembly for loose or missing grip screws. Check trigger guard for bends or cracks. Check trigger spring pin for bends, and or broken or missing spring.

f. **Cover Assembly**. Pivot the feed lever back and forth to ensure it operates smoothly without binding. Push in on the cover latches to make sure the retaining clip is not weak or missing and that they do not bind in the housing. Push down on the cartridge guides and feed pawls to make sure the springs are not weak or missing. Inspect accessory mounting rail for nicks or burrs.

g. **Feed Tray**. Check for cracks, deformation, broken welds, or loose rivets.

h. **Handguard**. Check handguard for cracks, broken or missing parts.

i. **Receiver Assembly**. Check that the rear sight assembly is securely mounted to the receiver and operates properly. Check that the cocking handle operates the slide properly. Pull the cocking handle to the rear and allow it to slowly return forward, making sure that the slide does not bind in the receiver. Check for damaged or missing ejection port cover, spring, and pin. Lower and raise the bipod legs, ensuring they move freely without binding. Check bipod legs for cracks, or twisted or incomplete assembly. Check the exterior surface of the M240B for the exterior protective finish.

j. **Machine Gun**. Assemble the weapon. Be sure parts are installed correctly and are in good working condition. When installing the barrel, move the barrel release slowly to the right and count the number of clicks. Fewer than two and more than seven clicks indicate defective parts. Check both barrels. Check weapon functioning with belted dummy ammunition by performing a function check. If weapon does not function properly and the cause cannot be determined using troubling shooting procedures, notify direct-support maintenance.

k. **M122A1 Mount.** The traversing and elevating mechanism should not bind. The numbers on the scales and dials must be legible.

(1) Distinct clicks must be heard when the handwheels are turned. Index lines should be calibrated with the indicator pointer.

(2) The pintle should fit snugly in the pintle bushing, and the pintle lock should hold the pintle securely.

(3) The sleeve latch should function properly, and the traversing bar should be tight when the tripod legs are extended and latched.

l. **Carrying Case**. Maintenance tools and equipment should be complete and serviceable. The case should be serviceable. Frequent washing of the case should be avoided. Such washing may destroy the waterproofing and shrink the case.

3-9. CLEANING, LUBRICATION, AND PREVENTIVE MAINTENANCE

The M240B machine gun should be cleaned immediately after firing. At a minimum, the M240B should be cleaned after firing a basic load of 900 to 1,200 rounds. The gunner disassembles the M240B into its major groups for cleaning. All metal components and surfaces that have been exposed to powder fouling should be cleaned using CLP on a bore-cleaning patch. CLP is used on the bristles of the receiver brush to clean the receiver. After the M240B is cleaned and wiped dry, a thin coat of CLP is rubbed on using a cloth. This lubricates and preserves the exposed metal parts during all normal temperature ranges.

CAUTION

When using CLP, do not use other cleaners. Never mix CLP with RBC or LSA. When cleaning the barrel, avoid getting CLP or RBC in the gas regulator. Damage could occur to the weapon.

a. When cleaning the weapon, any of the previously mentioned cleaning lubricating agents can be used. As soon as possible after firing the M240B, the gunner disassembles the weapon into its eight major assemblies and cleans them as follows. Before the weapon is disassembled, ensure it is clear:

(1) Clean the bore using CLP or RBC and a bore brush with a cleaning rod. Do not reverse direction of the bore brush while it is in the bore.

(a) Run the brush through the bore several times until most of the powder fouling and other foreign matter has been removed.

(b) Swab out the bore several times using a cleaning rod and a swab wet with CLP.

(c) Swab out the bore several times using a cleaning rod and a dry swab.

(2) Clean the chamber using CLP and a chamber brush attached to a cleaning rod.

(a) Run the brush through the chamber several times until most of the powder fouling and other foreign matter has been removed.

(b) Swab out the chamber several times using a cleaning rod and a swab wet with CLP.

(c) Swab out the chamber several times using a cleaning rod and a dry swab.

(3) Clean the receiver using a receiver brush and CLP.

(a) Brush the receiver until most of the powder fouling and other foreign matter is removed.

(b) Swab out the receiver several times using a cleaning rod section and a swab wet with CLP.

(c) Swab out the receiver several times using a cleaning rod section and a dry swab.

(4) Clean the gas regulator plug with special tools (cleaning reamers and combination regulator scraper). Remove *all* carbon dust. Do not use CLP on the collar, gas block, or body.

(a) Clean each gas inlet hole of the gas regulator plug. Insert the small reamer into each hole and twist back and forth to remove the carbon (apply hand pressure only) (Figure 3-20).

(b) Clean the central hole of the gas plug by inserting the scraper tool down to the bottom of the hole and twisting firmly (Figure 3-21).

(c) Clean the two grooves by inserting the scraper tool into the grooves and applying pressure as firmly as possible (Figure 3-21).

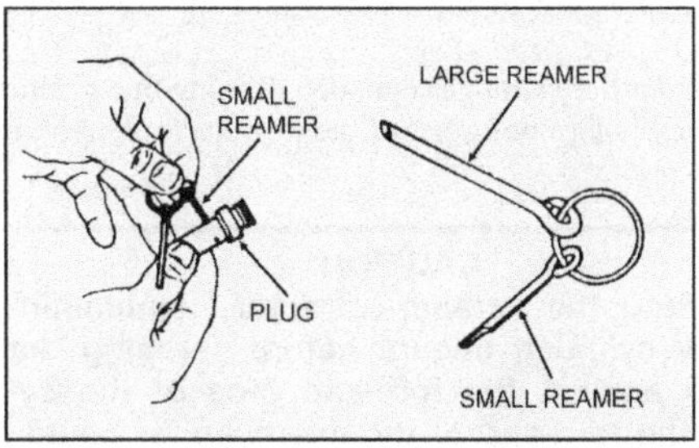

Figure 3-20. Tools for cleaning the gas regulator plug inlet holes.

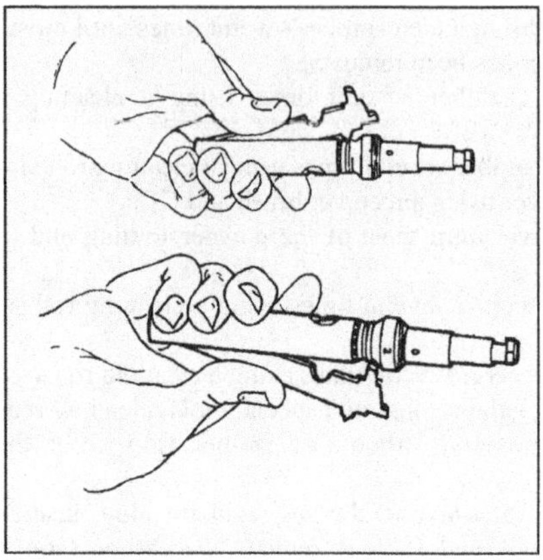

Figure 3-21. Cleaning of the gas regulator plug grooves.

(5) Clean the gas cylinder with the special tool scraper-extraction combination tool.

(a) Clean the front interior of the gas cylinder by carefully inserting the combination tool, with the handle upward. Be sure the tool is fully inserted and seated against the gas cylinder. Apply slight pressure to the handles and turn clockwise to remove carbon (Figure 3-22).

(b) Clean gas cylinder bore with gas cylinder cleaning brush dampened with CLP.

(c) Brush the gas cylinder until most of the powder fouling and other foreign matter are removed.

CAUTION

When inserting the scraper-extractor combination tool into the gas cylinder, ensure before scraping that it is fully seated against the fore-end face of the cylinder. Damage to the fore-end of the gas cylinder could cause gas leakage and subsequent weapon stoppage.

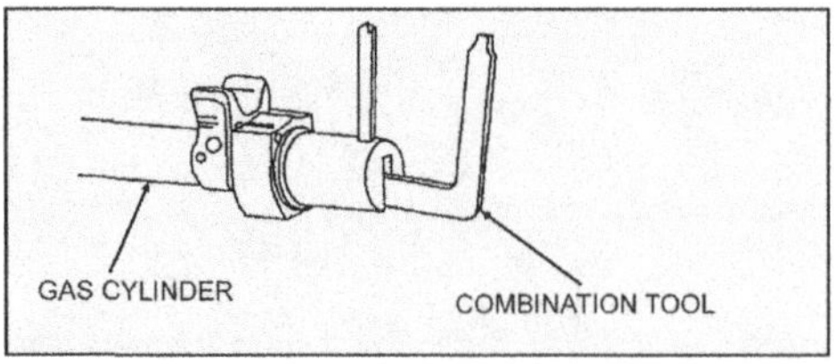

Figure 3-22. Cleaning tool for the gas cylinder.

(6) Clean the bolt and operating rod with the special scraper-extraction combination tool.

(a) Clean the piston head cavity by inserting the combination tool into the piston bottom of the operating rod. Squeeze handles firmly and twist the tool to remove carbon (Figure 3-23).

(b) Insert the screwdriver end of the tool into the piston to remove carbon residue on the bottom.

(c) Clean the bolt and operating rod with rag and CLP.

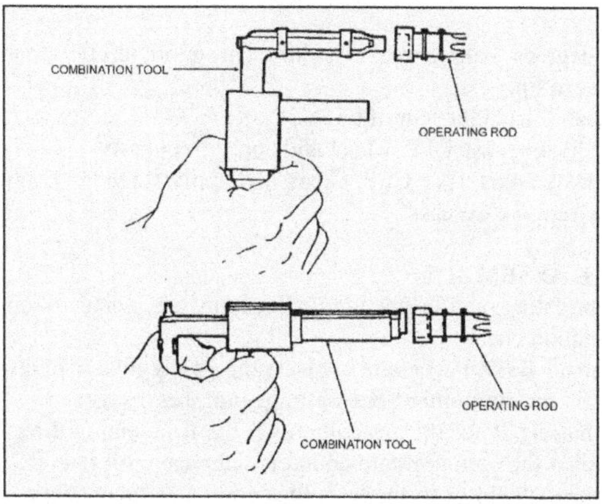

Figure 3-23. Cleaning tool of the piston head cavity.

(7) Remember the following important points during cleaning:

- Do not use gasoline, kerosene, benzene, shaving cream, high-pressure water, steam, or air for cleaning.
- Keep the gas hole bushing free of CLP or RBC. It must remain dry.
- During sustained firing, especially when using blank ammunition, the extractor assembly must be stripped and cleaned periodically.
- Improper cleaning of the gas cylinder and gas regulator plug will result in the two temporarily welding themselves together during firing.

(8) Lubricate the following parts with CLP as instructed:

- Driving spring rod assembly.
- Bolt.
- Receiver inner walls.
- Cover assembly (springs, and feed pawls).
- Trigger housing (inside only).

b. After lubricating, the components are cycled by hand to spread the CLP. Weapons fired infrequently or stored for prolonged periods should have a light film of CLP. This should be applied to the interior of the gas cylinder and the gas piston immediately after cleaning or after inspection. Preventive maintenance is performed every 90 days, unless inspection reveals more frequent servicing is necessary. The use of the lubricant does not eliminate the requirement for cleaning and inspecting to ensure that corrosion has not formed. Before the weapon is used, the gas system and components must be cleaned and free of oil and lubricants.

c. All exposed surfaces of the M122A1 tripod, flex-mount assembly, complete pintle and T&E mechanism are cleaned by wiping them down with a clean rag. For T&E and pintles that have stubborn areas with hard-to-remove dirt, a steel brush or bore brush is used to loosen the dirt (do not use on the flex-mount itself). A clean rag is then used to wipe them down and CLP is used to lubricate them.

d. The following procedures apply to cleaning and lubricating the M240B during unusual conditions:

(1) Below 0 degrees Fahrenheit—use lubricating oil, arctic weather (LAW). Oil lightly to avoid freeze-up.

(2) Extreme heat—use light coat of CLP.

(3) Damp or salty air—use CLP. Clean and apply frequently.

(4) Sandy or dusty areas—use CLP. Clean and apply frequently. Wipe with rag after each application to remove excess.

3-10. GENERAL ASSEMBLY

After cleaning, lubricating, and inspecting the weapon, the gunner assembles the weapon and performs a function check.

a. **Replacing the Barrel Assembly**. Insert the gas regulator plug into the gas hole bushing so that it is on the number one setting. (number 1 gas setting on the regulator faces towards the barrel). Place the gas collar over the front end of the gas regulator plug, while pushing against the spring, rotate counterclockwise until it stops. Insert one of the metal tabs of the heat shield in to the hole located of the sides of the gas hole bushing, than rotate it so that the other tab locks in place. Then push down on the heat shield so that it snaps onto the barrel. With gas regulator downward and carrying handle in the vertical position, place barrel on the barrel support (located on the gas cylinder). Keeping the gun upright, pull the barrel to the rear ensuring the gas regulator is guided into the gas cylinder. Pull the barrel fully into the receiver and rotate the carrying handle completely to the right, ensuring to count the number of clicks. If the number of clicks fall between 2 to 7, the headspace is set correctly. If the number falls outside 2 to 7, turn it in to the unit armor (make sure that the threads on the barrel are located on top and bottom and on the inside of the receiver make sure that the threads are located on the left and right) (Figure 3-24).

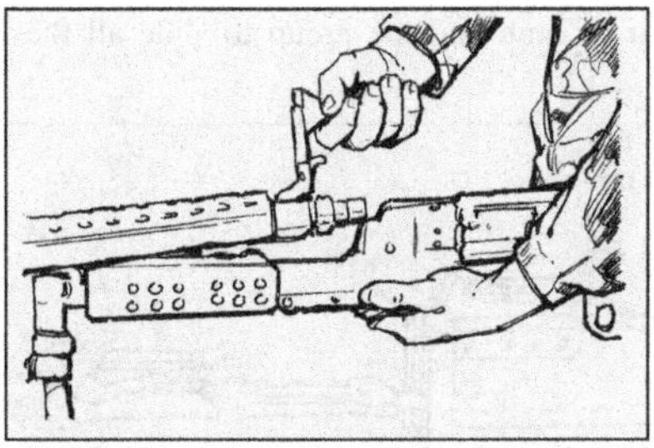

Figure 3-24. Replacement of the barrel assembly.

b. **Replacing the Cover Assembly and Feed Tray**. Position the feed tray on the receiver so that the feed tray guides are aligned with the receiver brackets. Place the cover assembly onto the receiver aligning its mounting holes with the mounting brackets on the receiver, close the cover assembly. Then, insert the spring pin into the holes to affix the cover and feed tray to the receiver (insert the spring of the spring pin into the hole than push in from right to left).

c. **Replacing the Trigger Housing Assembly**. Insert the holding notch on the front of the trigger housing into the forward recess on the bottom of the receiver. Rotate the rear of the trigger housing upwards and align the holes of the trigger housing with the mounting bracket on the receiver. Hold the trigger housing assembly and insert the spring pin into the hole, securing the assembly to the receiver. (insert the spring of the spring pin into the hole than push in from right to left) (Figure 3-25).

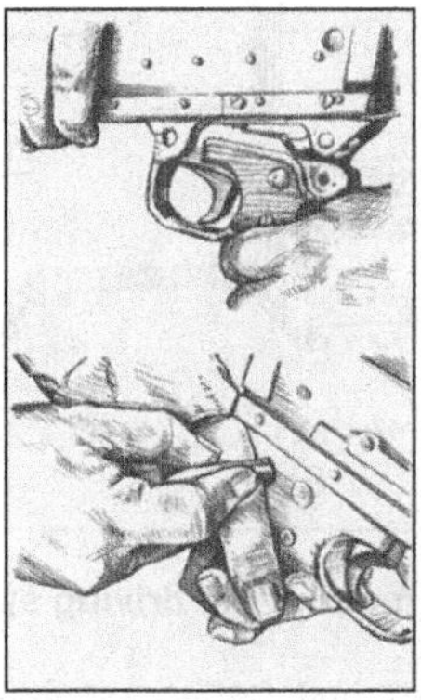

Figure 3-25. Replacement of the trigger housing assembly.

d. **Replacing the Bolt and Operating Rod Assembly**. Make sure the bolt and operating rod are fully extended (unlocked position). Insert the bolt and operating rod into the rear of the receiver (bolt upward, operating rod beneath bolt) ensuring the bolt is on top of the rails located on the left and right inner walls of the receiver. Push the entire bolt and operating rod assembly into the receiver as far forward as possible. Pull the trigger to allow the sear to drop and the group to slide all the way into the receiver (Figure 3-26).

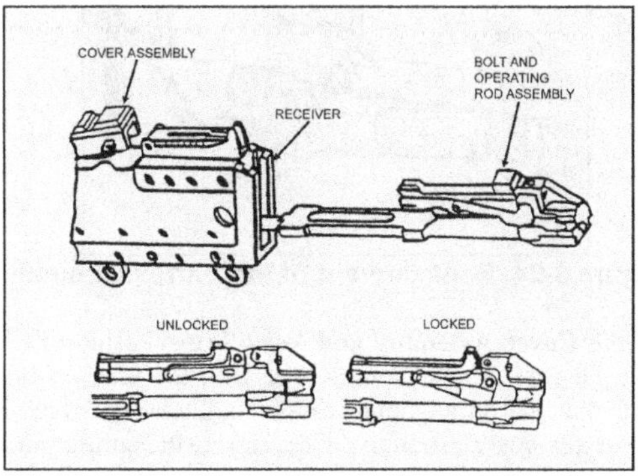

Figure 3-26. Replacement of the bolt and operating rod assembly.

e. **Replacing the Driving Spring Rod Assembly**. Insert the driving spring rod assembly into the receiver, sliding it all the way forward against the recess in the rear of the operating rod. Push in and lower the driving spring rod assembly to engage the retaining stud into the hole located on the bottom of the receiver (Figure 3-27).

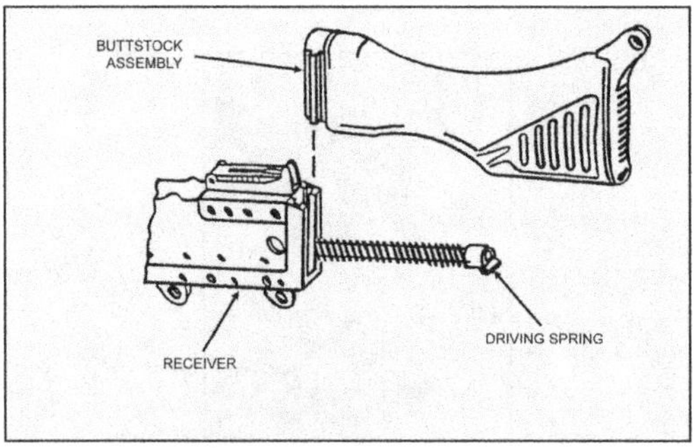

Figure 3-27. Replacement of the driving spring rod assembly.

f. **Replacing the Butt stock and Buffer Assembly**. Position the bottom recess grooves of the butt stock onto the top of the receiver recess grooves. Slide the butt stock down until it locks in place on the receiver. Ensure the butt stock is secure.

g. **Replacing the Handguard**. Line the handguard on the bottom of the gas cylinder and push upwards. The handguard snaps in place.

3-11. FUNCTION CHECK

The gunner must perform a function check to ensure that the M240B is correctly assembled. The procedures, in order, are—

a. Places the safety on "F."

b. Pulls the cocking handle to the rear, locking the bolt to the rear of the receiver.

c. Returns the cocking handle to the forward position.

d. Places the safety on "S," and closes the cover.

e. Pulls the trigger. (Bolt should not go forward).

f. Places the safety on "F."

g. Pulls the cocking handle to the rear, pulls the trigger, and rides the bolt forward.

h. Closes the ejection port cover.

3-12. MAINTENANCE PROCEDURES

Maintenance of the M240B requires certain actions to be taken before, during, and after firing.

a. Before firing—

(1) Wipe the bore dry.

(2) Inspect the weapon as outlined in operator's TM.

(3) Inspect the spare barrel.

(4) Lubricate the weapon.

b. During firing—

(1) Change the barrels. Changing the barrel prolongs the life of both barrels.

(2) Periodically inspect the weapon to ensure that it is properly lubricated.

(3) When malfunctions or stoppages occur, follow the procedures in Section IV.

c. After firing—

(1) Clear and clean the weapon immediately.

(2) Every 90 days during inactivity, clean and lubricate the weapon unless inspection reveals more frequent servicing is necessary (TM 9-1005-313-10).

3-13. MAINTENANCE DURING NUCLEAR, BIOLOGICAL, CHEMICAL CONDITIONS

If the M240B is contaminated by chemical, biological, or radiological agents, the gunner takes appropriate action to reduce exposure and minimize penetration.

a. **Chemical.** The gunner uses towelettes from the M258A1 kit to wipe off the weapon. If these are not available, he washes the weapon with soap and water.

b. **Biological**. The gunner uses towelettes or soap and water as previously described.

c. **Radiological**. The gunner wipes the weapon with warm soapy water if it is available. If not, he uses towelettes or rags. (For more details, see FM 3-5.)

Section III. OPERATION AND FUNCTION

This section discusses the operation of the M240B machine gun. This includes loading, unloading, cycle of functioning, adjustment of the sights, and use of both the bipod and tripod.

3-14. OPERATION

The M240B machine gun is loaded from the closed bolt position. The M240B is fired, unloaded, and cleared from the open bolt position. The safety must be placed on "F" before the bolt can be pulled to the rear. Before belted ammunition can be used, it must be linked with the double link at the open end of the bandoleer. It must be free of dirt and corrosion. In almost all cases, the M240B machine gun can be best used when fired from a tripod; the M240B's potential for continuous, accurate fire and control manipulation is maximized. However, in some circumstances, the gunner may use the bipod mount.

3-15. LOADING

The gunner makes sure the weapon is cleared. He place the safety on "F." With his palm facing up, he pulls the cocking handle to the rear. This puts the bolt assembly in the rear position. When the bolt is held to the rear by the sear, he manually returns the cocking handle to the forward position, places the safety on "S." He raises the cover assembly and ensures the feed tray, receiver assembly, and chamber are clear. He lowers the feed tray, places the safety on "F," and pulls the cocking handle to the rear. While maintaining rearward pressure, he pulls the trigger and eases the bolt assembly forward. He places the first round of the belt in the feed tray groove, double link leading, with open side of links face down. He holds the belt about six rounds from the loading end, while closing the cover assembly. *Ensure that the round remains in the feed tray groove, and close the cover assembly* (Figure 3-28).

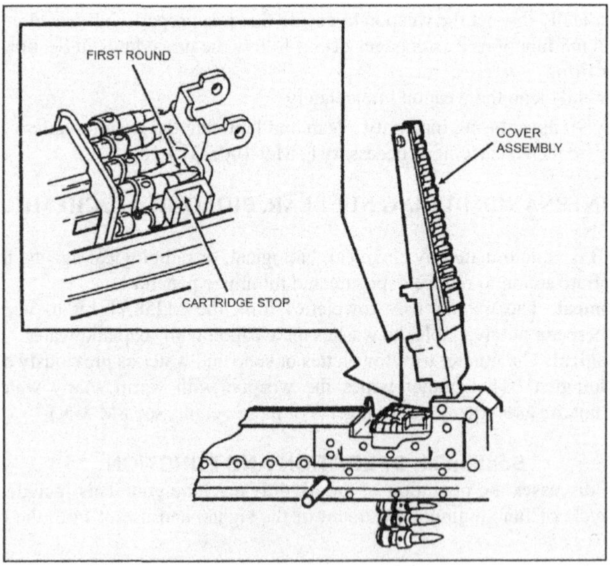

Figure 3-28. Loading.

WARNING

The M240B is carried loaded with the bolt locked to the *rear* in all *tactical situations* where noise discipline is critical to the success of the mission. Trained gun crews are the only personnel authorized to load the M240B and only when command directs the crew to do so. During *normal training exercises*, the M240B is loaded and carried with the bolt in the *forward position*.

3-16. UNLOADING

The gunner unloads the M240B by pulling and locking the bolt to the rear position, if it is not already there. He manually returns the cocking handle to its forward position. He places the safety on "S." He raises the cover assembly and removes any ammunition or links from the feed tray. He performs the four-point safety check (see Section III).

3-17. CYCLE OF FUNCTIONING

Crew members can recognize and correct stoppages when they know how the weapon functions. The weapon functions automatically as long as ammunition is fed into it and the trigger is held to the rear. Each time a round is fired, the parts of the weapon function in a cycle or sequence. Many of the actions occur at the same time and are separated only for teaching purposes. The sequence of functioning is known as the "cycle of functioning."

a. The cycle starts when the first round of the belt is placed in the tray groove. Then the trigger is pulled, releasing the sear from the sear notch. When the trigger is pulled to the rear, the rear of the sear lowers and disengages from the sear notch. This allows the bolt and operating rod assembly to be driven forward by the expansion of the driving spring rod assembly. The cycle stops when the trigger is released and the sear again engages the sear notch on the bolt and operating rod assembly.

b. The details of the cycle of functioning follows:

(1) *Feeding*. The actuating roller moves the feed lever side to side, which in turn moves the feed pawls. The forward movement of the bolt forces the outer pawls to the right, fully feeding the round. The inner pawl rides over the round and settles behind it. The rearward movement forces the inner pawl to the right, fully feeding the round. The action of fully feeding a round pushes the link of a fired round out of the side of the gun. The last link in a belt cannot be pushed out and is cleared during the unloading.

(2) *Chambering*. The first round is positioned in line with the chamber and is held in position by the cartridge stop and cartridge guide pawl. On trigger squeeze, the nose of the sear is depressed thus freeing the piston rod extension. The driving spring rod assembly pushes the working parts forward. The feed horn strikes the base of the round. The bolt strips the round from the belt link. The chambering ramp angles downward and, along with the spring tension of the cartridge guide pawl, forces the round toward the chamber. The cartridge guide pawl also holds back the belt link. When the round is fully seated in the chamber, the extractor snaps over the extractor rim of the cartridge, and the ejector is depressed.

(3) *Locking*. During chambering, as soon as the piston begins to move, the firing pin is withdrawn into the bolt block. The breech remains locked during the primary movement. The bolt enters the barrel breech as the operating rod is driven forward by the drive spring, and as the locking lever, which the bolt is riding on, swings forward, pushing the bolt forward and locking it to the barrel breech. Although the term "locking" is used here, in the M240B, the bolt and barrel do not physically interlock. This way, the barrel can be removed when the bolt is forward.

(4) *Firing*. As the working parts come forward and the round is fed into the chamber, the locking lever is forced down by the locking cams. This slows down the forward movement of the bolt assembly. The piston rod extension, still moving forward, causes the locking lever link to rotate downward and back. This forces the arms down to their fullest extent in front of the locking shoulder. The extractor rises over the base of the round and the ejector is compressed. The round is now fully home with the breech locked. The final forward movement of the piston extension drives the firing pin through the bolt assembly onto the cartridge primer and fires the round. The working parts are now fully forward.

(5) *Unlocking*. When the round is fired, some of the gases pass through the gas plug regulator into the gas cylinder. The rapidly expanding gases enter the hollow end cap of the gas piston and force the operating assembly to the rear. This powers the last four steps in the cycle of functioning. During the primary movement of the operating rod assembly, it moves independently of the bolt for a short distance. At this point, the locking lever begins to swing toward the rear, carrying the bolt with it into its unlocked position, and clearing the barrel breech. When the bolt assembly has been jerked back, slightly enough to unlock the breech, the primary effort is extraction of the empty case.

(6) *Extraction*. When the breech is fully unlocked and the bolt assembly starts its rearward movement, the extractor withdraws the empty case from the chamber.

(7) *Ejecting*. As the cartridge case is withdrawn from the chamber, the ejector pushes from the top, and the extractor pulls from the bottom. The casing falls down from the face of the bolt as soon as it reaches the cartridge-ejection port. The empty belt links are forced out the link ejection port as the rearward movement of the bolt causes the next round to be positioned in the tray groove.

(8) *Cocking*. As the working parts continue toward the rear, the return spring is compressed; the trigger is kept squeezed; sufficient is gas made available by the gas-regulator adjustment, which causes the working parts to rebound off the buffer; and the action of feeding and firing continues. In releasing the trigger, the sear remains down, but the tripping lever rises. As the working parts come to the rear, the end of the piston rod extension hits the tripping lever, which, in turn, allows the sear to rise and engage the sear notch, which holds the working parts to the rear.

3-18. SIGHTS

This paragraph provides information on how to make corrections if the initial setting is not accurate. At a 10-meter target, each paster is 1 cm. Therefore, ten clicks on the adjusting screw (windage) of the front sight assembly in either direction moves the strike of the round left or right 1 cm. One complete turn on the front sight blade moves the strike of the round up or down 1 cm.

a. **Elevation Correction.** If the shot group is above or below the point of aim, the front sight posts must be adjusted using the front sight-adjusting tool. Unlock the front-sight retaining strap and rotate it up. If the shot group is above the point of aim, rotate the sight post counterclockwise. If the shot group is below the point of aim, rotate the sight post clockwise. Rotating the front sight post counterclockwise brings the point of impact *down* on the target. Rotating the front sight post clockwise brings the point of impact *up* on the target. At a range of 10 meters, one-half turn of the front sight post blade will move the point of impact by 5 mm or .5 cm. One full turn of the front sight post blade moves the point of impact by 1 cm. After rotating the front sight post blade the desired amount, lower the retaining strap, but *do not* lock it down until elevation is confirmed. If the front sight post blade must be rotated counterclockwise to a point where its base is past flush (Number 2 blade), it should be replaced with a Number 1 front sight blade, which is smaller than a Number 2 blade. If the front sight post blade must be rotated counterclockwise to a point where its base is more than one full turn past flush (Number 1 blade), it should be replaced with a Number 2 front sight blade which is taller than a Number 1 blade (Table 3-3, page 3-32).

b. **Windage Correction.** If the shot group is to the left of the point aim, move the front sight assembly to the right to shift the point of impact to the left (towards the point of aim). Using the front sight adjusting tool, loosen (turn counterclockwise) the adjusting screw on the front sight assembly the desired amount. Then tighten (turn clockwise) the opposite side screw on the left *exactly* the same number of clicks. At a range of 10 meters, one complete rotation of the adjusting screws will move the point of impact 8 mm or .8 cm. As the adjusting screws are turned, noticeable clicks (eight per revolution) should be detected. Each click is 1 mm or .1 cm. If this is not the case, have your armorer repair it. The front sight windage adjusting procedure is the combination of creating slack on one side, and then taking up that slack from the opposite side. The front sight protector assembly should always be clamped between the heads of the two opposing screws. *Remember, each time one screw is loosened or backed off, the opposite screw must be turned exactly the same amount.* Check for play in the front sight assembly by lightly clamping it between finger and thumb and attempting to move the sight assembly laterally. If you feel no play, the windage adjustment is completed. If evident, *carefully* check both screws for looseness (Table 3-4, page 3-32).

c. **10-Meter Zeroing (Mechanical Zero).** Ten-meter zero (mechanical zero) is the standardized starting point for all weapons in the United States Army. The gunner places the range scale on a range of 500 meters on the rear sight. He gets the front sight post blade approximately centered for both elevation and windage. The gunner identifies what number blade is on the weapon for elevation.

(1) *Number 1 blade (low 9.8mm).* Unlock the retaining strap and unscrew (counterclockwise) until the base of the blade is flush with the front sight protector surface, then make one full turn (counterclockwise). This should put the base of the blade past the base of the protector. Screw in (clockwise), counting the number of turns it takes until it stops, making sure the blade is on line with the barrel. If needed, back off until the blade is on line. Unscrew (counterclockwise) half the number of turns. This brings the blade to about the center.

(2) Number 2 blade (high 11.8mm). Unlock the retaining strap and unscrew (counterclockwise) until the base of the blade is flush with the front sight protector

surface. Screw in (clockwise), counting the number of turns it takes until it stops, making sure the blade is on line with the barrel, if needed back off until the blade is on line. Unscrew (counterclockwise) half the number of turns. This procedure brings the blade to about the center. Assume the prone position and sight on the target. Ensure windage is accomplished by making sure the front sight protector is centered left and right on its base.

100 meters—one full turn moves strike 10.8 cm (4.25 inches)
200 meters—one full turn moves strike 21.6 cm (8.5 inches)
300 meters—one full turn moves strike 32.4 cm (12.75 inches)
400 meters—one full turn moves strike 43.2 cm (17 inches)
500 meters—one full turn moves strike 54 cm (21.25 inches)
600 meters—one full turn moves strike 64.8 cm (25.5 inches)
700 meters—one full turn moves strike 75.6 cm (29.75 inches)
800 meters—one full turn moves strike 86.4 cm (34 inches)
900 meters—one full turn moves strike 97.2 cm (38.25 inches)

Table 3-3. Elevation correction chart.

100 meters—one full turn moves strike 8 cm (3.15 inches).
200 meters—one full turn moves strike 16 cm (6.3 inches).
300 meters—one full turn moves strike 24 cm (9.45 inches).
400 meters—one full turn moves strike 32 cm (12.6 inches).
500 meters—one full turn moves strike 40 cm (15.75 inches).
600 meters—one full turn moves strike 48 cm (18.9 inches).
700 meters—one full turn moves strike 56 cm (22 inches).
800 meters—one full turn moves strike 64 cm (25.2 inches).
900 meters—one full turn moves strike 72 cm (28.35 inches).

Table 3-4. Windage correction chart.

3-19. M122A1 TRIPOD

The M122A1 tripod provides a stable mount for the M240B, and it permits a higher degree of accuracy and control. The tripod is recommended for marksmanship training and defensive employment. The M122A1 tripod consists of the tripod,and flex-mount with T&E mechanism.

a. **Mounting the M240B on the Tripod.** The tripod assembly provides a stable and relatively lightweight base that is far superior to the bipod. The tripod may be extended and collapsed without difficulty. It consists of a tripod head, one front leg and two rear legs, and traversing bar. The traversing bar connects the two rear legs. It is hinged on one side, and has a sleeve and sleeve latch on the other that allows the tripod to collapse to a closed position for carrying or storage, or to lock in an open, extended position for use. The traversing bar also supports the T&E mechanism. The increments are numbered every 100 mils to 425 mils right of center. On the bar, there is a scale that measures direction in mils. It is graduated in 5-mil increments and numbered every 100 mils to 450 mils left of the center.

(1) The T&E mechanism provides controlled manipulation and the ability to engage predetermined targets.

(a) The traversing portion of the mechanism consists of the traversing handwheel and traversing slide-lock-lever. As the traversing handwheel is turned, the muzzle of the weapon turns to the left or right depending on the direction it is turned. Each click of the traversing handwheel indicates a 1-mil change in direction of the muzzle: 1 click equals 1 mil. There is a total of 100-mils traverse (50 mils right and 50 mils left of center).

(b) The elevating portion of the mechanism consists of the elevating handwheel. The elevating handwheel has a mil-click device built into it (1 click equals 1 mil). Engraved into the handwheel is a scale divided into 5-mil divisions and 1-mil subdivisions for a total of 50-mil increments. There are 200 mils above and 200 mils below the zero mark for a total of 400 mils in elevation change. Elevation readings are taken in two parts. First, the major reading is taken from the elevation screw plate. The second, minor reading is from the handwheel. The two readings are separated by a slash (/) when they are recorded.

(c) The traversing slide-lock-lever allows rapid lateral adjustments along the traversing bar. Direction readings are taken from the scale on the traversing bar, using the left side of the traversing slide as an index. The direction of the reading comes from the position of the muzzle, not the position of the slide.

(2) The flex-mount consists of the mount itself and the traversing and elevating mechanism. It joins the weapon and the T&E mechanism to the tripod. The flex-mount enhances the stability of the tripod platform and dampens the recoil of the weapon.

(3) To setup the tripod, the gunner unfolds the front leg and spreads the rear legs until the leg lock engages (Figure 3-29, page 3-34).

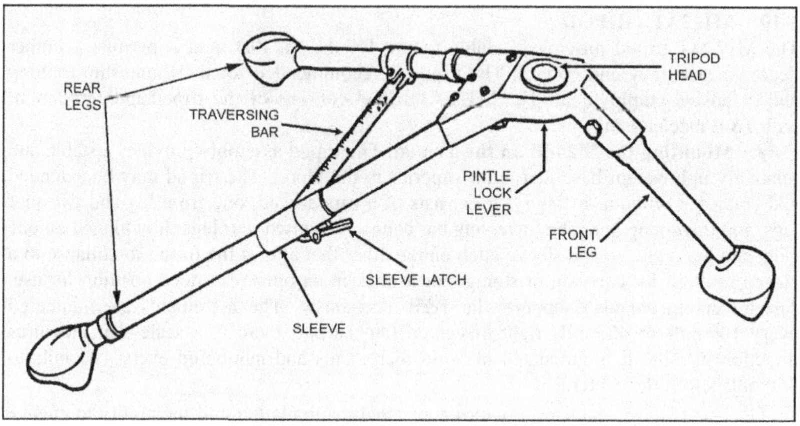

Figure 3-29. M122A1 tripod extended.

(4) Prepare the T&E mechanism for mounting. First center the elevating and traversing handwheel. To do this, he rotate the elevation handwheel until aboout 1 1/2 inches (two fingers) are visible on the upper elevating screw; rotating the traversing slide until approximately two fingers are visible on the lower elevating screw. The gunner rotates the traversing handwheel towards his body as far as it will go, then he turns it away two complete revolutions. He checks the traversing handwheel scale to ensure the "0" on the scale is aligned with the "0" index line before and after the two revolutions. The T&E should be centered now (Figure 3-30).

(5) Mount the T&E mechanism, pintle assembly, and fork assembly to the M122A1 tripod. With the T&E roughly centered, place the pintle assembly (1) into the sleeve bushing on the tripod leg assembly (2). Release the pintle lock (3) on the tripod leg assembly to secure the pintle assemblly to the tripod (Figure 3-30).

NOTE: The deflector on the fork assembly should deflect to the right.

(6) Align the holes in the fork assembly (4) with the holes in the T&E (5). Insert the pin (6) through the fork assembly and the T&E and secure with "C" clamps (7) (Figure 3-30).

(7) Mount the weapon on the M122A1 tripod assembly (Figure 3-30). Tilt the muzzle down slightly and insert the weapon's front receiver bushing (1) into the slots in the pintle assembly (2). Insert the quick-release pin (3) through the pintle (2) and front receiver bushing (1). Place the T&E assembly (5) (with fork assembly attached) onto the traverse bar (8) of the tripod leg assembly (2). Lock the T&E mechanism into place by turning the lock lever (9) clockwise. Lower the rear of the weapon into the fork assembly (4). Align the mounting holes (5) in the trigger housing with the hole in the fork assembly (4). Insert the spring pin (6) through the holes in the trigger assembly and fork but make sure the weapon is securely attached.

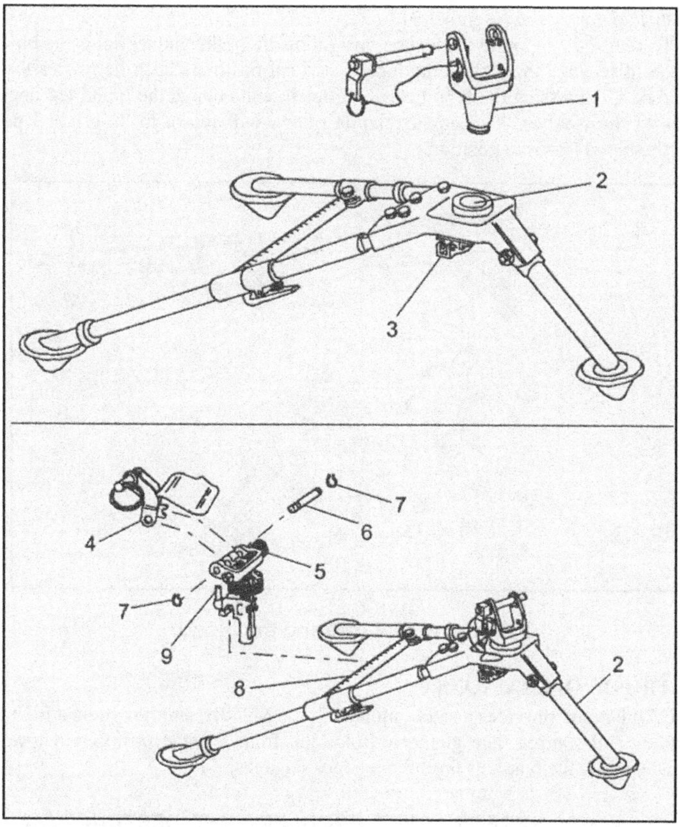

Figure 3-30. Mounting the M240B on the M122A1.

b. **Dismounting the M240B From the M122A1 Tripod.** The gunner dismounts the M240B from the M122A1 tripod by first removing the spring pin from the fork assembly, then he disengages the quick-release pin from the pintle and the front receiver bushing. Now, he raises the weapon up and off the tripod assembly.

3-20. BIPOD OPERATIONS

The bipod assembly is used to fire from the prone position. The buttstock in conjunction with the gunners nonfiring hand provides support for the weapon when firing in the bipod mode. The gas cylinder holds the bipod in place.

a. To lower the bipod legs, the gunner depresses the bipod retaining latch, while holding the bipod legs together to disengage from slots in the receiver. Then rotate the bipod legs down and release them so they lock in the vertical position. The bipod legs of the M240B do not extend (Figure 3-31).

b. To return the bipod to the locked upright position, the gunner holds the bipod legs together to disengage them from the locked vertical position. Then he rotates the bipod legs rearward, depressing the bipod retaining latch, and engage the bipod leg hooks into the slots of the receiver. The bipod retaining latch will return to its original position, locking the bipod legs into position.

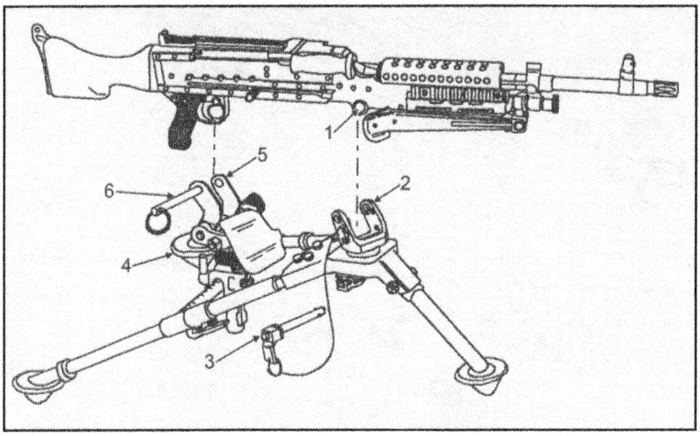

Figure 3-31. Lowering the bipod.

3-21. TRIPOD OPERATIONS

The M122A1 tripod provides a stable mount for the M240B, and it permits a high degree of accuracy and control. The gunner unfolds the front leg and positions it toward the target and spreads the rear legs until the leg lock engages.

Section IV. PERFORMANCE PROBLEMS AND DESTRUCTION

This section identifies some of the problems that cause the weapon to perform improperly. It also explains how to identify unserviceable parts, and how to destroy the M240B when authorized to do so.

3-22. MALFUNCTIONS

A malfunction occurs when a *mechanical failure* causes the *weapon to fire improperly*. Defective ammunition or improper operation by the gunner is not considered a malfunction. Sluggish operation and uncontrolled fire are the most common malfunction. If cleaning and or lubricating the weapon does not fix the problem, then the gunner turns it in to the unit armorer. Table 3-5 shows malfunctions, their probable causes, and the corrective actions.

MALFUNCTIONS	PROBABLE CAUSES	CORRECTIVE ACTIONS
Sluggish operation on gas regulator.	Carbon build-up.	Clean gas regulator.
Uncontrolled fire (runaway gun).	Broken or stuck Trigger.	
	Stuck sear.	Replace trigger.
	Broken of damaged sear spring	Replace sear spring

Table 3-5. Malfunctions.

a. **Uncontrolled Fire (Runaway Gun).** Uncontrolled fire (the weapon continues to fire after the trigger is released). This is usually caused by, the gunner not pulling and holding the trigger all the way to the rear. The following are immediate actions for uncontrolled fire:

(1) The gunner holds the weapon on target and fires the remaining ammunition.

(2) The assistant gunner stops the weapon from firing by breaking the belt of ammunition.

(3) The gunner as a last resort pulls the cocking handle to the rear thus, locking the bolt to the rear of the receiver.

b. **Sluggish Operation**. Sluggish operation is due to excessive friction caused by carbon build-up, improper lubrication, or burred parts. Corrective action includes cleaning, lubricating inspecting, and replacing worn parts. The gunner may adjust the gas regulator to maintain the rate of fire until he has a chance to clean the machine gun.

3-23. STOPPAGES

A stoppage is any *interruption* in the cycle of functioning caused by *faulty action* of the weapon or *faulty ammunition*. Stoppages are classified by their relationship to the cycle of functioning. Table 3-6 shows types of interruptions or stoppages, their probable causes, and the corrective actions.

STOPPAGE	PROBABLE CAUSE	CORRECTIVE ACTION
Failure to feed.	Insufficient gas pressure.	Clean gas port, inserts, and gas plug.
	Improper lubrication.	Lubricate as required.
	Defective links or ammunition.	Insert new link or ammunition.
	Ammunition belt installed wrong.	Reverse belt with open side of link down.
	Damaged or weak feed pawls and springs or feed lever.	Replace.
	Obstruction in receiver.	Remove obstruction; clean and lubricate as required.
	Damaged or weak feed pawls	Send to DS maintenance.
	Defective links or ammunition.	Insert new link or ammunition.
	Ammunition belt installed wrong.	Reverse belt with open side of link down.
Failure to chamber.	Ruptured cartridge case.	Remove IAW TM 9-1005-313-10.
	Damaged driving spring rod assembly.	Replace driving spring rod assembly.
	Damaged gas plug/collar (cracks/burrs).	Replace gas plug/collar.
	Built-up carbon on gas plug/collar, gas cylinder, piston or dirty chamber.	Remove carbon and clean IAW TM 9-1005-313-10.
Failure to extract.	Broken extractor or spring. Chipped or broken extractor. Defective extractor plunger. Insufficient gas pressure.	Replace. Replace. Replace. Clean and lubricate as required.
Failure to lock.	Dirty Chamber.	Clean IAW TM 9-1005-313-10.
	Dirty receiver or lack of lubrication.	Clean and lubricate IAW TM 9-1005-313-10.
	Insufficient gas pressure.	Clean gas regulator.

Table 3-6. Stoppages.

STOPPAGE	PROBABLE CAUSE	CORRECTIVE ACTION
Failure to fire.	Faulty ammunition. Broken or damaged firing pin or defective trigger. Insufficient gas pressure.	Replace. Replace or send to DS maintenance. Clean gas port, inserts, and gas plug.
Failure to cock.	Broken sear. Worn operating rod sear notch. Broken, defective, or missing sear plunger and or spring.	Send to DS maintenance. Send to DS maintenance. Send to DS maintenance.
Short to recoil.		Clean gas port and operating rod tube. Lubricate as required. Replace operating rod spring. See runaway gun (Malfunctions).

Table 3-6. Stoppages, (continued).

DANGER

1. **IF NOTHING IS EJECTED AND THE WEAPON IS HOT (200 OR MORE ROUNDS FIRED IN LESS THAN 2 MINUTES), DO NOT OPEN THE COVER. MOVE THE SAFETY TO "S," WHICH PLACES THE WEAPON ON "S." KEEP THE WEAPON POINTED DOWNRANGE AND KEEP AWAY FROM THE WEAPON FOR 15 MINUTES, THEN CLEAR THE WEAPON.**
2. **BE CAREFUL IN CLEARING THE WEAPON WHEN THE BARREL IS HOT, A ROUND MAY FIRE (COOK OFF) DUE TO THE BARREL'S HEAT INSTEAD OF DUE TO THE FIRING MECHANISM. DURING TRAINING OR ON A FIRING RANGE, AFTER THE WEAPON HAS FIRED 200 ROUNDS, ITS BARREL IS CONSIDERED A HOT BARREL.**
3. **DURING COMBAT, WAIT 5 SECONDS, BECAUSE OF THE POSSIBILITY OF A "HANGFIRE" BEFORE APPLYING IMMEDIATE OR REMEDIAL ACTION. DURING TRAINING, WAIT 15 MINUTES BEFORE CLEARING A HOT WEAPON AND APPLYING IMMEDIATE OR REMEDIAL ACTION.**

NOTE: When applying immediate or remedial action on a cold or hot gun, the gunner checks to see if any part of the round (ranging from the tip of the bullet to the rim) is in the chamber. The gunner removes the ammunition from the feed tray only, then closes the cover and attempts to fire. If the weapon fires, he reloads and continues firing. If it does not fire, he clears the weapon, and he inspects the weapon and ammunition.

3-24. IMMEDIATE ACTION

Immediate action is action taken to *reduce a stoppage without looking for the cause.* Immediate action should be taken in the event of either a misfire or a cook off. A *misfire* is the failure of a chambered round to fire. Such failure can be due to an ammunition defect or faulty firing mechanism. A *cook off* is the firing of a round by the heat of a hot barrel and not by the firing mechanism. Cookoffs can be avoided by applying immediate action within 10 seconds after a failure to fire. If the M240B stops firing, the gunner performs the following immediate actions are taken. (An effective memory aid is POPP, which stands for pull, observe, push, and press.)

a. Pulls and locks the cocking handle to the rear while observing the ejection port to see if a cartridge case, belt link, or round is ejected. Ensures that the bolt remains to the rear to prevent double feeding if a round or cartridge case is not ejected.

b. If a cartridge case, belt link, or a round is ejected, returns cocking handle to forward position, aim on the target, and presses the trigger. If the weapon still does not fire, takes remedial action. If a cartridge case, belt link, or round is not ejected, takes remedial action.

3-25. REMEDIAL ACTION

Remedial action is any action taken to determine the cause of a stoppage and to restore the weapon to an operational condition. This action is taken only after immediate action did not remedy the problem.

a. **Cold Weapon Procedures**. When a stoppage occurs with a cold weapon and immediate action has failed, the gunner uses the following procedures.

(1) Pulls the cocking handle to the rear, locking the bolt. Returns the cocking handle and places the safety to SAFE.

(2) Places the weapon on the ground or away from his face and opens the cover, performs the four-point safety check. Reloads and continues to fire.

(3) If the weapon does not fire, clears the weapon and inspects it and the ammunition.

b. **Hot Weapon Procedures**. If the stoppage occurs with a hot weapon (200 rounds or more in 2 minutes or as noted above for training), the gunner moves the safety to SAFE, waits 5 seconds (during training, lets the weapon cool for 15 minutes), uses the same procedures as outlined for cold weapon procedures.

c. **Jammed Cocking Handle**. If a stoppage occurs and the cocking handle cannot be pulled to the rear by hand (the bolt may be fully forward and locked or only partially forward), the gunner takes the following steps.

(1) Tries once again to pull the cocking handle *by hand.*

WARNING

Do not try to force the cocking handle to the rear with your foot or a heavy object. This could damage the weapon.

(2) If the weapon is hot enough to cause a cook off, moves all soldiers a safe distance from the weapon and keeps them away for 15 minutes.

(3) After the gun has cooled, the gunner will pull the cocking handle to the rear. Ensures rearward pressure is kept on the cocking handle until the driving spring rod

assembly is removed. Opens the cover and disassembles the gun. (The assistant gunner helps the gunner do this.)

(4) Removes the round or fired cartridge. Uses cleaning rod or ruptured cartridge extractor if necessary.

(a) In a training situation, after completing the remedial action procedures, the gun should not be fired until an inspection by an ordnance specialist has been made.

(b) In a combat situation, after the stoppage has been corrected, the gunner changes the barrel and tries to fire. If the weapon fails to function properly, the gunner sends it to the unit armorer.

3-26. STUCK BARREL

Stuck barrel is the result of the machine gun crew not properly cleaning the gas cylinder and gas regulator plug. During training or range firing the M240B should be cleared, disassemble and cleaned immediately. In combat, the M240B should be cleaned as soon as possible. The gun crew performs the following actions, only if the weapon can not be properly cleaned at that time.

a. Pulls the cocking handle to the rear, locking the bolt. Returns the cocking handle and places the safety to SAFE.

b. Places the weapon on the ground or away from his face and opens the cover, performs the four-point safety check.

c. The gunner ensures that the barrel is still locked to the receiver with the carrying handle to the right.

d. The assistant gunner places the heat protective mitten on his right hand. With the mitten on he will remove the gas regulator collar from the barrel that is secured to the receiver.

e. With the gas regulator collar removed, the gunner and assistant gunner remove the barrel as outlined in Section II.

f. After removal of the barrel, the assistant gunner will remove the gas regulator collar and gas regulator plug from the spare barrel.

g With these removed the gunner and assistant gunner inserts the barrel into the socket of the receiver ensuring that the gas regulator plug is going into the gas hole bushing. Once the barrel is secured to the receiver the assistant gunner secures the gas regulator collar on the gas regulator plug.

h. The gunner, after ensuring the barrel is secured to the receiver (2 to 7 clicks) and the collar is secure, will reload and continue firing.

3-27. DESTRUCTION PROCEDURES

Destruction of any military weapon is only authorized as a last resort to prevent enemy capture or use. This paragraph discusses the field-expedient means of this destruction; it does not replace published policies. In combat situations, the commander has the authority to destroy weapons, but he must report this destruction through channels.

a. Disassemble the weapon as completely as time permits. Use the barrel or tripod mount to destroy the bolt and operating rod assembly, barrels, rear and front sights, and mounts.

b. Bury the disassembled weapon or dump the parts into a stream, a sump, or a latrine.

c. Burn the weapon by placing an incendiary grenade on the receiver group over the bolt (with the cover resting on the grenade) and detonating the grenade.

d. Smash the traversing and elevating mechanism and pintle assembly. Bend the tripod legs.

CHAPTER 4

MACHINE GUN MARKSMANSHIP TRAINING

This chapter aids trainers in preparing and conducting machine gun marksmanship training for the machine gun. Machine gun marksmanship training is conducted in three phases: preliminary gunnery, basic gunnery, and advanced gunnery in Chapter 5.

Section I. INTRODUCTION

Marksmanship begins with nonfiring individual skill proficiency and concludes with collective proficiency firing under demanding conditions.

4-1. OBJECTIVES

The objectives of machine gun marksmanship training are to produce gunners that are thoroughly capable of the following:

a. **Accurate Initial Burst**. Obtaining an accurate initial burst of fire on the target is essential to good marksmanship. This requires the gunner to estimate range to the target, set the sights, and apply the fundamentals of marksmanship while engaging targets.

b. **Adjustment of Fire**. The gunner must observe the strike of the rounds when the initial burst is fired. If not on target, he manipulates the T&E mechanism until the rounds do strike the target. The assistant gunner must be proficient in observing the strike of rounds and in observing and using tracers so the gunner can rapidly relay the machine gun on the target for engagement.

c. **Speed**. Speed is also essential to good marksmanship; it is attained by practice in both dry-fire and live-fire exercises. It is an acquired skill gained through extensive training that combines other skills when delivering fire. Speed should not be stressed to the detriment of accuracy.

4-2. TRAINING PHASES

Marksmanship training for the machine gun is progressive in nature. It begins with nonfiring individual skill proficiency and concludes with collective proficiency firing under demanding conditions. Gunners and leaders must master the fundamentals before attempting individual and collective firings. More effective and efficient marksmanship occurs if live firing is preceded with good preliminary marksmanship training. Likewise, proficient individual firing will achieve more proficient collective firing.

a. **Preliminary Gunnery.** In this phase, the gunner learns and demonstrates proficiency on individual skills that prepare him to fire live ammunition. This includes mastering mechanical training, the four fundamentals of marksmanship, T&E manipulation, sight adjustments, crew drill, and fire commands.

b. **Basic Gunnery**. In this phase, the gunner applies the fundamentals in live-fire exercises during day and night conditions. This includes zeroing, 10-meter firing with crew drill, field zeroing, and transition firing with crew drill.

c. **Advanced Gunnery.** In this phase, gunners are trained on combat techniques of fire, techniques of employment, and live-fire exercise during NBC conditions.

4-3. TRAINING STRATEGY

Training strategy involves the overall concept for integrating resources into a program that trains individual and collective skills needed to perform a wartime mission. The goal of a marksmanship program is to produce well-trained gunners who can win and survive on the battlefield.

a. Leaders implement training strategies for machine gun marksmanship in TRADOC institutions (IET, NCOES, IOBC, and IOAC) and in units. The overall training strategy is multifaceted and is inclusive of the specific strategies used in institution and unit programs. Also included are the supporting strategies that use resources such as publications, ranges, ammunition, training aids, devices, simulators, and simulations. These strategies focus on developing critical soldier skills and leader skills that are required for the intended outcome.

b. The training strategies contain two components: initial training and sustainment training. Both may include individual and collective skills. Initial training is critical because a task that is taught correctly and learned well is retained longer. When an interim of nonuse occurs, well-trained skills are more quickly regained and sustained. The more difficult and complex the task, the harder it is to sustain the skill. Personnel turnover plays a major factor in the decay of collective skills, since the loss of critical team members requires retraining to regain proficiency. Retraining becomes necessary when a long period elapses between initial and sustainment training sessions or when the training doctrine is altered.

c. The training strategy for machine gun marksmanship begins in the institutions and continues in the unit. Figure 4-1, illustrates an example of this overall process, which provides a concept of the flow of unit sustainment training. Combat arms IET provides field units with soldiers who are familiar with standards in basic marksmanship tasks. The soldiers graduating from these courses have been trained to maintain their machine guns and to hit a variety of targets. They have learned range determination, target detection, application of marksmanship fundamentals, and other skills needed to engage a target.

d. Additional skills trained in the institution include techniques for employment, classes of fire, and fire commands. These skills must then be reinforced in the unit. Related soldier skills of camouflage, cover and concealment, maneuver, and preparation and selection of a fighting position are addressed in STP 21-24-SMCT, which must be integrated into tactical training.

e. Training continues in units on the basic skills taught in combat arms IET. Additional skills, such as suppressive fire and supporting fire, are trained and then integrated into collective training exercises, which include squad and platoon live-fire exercises. (A unit machine gun marksmanship training program is explained in Appendix B.) The strategy for sustaining the basic marksmanship skills that is taught in combat arms IET involves periodic preliminary gunnery, followed by 10-meter, transition firing, and qualification range firing. However, a unit must establish a year-round program to sustain skills. Key elements include training the trainers and refresher training of nonfiring skills.

f. In the unit, individual proficiency and leader proficiency of marksmanship tasks are integrated into collective training that includes squad, section, and platoon drills and STXs. The collective tasks in these exercises, and how they are planned and conducted, are in ARTEP 7-8-MTP and ARTEP 7-8-DRILL. Collective tasks are evaluated to standard and discussed during leader and trainer after-action reviews. Objective evaluations of both individual and unit proficiency provide readiness indicators and future training requirements.

g. A critical step in the Army's overall marksmanship training strategy is to train the trainers and leaders first. Leader courses include limited machine gun training, but unit publications will help develop officer and NCO proficiency necessary to plan and conduct gunnery training and to evaluate the effectiveness of their programs. Proponent schools provide training support materials to include field manuals, training aids, devices, simulators, and programs that are doctrinal foundations and guidance for training the force.

h. Once the soldier understands the weapon, knows how to zero, and has demonstrated proficiency at 10-meter and transition ranges, he should be exposed to more difficult ranges and scenarios.

i. IET culminates in the soldier's proficiency assessment, which is conducted on the 10-meter and transition and record fire ranges. Unit training culminates in a collective, live-fire, tactical exercise that provides an overview of unit proficiency and training effectiveness.

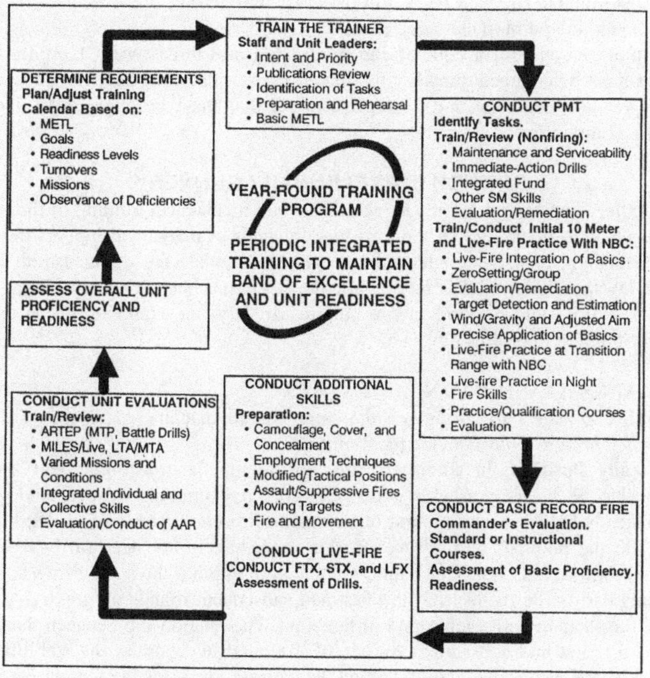

Figure 4-1. Unit marksmanship sustainment strategy.

4-4. TRAINING FOR COMBAT CONDITIONS

The trainer must realize that qualification is not an end but a step towards reaching combat requirements. To reach this goal, the gunner not only considers his position and the use of his weapon, but also some of the following combat conditions as well.

a. Most engagements will be within 300 meters; however, the gunner must still engage targets out to the maximum range of the machine gun.

b. Enemy personnel are seldom visible except when assaulting.

c. Most combat fire must be directed at an area where the enemy has been detected or where he is suspected of being but cannot be seen. Area targets consist of objects or outlines of men irregularly spaced along covered and concealed areas (ground folds, hedges, borders of woods).

d. Most combat targets can be detected by smoke, flash, dust, noise, or movement, but the targets are only visible for a moment.

e. Some combat targets can be engaged by using reference points, predetermined fire, or range card data.

f. The nature of the target and irregularities of terrain and vegetation may require a gunner to move from one position to another to place effective fire on the target. The most stable position for the gunner is the prone tripod-supported position.

g. Most combat targets have a low contrast outline and are obscured. Therefore, choosing an aiming point in elevation is difficult.

h. Time-stressed fire in combat can be divided into three types: a single, fleeting target that must be engaged quickly; distributed targets that must be engaged within the time they remain available; and a surprise target that must be engaged at once with instinctive, accurate fire.

SECTION II. PRELIMINARY GUNNERY

Once a soldier is proficient in the characteristics and mechanical training of the machine gun, he is ready to be trained on the four fundamentals of marksmanship. As the gunner learns the fundamentals, he should be required to manipulate the sights, use his body to shift and lay the sights on the target, use the T&E mechanism to lay on the target, conduct crew drill, and respond to fire commands. Dry-fire exercises are an excellent method for training to proficiency.

4-5. MARKSMANSHIP FUNDAMENTALS

The four fundamentals for firing are the same for all machine guns, they are *steady position*, *aim*, *breath control*, and *trigger control*.

a. **Steady Position**. In automatic fire, position is the most important aspect of marksmanship. If the gunner has a good zero, correctly aims his weapon, and properly applies a steady hold in firing a burst of automatic fire, the first round of that burst hits the target at the point of aim. However, this procedure is not necessarily true of the second and third rounds. The first round hits the aiming point the same as when a round is fired singularly. The recoil from the first and subsequent rounds progressively disturb the lay of the weapon with each round of the burst. The relationship between the point of impact of the first and subsequent rounds of the burst depends on the stability of the gunner's position. His body, directly behind the weapon, serves as the foundation, and his grip serves as a lock to hold the weapon against the foundation. The better the body alignment and the steadier the grip, the less dispersed the rounds of a burst of automatic fire will be.

b. **Aim.** To aim the machine gun, the gunner must align the sights, focus his eye, obtain a correct sight picture, control his breathing, and maintain trigger control.

(1) *Sight Alignment.* To obtain correct align, the gunner centers the front sight post in the aperture of the rear sight. For a correct sight picture, the gunner centers the target over the front sight post so that it appears to rest lightly on top of the sight. The aspects of obtaining an accurate initial burst through sight alignment and sight picture, trigger manipulation, and zeroing are the same for tripod training as for bipod training

(2) *Focus of the Eye.* A good firing position places the eye directly on line with the center of the rear sight. The gunner must focus on the tip of the front sight post. The natural ability of the eye to center objects in the rear sight and to seek the point of greatest light aids in providing correct sight alignment.

(3) *Sight Picture.* A correct sight picture has the target, front sight post, and rear sight aligned. The sight picture consists of sight alignment and placement of the aiming point on the target. The gunner aligns the front sight post in the center of the rear sight and then aligns the sights with the target. The top of the front sight post is aligned on the center base of the target (Figure 4-2).

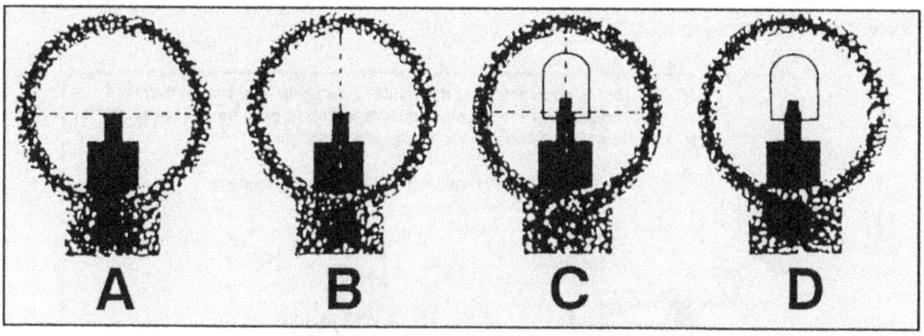

Figure 4-2. Sight picture.

c. **Breath Control.** When firing in bipod-mounted mode, two types of breath control are used. When firing single shots, as in zeroing, the gunner stops breathing after most of the air has been exhaled during the normal breathing cycle. He fires before he feels any discomfort. During automatic fire, ideally, the gunner exhales and stops his breath when pressing the trigger. He does not have time to take deep breaths between bursts. He must hold his breath before each burst or adapt his breathing by taking quick shallow breaths or taking deeper breaths between several bursts.

d. **Trigger Control.** Pressing the trigger straight to the rear and releasing it helps control the number of rounds in each burst and prevents disturbing the lay of the weapon. For this the gunner must learn how to manipulate the trigger so, that he may get the desired burst he wishes to obtain.

4-6. FIRING POSITIONS

The bipod-supported prone and fighting positions and the tripod-supported prone and fighting positions are covered in preliminary gunnery.

a. **Prone Position, Bipod-Supported.**

(1) Assume a prone position to the rear of the weapon (place the shoulder rest on your firing shoulder for the M249 and M60 only). An imaginary line drawn through the weapon should bisect the firing shoulder and buttock, and continue through the heel of your foot.

(2) Spread your legs a comfortable distance apart with your heels as close to the ground as possible, yet comfortable.

(3) Grasp the pistol grip with your firing hand. Place the fleshy end of the index finger resting lightly on the trigger. Place your nonfiring hand on the small of the stock with your thumb is curled underneath. Then slide your nonfiring hand forward until your little finger touches the receiver, so your aiming point will always be the same.

(4) Place your cheek against the forefinger of your nonfiring hand to form a stock weld. Try to position your nonfiring hand and cheek at the same spot on the stock each time you fire the weapon. The stock weld should provide for a natural line of sight through the center of the rear sight aperture to the front sight post and to the target. Relax your neck so that your cheek rests on your forefinger naturally.

(5) Apply a firm, steady pressure rearward and down, holding the weapon tightly into the hollow of your shoulder while aiming and firing.

(6) Keep your shoulders level and elbows about an equal distance from the receiver of the weapon (Figure 4-3).

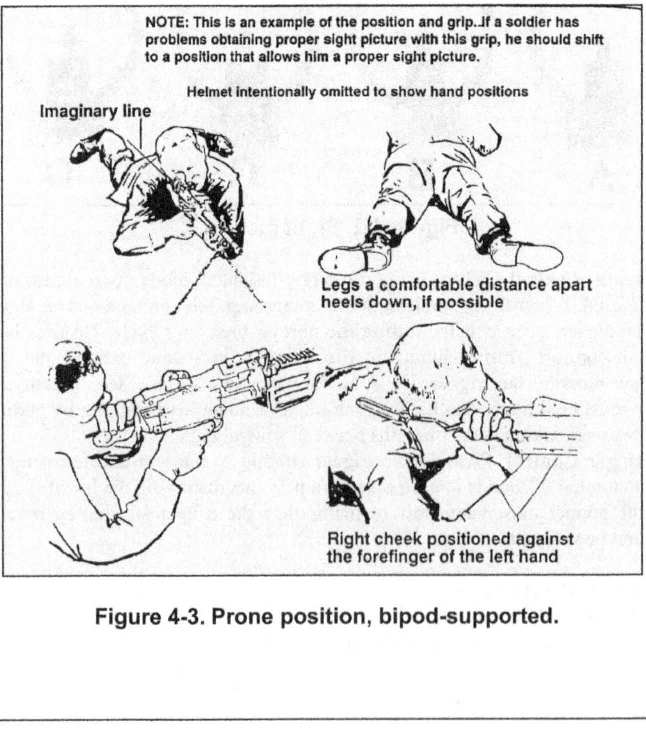

Figure 4-3. Prone position, bipod-supported.

NOTES: 1. The assistant gunner assumes a prone position along the left side of the gunner to load ammunition and observe.
2. Left-handed firing with the M249 and M60 is discouraged because the ejection pattern of some weapons is almost directly to the rear. When firing any machine gun using the tripod, the gunner must use his left hand to manipulate the T&E mechanism, therefore precluding the gunner from firing the machine gun left handed.
3. If a gunner has problems obtaining a proper sight picture, he should shift to a position that allows him to do so.

b. **Fighting Position, Bipod-Supported.** This is an excellent position that provides a stable firing platform. The depth of the fighting position and the support should be adjusted for the height and arm length of the gunner. This allows for a steadier position.

(1) Extend the bipod legs and place the machine gun in front of the position.

(2) Place your right (firing side foot) foot sideways against the rear of the fighting position and lean forward until your chest is squarely against the forward wall.

(3) Raise the folding shoulder rest and place it on your firing shoulder (M249 and M60 only). Keep your shoulders level or parallel to the ground.

(4) Grasp the pistol grip with your firing hand. place the fleshy end of the index finger resting lightly on the trigger. Place your nonfiring hand on the small of the stock and ensure that your thumb is curled underneath.

(5) Place your cheek against the forefinger of your nonfiring hand to form a stock weld. Try to position your nonfiring hand and cheek at the same spot on the stock each time you fire the weapon. The stock weld should provide for a natural line of sight through the center of the rear sight aperture to the front sight post and to the target. Relax your neck so that your cheek rests on your forefinger naturally.

(6) Apply a firm, steady pressure rearward and down, holding the weapon tight into the hollow of your shoulder while aiming and firing.

(7) Keep your shoulders level and elbows about an equal distance from the receiver of the weapon (Figure 4-4).

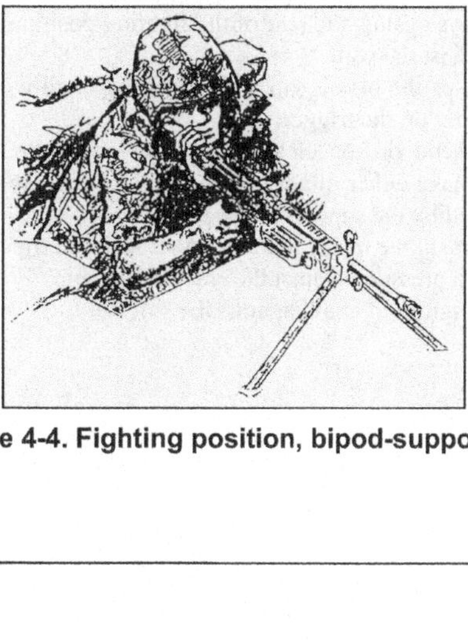

Figure 4-4. Fighting position, bipod-supported.

c. **Prone Position, Tripod-Supported**. The gunner assumes a prone position to the rear of the weapon (place the shoulder rest on your firing shoulder for the M249 and M60 only). An imaginary line drawn through the weapon should bisect the right shoulder and buttock and continue through the heel of his foot. When using the tripod, the assistant gunner assumes a prone position along the left side of the gunner to load ammunition and observe.

(1) The gunner, spreads his legs a comfortable distance apart with his heels as close to the ground as possible and still be comfortable.

(2) Grasps the pistol grip with his right hand with the fleshy end of his index finger resting lightly on the trigger. (The machine gun is not fired left-handed with the tripod because turning the traverse handwheel with the right hand is difficult.)

(3) Grasps the elevating handwheel with his left hand, palm down. Exerts a firm downward pressure with both hands while aiming and firing.

(4) Places both elbows on the ground between the tripod legs and his body. The position of his elbows raises or lowers his body in relation to the machine gun.

(5) Places his shoulder lightly against the stock without applying any pressure.

(6) Rests his cheek lightly (if at all) against the stock (Figure 4-5).

PRONE POSITION TRIPOD

Figure 4-5. Prone position, tripod-supported.

d. **Fighting Position, Tripod-Supported.** (Figure 4-6.) The gunner places his right (firing side) foot sideways against the rear of the fighting position and leans forward until his chest is squarely against the wall.

(1) The gunner, grasps the pistol grip with his firing hand with the fleshy end of his index finger resting lightly on the trigger.

(2) Places his left hand on the elevating handwheel, palm down, exerting a firm downward pressure to make either minor or major adjustments in deflection or elevation. (The weapon is stabilized by the support of the tripod.)

(3) Places his elbows on the inside and does not touch the tripod.

(4) Places little or no pressure against the stock of the gun.

(5) Rests his cheek lightly, if at all, against the stock.

Figure 4-6. Fighting position, tripod-supported.

4-7. NIGHT FIRE

Although the same four fundamentals of marksmanship are used for night firing, adjustments must be made to accommodate the night vision devices.

a. **Bipod**.

(1) *Steady Position.* When firing unassisted, changes in head position and stock weld are necessary especially when using weapon-target alignment techniques. Normally, the gunner positions his head so that he can align the weapon on the target and look over the sights. In some cases, the lower part of his jaw makes firm contact with his nonfiring hand on the stock, with his eyes an inch or so above the sights. The key is to use the natural pointing ability to align the machine gun on the target. When using NVDs, the head position and stock weld must be altered to be able to use the device. Sometimes height of the NVD may make this impossible. NVDs alter the machine gun's weight and center of gravity. The gunner must compensate by exerting greater pressure and control with his firing hand on the pistol grip and his nonfiring hand on the stock.

(2) *Aim.* Various modifications are necessary when aiming the machine gun at night. When firing unassisted, the gunner uses off-center vision instead of pinpoint focus. Both eyes are open and focused downrange on the target and not on the sights. Rather than aim using the sights, the gunner looks over the sights and points the machine gun where he is looking. The normal tendency is to fire high so the gunner must improve weapon-target alignment by pointing slightly low to compensate. When using NVDs, the gunner uses the necessary aiming process to use the device.

(3) *Breath Control.* This fundamental is not affected by night firing conditions; however, wobble is more pronounced when using NVDs, because they magnify the field of view.

(4) *Trigger Control.* There is no change to this fundamental during night firing. The objective is to not disrupt alignment of the weapon with the target.

b. **Tripod.**

(1) *Steady Position.* When firing at predetermined targets with the weapon laid on each target, there are not differences in steady position at night as compared to day. However, firing at night at targets of opportunity requires modifications. The gunner is required to use weapon-target alignment techniques. He must align the weapon on the target and look over the sights. His head is higher and his lower jaw is lightly on the stock if at all. With night vision devices, the gunner must position his head so that his firing eye is in line with the device.

(2) *Aim.* For targets of opportunity, the gunner uses the same techniques as with a bipod during night firing except weapon-target alignment is achieved with the T&E mechanism.

(3) *Breath Control.* There are no changes in this fundamental.

(4) *Trigger Control.* There are no changes in this fundamental.

4-8. NUCLEAR, BIOLOGICAL, CHEMICAL FIRE

The four fundamentals remain valid in an NBC environment, although some modifications may be needed to accommodate the equipment.

a. **Bipod**

(1) *Steady Position.* The bulk of overgarments may require adjustments to the position for stability and comfort. A consistent stock weld is difficult to maintain because of the shape of the protective masks. The gunner has to hold his head in an awkward position to see through the sight. If necessary, he may cant the weapon to overcome this situation. This procedure relieves the neck muscles and places the eye on line with the center of the rear sight.

(2) *Aim.* The gunner may have to rotate (cant) the machine gun to see through the rear sight aperture. He should rotate only enough to align the sights, and only if necessary. Ballistics cause rounds to impact low in the direction of the cant at long ranges. If canting at targets beyond 175 meters, the gunner must adjust his point of aim. The best technique is to aim at center base of the target initially and then make adjustments based on the strike of the rounds. Right-handed firers adjust point of aim to the right and high; left-handed firers to the left and high.

(3) *Breath Control.* Although breathing is somewhat restricted and more difficult while wearing the protective mask, the impact is negligible. Care must be taken, to avoid hyperventilating during burst fire. The amount of oxygen inhaled by taking quick shallow breaths or deeper breaths between bursts is significantly reduced.

(4) *Trigger Control.* Trigger control is affected when the gunner wears gloves. The effect cannot be accurately predicted for each soldier; therefore, practice and training under these conditions are required.

b. **Tripod.**

(1) *Steady Position.* Modifications are similar to those in bipod firing. There are two other points of importance. Manipulating the T&E with gloves on is more difficult because the feel of the handwheel differs. The gunner may not sense the same control as without gloves. Second, hearing is impaired. Together, reduced sense of touch and hearing impairment make T&E manipulations especially difficult. For these reasons, adjustments may be considerably slower.

(2) *Aim.* Unlike the bipod, the tripod does not allow the machine gun to be canted. This requires the gunner to position his head behind the stock to use the sight. Skilled gunners who make adjustments to the T&E quickly can confirm their sight picture and then look over the sights to observe the strike of the round while firing. This not only provides relief for the neck muscles but aids in making adjustments.

(3) *Breath Control.* Some considerations apply in the same way as with of the bipod; however, the stable platform of the tripod negates movement associated with breathing.

(4) *Trigger Control.* Like the bipod, control is different because the trigger feels different. Training familiarizes the gunner with the changes he must make while wearing gloves.

4-9. ENGAGEMENT OF MOVING TARGETS

The fundamentals used to hit moving targets are the same as those needed to hit stationary targets. However, the procedures to engage moving targets vary as the angle, speed, and range of the target varies. Targets moving directly at the gunner are engaged the same as a stationary target; there is no change in the application of the fundamentals. But fast-moving targets at varying ranges and angles do require changes in the application of steady position and aiming. (For aerial targets engagements, see Appendix C.)

a. **Leads.** To hit a moving target, the machine gun must be aimed ahead of the target far enough to cause the bullet and target to arrive at the same time at the same point. This distance is measured in target lengths. One target length as seen by the gunner is one lead. Leads are measured from the center of mass. Table 4-1 gives the amount of lead needed to hit a target moving at right angles, to the gunner, and at speeds and ranges indicated. The gunner makes adjustments as conditions change. If target speed is 7 1/2 mph, the amount of lead is half that shown on the table; at 30 mph, double that shown. The angle at which the target moves also changes the lead. If the target is moving on an oblique angle, only half the lead is required. For a target moving directly at the gunner, the aiming point is below the center base of the target depending on range and slope of the ground. For a target moving directly away from a gunner, the aiming point is above the center base of the target (Figure 4-7). Too much lead is better than too little because the target moves into the beaten zone, and observation of the strike of the rounds is easier in relation to the target.

SPEED	RANGE OF TARGET		
	300 meters	500 meters	900 meters
15 mph	1/2 X Target length	1 X Target length	2 X Target lengths

Table 4-1. Vehicle lead table.

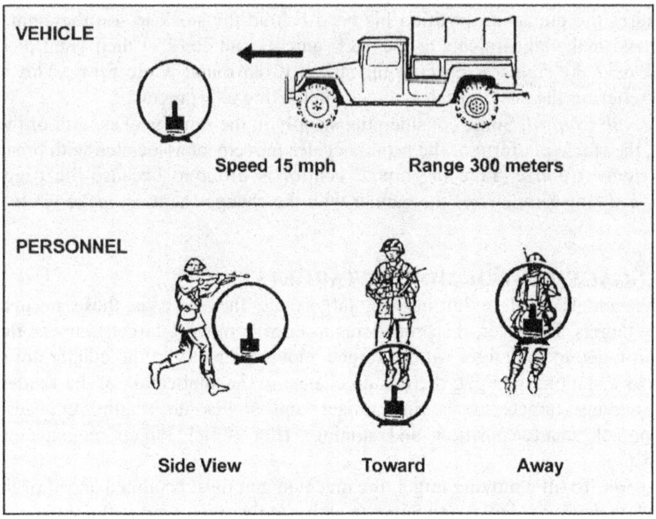

Figure 4-7. Moving-target aiming points.

NOTE: A soldier carrying a full combat load can run as fast as 8 mph for short distances on the battlefield.

b. **Tracking Techniques**. The gunner aims at a point ahead of the target equal to the estimated number of leads, maintains this lead by tracking the target (manipulates the weapon at the same angular speed as that of the target), and then fires. Tracking allows the gunner in position for a second burst if the first one misses.

c. **Trapping Techniques.** The gunner establishes an aiming point forward of the target and along the target path. He pulls the trigger as the target reaches the appropriate point in regard to lead.

d. **Position and Aim**.

(1) *Steady Position*. The gunner makes no change in position for targets moving directly toward or away from him. He manipulates the T&E mechanism to obtain the proper lead and sight picture. Some targets at varying speeds, angles, and ranges may require the gunner to reposition when in the prone position. The gunner redistributes his weight to his elbows and toes raising his body directly behind the weapon. He uses the T&E mechanism to traverse on to the target.

(2) *Aim*. The gunner uses the T&E mechanism to acquire the appropriate sight picture in relation to leading the target. He must quickly determine speed, angle, and range to the target, decide whether to track or trap, acquire lead, and engage the target. He uses the traversing handwheel to maintain lead.

(3) *Breath Control*. The gunner makes no change, but he must be quick to hold his breath because of the fleeting nature of moving targets.

(4) *Trigger Control.* The gunner makes no change in applying this fundamental.

e. **Bipod Techniques.** For targets moving to or from a gunner using a bipod, the same procedures are used. From a prone position, the gunner may be required to adjust his position quickly depending on range, angle, and speed of the target.

(1) *Steady Position.* If appropriate lead cannot be achieved by shifting his shoulders right or left (traverse) or by moving his elbows closer or farther a part (search), the gunner redistributes his weight to his elbows and toes and raises his body off the ground. Using his toes, the gunner shifts his body right or left in the opposite direction of the target and pivots on his elbows until the aiming point is well ahead of the target. The gunner rapidly assumes a steady position, obtains the sight picture, leads and engages the target. Trapping is the preferred technique. In order to apply this method, the bipod legs must move freely. When firing from a fighting position, the gunner must be flexible enough to track any target in his sector. If lead cannot be achieved, he slides the bipod legs in the appropriate direction (left or right) ahead of the target and continues as in the prone position. Trapping is still the preferred technique. If the terrain does not permit sliding the weapon left or right, the gunner lifts the bipod legs off the ground and places them where he can aim ahead of the target, reestablishes a steady position, and continues as before.

(2) *Aim.* The gunner determines angle, speed, and range quickly; acquires the appropriate lead; and engages the target. He aligns the front sight post in the proper position to lead the target. For targets moving directly away, he places the front sight post above center of mass. For targets moving directly at him, he aligns the front sight post below center of mass. For all other targets, he aligns the front sight with center base of the target applying the appropriate lead.

(3) *Breath Control.* The gunner must hold his breath quickly because of the fleeting nature of moving targets.

(4) *Trigger Control.* This is the same as for engaging stationary targets.

4-10. TRAVERSE AND SEARCH

The traverse technique moves the muzzle of the weapon to the left or right to distribute fire laterally. Search moves the muzzle up or down to distribute fire in depth.

a. **Tripod.**

(1) *Traverse.* To move the muzzle to the right, the gunner places his left hand on the traversing handwheel, thumb up, and pushes his thumb away from his body (right). To move the muzzle to the left, he pulls his thumb towards his body (left).

(2) *Search.* To move the muzzle up, the gunner grasps the elevating handwheel with his left hand and pushes his thumb away from his body (add). To move the muzzle down, he pulls his thumb towards his body (drop).

b. **Bipod.**

(1) *Traverse.* To make minor changes in direction, the gunner shifts his shoulders to the right or left to select successive aiming points in the target area. Major changes require him to redistribute his weight to his elbows and toes and raise his body off the ground. Using his toes, he shifts his body to the right or left to be in the opposite direction of the target, and pivots on his elbows until he is aligned with the target. The gunner rapidly assumes a steady position, obtains the proper sight picture, and engages the target.

(2) *Search.* To make changes in elevation, the gunner moves his elbows closer together to lower the muzzle or farther apart to raise the muzzle. He corrects gross errors in range by adjusting the range setting.

4-11. DIRECT LAY

The simplest, quickest, and most effective technique of delivering fire with the machine gun is to align the sights on the target and properly apply fire. This technique of fire is called *direct lay*.

4-12. APPLICATION OF FIRE

The gunner must aim, fire, and adjust on a certain point of the target. He always keeps the center of his beaten zone at the center base of the target for maximum effect from each burst of fire. When this procedure is done, bullets in the upper half of the cone of fire run through the target if it has height, and the bullets in the lower half of the beaten zone ricochet into the target.

4-13. FIRE ADJUSTMENT

The gunner initially sets his sights with the range to the target, lays on the target (sight alignment and sight picture on the center base of the target), fires a burst, and observes the strike of the rounds or flight of the tracers. When the initial burst is correct, he continues to fire until the target is covered. He must regain a good sight picture before each burst when using the bipod. When using the tripod, the gunner makes a rapid check of the sight picture after each traverse and search adjustment.

a. **Sight Corrections Method**. A gunner must observe and adjust fire rapidly to be effective. He observes bursts of fire by noting the strike of the rounds in the target area and the tracers in flight. The technique to adjust fire depends on time, range, and amount of adjustment. These factors assist the gunner in determining whether or not to make sight corrections or adjust position and point of aim. When the initial burst is not correctly placed, the gunner may change the elevation and windage on the sights and fire another burst on the target. This method is time-consuming, even for the well-trained soldier.

b. **Adjusted Aiming Point Method.** In this method of fire adjustment, the gunner uses his sight but does not make sight corrections. This method is quick. If the gunner misses the target with his initial burst, he must rapidly select a new aiming point the same distance from the target as the center of impact of the initial burst, but in the opposite direction. For example, if the initial burst is 20 meters beyond and 10 meters to the right of the target, the gunner rapidly selects an aiming point about 20 meters short and 10 meters to the left of the target, lays on that aiming point, and fires (Figure 4-8).

(1) When selecting a new aiming point from bipod mode, he may have to shift his shoulders slightly to the left or right for windage corrections. For elevation changes, he moves his elbows closer together (lowers the impact) or farther apart (raises the impact). For large corrections, he must move his elbows and realign his body to remain directly behind the weapon. He does this by redistributing weight to his elbows and toes and raises his body off the ground. He shifts his body using his toes, to the right or left, pivoting on his elbows until he is on line with the target. Then he assumes a steady position, obtains the sight picture, and engages the target.

(2) When selecting a new aiming point from tripod mode, the gunner may have to manipulate the T&E mechanism.

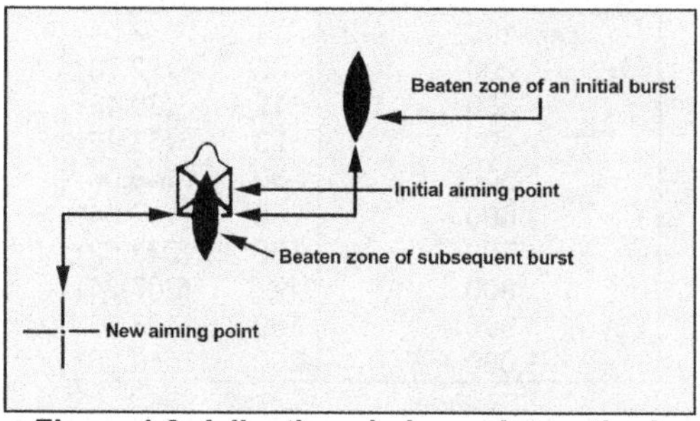

Figure 4-8. Adjusting aiming point method.

4-14. EFFECTS OF WIND

The effects of wind vary depending on changes in speed and direction. Wind is classified by the direction it is blowing in relationship to the firer and target line. The *clock system* is used to indicate wind direction and value (Figure 4-9).

a. **Clock System**. Winds that blow from the left (9 o'clock) or right (3 o'clock) are called *full-value winds*, because they have the most effect on the round. Winds that blow at an angle from the front or rear area are called *half-value winds*, because they have about one-half the effect on the round as full-value winds. Winds that blow straight into the gunner's face or winds that blow straight into the target are termed *no-value winds*, because their effect on the round is too small to be a concern. Effects of the wind increase as the range increases. (Figure 4-10) shows the effects of a 10-mph wind at varying ranges. A 20-mph wind doubles the effect. Winds at other than right angles have less effect. As indicated in Figure 4-10, wind has almost no effect up to 300 meters.

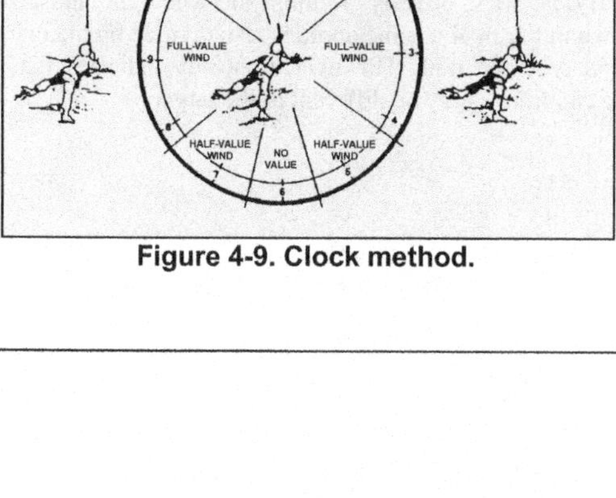

Figure 4-9. Clock method.

10-MILES PER HOUR WIND DRIFT		
RANGE IN METERS	INCHES (CM)	
100	1	(2.54)
200	5	(12.70)
300	12	(30.48)
400	23	(53.42)
500	39	(49.06)
600	60	(152.04)
700	88	(223.52)
800	121	(307.34)
900	159	(403.86)
1,000 +	202	(513.08)

Figure 4-10. Effects of winds.

NOTE: When in doubt, the gunner aims the initial burst directly at the center base of the target and, using the techniques of observation and adjustment of fire, adjusts the fire onto the target.

b. **Wind Measurement**. Wind is highly variable and sometimes quite different at the firing position than at the target position. Even though the wind is blowing hard at the firing position, trees, brush, or terrain could protect the path of the round. The wind can vary by several miles per hour between the time a measurement is taken and when the round is fired. Therefore, training time should not be wasted trying to teach gunners an exact way to measure wind speed. They should know that even though wind can affect trajectory, it can be overcome by adjusting fire. A wind gauge can be used for precise measurement of wind velocity. When a gauge is not available, velocity is estimated by one of the following methods.

(1) *Observation Method.* The following information can assist in determining wind velocities.

(a) Winds under 3 mph can barely be felt, but the presence of slight wind can be determined by drifting smoke.

(b) Winds of 5 to 8 mph constantly move the leaves of trees.

(c) Winds of 8 to 12 mph raise dust and loose paper.

(d) Winds of 12 to 15 mph cause small trees to sway.

(2) *Pointing Method.* A piece of paper or other light material can be dropped from shoulder height. By pointing directly at the spot where it lands, the angle can be estimated. As shown in Figure 4-11, the angle is also divided by the constant number 4 to determine the wind speed in mph. However, this only indicates the conditions at the firing position; the conditions may be different at the target.

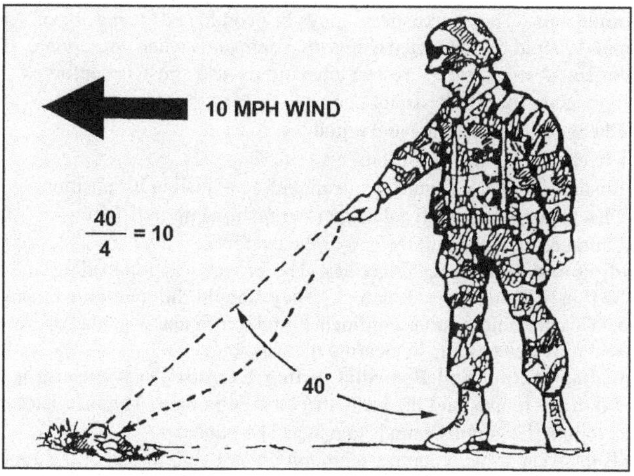

Figure 4-11. Pointing method.

4-15. FIRE COMMANDS

The standard fire commands are used as means of control during preliminary, basic, and advanced gunnery. The fire command must be explained to the gunner. The elements are given (as appropriate) before each dry-fire or live-fire exercise. The gunner takes action as directed and repeats each element as it is announced. (For a detailed explanation of fire commands, see Chapter 5.) When using the basic 10-meter range target, the fire command elements are as follows:

a. **Alert.** The alert is given as "Fire mission." Upon hearing the alert, the gunner loads his weapon and places the safety on "F".

b. **Direction**. Direction is given as FRONT since the targets appear to the gunner's front on the basic range.

c. **Description**. Description is given as PASTER NUMBER (pasters 1 through 8 as appropriate), at which time the gunner lays his weapon on the announced paster.

d. **Range**. The elevation on the rear sight assembly is always used on the basic range. This is announced as FIVE HUNDRED or SEVEN HUNDRED, at which time the gunner must ensure that his rear sight assembly has the correct elevation setting.

e. **Method of Fire**. Firing on the basic range is at a point target, so the method of fire is announced as FIXED. The gunner fires either single rounds or bursts at a rate slower than the sustained rate; therefore, the rate-of-fire element is omitted.

f. **Command to Open Fire**. This is announced as AT MY COMMAND. When the gunner is ready, he announces "Up." When all gunners are ready to fire, the command FIRE is given.

4-16. DRY-FIRE EXERCISES

Dry-fire exercises train the techniques of loading, unloading, immediate action, remedial action, fundamentals of marksmanship, sight settings, and T&E manipulation.

a. **Ammunition**. These exercises may be conducted using blank or dummy ammunition and should be conducted using fire commands when appropriate. If the blank firing attachment is used, safety restrictions for its use must be enforced. While the gunner performs the tasks, the assistant gunner—

- Checks the sight setting and initial lay.
- Checks the gunner's position.
- Ensures the gunner simulates firing before adjusting his position.
- Checks for proper body adjustment or manipulation of T&E.
- Critiques the gunner at the end of the exercise.

b. **Loading and Unloading Exercises**. The procedures for loading and unloading are prescribed in Chapter's 1, 2 and 3. They should be reinforced using dummy ammunition. This training instills confidence and proficiency in the operation of the weapon. It also provides training in clearing the weapon.

c. **Immediate Action and Remedial Action Exercise**. This exercise is conducted using linked dummy rounds and the basic machine gun target. The instructor should use salvage links to link the dummy rounds together. The gunner—

(1) Loads the weapon with dummy ammunition and aims at one of the aiming pasters on the basic machine gun target.

(2) Being conscious of the sight picture, pulls the trigger and the bolt goes forward (simulate firing the weapon). If the sight picture is disturbed, checks his position and grip, and maintains better control of the weapon.

(3) If he has a stoppage, applies immediate action procedures and continues to fire.

(4) If immediate action has failed, applies remedial action procedures and continue to fire.

d. **Operational Exercise**. The gunner aims and simulates firing each dummy round at the aiming paster on the basic machine gun target.

(1) Observes the sight picture through the feeding, locking, and firing cycle. (This provides feedback on his ability to maintain and hold the sight picture.)

(2) If at the completion of the firing cycle there is significant movement of the sight picture, his position is not steady enough or the tripod is not stable.

(3) Applies immediate action after firing each shot to extract and eject the dummy cartridge, and returns the bolt to the cocked position. Returns the cocking handle to the forward position.

WARNING

The M240B is carried loaded with the bolt locked to the *rear* in *tactical situations* where noise discipline is critical to the success of the mission. Trained gun crews are the only personnel authorized to load the M240B and only when command directs the crew to do so. During *normal training exercises*, the M240B is loaded and carried with the bolt in the *forward position*.

e. **Sight Setting and Sight Change Exercises**. These exercises are designed to train the gunner in the operation and adjustment of the rear sight, and making corrections in elevation and windage on the machine gun.

(1) For large adjustments in elevation (range), the gunner manipulates the rear sight to achieve different range settings. For fine adjustments in elevation, the gunner rotates the elevation knob for the machine gun.

(2) To make adjustments for windage, the gunner traverses the rear sight across the windage scale for the machine gun.

f. **Practice**. Before the dry-fire proficiency examination, soldiers should practice the tasks until they become proficient.

g. **Traversing and Searching Exercise**. After the gunner knows the principles of sighting and aiming and can assume a satisfactory firing position, he learns how to make minor and major body position changes to obtain an accurate initial lay. He practices shifting the direction of the weapon to successive points by moving his body. The basic machine gun target is placed 10 meters from the weapon for this exercise.

(1) Makes adjustments for large shifts in direction by using his elbows-and-toes technique described earlier. Makes small changes in direction by adjusting his shoulders.

(2) Makes major elevation changes by adjusting the range setting on the rear sight. Makes minor elevation changes by adjusting his elbows.

(3) Traverses and searches the target by sighting on the initial aiming paster (number 5 or 6) and then shifting to each of the other pasters in order (5 through 6 or its reverse).

(4) Upon receiving a fire command, the gunner repeats the instructions, sets the sights, lays the weapon on the designated paster, assumes the correct position, and reports *up*.

(5) At the command FIRE, the gunner simulates firing two single shots, then shifts to the next paster and simulates firing until the exercise is complete.

h. **T&E Manipulation Exercise**. After the gunner understands the principles of sighting and aiming and can assume a satisfactory firing position, he is instructed in manipulating the tripod-mounted machine gun to obtain an accurate initial lay. He is taught to shift the direction of the weapon to successive points with proficiency. The basic machine gun target is placed 10 meters from the weapon for this exercise.

(1) Makes large shifts in direction by releasing the traversing slide lock lever and moving the slide to the right or left. Makes minor changes in direction by using the traversing handwheel. (One click on the handwheel moves the strike of the round 1 cm on the target.)

(2) Adjusts for elevation by rotating the elevating handwheel with his left hand.

(3) Traverses and searches the target by laying on the initial aiming paster (number 5 or 6) and then shifts to each of the other pasters in order (5 through 6 or its reverse). (All major shifts in traverse are accomplished by loosening the traversing slide lockding lever.) When shifting from pasters number 7 through 8 or 8 through 7, uses the traversing handwheel.

(4) Upon receiving the command, the gunner repeats the instructions, sets the sights, lays the weapon on the designated paster, assumes the firing position, and reports UP.

(5) At the command FIRE, the gunner repeats the command, simulates firing two single shots, then shifts to the next paster and simulates firing until the exercise is completed.

i. **Dry-Fire Proficiency (Performance) Examination**. A gunner must demonstrate skill in all the tasks of the dry-fire proficiency examination before he is allowed to progress to 10-meter live firing. This examination emphasizes learning by doing. Proficiency is tested on a pass or fail basis. (The proficiency test is in Appendix B.)

j. **Remedial Training**. Remedial training must be given to soldiers who fail the performance objectives. Gunners who have passed the proficiency test may be used to assist in the training of soldiers having difficulty. Following retraining, the soldiers are retested in those tasks.

4-17. MULTIPURPOSE MACHINE GUN RANGE LAYOUT

The multipurpose machine gun range is used for conducting the 10-meter course as well as transition day, night, and integrated NBC firing. The firing area has 10 lanes. (Detailed setup and target configurations are described in TC 25-8. The layout is shown in Figure 4-12.) Personnel required for conducting the 10-meter range, as well as the transition firing, are the same, and they should perform the same duties for each training period. Local policy may dictate personnel requirements. The following are the minimum required personnel: OIC, NCOIC, safety officer or NCO, ammunition NCO, tower operator, lane NCOs, trainer and assistant gunners, or IAW TC 25-8. All personnel must adhere to safety rules IAW AR 385-63, local regulations, and Appendix D.

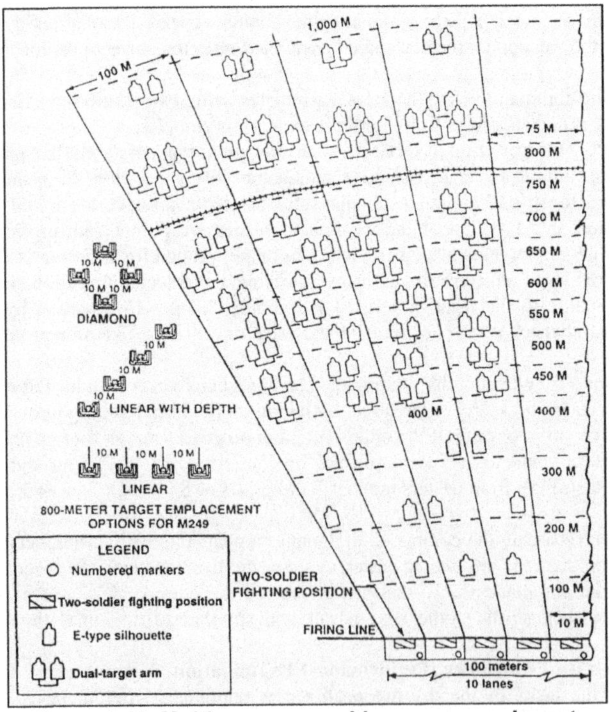

Figure 4-12. Multipurpose machine gun range layout.

NOTE: Targets beyond 800 meters are to be used with a machine gun optic and are not to be used during qualification without a machine gun optic.

4-18. BASIC MACHINE GUN TARGET

The basic machine gun target (FSN 6920-078-5128 and NSN 6920-00-078-5123) is used for the 10-meter firing exercise (Figure 4-13). The following explanation of the target, including the size of the aiming pasters and scoring spaces, aids in zeroing the machine guns and facilitates control during the 10-meter firing exercises. The target consists of four sections lettered A, B, C, and D. Each section has four point targets numbered 1, 2, 3, and 4; and two sets of area targets numbered 5 through 6 and 7 through 8. Each space is 4 cm wide and 5 cm high. The black aiming paster within the numbered scoring spaces is 1 cm square. The target is used to score one gunner—with one refire. Each gunner uses sections A, B and C. Sections C for qualification and section D for refire.

a. **Point Targets**. Point targets on the basic machine gun target are pasters 1 through 4 of sections A, B, C, and D. Firing at point targets exposes the gunner to zeroing techniques and controlled-burst fire techniques. Targets 1 through 4 can also be used for qualification.

b. **Area Targets**. Area targets on the basic machine gun target consist of pasters 5 through 6, and 7 through 8 of sections A, B, C, and D. Target group 5 through 6 provides the gunner with targets in depth and allows him to use a series of aiming points to disburse fire across the target by using the T&E mechanism. Target group 7 through 8 provides the gunner with linear targets with depth. This series of targets uses a series of aiming points to disburse fire across the target and in depth by using the T&E mechanism.

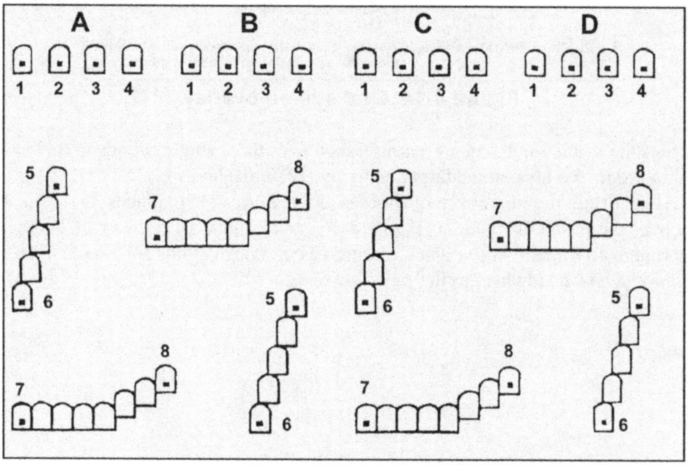

Figure 4-13. Basic machine gun target.

c. **Grid Square Overlay**. This device assists the gunner in zeroing his weapon at 10 meters, while using the basic machine gun target (Figure 4-14). The grid square overlay is used the same as an M16 25-meter zero target, except the material can be made of plastic or view graph transparency. Each square is equal to 1 cm.

1 CLICK = 1 CM. Turn the traversing handwheel to move the strike of the round left or right.
1 CLICK = 1 CM. Turn the elevation handwheel to move the strike of the round up or down.

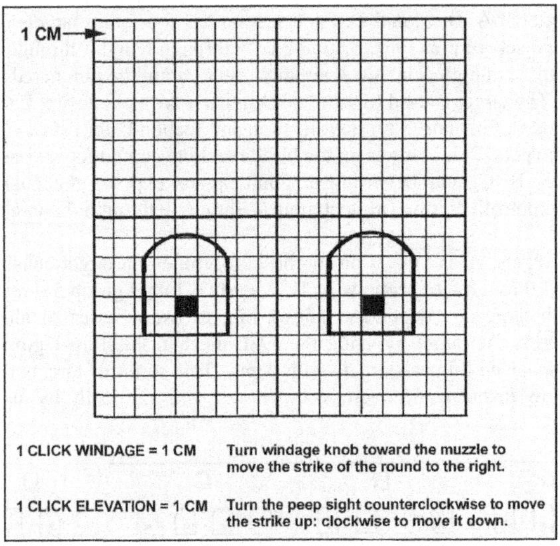

Figure 4-14. Grid square overlay.

(1) Sets the sights for 10-meter zeroing, then fires three single rounds to form a three-round shot group. Re-lays on the target using the T&E mechanism.

(2) After firing the three-round shot group (Figure 4-15), places the grid square overlay over the pasters (1 and 2) (Figure 4-16) and counts the number of clicks it will take for rounds to impact on the black aiming paster. (Corrections for Figure 4-16 would be turn the traverse handwheel to the right one click.)

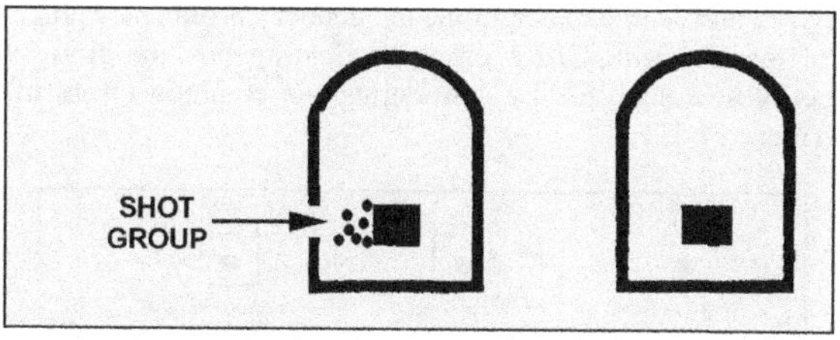

Figure 4-15. Shot group on basic machine gun target.

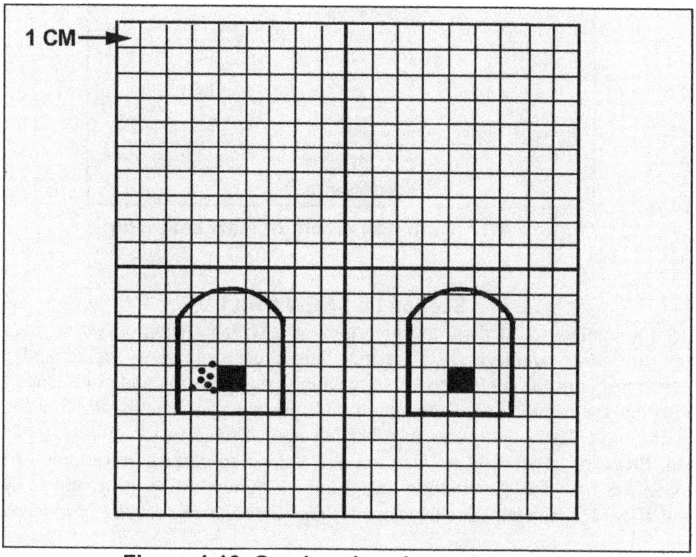

Figure 4-16. Overlay placed over pasters.

4-19. TARGET ANALYSIS

Targets are analyzed and scored to determine the gunner's proficiency and to reinforce the fundamentals of marksmanship. In a prone or fighting position firing with a zeroed weapon, a target is best analyzed by considering the common errors of machine gun marksmanship (Figure 4-17).

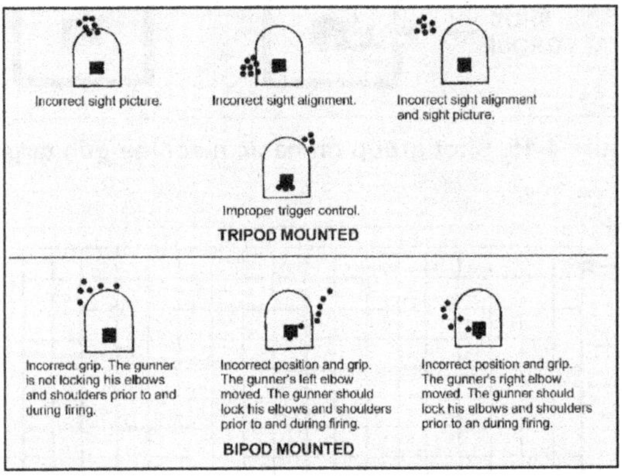

Figure 4-17. Common errors of marksmanship.

Section III. CREW DRILL

This section applies to all three machine guns and will be incorporated in *preliminary gunnery* and *basic gunnery*. The machine gun crew drill gives squad and platoon members training in the fundamentals of machine gun operation and confidence in their ability to put the machine gun into action with precision and speed. Rotation of duties during training ensures that every member becomes trained in the duties of each crew position. Precision is attained by learning and practicing correct procedures to include inspecting the machine gun before firing and observing safety procedures. Speed is acquired after precision has been developed. *Precision is never sacrificed for speed.*

4-20. PREPARATION

The crew drill will be conducted with preliminary gunnery and will be part of the 10-meter and transition firing practice and qualification, concurrently during other courses of fire, or anytime at the discretion of the unit commander. The organization for crew drill described in this section is for training crews in the fundamentals of machine gun operation; it is not the organization to be employed in every tactical situation.

a. To instill realism and relate the crew drill to actual situations, the unit leader should vary his method of instruction. Possible approaches to this method of instruction include the following:

- Conduct the crew drill from the prone position.
- Initiate the crew drill from all types of tactical formations.
- Perform the crew drill in simulated tactical situations.

b. The crew drill, as discussed here, involves the leader and one machine gun crew. The machine gun crew consists of three members (a gunner, assistant gunner, and an ammunition bearer). There are two complete machine gun crews in the light headquarters section of infantry, air assault infantry platoons, and airborne infantry platoons.

c. All commands are given by a leader. This leader may be a team leader, squad leader, or someone placed in charge of the crew. The gunner and assistant gunner repeat all commands. After the machine gun is mounted, the assistant gunner transmits all signals from the leader to the gunner and from the gunner to the leader.

4-21. CREW EQUIPMENT

In addition to individual weapons and equipment, crew members carry equipment for both bipod and tripod training. The following is a suggested assignment of the equipment to the machine gun crew members:

a. **Day Time Equipment**.

(1) Leader (designated)—binoculars, compass.

(2) Gunner—machine gun, compass, MGO or AN/PAS-13, two bandoleers (with dummy ammunition).

(3) Assistant Gunner—binoculars, spare barrel case (spare barrel and accessories), traversing and elevating mechanism, pintle assembly, and three bandoleers (with dummy ammunition).

(4) Ammunition bearer—compass, tripod and four bandoleers (with dummy ammunition).

b. **Night Time Equipment**.

(1) Leader (designated)—AN/PVS-7B with 3XMAG, compass.

(2) Gunner—machine gun, compass, AN/PVS-4 or AN/PAS-13, two bandoleers (with dummy ammunition).

(3) Assistant Gunner—AN/PVS-14 with 3XMAG, spare barrel case (spare barrel and accessories), traversing and elevating mechanism, pintle assembly, and three bandoleers (with dummy ammunition).

(4) Ammunition bearer—AN/PVS-7B with 3XMAG, compass, tripod and four bandoleers (with dummy ammunition).

4-22. FORMATION (BIPOD OR TRIPOD)

The leader commands FORM FOR CREW DRILL. The crew forms in a file with five steps between each crew member in this order: gunner, assistant gunner, and ammunition bearer. The gunner is five steps from and facing the leader. When the crew members reach their positions, each assumes the prone position and is ready for the crew drill. (Figure 4-18, page 4-26.)

Figure 4-18. Crew in ready position.

4-23. CROSS-TRAINING PROCEDURES

Duties are rotated during the crew drill to train each soldier in the duties of all crew members. The command to rotate duties is FALL OUT, GUNNER. At this command, the gunner becomes the ammunition bearer, the assistant gunner becomes the gunner, and the ammunition bearer becomes the assistant gunner. When crew members have assumed their new positions, they call out their new duties in order: AMMUNITION BEARER, ASSISTANT GUNNER, GUNNER.

4-24. INSPECTION FOR BIPOD FIRE

An inspection of equipment is made at the beginning of each exercise.

a. **Command**. After the crew is formed for crew drill, the leader commands INSPECT EQUIPMENT BEFORE FIRING, BIPOD. At the command, each crew member inspects his equipment as explained below.

(1) *Inspection by Gunner*. The gunner inspects the ammunition first. He ensures that the ammunition is properly linked and free of dirt and corrosion, and that the double link is up (ready for loading). After he inspects the ammunition, he places the cloth slings over his shoulder (except for one bandoleer, which he prepares for loading). He then inspects the machine gun and takes his position parallel to the machine gun (his head on line with the feed tray). The night personnel also check the AN/PVS-4 or AN/PAS-13.

(a) Holding the machine gun with his left hand, using his right hand he lowers the bipod legs and then rest the machine gun on the bipod.

(b) Attaches the bandoleer to the machine gun.

(c) Places the safety on "F", pulls the cocking handle to the rear, places the safety on "S", returns the cocking handle to the forward position, raises the cover assembly.

(d) Calls for the cleaning rod and receives it from the assistant gunner.

(e) Crawls forward, then runs the cleaning rod through the barrel to ensure it is clear.

(f) Checks the flash suppressor for cracks.

(g) Checks the front sight for tightness and for damage to the blade.

(h) Checks the carrying handle to ensure that it can be positioned so it will not be in the way during aiming and firing.

(i) Ensures that the barrel is securely locked to the receiver.

(j) Returns the cleaning rod to the assistant gunner.

(k) Moves to the rear of the machine gun and checks the moving parts in the feed cover.

- Ensures that the feed cam is clean and properly lubricated.
- Pushes back and forth on the feed cam to check for freedom of movement.
- Pushes on the belt feed pawl to ensure that it has spring tension.
- Pushes on the cartridge guides to ensure that they a have spring tension.

(l) Pushes the belt holding pawl to ensure that it has spring tension.

(m)Lowers and latches the cover (without inserting the belt).

(n) Pulls the trigger to check the functioning of the safety.

(o) Places the safety on "F", pulls the cocking handle to the rear, pulls the trigger, eases the bolt forward manually with the cocking handle.

(p) Checks the rear sight.

(2) *Inspection by Assistant Gunner*. Remaining in a prone position, the assistant gunner begins by inspecting his ammunition. He takes the cleaning rod from the carrying case and assembles the cleaning rod. He then takes the traversing and elevating mechanism from the case and prepares it as follows. Night will also check AN/PVS-14 with 3X MAG.

(a) Rotates the elevating handwheel, exposing 1 1/2 inches or the width of two fingers) of threads above the elevating handwheel.

(b) Rotates the traversing slide sleeve, exposing 1 1/2 inches or the width of two fingers) of threads below the elevating handwheel.

(c) Centers the traversing mechanism.

(d) Checks the to ensure that the locking mechanism that attach to the machine gun are present and in working order.

(e) Replaces the traversing and elevating mechanism in on the case and removes the spare barrel from the spare barrel case.

(f) Checks the barrel.

(g) Checks the flash suppressor for cracks.

(h) Checks the front sight for tightness and for damage to the blade.

(i) Checks the pintle assembly for proper functioning.

(j) Places the spare barrel its case; disassembles the cleaning rod and returns it accessory pocket; and checks the ruptured cartridge extractor, bore brush, chamber brush, receiver brush, and heat protective mitten for serviceability.

(3) *Inspection by Ammunition Bearer*. Remaining in a prone position, the ammunition bearer inspects his ammunition as described above for gunner and assistant gunner. He then inspects the tripod, pintle assembly and T&E mechanism. Night personnel also check the AN/PVS-7 with the 3XMAG.

(a) Ensures that the front leg will unfold properly and the rear legs unfold and lock securely in place with the sleeve latch.

(b) Checks the sleeve latch to ensure that it has spring tension and will function.

(c) Checks the pintle assembly to ensure that it is locked into the pintle bushing and that the pintle rotates freely within the bushing.

(d) Checks to ensure that the T&E mechanism will lock on the traversing bar and move freely when unlocked for major changes in direction.

(e) Unlocks the pintle and T&E mechanism from the tripod and return to the assistant gunner.

(f) Folds the rear legs by unlocking the sleeve latch and folds the front leg so that the tripod is in the carrying position.

b. **Report**. When crew members have completed their inspection of the equipment, they call out their report, without command, starting from the rear.

(1) AMMUNITION BEARER CORRECT (or reports deficiencies).

(2) AMMUNITION BEARER AND ASSISTANT GUNNER CORRECT (or reports the ammunition deficiencies.

(3) GUNNER ALL CORRECT (or deficiencies found during the inspections).

4-25. PLACEMENT INTO ACTION (BIPOD)

To place the machine gun into action, the leader commands and signals MACHINE GUN TO BE MOUNTED HERE (pointing to the position where the machine gun is to be mounted), FRONT (pointing in the direction of fire), ACTION (raising fist to shoulder level and thrusting it several times in the direction of the selected position).

a. At the command ACTION, the gunner stands, grasps the carrying handle with his left hand, grasps the top of the stock with his right hand, raises the machine gun to a carrying position (muzzle to the front) and moves to the selected position.

b. Upon arrival at the position, the gunner places the machine gun on the ground. He then assumes a prone position to the rear of the machine gun, positions the carrying handle so that it will not interfere during aiming and firing, aligns the machine gun in the direction of fire, and set the rear sight. He places the safety on "F", pulls the bolt to the rear, places the safety on "S", and returns the cocking handle to the forward position. He then raises the feed cover, places the first round of ammunition in the cartridge feed tray groove, and closes the feed cover ensuring that the round does not slip out of the cartridge feed tray groove. He then places the machine gun to his shoulder and puts the safety on "F".

WARNING

The M240B is carried loaded with the bolt locked to the *rear* in *tactical situations* where noise discipline is critical to the success of the mission. Trained gun crews are the only personnel authorized to load the M240B and only when command directs the crew to do so. During *normal training exercises*, the M240B is loaded and carried with the bolt in the *forward position*.

c. The assistant gunner times his movements so that he arrives at the position as the gunner is assuming the prone position. He lies prone on his left hip, feet to the rear, and on the left side of the gunner. He places the spare barrel case parallel to the gun with the zippered side towards the machine gun. He opens the case and removes the spare barrel. He places the spare barrel on the case, muzzle to the front and even with the muzzle of the machine gun. (Figure 4-19.)

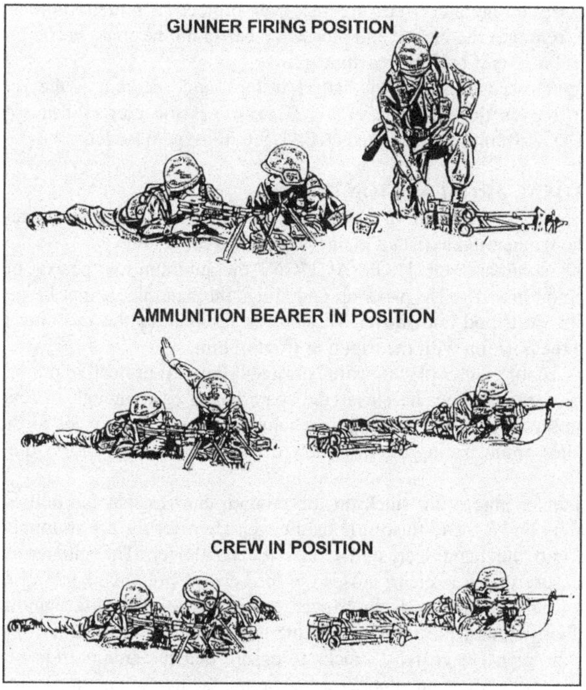

Figure 4-19. Crew members in firing position.

d. The ammunition bearer times his movements so that he arrives at the position as the assistant gunner is assuming the prone position. He places the folded tripod one step to the left of the muzzle of the machine gun and on line with the machine gun. He unslings his bandoleers and places them next to the folded tripod legs. He then lies prone 10 meters to the left and on line with the position, provides security, and prepares to fire into the target area with his rifle.

e. When ready to fire, the gunner puts the safety lever on "F" and reports UP. The assistant gunner signals READY to the leader.

4-26. PROCEDURES FOR CHANGING THE BARREL (BIPOD)

To ensure proficiency and speed in changing barrels, the barrel changing process is included in crew drill. When the gunner has reported UP and the assistant gunner has signaled READY, the leader commands CHANGE BARRELS.

a. The gunner ensures that the bolt is to the rear, puts the safety on "S", and puts the stock on the ground. Next, he moves his left hand to the top of the stock to ensure the weapon stays parallel to the ground. He puts his right hand under the handguard/forearm assembly help support the machine gun when the assistant gunner removes the barrel.

b. The assistant gunner (wearing the heat protective mitten) unlocks the barrel locking lever, removes the barrel, and places the barrel on the spare barrel case. He holds the spare barrel inserts it into the machine gun.

c. The gunner ensures that the barrel is lock and secured in the receive of the machine gun, moves the safety lever to "F", assumes the correct firing position, and reports UP. The assistant gunner signals READY to the squad leader.

4-27. REMOVAL FROM ACTION (BIPOD)

To take the machine gun out of action, the leader commands and signals OUT OF ACTION. The gunner and assistant gunner repeat the command.

a. At the command OUT OF ACTION, the ammunition bearer moves to the position, slinging his rifle. He picks up and slings the bandoleers that he previously left there. He gets the tripod and moves 15 steps to the rear of the machine gun. He lies prone, facing the position with the tripod in front of him.

b. The assistant gunner places spare barrel and the heat protective mitten in the spare barrel case. Before standing, he closes the spare barrel case enough to retain the spare barrel and the traversing and elevating mechanism. He moves 10 steps to the rear of the position and lies prone, facing the position. At this time, he fully closes the spare barrel case.

c. The gunner places the stock on the ground, ensures that the bolt is to the rear, places the safety on "S", and raises the feed cover. He removes the ammunition from the tray, puts it into the bandoleer, and closes the bandoleer. The gunner examines the chamber to ensure that it is clear; closes the feed cover; pulls the cocking handle to the rear; puts the safety on "F"; pulls the trigger, easing the bolt forward. Standing, he pivots on his right foot; without turning the machine gun, he raises it to his left hip and moves five steps to the rear. He visually checks to ensure that the ammunition bearer and the assistant gunner are in their positions. He lies prone, facing the position with the machine gun on his right. He folds the bipod legs alongside the barrel and reports UP to the squad leader.

4-28. INSPECTION FOR TRIPOD FIRE

The inspection of equipment for tripod training is the same as for bipod training except that the leader's command to start the inspection of equipment is INSPECT EQUIPMENT BEFORE FIRING TRIPOD. Also, the gunner inspects the bipod legs and folds them to their position alongside the barrel.

4-29. PLACEMENT INTO ACTION (TRIPOD)

The leader commands and signals MACHINE GUN TO BE MOUNTED HERE, FRONT, ACTION. (Figure 4-20.)

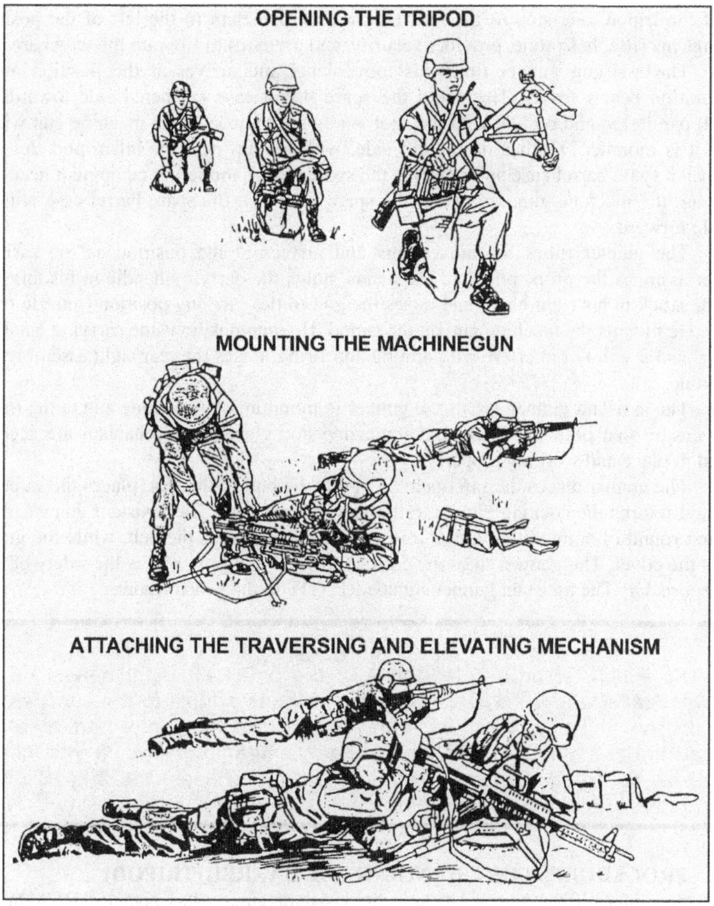

Figure 4-20. Placing the machine gun into action.

a. Upon the command ACTION, the ammunition bearer stands, holds the tripod with his right hand, and moves forward to the position. He kneels on his right knee and rests the shoes of the rear tripod legs on the ground, with the mount in a vertical position. Steadying the mount with his right hand near the tripod head, he raises the front leg with his left hand. He grasps the right shoe with his right hand and the left shoe with is left hand, and raises the tripod chest high. He separates the tripod legs with a quick jerk. Ensuring that the sleeve latch engages the sleeve, he places the tripod on the ground with the front leg pointing in the direction of fire. He rises to his feet and stamps the rear shoes into the ground. He then unslings his bandoleers and places them on line with the front leg of the tripod, one step to the left. He moves 10 meters to the left of the position, unslings his rifle, lies prone, provides security, and prepares to fire into the target area.

b. The assistant gunner times his movements and arrives at the position as the ammunition bearer leaves. He places the spare barrel case (zippered side towards the tripod) parallel to and on line with the spot where the muzzle of the machine gun will be when it is mounted. He lies on his left side, with his hip near the left tripod shoe. He unzips the spare barrel case and removes the spare barrel, mounting equipment necessary to mount the machine gun. He places the spare barrel on the spare barrel case with the muzzle forward.

c. The gunner times his movements and arrives at the position as the assistant gunner assumes the prone position. He stands, holds the carrying handle in his left hand and the stock in his right hand, and raises the gun to the carrying position (muzzle to the front). He mounts the machine gun on the tripod. He then positions the carrying handle to the right so it will not interfere with aiming and firing, raises the rear sight assembly, and lies prone.

d. The assistant gunner assists the gunner in mounting the machine gun to the tripod. They ensure that both the pintle and traversing and elevating mechanism are securely locked in place and working properly.

e. The gunner places the safety on "F", pulls the bolt to the rear, places the safety on "S", and returns the cocking handle to the forward position. The assistant gunner places the first round of ammunition in the tray groove and supports the belt, while the gunner closes the cover. The gunner takes the correct position and grip, places the safety on "F", and reports UP. The assistant gunner signals READY to the squad leader.

WARNING

The M240B is carried loaded with the bolt locked to the *rear* in *tactical situations* where noise discipline is critical to the success of the mission. Trained gun crews are the only personnel authorized to load the M240B and only when command directs the crew to do so. During *normal training exercises*, the M240B is loaded and carried with the bolt in the *forward position*.

4-30. PROCEDURES FOR CHANGING THE BARREL (TRIPOD)

When the gunner has reported UP and the assistant gunner has signaled READY, the leader commands CHANGE BARRELS.

a. The gunner ensures that the bolt is to the rear, puts the safety on "S". He also assists the assistant gunner in changing the barrel, if needed.

b. The assistant gunner (wearing the heat protective mitten) unlocks the barrel locking lever, removes the barrel, and places the barrel on the spare barrel case. He holds the spare barrel inserts it into the machine gun.

c. The gunner ensures that the barrel is lock and secured in the receiver of the machine gun, moves the safety lever to "F", assumes the correct firing position, and reports UP. The assistant gunner signals READY to the squad leader.

4-31. REMOVAL FROM ACTION (TRIPOD)

At the command OUT OF ACTION, the gunner ensures that the bolt is to the rear, places the safety on "S", and raises the cover. The assistant gunner removes the ammunition from the tray, returns it to the bandoleer, and closes the bandoleer. The gunner inspects the chamber to ensure that it is clear; closes the cover; pulls the cocking handle to the rear; puts the safety on "F"; pulls the trigger, easing the bolt forward. The gunner unlocks the rear of the machine gun from the tripod.

a. The assistant gunner will assist the gunner in dismounting the rear of the machine gun. He puts the spare barrel and heat protective mitten into the case and closes it enough to hold the contents. He stands, moves 10 steps to the rear of the position, and lies prone, facing to the front. After receiving all mounting equipment from the ammunition bearer, he puts it in the spare barrel case and fully closes the spare barrel case.

b. After the assistant gunner leaves, the gunner stands, lowers the rear sight, and holds the carrying handle with his left hand. With his right hand, he dismounts the front of the machine gun from the tripod. Holding the stock with his right hand, he pivots to his right as he raises the machine gun to the carrying position. He then moves five steps to the rear of the position and lies prone, facing to the front.

c. The ammunition bearer rises, slings his rifle, moves to the machine gun, and secures his bandoleers, timing his arrival so that the gunner and assistant gunner will be clear of the tripod. He grasps the tripod with his left hand and moves five steps to the rear of the position. He turns, facing the front, and kneels on his right knee. He places the tripod in a vertical position with the rear shoes on the ground and supports it with his right hand near the head of the tripod. At this time, he hands the assistant gunner all mounting equipment. He reaches up with his right hand down the right leg, and releases the sleeve latch. He then grasps the shoes and closes the tripod legs. He lowers the tripod to the ground, head to the left, lies prone behind it, and reports UP.

4-32. PRONE POSITION

Machine gun crew drill, as it is described in the preceding paragraphs, is an excellent training vehicle for the machine gun crew. A continuation or second phase of the crew drill is outlined in this paragraph. It should be used only as a technique for adding realism to training.

a. **Inspecting Equipment Before Firing**. The inspection of equipment for crew drill from the prone position is the same as that for bipod training and tripod training.

b. **Placing the Machine Gun Into Action**. The leader commands and signals MACHINE GUN TO BE MOUNTED HERE, FRONT, ACTION in the same manner as for bipod training. The procedures for bipod training are the same with one exception—crew members do not get to their feet and movements are executed in the low crawl. Once in position, all actions are performed from the prone position.

c. **Training With the Tripod**. Upon the command ACTION, the ammunition bearer crawls forward to the designated position and extends the front leg of the tripod. Grasping the rear legs firmly, he emplaces the front leg. Applying downward pressure, he emplaces the rear legs. He then crawls to a position about 10-meters to the left of the machine gun and gets into a good firing position with his rifle.

(1) The assistant gunner crawls forward, timing his movement to arrive as the ammunition bearer leaves. Positioning himself on the left side and facing the tripod, he

places the spare barrel case alongside the tripod, unzips the case, and removes the spare barrel and mounting equipment.

(2) The procedures for mounting the machine gun on the tripod remain the same except all are performed in a prone position and all movements are in the low crawl.

d. **Taking the Machine Gun Out of Action**. The procedures for taking the machine gun out of action remain the same except all are performed in a prone position and all movements are in the low crawl.

Section IV. BASIC GUNNERY

In basic marksmanship, the gunner applies the fundamentals in live-fire exercises during day and night. This includes 10-meter zeroing, 10-meter firing, field zeroing, transition firing, and record firing.

4-33. ZERO

Zeroing aligns the sights with the barrel so that the point of aim equals the point of impact. Ten-meter zeroing is for conducting 10-meter fire only and has no further application. (Zeroing at range or field zeroing is the gunner's battlesight zero and must be recorded.) Remember to zero both barrels of the machine gun.

a. **10-Meter Zero, Set the Sights (Mechanical Zero).** The gunner indexes or places the range scale on a range of 500 meters. He assumes a prone position and sights on the target.

b. **Three-Round Group**. The gunner fires three single rounds loaded individually at the center base of the aiming points on the basic machine gun marksmanship target. He fires the three rounds without making any adjustments to the sights. The shot group must be about a 4-cm circle or smaller to establish the center of the group in relation to the center base of the aiming paster.

c. **Grid Square Overlay**. For a more accurate adjustment, the gunner moves downrange and places the grid square overlay over pasters 1 and 2. He ensures that he aligns the overlay with the pasters and squares.

(1) Counts the number of squares it will take to move the shot group to the aiming paster.

(2) Upon completion, returns to the firing line to make corrections to the weapon. (Figure 4-21 illustrates a zero group size on which adjustments can be made and a group that is too loose for adjustments [bipod mode].) If a group is too loose, the gunner checks his position and group.

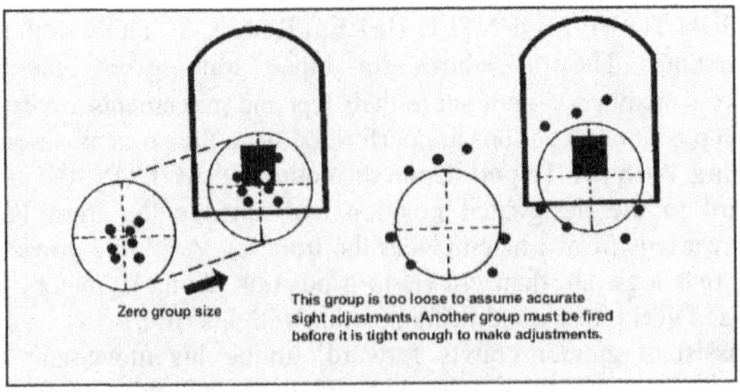

Figure 4-21. Zero group size.

d. **Windage Correction**. If the center of the group is to the left or right of the black aiming paster, the gunner must correct for windage.

e. **Elevation Correction**. If the center of the shot group is above or below of the black aiming paster, the gunner must correct for elevation.

f. **Confirmation**. The gunner fires another three-shot group (loaded singly) after making his corrections for windage and elevation. If the center of the group is still off the aiming point, he adjusts further until the group is centered on the point of aim.

g. **Recording of Zero**. There is no reason to record the 10-meter zero, because it applies only to firing at the 10-meter basic machine gun target.

NOTE: Remember to zero both barrels.

WARNING

The M240B is carried loaded with the bolt locked to the *rear* in *tactical situations* where noise discipline is critical to the success of the mission. Trained gun crews are the only personnel authorized to load the M240B and only when command directs the crew to do so. During *normal training exercises*, the M240B is loaded and carried with the bolt in the *forward position*.

4-34. FIELD ZERO

A gunner must know how to zero the machine gun at distance. He should select a known distance target between 300 and 700 meters. As the range increases, it becomes more difficult to determine where the center of the beaten zone is in relation to the target. Therefore, the 500-meter target on the transition range is recommended because of the ease of determining adjustments.

a. **Setting of the Sights**. The gunner uses the same procedures as for 10-meter zeroing except that he places the rear sight on the range to the target. The recommended range is 500 meters.

b. **Burst**. Fire a burst of 5 to 7 rounds for the M249 or 7 to 9 for the M60/M240B. The gunner assumes a good stable position and fires bursts of 5 to 7 rounds for the M249 or 7 to 9 for the M60/M240B at the center base of the target and notes where the burst strikes.

c. **Correction for Windage**. If the center of the beaten zone is to the left or right of the target, he corrects for windage. He adjusts the windage accordingly.

d. **Correction for Elevation**. If the center of the beaten zone is high or low in relation to the target, he corrects for elevation. Because determining that relationship is difficult, the gunner relies on trial and error to gain sufficient experience in making reliable estimates. He makes corrections in the same manner as 10-meter zeroing.

e. **Confirmation**. After making corrections for windage and elevation, the gunner fires confirming bursts of 5 to 7 rounds for the M249 or 7 to 9 rounds for the M60/M240B. If the target is not hit, he repeats the procedures.

f. **Recording of Zero**. Upon confirming the zero, the gunner records it by counting the number of clicks he moved the sight for windage and elevation from the initial setting.

4-35. 10-METER FIRING

The 10-meter firing trains the gunner to apply the fundamentals of machine gun marksmanship in live-fire exercises. It familiarizes the soldier with the weapon's characteristics, noise, and recoil. It instills in the soldier confidence in his weapon. Each gunner learns to zero his machine gun, conducts crew drill, controlled-burst fire at point targets, and uses traverse and search techniques of fire at area targets. The 10-meter firing is conducted on a 10-meter range or a multipurpose range using the basic machine gun target. These exercises are fired with the machine gun on the bipod from both the prone position and the fighting position and with the tripod from prone and fighting positions. The 10-meter firing exercises are for practice as well as part of record qualification. All 10-meter firing exercises are recorded and scored to provide the gunner an assessment of his performance. The 10-meter firing is conducted IAW Firing Table I (Table 4-2, page 4-42). There are ten tasks.

a. **Task 1—Zero Bipod.** The gunner fires single shots to determine his weapon's zero for 10 meters. This task reinforces the dry-fire experience and allows the gunner to practice loading, while providing the most accurate and tight shot group obtainable. (A1 and A2)

b. **Task 2—Controlled-Burst Fire Bipod Fixed**. Using point targets, the gunner fires bursts of 5 to 7 rounds for the M249 or 7 to 9 rounds for the M60/M240B. This task exposes the gunner to automatic fire and the action of the weapon and at the same time introduces trigger control. (A3 and A4)

c. **Task 3—Controlled-Burst Fire Bipod Fixed**. This task requires the gunner to make body position to engage area targets in depth, to use controlled-burst firing, and to use a series of aiming points to disburse fire across the target. (A5 and A6)

d. **Task 4—Controlled-Burst Fire Bipod Fixed**. This task requires the gunner to make position changes to engage linear targets with depth, to use controlled-burst firing, and to use a series of aiming points to disburse fire across the target. (A7 and A8)

e. **Task 5—Zero Tripod**. The gunner fires single shots to determine his weapon's zero for 10 meters. This task reinforces the dry-fire experience and allows the gunner to practice loading, while providing the most accurate and tight shot group obtainable. (B1 and B2)

f. **Task 6—Controlled-Burst Fire Tripod**. Using point targets, the gunner fires bursts of 5 to 7 rounds for the M249 or 7 to 9 rounds for the M60/M240B. This task exposes the gunner to automatic fire and the action of the weapon and at the same time introduces trigger control. (B1 through B4)

g. **Task 7—Traverse and Search Fire.** This task requires the gunner to make position changes or manipulate the T&E mechanism to engage linear targets with depth, to use controlled-burst firing, and to use a series of aiming points to disburse fire across the target. (B7 through B8)

h. **Task 8—Traverse and Search Fire**. This task requires the gunner to make body position changes or manipulate the T&E mechanism to engage area targets in depth, to use controlled-burst firing, and to use a series of aiming points to disburse fire across the target, while wearing a protective mask and gloves. (B5 through B6)

i. **Task 9—Search and Traverse Fire Qualification**. This task requires the gunner to make position changes or manipulate the T&E mechanism to engage area targets in depth during timed conditions. (C5 through C6)

j. **Task 10—Traverse and Search Fire Qualification**. This task requires the gunner to engage area targets with width and depth, while making position changes or manipulating the T&E mechanism during timed conditions. (C7 through C8)

4-36. 10-METER CONDUCT OF FIRE

The gunners are instructed on the objectives and fundamentals of firing from the bipod and tripod-supported prone or fighting positions, on fire commands used on the basic range, on the basic machine gun marksmanship target, and on analyzing and scoring the target. The unit is organized in firing orders based on range constraints. Each firing order should consist of a gunner and an assistant gunner. The assistant gunner assists the gunner during prefire checks and zeroing. The assistant gunner also relays signals to the tower operator, checks the gunner's position, and assists him. During qualification, an assistant gunner is not used. The ten tasks are fired in the following manner:

a. **Task 1—Bipod, Zero.**

(1) The tower operator gives the command MACHINE GUN TO BE MOUNTED HERE (weapon squad leader's pointing to the firing points on the 10-meter line), FRONT (weapon squad leader's pointing to the 10-meter targets), ACTION.

(2) At the command ACTION the machine gun crew conducts, placing the machine gun into action (bipod mode)

(3) The gunner prepares the rear sight for zeroing and checks the front sight.

(4) The gunner assumes a good position.

(5) The tower operator instructs the gunner to prepare a single round.

(6) The following fire command is given. The gunner and assistant gunner repeat each element of the fire command as it is given.

FIRE MISSION (The gunner loads and moves the safety to "F".)
FRONT (The gunner focuses on the target or target area.)
PASTER A ONE (The gunner locates target.)
FIVE HUNDRED (The gunner adjusts sights and acquires the sight picture.)
FIXED, ONE ROUND (The gunner is given the method of fire.)
COMMENCE FIRING (The gunner fires on command from tower operator, but when ready.)

NOTE: Throughout all firing exercises, the gunner performs the appropriate tasks during each element of the fire command. The number of rounds fired is used instead of the rate for METHOD OF FIRE. This is for control. (Omitting the rate specifies RAPID fire, which is not desirable for the tasks.)

WARNING

The M240B is carried loaded with the bolt locked to the *rear* in *tactical situations* where noise discipline is critical to the success of the mission. Trained gun crews are the only personnel authorized to load the M240B and only when command directs the crew to do so. During *normal training exercises*, the M240B is loaded and carried with the bolt in the *forward position*.

(7) The gunner loads one round, obtains the proper sight picture, and gives an UP to the assistant gunner.

(8) The assistant gunner relays the READY signal to the tower operator.

(9) The tower operator gives the command COMMENCE FIRING.

(10) The gunner engages paster A1 with three-single shots when he is ready.

(11) The gunner moves downrange to observe, mark, and triangulate the shot group. He makes adjustments as needed.

(12) Steps 3 through 10 are repeated, but the gunner fires at paster A2 firing a single round, then he adjusts.

NOTE: If the gunner should zero his weapon using 9 rounds, he uses the remaining 3 rounds to confirm his zero. If he is unable to zero with 12 rounds, he is removed from the firing line for remedial training.

WARNING

The M240B is carried loaded with the bolt locked to the *rear* in *tactical situations* where noise discipline is critical to the success of the mission. Trained gun crews are the only personnel authorized to load the M240B and only when command directs the crew to do so. During *normal training exercises*, the M240B is loaded and carried with the bolt in the *forward position*.

b. **Task 2—Bipod, Controlled-Burst Fire, Fixed.**

(1) The tower operator instructs the gunner to prepare two 7-round belts (M249) or two 9-round belts (M60/M240B).

(2) When the fire command is given, the gunner and assistant gunner repeat each element as it is given.

FIRE MISSION
FRONT
PASTER A THREE
FIVE HUNDRED
FIXED, FIVE- TO SEVEN-ROUND BURSTS (M249) or SEVEN- TO NINE-ROUND BURSTS (M60/M240B).
AT MY COMMAND

(3) The gunner acquires the proper sight picture and gives an UP to the assistant gunner.

(4) The assistant gunner relays the READY signal to the tower operator.

(5) The tower operator gives the command to FIRE.

(6) The gunner fires the first burst of 5 to 7 rounds (M249) or 7 to 9 rounds (M60/M240B) at paster A3.

(7) Steps 2 through 6 are repeated, but the gunner fires at paster A4.

c. **Task 3—Bipod, Controlled-Burst Fire, Fixed.**

(1) The tower operator instructs the gunner to prepare a 14-round belt (M249) or 18-round belt (M60/M240B).

(2) When the fire command is given, the gunner and assistant gunner repeat each element as it is given.

FIRE MISSION
FRONT
PASTER A FIVE
FIVE HUNDRED
TRAVERSE AND SEARCH, FIVE- TO SEVEN-ROUND BURSTS (M249) OR SEVEN- TO NINE-ROUND BURSTS (M60/M240B)
AT MY COMMAND

(3) The gunner acquires the proper sight picture and gives an UP to the assistant gunner.

(4) The assistant gunner relays the READY signal to the tower operator.

(5) The tower operator gives the command to FIRE.

(6) The gunner fires the first burst of 5 to 7 (M249) OR 7 TO 9 (M60/M240B) rounds at paster A5.

(7) Steps 2 through 6 are repeated, but the gunner fires at paster A6.

d. **Task 4—Bipod, Controlled-Burst Fire, Fixed.**

(1) The tower operator instructs the gunner to prepare 14-round belt (M249) or 18-round belt (M60/M240B).

(2) When the fire command is given, the gunner and assistant gunner repeat each element as it is given.

FIRE MISSION
FRONT
PASTER A SEVEN
FIVE HUNDRED
TRAVERSE AND SEARCH, FIVE- TO SEVEN-ROUND BURSTS (M249) or SEVEN- TO NINE-ROUND BURSTS (M60/M240B)
AT MY COMMAND

(3) The gunner acquires the proper sight picture and gives an UP to the assistant gunner.

(4) The assistant gunner relays the READY signal to the tower operator.

(5) The tower operator gives the command to FIRE.

(6) The gunner fires the first burst of 5 to 7 rounds (M249) or 7 to 9 rounds (M60/M240B) at paster A7.

(7) Steps 2 through 6 are repeated, but the gunner fires at paster A8.

(8) The gunner and assistant gunner moves downrange to observe and analyze his target and shot groups.

(9) The Tower operator gives the following command when the gunner and assistant gunner return from downrange. OUT OF ACTION.

(10) At the command OUT OF ACTION the machine gun crew conducts, taking the machine gun out of action (bipod mode)

e. **Task 5—Tripod, Zero.** If the gunner should zero his weapon in 9 rounds, he uses the remaining 3 rounds to confirm his zero. If he is unable to zero in 12 rounds, he is removed from the firing line for remedial training.

(1) The tower operator gives the command MACHINE GUN TO BE MOUNTED HERE (weapon squad leaders point to the firing points on the 10-meter line), FRONT (weapon squad leader points to the 10-meter targets), ACTION.

(2) At the command ACTION the machine gun crew conducts, placing the machine gun into action (tripod mode)

(3) The gunner prepares the rear sight for zeroing and checks the front sight.

(4) The gunner assumes a good tripod position.

(5) The tower operator instructs the gunner to prepare a single round.

(6) The following fire command is given. The gunner and assistant gunner repeat each element of the fire command as it is given.

FIRE MISSION (The gunner loads and moves the safety to "F".)
FRONT (The gunner focuses on the target or target area.)
PASTER B ONE (The gunner locates target.)
FIVE HUNDRED (The gunner adjusts sights and acquires the sight picture.)
FIXED, ONE ROUND (The gunner is given the method of fire.)
COMMENCE FIRING (The gunner fires on command from tower operator, but when ready.)

(7) The gunner loads one round, obtains the proper sight picture, and gives an UP to the assistant gunner.

(8) The assistant gunner relays the READY signal to the tower operator.

(9) The tower operator gives the command COMMENCE FIRING.

(10) The gunner engages paster B1 with three single shots when he is ready.

(11) The gunner moves downrange to observe, mark, and triangulate the shot group. He makes adjustments as needed.

(12) Steps 3 through 10 are repeated, but the gunner fires at paster B2 firing a single round, then he adjusts.

f. **Task 6—Tripod, Controlled-Burst Fire, Traverse.**

(1) The tower operator instructs the gunner to prepare a 28-round belt (M249) or 36-round belt (M60/M240B).

(2) When the fire command is given, the gunner and assistant gunner repeat each element as it is given.

FIRE MISSION
FRONT
PASTERS B ONE THROUGH B FOUR
FIVE HUNDRED
FIXED, FIVE TO SEVEN-ROUND BURSTS (M249) or SEVEN- TO NINE-ROUND BURSTS (M60/M240B)
AT MY COMMAND

(3) The gunner acquires the proper sight picture and gives an UP to the assistant gunner.

(4) The assistant gunner relays the READY signal to the tower operator.

(5) The tower operator gives the command to FIRE.

(6) The gunner engages pasters B1 through B4, firing 5- to 7-round bursts (M249) or 7 to 9 round bursts (M60/M240B) at each paster, using traverse technique.

g. **Task 7—Tripod, Controlled-Burst Fire, Traverse and Search**.

(1) The tower operator instructs the gunner to prepare a 56-round belt (M249) or 63-round belt (M60/M240B).

(2) When the fire command is given, the gunner and assistant gunner repeat each element as it is given.

FIRE MISSION
FRONT
PASTERS B SEVEN THROUGH B EIGHT
FIVE HUNDRED
TRAVERSE AND SEARCH, FIVE- TO SEVEN-ROUND BURSTS (M249) or SEVEN- TO NINE-ROUND BURSTS (M60/M240B)
AT MY COMMAND

(3) The gunner acquires the proper sight picture and gives an UP to the assistant gunner.

(4) The assistant gunner relays the READY signal to the tower operator.

(5) The tower operator gives the command to FIRE.

(6) The gunner engages pasters B7 through B8, firing a 5- to 7-round bursts or 7- to 9-round bursts at each paster, using traverse and search technique.

h. **Task 8—Tripod, Controlled-Burst Fire, Search and Traverse.**

(1) The tower operator instructs the assistant gunner to prepare a 35-round belt (M249) or 45-round belt (M60/M240B).

(2) When the fire command is given, the gunner and assistant gunner repeat each element as it is given

FIRE MISSION
FRONT
PASTERS B FIVE THROUGH B SIX
FIVE HUNDRED
TRAVERSE AND SEARCH, FIVE- TO SEVEN-ROUND BURSTS (M249) or SEVEN- TO NINE-ROUND BURSTS (M60/M240B)
AT MY COMMAND

(3) The gunner acquires the proper sight picture and gives an UP to the assistant gunner.

(4) The assistant gunner relays the READY signal to the tower operator.

(5) The tower operator gives the command to FIRE.

(6) The gunner engages pasters B5 through B6, firing a three round burst at each paster, using search and traverse technique.

(7) The gunner and assistant gunner moves downrange to observe and analyze his targets.

i. **Task 9—Tripod, Qualification, Search and Traverse Fire.** On completion of all firing, the firing line is cleared and the instructors or safeties move downrange and score the targets. The firer will not score his own target.

(1) The tower operator instructs the assistant gunner to prepare a 35-round belt (M249) or 45-round belt (M60/M240B).

(2) When the fire command is given, the gunner and assistant gunner repeat each element as it is given

FIRE MISSION
FRONT
PASTERS C FIVE THROUGH C SIX
FIVE HUNDRED
TRAVERSE AND SEARCH, FIVE- TO SEVEN-ROUND BURSTS (M249) or SEVEN- TO NINE-ROUND BURSTS (M60/M240B)
AT MY COMMAND

(3) The gunner acquires the proper sight picture and gives an UP to the assistant gunner.

(4) The assistant gunner relays the READY signal to the tower operator.

(5) The tower operator gives the command to FIRE.

(6) The gunner engages pasters B5 through B6, firing a 5- to 7- round bursts (M249) or 7- to 9-round bursts (M60/M240B) at each paster, using search and traverse technique. The gunner has 30 seconds to engage as many pasters as he can during the time allowed.

j. **Task 10—Tripod, Qualification, Traverse and Search.** On completion of all firing, the firing line is cleared and the instructors or safeties move downrange and score the targets. The firer will not score his own target.

(1) The tower operator instructs the gunner to prepare a 56-round belt (M249) or 72-round belt (M60/M240B).

(2) When the fire command is given, the gunner and assistant gunner repeat each element as it is given.

FIRE MISSION
FRONT
PASTERS C SEVEN THROUGH C EIGHT
HUNDRED
TRAVERSE AND SEARCH, FIVE- TO SEVEN-ROUND BURSTS (M249) or SEVEN- TO NINE-ROUND BURSTS (M60/M240B)
AT MY COMMAND

(3) The gunner acquires the proper sight picture and gives an UP to the assistant gunner.

(4) The assistant gunner relays the READY signal to the tower operator.

(5) The tower operator gives the command to FIRE.

(6) The gunner engages pasters C7 through C8, firing 5- to 7-round bursts (M249) or 7- to 9-round bursts (M60/M240B) at each paster, using traverse and search technique. The gunner has 45 seconds to engage as many pasters as he can during the time allowed.

(7) The tower operator gives the following command when the gunner and assistant gunner return from downrange. OUT OF ACTION.

(8) At the command OUT OF ACTION the machine gun crew conducts, taking the machine gun out of action (tripod mode)

BASIC (10-METER) FIRE PRONE OR FIGHTING POSITION, BIPOD OR TRIPOD, PRACTICE AND QUALIFICATION						
TASK	**TIME**	**RDS M249**	**RDS M60/ M240B**	**TYPE**	**TARGET**	**TYPE FIRE**
1	No limit	12	12	Ball	Pasters A1 and A2	12 single rd (zero).
2	No limit	14	18	Ball	Pasters A3 and A4	5- to 7-rd bursts (M249) or 7- to 9-rd bursts (M60/ M240B) for each paster.
3	No limit	14	18	Ball	Pasters A5 and A6	5- to 7-rd bursts (M249) or 7- to 9-rd bursts (M60/ M240B) for each paster.
4	No limit	14	18	Ball	Pasters A7 and A8	5- to 7-rd bursts (M249) or 7- to 9-rd bursts (M60/ M240B) for each paster.
5	No limit	12	12	Ball	Pasters B1 and B2	12 single rd (zero).
6	No limit	28	36	Ball	Pasters B1 thru B4	5- to 7-rd bursts (M249) or 7- to 9-rd bursts (M60/ M240B) for each paster.
7	No limit	56	72	Ball	Pasters B7 thru B8	5- to 7-rd bursts (M249) or 7- to 9-rd bursts (M60/ M240B) for each paster traverse and search.
8	No limit	35	45	Ball	Pasters B5 thru B6	5- to 7-rd bursts (M249) or 7- to 9-rd bursts (M60/ M240B) for each paster.
*9	30 seconds	35	45	Ball	Pasters C5 thru C6	5- to 7-rd bursts (M249) or 7- to 9-rd bursts (M60/ M240B) for each paster.
*10	45 seconds	56	72	Ball	Pasters C7 thru C8	5- to 7-rd bursts (M249) or 7- to 9-rd bursts (M60/ M240B) for each paster.
NOTE: The unit commander determines the position to be used. A summary of the ammunition requirements is on page 4-54. *Indicates qualification tasks.						

Table 4-2. Firing Table I—Basic (10-meter) fire.

WARNING

The M240B is carried loaded with the bolt locked to the *rear* in *tactical situations* where noise discipline is critical to the success of the mission. Trained gun crews are the only personnel authorized to load the M240B and only when command directs the crew to do so. During *normal training exercises*, the M240B is loaded and carried with the bolt in the *forward position.*

4-37. 10-METER FIRING, QUALIFICATION

The first phase of qualification consists of firing tasks 2 through 8 of Firing Table I for practice, and tasks 9 and 10 of Firing Table I for record. Before firing, all soldiers must be familiar with the tasks, the time allowed, the ammunition allowances, the procedures to follow in the event of a stoppage, and the penalties imposed.

a. **Time and Ammunition**. Each gunner completes zeroing before record firing. Individual fire commands are given for each task. Task 9 is fired in 30 seconds, and task 10 is fired in 45 seconds.

b. **Stoppages**. If a stoppage occurs, the gunner must apply immediate action. If the stoppage is reduced, he continues to fire the course.

(1) If a stoppage occurs that cannot be reduced by immediate action, the gunner raises his hand and awaits assistance.

(2) Once the stoppage is reduced, the gunner completes firing beginning with the next task.

(3) If a stoppage is caused by an error on the part of the gunner, additional time is not permitted. The gunner receives the score he earned before the stoppage occurred.

(4) If it is necessary to replace the machine gun, the gunner must zero the new weapon. The gunner can fire the exercise again.

(5) Gunners who cannot fire a task or cannot complete firing in the time allowed (because of malfunctions) can finish the exercise in an *alibi run* after all other gunners complete firing. They fire only those tasks they failed to engage because of the malfunction.

c. **Penalties**. Five points are deducted from the score of any gunner who fails to stop firing at the command or signal to cease fire. If a gunner fires at the wrong target or exercise, he loses the points for those rounds. A gunner whose target was fired upon by another gunner is permitted to refire the exercise.

d. **Scoring**. When scoring the 10-meter target, the trainer scores all scoring pasters (C5 through C6 and C7 through C8). One point is given for each round impacting within the scoring space. The maximum point value is 7 points (M249) or 9 points (M60/M240B) for each paster. Rounds touching the line on the paster are considered a HIT. When firing C5 though C6, the gunner engages 5 scoring pasters with 35 rounds. (M249) or 45 rounds (M60/M240B) The maximum possible is 35 points (M249) or 45 points (M60/M240B). When firing pasters C7 through C8, the gunner engages 8 scoring pasters with 56 rounds (M249) or 72 rounds (M60/M240B). The maximum possible is 56 points (M249) or 72 points (M60/M240B). Gunners do not score their own targets when firing for qualification. During qualification firing, at least 63 points (M249) or 81 points

(M60/M240B) must be achieved on Firing Table I. DA Form 85-R is used to record scores (Figure 4-23). (See paragraph 4-42).

e. **Position**. Based on his METL, the commander selects either the bipod-supported prone or fighting position for table A only. For qualification the position will be either tripod-supported prone or tripod-supported fighting position table B for practice and table C for qualification.

4-38. TRANSITION FIRE

Transition firing provides the gunner the experience necessary to progress from 10-meter firing to field firing at various types of targets at longer ranges. The gunner experiences and learns the characteristics of fire, field zeroing, and range determination, and engaging targets in a timed scenario. He uses the adjusted aiming-point method of fire adjustment. Transition firing is conducted on a machine gun transition range or the MPRC. These exercises are fired with the bipod prone or fighting position. Transition firing is fired and scored for practice and qualification to provide feedback to the gunner. Firing Table II consists of eight tasks (Table 4-3, page 4-47).

a. **Range Facilities**. The transition range should consist of several firing lanes. Each lane should be 10 meters wide at the firing line and 100 meters wide at a range of 800 meters. Ideally, each lane has a fighting position with an adjacent prone firing position.

b. **Targets** (Card board: NSN 6920-00-795-1806 and plastic: 6920-00-071-4780). The E-type silhouette targets are used—single and double are needed for qualification. The double represents an enemy automatic weapon, which for the gunner is a priority target (Figure 4-22). The targets are at various ranges that a gunner might engage. All targets should be plainly seen from the firing positions. Electrical targets are desirable.

c. **Stoppage.** The same procedures used in Firing Table I qualification firing are used (page 4-42).

d. **Penalties**. The same procedures used in Firing Table I qualification firing are used (page 4-42).

e. **Scoring**. Ten points are given for each target hit, whether hit on the first or second burst. The total possible points are 110. The gunner must hit at least 7 (70 points) targets out of 11 exposures to qualify. DA Form 85-R is used to record scores Figure 4-23 (see paragraph 4-42).

f. **Position.** Based on his METL, the commander selects either the bipod-supported prone or bipod-supported fighting position for qualification.

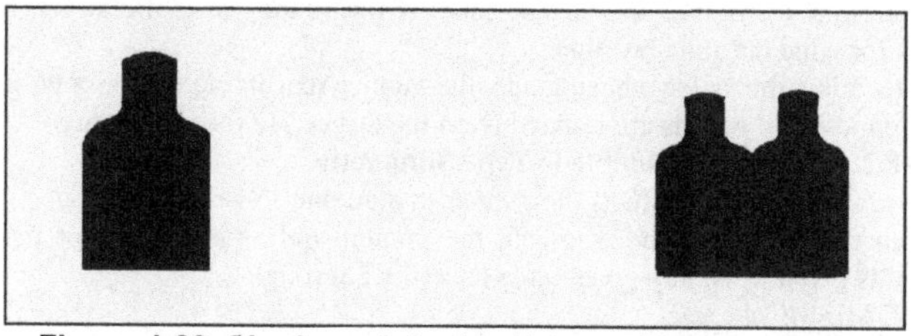

Figure 4-22. Single and double E-type silhouette targets.

4-39. TRANSITION CONDUCT OF FIRE, BIPOD

The unit is organized in firing orders based on range constraints. Each firing order should consist of a gunner and an assistant gunner. The assistant gunner assists the gunner during prefire checks and zeroing. He also relays signals to the tower operator, checks the gunner's position, and assists him during qualification in target detection and adjustments. The bipod-supported prone or fighting positions are used. The eight tasks are fired in the following manner.

a. **Task 1—Field Zero, 500-Meter, Double E-Type Silhouette.**

(1) The tower operator gives the command MACHINE GUN TO BE MOUNTED HERE (weapon squad leader's pointing to the firing points on the transition line), FRONT (weapon squad leader's pointing to the targets), ACTION.

(2) At the command ACTION the machine gun crew conducts, placing the machine gun into action (bipod mode)

(3) The gunner prepares the rear sight for field zeroing and checks the front sight blade. He sets the range to the zero target on the range scale. The preferred range is 500 meters.

(4) The gunner assumes a good position.

(5) The tower operator instructs the assistant gunner to prepare a 28-round belt (M249) or 36-round belt (M60/M240B).

(6) When the fire command is given, the gunner and assistant gunner repeat each element as it is given.

FIRE MISSION
FRONT
TARGETS: TROOPS IN THE OPEN
THREE HUNDRED
FIXED, FIVE- TO SEVEN-ROUND BURST (M249) or SEVEN- to NINE-ROUND BURST (M60/M240B)
AT MY COMMAND

(7) The gunner loads one 28-round belt of ammunition (M249) or 36-round belt of ammunition (M60/M240B), obtains the proper sight picture, and gives an UP to the assistant gunner.

(8) The assistant gunner relays the READY signal to the tower operator.

(9) The tower operator gives the command FIRE.

(10) The gunner fires a 5- to 7-round burst (M249) or 7- to 9-round burst (M60/M240B) at the target when ready.

(11) The gunner observes the beaten zone. If the rounds miss the target, he makes adjustments for windage and elevation.

(12) After adjustments have been made, the gunner repeats steps 8 through 9 with the remaining rounds until rounds are impacting on the target. He records his zero.

b. **Task 2—400-Meter, Double E-Type Silhouette.**

(1) The tower operator instructs the gunner to load one 154-round belt.

(2) When the fire command is given, the gunner and assistant gunner repeat each element as it is given. It is only given once for tasks 2 through 8.

FIRE MISSION
FRONT
TARGET: TROOPS IN THE OPEN

ONE HUNDRED TO EIGHT HUNDRED METERS
FIXED, FIVE- TO SEVEN-ROUND BURST (M249) or SEVEN- TO NINE-ROUND BURST (M60/M240B)
AT MY COMMAND

(3) The gunner gives an UP to the assistant gunner.

(4) The assistant gunner gives the READY signal to the tower operator.

(5) The tower operator gives the command FIRE.

(6) The gunner scans the sector.

(7) A 400-meter, double E-type target is exposed for 10 seconds.

(8) The gunner determines the range, places the proper setting on the rear sight, assumes the proper position, obtains the correct sight alignment and sight picture, and fires a 5- to 7-round burst (M249) or 7- to 9-round burst (M60/M240B).

(9) If the gunner fails to hit the target, he fires another 5- to 7-round burst (M249) or 7- to 9-round burst (M60/M240B) using the adjusted aiming point method of fire adjustment.

c. **Task 3—500-Meter, Double E-Type Silhouette.**

(1) The gunner and assistant gunner continues to scan the sector.

(2) A 500-meter, double E-type target is exposed for 10 seconds.

(3) The gunner determines the range, places the proper setting on the rear sight, assumes the proper position, obtains the correct sight alignment and sight picture, and fires a 5- to 7-round burst (M249) or 7- to 9-round burst (M60/M240B).

(4) If the gunner fails to hit the target, he fires another 5- to 7-round burst (M249) or 7- to 9-round burst (M60/M240B) using the adjusted aiming point method of fire adjustment.

d. **Task 4—600-Meter, Single E-Type Silhouette.**

(1) The gunner and assistant gunner continues to scan the sector.

(2) A 600-meter, single E-type target is exposed for 20 seconds.

(3) The gunner determines the range, places the proper setting on the rear sight, assumes the proper position, obtains the correct sight alignment and sight picture, and fires a 5- to 7-round burst (M249) or 7- to 9-round burst (M60/M240B).

(4) If the gunner fails to hit the target, he fires another 5- to 7-round burst (M249) or 7- to 9-round burst (M60/M240B) using the adjusted aiming point method of fire adjustment.

e. **Task 5—800-Meter, Single E-Type Silhouette.**

(1) The gunner and assistant gunner continues to scan the sector.

(2) A 800-meter, single E-type target is exposed for 30 seconds (total of six targets).

(3) The gunner determines the range, places the proper setting on the rear sight, assumes the proper position, obtains the correct sight alignment and sight picture, and fires a 5- to 7-round burst (M249) or 7- to 9-round burst (M60/M240B).

(4) If the gunner fails to hit the target, he fires another 5- to 7-round burst (M249) or 7- to 9-round burst (M60/M240B) using the adjusted aiming point method of fire adjustment.

f. **Task 6—400-Meter, Single E-Type Silhouette; and 600-Meter, Double E-Type Silhouettes.**

(1) The gunner and assistant gunner continue to scan the sector.

(2) A 400-meter single E-type target and a 600-meter double E-type target are exposed for 30 seconds.

(3) The gunner determines the range, places the proper setting on the rear sight, assumes the proper position, obtains the correct sight alignment and sight picture, and fires a 5- to 7-round burst (M249) or 7- to 9-round burst (M60/M240B) at each target.

(4) If the gunner fails to hit the target, he fires another 5- to 7-round burst (M249) or 7- to 9-round burst (M60/M240B) using the adjusted aiming point method of fire adjustment at each target.

g. **Task 7—700-Meter and 800-Meter, Double E-Type Silhouettes.**

(1) The gunner and assistant gunner continue to scan the sector.

(2) A 700-meter and a 800-meter double E-type targets are exposed for 45 seconds (total of four targets at 700 meters and six targets at 800 meters).

(3) The gunner determines the range, places the proper setting on the rear sight, assumes the proper position, obtains correct sight alignment and sight picture, and fires a 5- to 7-round burst (M249) or 7- to 9-round burst (M60/M240B) at each target.

(4) If the gunner fails to hit the target, he fires another 7-round burst using the adjusted aiming point method of fire adjustment at each target.

h. **Task 8—400-Meter, Single E-Type Silhouette, and 500-Meter, 600-Meter, Double E-Type Silhouettes.**

(1) The gunner and assistant gunner continue to scan the sector.

(2) The 400-meter single E-type silhouettes, and 500- and 600-meter double E-type silhouettes are exposed for 45 seconds (total of two targets at 400 meters, four targets at 500 meters, and six targets at 600 meters).

(3) The gunner determines the range, places the proper setting on the rear sight, assumes the proper position, obtains correct sight alignment and sight picture, and fires a 5- to 7-round burst (M249) or 7- to 9-round burst (M60/M240B) at each target.

(4) If the gunner fails to hit the target, he fires another 5- to 7-round burst (M249) or 7- to 9-round burst (M60/M240B) using the adjusted aiming point method of fire adjustment at each target.

BIPOD TRANSITION FIRE PRONE OR FIGHTING POSITION PRACTICE AND QUALIFICATION							
TASK	TIME	RDS M249	RDS M60/ M240B	TYPE	TARGET	RANGE	TYPE FIRE
1	No limit	28	36	X 4:1	Double E	500	Fixed, 5- to 7-rd burst (M249) or 7- to 9-rd burst (M60/M240B) (field zero)
*2	5 seconds	14	18	X 4:1	Double E	400	Fixed, 5- to 7-rd burst (M249) or 7- to 9-rd burst (M60/M240B)
*3	10 seconds	14	18	X 4:1	Double E	500	Fixed, 5- to 7-rd burst (M249) or 7- to 9-rd burst (M60/M240B)
*4	5 seconds	14	18	X 4:1	Double E	600	Fixed, 5- to 7-rd burst (M249) or 7- to 9-rd burst (M60/M240B)
*5	10 seconds	14	18	X 4:1	Double E	800	Fixed, 5- to 7-rd burst (M249) or 7- to 9-rd burst (M60/M240B)
*6	20 seconds	28	36	X 4:1	Single E Double E	400 600	Fixed, 5- to 7-rd burst (M249) or 7- to 9-rd burst (M60/M240B)
*7	20 seconds	28	36	X 4:1	Double E Double E	700 800	Fixed, 5- to 7-rd burst (M249) or 7- to 9-rd burst (M60/M240B)
*8	25 seconds	42	54	X 4:1	Single E Double E Double E	400 500 600	Fixed, 5- to 7-rd burst (M249) or 7- to 9-rd burst (M60/M240B)
NOTE: The unit commander determines the firing position. A summary of the ammunition requirements is on page 4-53. *Indicates qualification tasks. **X** Indicates ball and tracer 4:1 mix.							

Table 4-3. Firing Table II—Bipod transition fire.

4-40. TRANSITION FIRE, LIMITED VISIBILITY

Night or limited visibility firing requires the soldier to apply the fundamentals of gunner marksmanship while using nightsights. This training instills confidence in the machine gunner. Each soldier learns how to engage targets using nightsight. He learns to mount the sight, boresight the weapon at 10-meters, and zero the aided vision devices (IAW Appendix G for that device) at 10-meters using a 10-meter (M16A2) zero target. Finally, he learns to detect and engage a series of undetermined targets at various ranges with the aided vision device. Night firing exercises can be conducted during daylight with the AN/PVS-4 when the daylight cover is used. These exercises are for instructional, practice and qualification purposes. The commander can use this training to assess his unit's METL. Night firing is conducted on the same 10-meter range and transition range or a multipurpose machine gun range used for Firing Tables I and II. The tasks and conduct of fire in Firing Table III are the same as in Firing Table II. Therefore, a conduct of fire is not necessary.

a. **Time and Ammunition**. Table 4-4, Firing Table III (page 4-51) outlines ammunition requirements.

b. **Stoppage**. The same procedures that are used in Firing Table II, page 4-49.

c. **Penalties**. No penalties are used.

d. **Scoring.** No points are given when the target is hit on the first or second hit, only a hit or miss. The gunner must hit 6 out of 11 targets in order to be a qualified gunner. The gunner must have qualified on both the 10-meter and transition in order to advance to this step. DA Form 85-R, provided in the back of this manual, can be used to record the number of hits.

e. **Conditions**. Table 4-3, Firing Table III is used for engaging targets out to 400 meters under ideal moonlight or during daylight conditions. If visibility is limited because of a lack of ambient light, commanders may use field-expedient means to identify targets.

NOTE: The commander may lower the ranges by 100 meters when the ambient conditions do not allow the gunners to engage targets at extended ranges.

f. **Targets**. Single E-type silhouette targets and double E-type silhouette targets are used.

g. **Position**. Based on his METL, the commander selects either the bipod-supported prone position or bipod-supported fighting position.

h. **Conduct of Fire.** Limited visibility is the same as Firing Table II. The only difference is time and distance of the targets to be engaged and firing the scanning, walking and IR discipline exercise.

4-41. AN/PVS-4 ZERO

Refer to Appendix G and the appropriate TM for this piece of equipment.

CAUTION

When mounting the AN/PVS-4 to the mounting bracket, make sure the hole for the screw in the AN/PVS-4 is aligned and flush against the bracket screw. If not, it will strip the threads on the screw, and the AN/PVS-4 cannot be used with the M249 machine gun.

TRANSITION FIRE, LIMITED VISIBILITY						
TASK	TIME	RDS M249	RDS M60/ M240B	TYPE	TARGET	TYPE FIRE
1	No limit	6	6	X 4:1	25-meter zero at 10 meters	6-single rounds
2	No limit	18	18	X 4:1	25-meter zero at 10 meters	18-single rounds
3	No limit	28	36	X 4:1	500	Fixed, 5- to 7- rd bursts (M249) or 7- to 9-rd bursts (M60/M240B)
*4	10 seconds	14	18	X 4:1	200	Fixed, 5- to 7- rd bursts (M249) or 7- to 9-rd bursts (M60/M240B)
*5	15 seconds	14	18	X 4:1	400	Fixed, 5- to 7- rd bursts (M249) or 7- to 9-rd bursts (M60/M240B)
*6	10 seconds	14	18	X 4:1	100	Fixed, 5- to 7- rd bursts (M249) or 7- to 9-rd bursts (M60/M240B)
*7	15 seconds	14	18	X 4:1	300	Fixed, 5- to 7- rd bursts (M249) or 7- to 9-rd bursts (M60/M240B)
*8	25 seconds	28	36		200 400	Fixed, 5- to 7- rd bursts (M249) or 7- to 9-rd bursts (M60/M240B)
*9	25 seconds	28	36		100 300	Fixed, 5- to 7- rd bursts (M249) or 7- to 9-rd bursts (M60/M240B)
*10	30 seconds	42	56	X 4:1	100, 200 400	Fixed, 5- to 7- rd bursts (M249) or 7- to 9-rd bursts (M60/M240B)
NOTE: The unit commander determines the position to be used. A summary of the ammunition requirements is on page 4-53. * Indicates qualification tasks. **X** Indicates ball and tracer 4:1 mix						

Table 4-4. Firing Table III—Transition fire, limited visibility.

4-42. QUALIFICATION STANDARDS

Qualification with the M249, M60/M240B machine gun consists of achieving the minimum standards for 10-meter day and transition day firing tables. One point is allowed for each round impacting within the scoring space (maximum of 7 points [M249] or 9 points [M60/M240B] for each space) for Firing Table I. For Firing Table II, 7 points (M249) or 9 points (M60/M240B) are allowed for each target hit whether the target is hit on the first or second burst. For Firing Table III, place an X in the hit column and place an O in the miss column. The maximum possible score for Firing Table I is 91 points (M249), 117 points (M60/M240B). A minimum of 63 points (M249), 81 points (M60/M240B) is required. The maximum score for Firing Table II is 110 points; at least

70 points must be scored on this table to qualify. The maximum possible score for Firing Table III is 11 hits. A minimum of 7 hits is required. The combined minimum total score is 133 (M249), 151 (M240B); the combined maximum total score is 201 points (M249), 227 points (M240B). The overall ratings are as follows:

	M249	M60/M240B
EXPERT	182 to 201	206 to 227
GUNNER Ist CLASS	158 to 181	180 to 205
GUNNER 2nd CLASS	133 to 157	151 to 179
UNQUALIFIED	0 to 132	0 to 150

The trainer uses DA Form 85-R (Scorecard for M249, M60/M240B Machine Gun) for recording the gunner's performance on the machine gun qualification range. The instructions for completing the scorecard are on its reverse side. For an example of a completed form, see Figure 4-23. A blank locally reproducible form is in the back of this manual. The instructions are on the back of the form explaining how to fill out the form:

Use the following procedures to fill out the M249, M60/M240B scorecard:

1. **NAME:** Enter the gunner's last name, first name, middle initial, and rank.
2. **SSN:** Enter the gunner's social security number.
3. **UNIT:** Enter the gunner's unit designation.
4. **DATE:** Enter the date of firing.
5. **LANE:** Enter the lane number for the gunner's firing point.
6. **RECORD:** Tasks used for record and qualification are Firing Table I, tasks 9 through 10; Firing Table II, tasks 2 through 8; and Firing Table III, tasks 4 through 10.
7. **HIT/MISS:** For Table I, tasks 9 through 10, enter the number of rounds impacting within target spaces (maximum of 7 [M249] or 9 [M60/M240B] per space). For Table II, tasks 2 through 8, and Table III, tasks 4 through 10, enter an X for a hit and an O for a miss (regardless of whether the target is hit on the first or second burst).
8. **TOTAL HITS/ POINTS:** For Table I, tasks 9 through 10, give 1 point for each round impacting within a scoring space.
   For Table II, tasks 2 through 8, give 10 points for each target hit.
   For Table III, tasks 4 through 10, enter the number of targets hit (no points awarded).
9. **TOTAL SCORE:** Add points from Tables IV and V. Use the following qualification levels*:

	M249	M60/M240B
EXPERT GUNNER	182-201	206-227
GUNNER 1ST CLASS	158-181	180-205
GUNNER 2D CLASS	133-157	151-179
UNQUALIFIED	0-132	0-150

* The gunner must score 63 points (M249), 81 points (M60/M240B) on Table I, 70 points on Table II, and 6 hits on Table III to meet the minimum score for each.

The following is a summary of ammunition required:

	M249	M60/M240B	TYPE
Table I, Practice	185	231	Ball
Table I, Record	91	117	X4:1
Table II, Practice	182	236	X4:1
Table II, Record	154	200	X4:1
Table III, Practice	360	460	X4:1
Table III, Record	154	200	X4:1

NOTE: See DA Pam 350-38 for STRAC ammunition requirements.

SCORECARD FOR M249, M60/M240B MACHINE GUNS
For use of this form, see FM 3-22.68; the proponent agency is TRADOC.
See back of this form for instructions.

PRIVACY ACT STATEMENT

AUTHORITY: 10 USC 30129(g) Executive Order 9397.

PRINCIPAL PURPOSE: Records individual's performance on record fire range.

ROUTINE USES: Evaluate individual's proficiency and basis for determination of award of proficiency badge; SSN is used for positive identification purposes only.

DISCLOSURE: Voluntary, individuals not providing information cannot be rated/scored on a mass basis.

NAME	SSN	UNIT	DATE (YYYYMMDD)	LANE
PVT DAVID JONES	123-45-6789	C Co, 2/45TH IN	12 DEC 02	3

TABLE I (10-METERS)

TASK	RANGE (meters)	TIME	HITS
1*	10	N/A	N/A
2*	10	N/A	N/A
3*	10	N/A	N/A
4*	10	N/A	N/A
5*	10	N/A	N/A
6*	10	N/A	N/A
7*	10	N/A	N/A
8*	10	N/A	N/A
9	10	40 SEC	30
10	10	50 SEC	52

TOTAL HITS (POINTS) 82

TABLE II (DAY TRANSITION)

TASK	RANGE (meters)	TIME	*PRACTICE HIT	*PRACTICE MISS	**QUALIFY HIT	**QUALIFY MISS
1*	500	N/A	N/A	N/A	N/A	N/A
2**	400	10 SEC	X		X	
3**	500	15 SEC	X		X	
4**	600	20 SEC		0	X	
5**	800	30 SEC	X			0
6**	400 600	30 SEC	X X		X X	
7**	700 800	45 SEC	X	0	X X	
8**	400 500 600	45 SEC	X X X		X X X	

**TOTAL POINTS 100

TABLE III (LIMITED VISIBILITY)

TASK	RANGE (meters)	TIME	*PRACTICE HIT	*PRACTICE MISS	**QUALIFY HIT	**QUALIFY MISS
1*	10	N/A	N/A	N/A	N/A	N/A
2*	10	N/A	N/A	N/A	N/A	N/A
3*	500	N/A	N/A	N/A	N/A	N/A
4	200	10 SEC	X		X	
5	400	15 SEC	X			0
6	100	10 SEC		0	X	
7	300	15 SEC	X		X	
8	200 400	25 SEC	X	0	X X	
9	100 300	25 SEC	X X		X X	
10	100 200 400	30 SEC	X X	0	X X X	

TOTAL HITS 9

TOTAL SCORE ____________

* NONSCORED TASKS
**10 POINTS PER HIT

CHECK APPROPRIATE WEAPON

EXPERT		FIRST CLASS		SECOND CLASS	
☑ M249 182-201	☐ M60/M240B 206-227	☐ M249 158-181	☐ M60/M240B 180-205	☐ M249 158-181	☐ M60/M240B 180-205

OIC SIGNATURE	GRADER	RATING
John Smith	Paul Davis	EXPERT

DA FORM 85-R, OCT 2002 — DA FORM 85, SEP 62, IS OBSOLETE. — Page 1 of 2

USAPA V1.00ES

Figure 4-23. Example of a completed DA Form 85-R.

CHAPTER 5

COMBAT TECHNIQUES OF FIRE

Technique of fire is the method of delivering and controlling effective fire. The machine gunners must be trained in the standard methods of applying fire. This chapter discusses combat techniques of fire, application of fire on the battlefield, and advanced gunnery. Before the machine gun can be employed to its full potential, the soldier must know and be trained on characteristics of fire, classes of fire, types of targets, and application of fire.

Section I. CHARACTERISTICS OF FIRE

Each gunner must know the effects of rounds when fired. Factors influencing the path and strike of rounds are not limited to applying the fundamentals. They include the velocity of the round, gravity, terrain, atmospheric conditions, and the innate differences between each round.

5-1. TRAJECTORY

The trajectory is the path of the round in flight (Figure 5-1). The gunner must know the machine gun trajectory to effectively fire the weapon throughout its full range. The path of the round is almost flat at ranges up to 300 meters; then it begins to curve, and the curve becomes greater as the range increases.

5-2. MAXIMUM ORDINATE

Maximum ordinate is the highest point the trajectory reaches between the muzzle of the weapon and the base of the target. It always occurs about two-thirds of the distance from the weapon to the target. The maximum ordinate increases as the range increases (Figure 5-1).

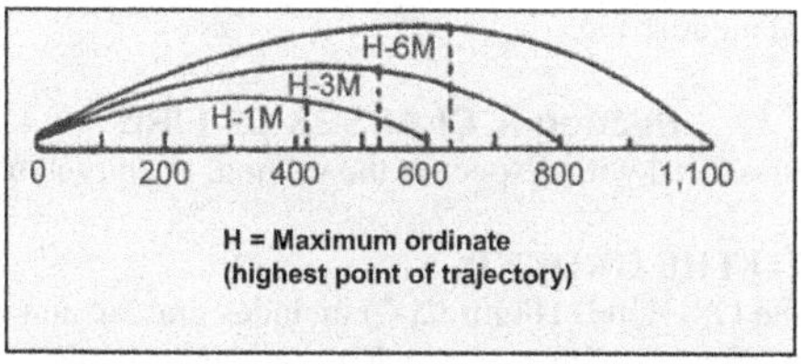

Figure 5-1. Trajectory and maximum ordinate.

5-3. CONE OF FIRE

When several rounds are fired in a burst from any machine gun, each round takes a slightly different trajectory. The pattern these rounds form on the way to the target is called a cone of fire (Figure 5-2, page 5-2). This pattern is caused primarily by vibration of the machine gun and variations in ammunition and atmospheric conditions.

5-4. BEATEN ZONE

The beaten zone (Figure 5-2) is the elliptical pattern formed by the rounds striking the ground or the target. The size and shape of the beaten zone changes when the range to the target changes or when the machine gun is fired on different types of terrain. On uniformly sloping or level terrain, the beaten zone is long and narrow. As the range to the target increases, the beaten zone becomes shorter and wider. When fire is delivered on terrain sloping down and away from the machine gun, the beaten zone becomes longer. When fire is delivered on rising terrain, the beaten zone becomes shorter. The terrain has little effect on the width of the beaten zone.

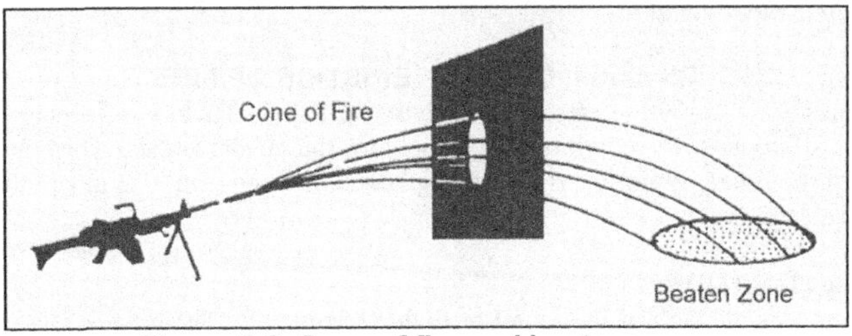

Figure 5-2. Cone of fire and beaten zone.

5-5. DANGER SPACE

The danger space is the space between the machine gun and the target where the trajectory does not rise above 1.8 meters (the average height of a standing soldier). This space includes the area of the beaten zone. When the machine gun is fired on level or uniformly sloping terrain at a target less than 700 meters away, the trajectory does not rise above the average height of a standing soldier. When targets are engaged on level or uniformly sloping terrain at ranges greater than 700 meters, the trajectory rises above the average height of a standing soldier, therefore, not all the distance between the machine gun and the target is danger space.

Section II. CLASSES OF FIRE

Machine gun fire is classified with respect to the ground, the target, and the weapon.

5-6. RESPECT TO THE GROUND

Fire with respect to the GROUND (Figure 5-3) includes grazing and plunging fires.

a. **Grazing Fire**. Grazing fire occurs when the center of the cone of fire does not rise more than 1 meter above the ground. When firing on level or uniformly sloping terrain, the gunner can obtain a maximum of 600 meters of grazing fire.

b. **Plunging Fire**. Plunging fire occurs when the danger space is confined to the beaten zone. Plunging fire also occurs when firing at long ranges, from high ground to low ground, into abruptly rising ground, or across uneven terrain, resulting in a loss of grazing fire at any point along the trajectory.

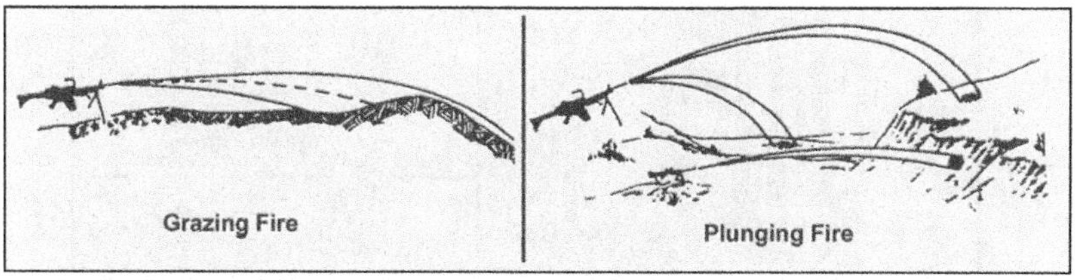

Figure 5-3. Classes of fire with respect to the ground.

5-7. RESPECT TO TARGET

Fire with respect to the TARGET includes frontal, flanking, oblique, and enfilade fires.

a. **Frontal Fire.**Frontal fire is when the long axis of the beaten zone is at a right angle to the front of the target. An example is when firing at the front of a target (Figure 5-4).

b. **Flanking Fire**. Flanking fire is firing at the side of a target (Figure 5-4).

Figure 5-4. Frontal fire and flanking fire.

c. **Oblique Fire**. Oblique fire is when the long axis of the beaten zone is at an angle other than a right angle to the front of the target (Figure 5-5).

d. **Enfilade Fire**. Enfilade fire is when the long axis of the beaten zone coincides or nearly coincides with the long axis of the target. This type of fire is either frontal or flanking. It is the most desirable type of fire with respect to a target, because it makes maximum use of the beaten zone (Figure 5-5, page 5-4).

Figure 5-5. Oblique fire and enfilade fire.

5-8. RESPECT TO MACHINE GUN

Fire with respect to the machine gun (Figure 5-6) includes fixed, traversing, searching, and traversing and searching, and free-gun fires.

a. **Fixed Fire**. Fixed fire is fire delivered against a point target when the depth and width of the beaten zone covers the target. Fixed fire also means only one aiming point is necessary to provide coverage of the target.

b. **Traversing Fire**. Traversing fire is fire distributed in width by successive changes in direction. The gunner selects successive aiming points throughout the width of the target. These aiming points must be close enough to ensure adequate coverage but not so close as to waste ammunition.

c. **Searching Fire**. Searching fire is fire distributed in depth by successive changes in elevation. The gunner selects successive aiming points in depth. The changes made in each aiming point will depend on the range and slope of the ground.

d. **Traversing and Searching Fire**. Traversing and searching fire is fire distributed in width and depth by successive changes in direction and elevation. Combining traversing and searching provides good coverage of the target. Adjustments are made in the same manner as described for traversing and searching fire.

e. **Free-Gun Fire**. Free-gun fire is fire delivered against targets requiring rapid major changes in direction and elevation that cannot be applied with the T&E mechanism. To deliver this type of fire, the gunner removes the T&E mechanism from the traversing bar on the tripod, allowing the weapon to be moved freely in any direction.

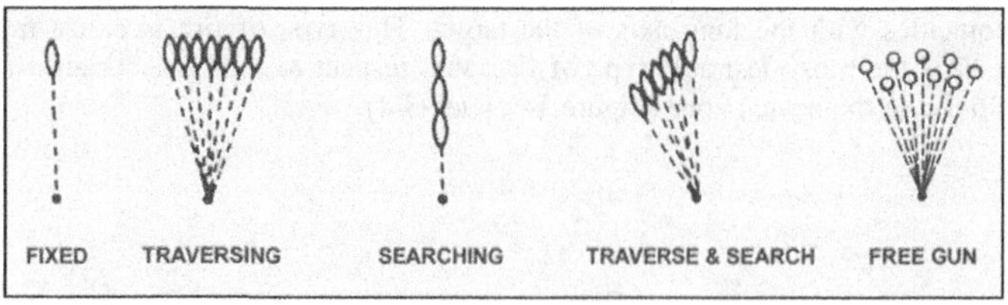

Figure 5-6. Classes of fire with respect to the machine gun.

Section III. APPLICATION OF FIRE

Application of fire consists of the methods the gunner uses to cover a target area. Training these methods of applying fire can be accomplished only after the soldiers have learned how to recognize the different types of targets they may find in combat, how to distribute and concentrate their fire, and how to maintain the proper rate of fire. Normally, the gunner is exposed to two types of targets in the squad or platoon sector: enemy soldiers and supporting automatic weapons. These targets have priority and should be engaged immediately.

5-9. TYPES OF TARGETS

Targets presented to the gunner in combat are usually enemy troops in various formations, which require distribution and concentration of fire. Targets with width and depth must be thoroughly covered by fire.

a. **Point Targets**. Point targets require the use of a single aiming point. Examples of point targets are enemy soldiers, bunkers, weapons emplacements, and lightly armored vehicles. Fixed fire is delivered at point targets.

b. **Area Targets**. Area targets may have considerable width and depth and may require extensive traversing and searching fire. These include targets in which the exact location of the enemy is unknown. The following are varieties of area targets likely to be engaged.

(1) *Linear Targets*. Linear targets have sufficient width to require successive aiming points (traversing fire). The beaten zone effectively covers the depth of the target area (Figure 5-7). Traversing fire is delivered at linear targets.

(2) *Deep Targets*. Deep targets require successive aiming points (searching fire) (Figure 5-8, page 5-6). Searching fire is delivered at deep targets.

(3) *Linear Targets with Depth*. Linear targets with depth have sufficient width requiring successive aiming points in which the beaten zone does not cover the depth of the target area. A combined change in direction and elevation (traversing and searching) is necessary to effectively cover the target with fire (Figure 5-9, page 5-6). Traversing and searching fire are delivered at linear targets with depth.

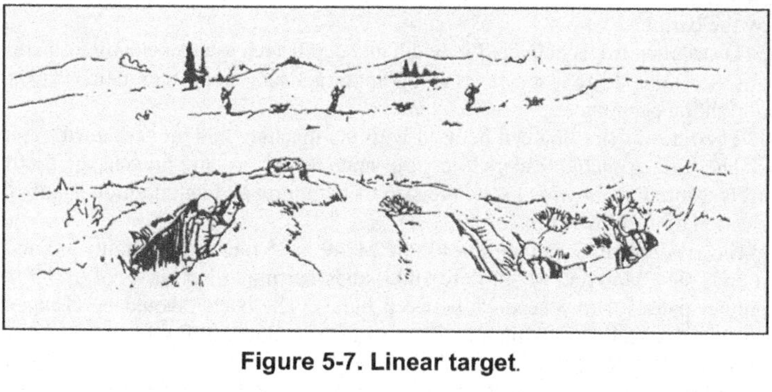

Figure 5-7. Linear target.

Figure 5-8. Deep target.

Figure 5-9. Linear targets with depth.

5-10. DISTRIBUTION, CONCENTRATION, AND RATE OF FIRE

The size and nature of the target determine how the gunner applies his fire. He must manipulate the machine gun to move the beaten zone throughout the target area. The rate of fire must be controlled to adequately cover the target but not waste ammunition or destroy the barrel.

a. Distributed fire is delivered in width and depth such as at an enemy formation.

b. Concentrated fire is delivered at a point target such as an automatic weapon or an enemy fighting position.

c. The rates of fire that can be used with the machine gun are sustained, rapid, and cyclic. These rates enable leaders to control and sustain fire and prevent the destruction of barrels. More than anything else, the size of the target and ammunition supply dictate the selection of the rate of fire.

(1) *Sustained Fire.* Sustained fire for the M249 is 85 rounds per minute in bursts of 3 to 5 rounds. The M60 and M240B are 100 rounds per minute in bursts of 6 to 9 rounds. The gunner pauses 4 to 5 seconds between bursts. The barrel should be changed after firing at sustained rate for 10 minutes. This is the normal rate of fire for the gunner.

(2) *Rapid Fire.* Rapid fire for the M249, M60, and M240B gunner is 200 rounds per minute in bursts of (6 to 8 M249) 10 to 12 rounds. The gunner pauses 2 to 3 seconds

between bursts. The barrel should be changed after firing at a rapid rate for 2 minutes. This procedure provides for an exceptionally high volume of fire, but for only a short period.

(3) *Cyclic Fire.* Cyclic fire uses the most ammunition that can be used in 1 minute. The cyclic rate of fire with the machine gun is achieved when the trigger is held to the rear and ammunition is fed into the weapon uninterrupted for one minute. Normal cyclic rate of fire for the M249 is 850 rounds, M60 is 550 rounds, and for the M240B it is 650 to 950 rounds. Always change the barrel after firing at cyclic rate for 1 minute. This procedure provides the highest volume of fire that the machine gun can fire, but this adversely affects the machine gun, and should only be fired in combat under emergency purposes only.

5-11. TARGET ENGAGEMENT

The gunner engages targets throughout his sector. He must know how to effectively engage all types of targets either by himself or in conjunction with another gunner.

a. **Single Gunner.**

(1) *Point Target.* When engaging a point target, the gunner uses fixed fire (Figure 5-10). If the target moves after the initial burst, the gunner adjusts fire onto the target by following its movement.

Figure 5-10. Engagement of point target.

(2) *Area Target.* When engaging an area target, the gunner fires in the center of mass, then traverses and searches to either flank (Figure 5-11, page 5-8). Upon reaching the flank, he reverses direction and traverses and searches in the opposite direction. A leader may indicate the width and depth of the target.

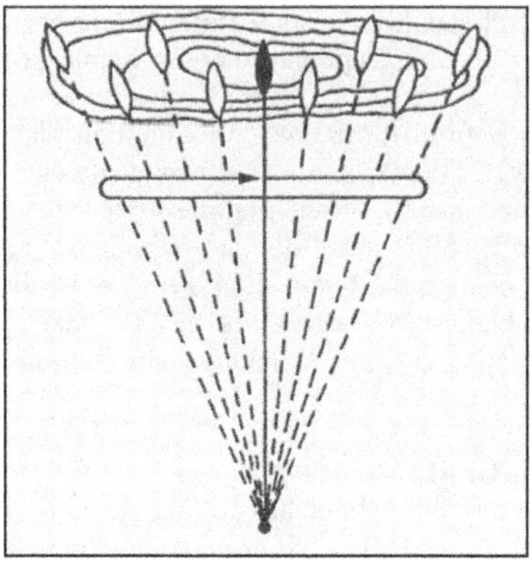

Figure 5-11. Engagement of area target.

(3) *Linear Target.* When engaging a linear target, the gunner traverses the machine gun to distribute fire evenly onto the target. He must cover the entire width of a linear target. The initial point of aim is on the midpoint. The gunner then manipulates to cover the rest of the target. If a linear target is hard to identify, a leader may designate the target by using a reference point (Figure 5-12). When this method is used, the leader determines the center of mass of the target and announces the number of meters from the reference point that will cause the gunner to aim on the center of mass. The reference point may be within or adjacent to the target (Figure 5-13). However, the reference point should be on line with the target for the best effect. After the command to fire has been given, the leader maintains and controls the fire by subsequent fire commands.

Figure 5-12. Engagement of hard-to-identify linear targets with a reference point outside the target area.

(4) *Deep Target.* When engaging a deep target, the gunner must use searching fire. If the range is announced, he initially aims on the midpoint of a deep target unless another portion of the target is more critical or presents a greater threat. The gunner then searches

down to one aiming point in front of the near end and back up to one aiming point beyond the far end. If a deep target is hard to identify, use the reference points to designate the center of mass. The extent (depth) of the target is always given in meters.

(5) *Linear Target With Depth.* When engaging a linear target with depth, the gunner uses traversing and searching fire. He begins engagement at the midpoint of the target unless another portion of the target is more critical or presents a greater threat. He traverses and searches to the near flank, then back to the far flank. When engaging hard-to-identify linear targets with depth, he designates the flanks and midpoint with rifle fire. The reference-point method is not used because at least two reference points are required to show the angle of the target.

Figure 5-13. Engagement of hard-to-identify targets with a reference point within the target area.

b. **Pair of Gunners.**

(1) *Area Targets.* When using a pair of machine guns to engage area targets, the gunner on the right fires on the right half, and the gunner on the left fires on the left half. The point of initial aim and adjustment for both gunners is on the midpoint. After adjusting fire on the center of mass, both gunners distribute fire by applying direction and elevation changes that give the most effective coverage of the target area. The right gunner traverses to the right, applies the necessary amount of search, and fires a burst. He traverses and searches up and down until the right flank of the area target has been reached. The left gunner traverses and searches to the left flank in the same way. Both gunners then reverse the direction of manipulation and return to the center of mass, firing a burst after each combined direction and elevation change (Figure 5-14, page 5-10).

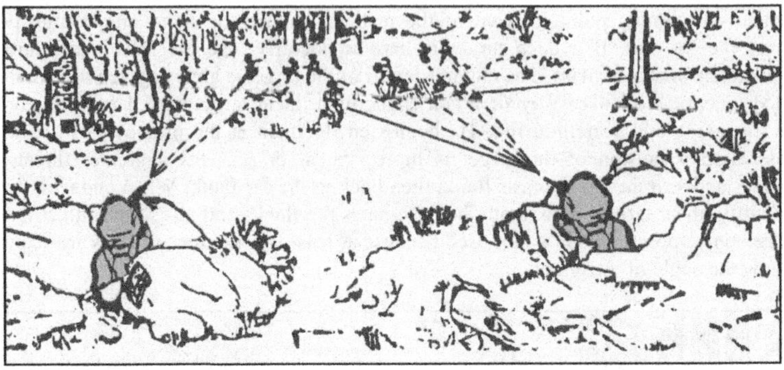

Figure 5-14. Engagement of area targets with a pair of gunners.

(2) *Linear Targets*. When using a pair of machine guns to engage a linear target, the target is divided at midpoint with the gunner on the right of the target firing on the right half, and the gunner on the left of the target firing on the left half (Figure 5-15).

(a) Both gunners aim on the midpoint initially. After adjusting on the midpoint, the gunner on the right traverses right, firing a burst after each change in direction until the rounds reach one aiming point beyond the right flank (this ensures complete target coverage). The gunner on the left traverses to the left flank in the same way the gunner on the right did. Both gunners then reverse their directions and return to the midpoint. The gunner must select aiming points for each burst rather than "spray" the target area.

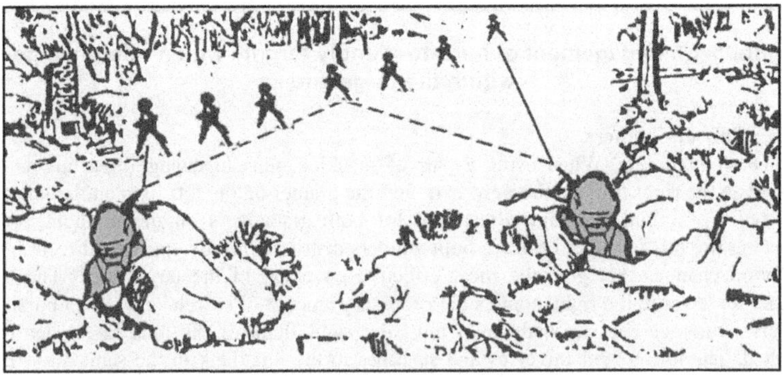

Figure 5-15. Engagement of linear targets with a pair of gunners.

(b) If one part of the target is a greater threat, fire can be concentrated on the greater threat by dividing the target unevenly. This special division of the target is done with fire commands. To preclude confusion, the gunners initially aim on the midpoint regardless of the special division to be made.

(3) *Deep Targets.* When using a pair of machine guns to engage a deep target, the initial point of aim is also on the midpoint for both gunners. Normally, the gunner on the right has the near half and the gunner on the left has the far half. Since enfilade fire is being used, they do not adjust on the midpoint of the target, because the long beaten zone compensates for any range errors. After the initial burst, the gunner on the right searches down to one aiming point in front of the near end of the target, and the gunner on the left searches up to one aiming point beyond the far end. Both gunners then reverse their direction of search and return to the midpoint (Figure 5-16).

Figure 5-16. Engagement of deep targets with a pair of gunners.

(4) *Linear Target With Depth.* When using a pair of machine guns to engage a linear target with depth, the initial point of aim and the extent of manipulation for both gunners is the same as those prescribed for linear targets (Figure 5-17).

Figure 5-17. Engagement of linear target with depth with a pair of gunners.

5-12. TARGET ENGAGEMENT DURING LIMITED VISIBILITY

Gunners have problems detecting and identifying targets during limited visibility. The leader's ability to control the fires of his weapons is also reduced, therefore, he may instruct the gunners to fire without command when targets present themselves.

a. Gunners should engage targets only when they can identify the targets, unless ordered to do otherwise. For example, if one gunner detects a target and engages it, the other gunner observes the area fired upon and adds his fire only if he can identify the target or if ordered to fire.

b. Tracer ammunition helps a gunner engage targets during limited visibility and should be used, if possible. If firing unaided, gunners must be trained to fire low at first and adjust upward. This overcomes the tendency to fire high.

c. When two or more gunners are engaging linear targets, linear targets with depth, or deep targets, they do not engage these targets as they would when visibility is good. With limited visibility, the center and flanks of these targets may not be clearly defined; therefore, each gunner observes his tracers and covers what he believes to be the entire target.

5-13. OVERHEAD FIRE

Fire delivered over the heads of friendly soldiers is called overhead fire. It is used during training ONLY AFTER SOLDIER SAFETY IS CHECKED AND VERIFIED. The terrain and visibility dictate when overhead fire can be delivered safely. (See AR 385-63 for a complete summary of training safety requirements.) Overhead fire is delivered with any machine gun mounted on a tripod because the machine guns provide greater stability and accuracy, and because vertical mil angles can be measured by using the elevating mechanism.

DANGER

OVERHEAD FIRE CANNOT BE SAFELY DELIVERED ON A TARGET AT GREATER THAN 850 METERS FROM THE MACHINE GUN, AND IT IS NOT DELIVERED OVER LEVEL OR UNIFORMLY SLOPING TERRAIN. IT CAN CAUSE DEATH OR INJURY.

a. Ideally, overhead fire is delivered when there is a depression in the terrain between the machine gun position and the target. The depression should place the gunner's line of aim well above the heads of friendly soldiers.

b. The squad leader normally controls overhead fire. He lifts or shifts the fire when the friendly soldiers reach an imaginary line, parallel to the target, where further fire would cause casualties to friendly soldiers. This imaginary line is called the "safety limit." The leader of the friendly soldiers may direct lifting of fire by prearranged signals transmitted by radio, wire, or visual means. The safety limit can be determined by observing the fire or by using the gunner's rule (Figure 5-18).

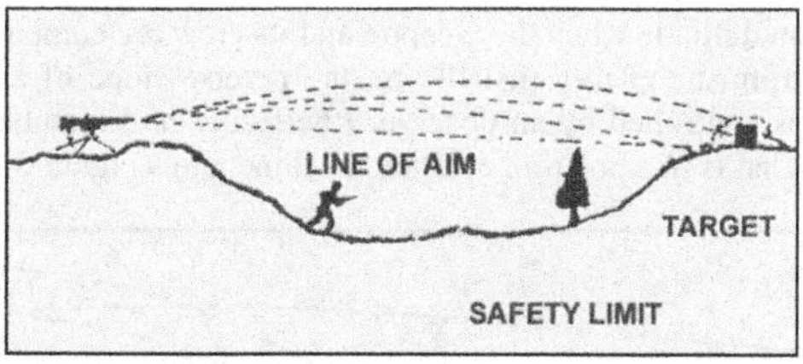

Figure 5-18. Overhead fire safety limit.

(1) To determine the safety limit by observation, the leader uses binoculars to see how close the fire is to advancing friendly soldiers.

(2) A safety limit can be selected by using the gunner's rule before the weapon is fired. The accuracy and safety of this method depends on the machine gun being zeroed and the range to the target being known. The gunner's rule is used only when the target is between 350 and 850 meters from the machine gun. The gunner's rule consists of the following procedure.

(a) Determine the range to the target and set the range on the rear sight.

(b) Aim the machine gun to hit the target.

(c) Set the rear sight to 1,000 meters.

(d) Depress the muzzle 10 mils by using the elevating handwheel (one click equals 1 mil).

(e) Look through the rear sight and note the point where the new line of aim strikes the ground. (An imaginary line drawn through this point and parallel to the target is the safety limit.)

(f) Reset the range to the target on the rear sight, aim on the target, and prepare to fire.

(g) Cease or shift fire when soldiers reach the safety limit.

c. The following safety measures MUST be applied when delivering overhead fire.

(1) Firmly emplace the tripod mount.

(2) Use field-expedient depression stops to prevent the muzzle from accidentally being lowered below the safety limit.

(3) Do not deliver overhead fire through trees.

(4) Inform commanders of friendly soldiers when fire is to be delivered over their heads.

(5) Ensure that all members of the crew are aware of the safety limit.

(6) Do not deliver overhead fire if the range from the machine gun to the target is less than 350 meters or more than 850 meters.

(7) Do not use a barrel that is badly worn.

(8) During training exercises, do not aim any machine gun where their trajectories will cross at a point directly over the heads of friendly soldiers. See AR 385-63 and local safety regulations concerning overhead fire.

5-14. DEFILADE POSITIONS

A machine gun is in defilade when the weapon and its crew are completely behind terrain that masks them from the enemy (usually on the reverse slope of a hill). Fire, from a defilade position, is controlled by an observer (the leader or a member of the crew who can see the target) that is in a position near the machine gun. (Figure 5-19.)

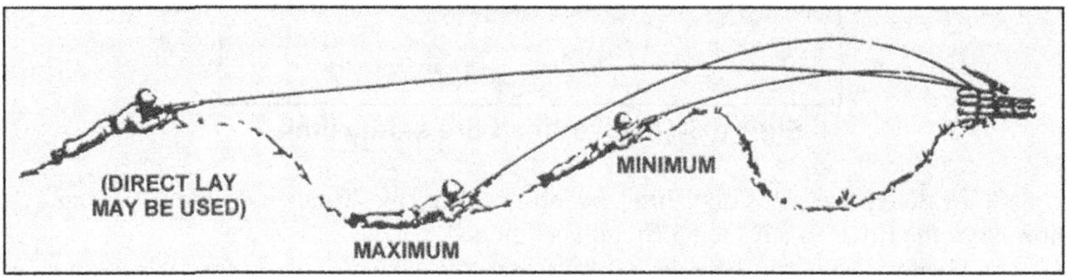

Figure 5-19. Defilade positions.

a. The machine gun must fire up and over the hill. Its fire must be observed and adjusted by a crewmember that can observe the target from a position on a flank or to the rear of the weapon (on higher ground). A defilade position allows little opportunity to engage new targets. The tripod mount is used when firing from defilade, because the gunner can measure vertical angles with it. This makes changes in elevation for adjusting fire easier, and if data is determined during daylight, the crew can fire from the same position after dark. A machine gun is in partial defilade when it is positioned just back of the crest of a hill, so that the crest provides some protection from enemy direct-fire and the machine guns are still able to engage its target by direct-lay techniques.

(1) *Advantages.*

- The crew has cover and concealment from enemy direct-fire weapons.
- The crew has some freedom of movement near the position.
- Control and supply are easier.
- The smoke and flash of the machine gun are hidden from the enemy.

(2) *Disadvantages.*

- Rapidly moving ground targets are hard to engage, because adjustment of fire must be made by using an observer.
- Targets close to the mask usually cannot be engaged.
- Final protective line is hard to understand.

b. The essential elements in the engagement of a target from a defilade position are mask clearance, direction, and adjustment of fire. If possible, a minimum mask clearance (minimum elevation) is determined for the entire sector of fire, however, a mask clearance for each target may be necessary (due to the slope of the mask). The elevation readings obtained using the methods below give the minimum elevation for the sector or target(s). The minimum elevation should be recorded on a range card.

(1) If the mask is 300 meters or less from the machine gun position, the gunner places a 300-meter range setting on the rear sight, aims on the top of the mask and adds 3 mils (clicks) of elevation with the elevating handwheel.

(2) If the mask is over 300 meters from the machine gun position, the gunner places the range setting to the mask on the rear sight, aims on the top of the mask, and adds 3 mils (clicks) of elevation.

c. The observer places himself to the rear of the machine gun on the gun-to-target line and in a position where he can see the machine gun and the target. He aligns the machine gun for general direction by directing the gunner to shift the machine gun until it is aligned on the target. A prominent terrain feature or landmark visible to the gunner through his sights is selected as an aiming point. This aiming point should be at a greater range than the target and at a higher elevation. When laying the machine gun on the aiming point, the range setting on the rear sight must correspond to the range to the target. (Figure 5-20.)

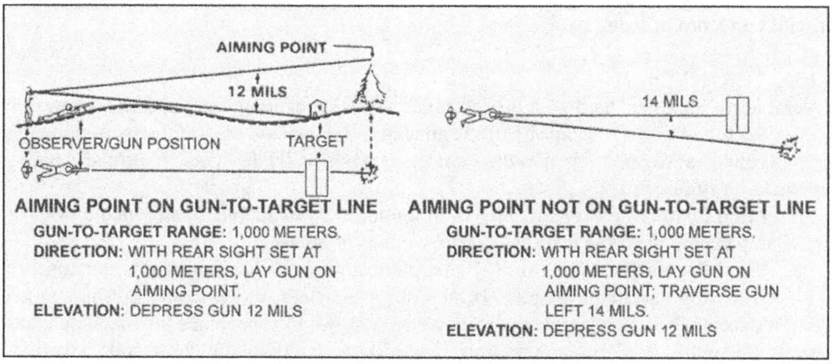

Figure 5-20. Observer adjusting fire.

(1) If the aiming point is on the gun-to-target line, the gun is laid on the aiming point and is thereby aligned for direction.

(2) If the aiming point is not on the gun-to-target line, the horizontal distance in mils is determined using the best means available (usually binoculars) and announced to the gunner. This measured distance is then set with the traversing handwheel.

d. The observer measures the vertical distance from the aiming point to the base of the target using the best means available and directs the gunner to depress the muzzle of the machine gun the number of mils measured. The machine gun should now be laid to hit the target.

Section IV. PREDETERMINED FIRES

Predetermined fires organize the battlefield for the gunners. They allow the leader and gunner to select potential targets or target areas that will most likely be engaged or that have tactical significance. This includes dismounted enemy avenues of approach, likely positions for automatic weapons, and probable enemy assault positions. The gunners do this by using sectors of fire, final protective lines, or a principal direction of fire and selected target areas. This preparation maximizes the effectiveness of the machine gun during good as well as limited visibility. It enhances fire control by reducing the time required to identify targets, determine range, and manipulate the weapon onto the target. Abbreviated fire commands and previously recorded data enable the gunner to aim or adjust fire on the target quickly and accurately. Selected targets should be fired on in daylight whenever practical to confirm data. The range card identifies the targets and provides a record of firing data.

5-15. TERMINOLOGY

Several terms are associated with predetermined fire that every gunner needs to know.

a. **Sector of Fire**. A sector of fire is an area to be covered by fire that is assigned to an individual, a weapon, or a unit. Gunners are normally assigned a primary and a secondary sector of fire.

b. **Final Protective Fire**. An FPF is an immediately available prearranged barrier of fire to stop enemy movement across defensive lines or areas.

c. **Final Protective Line**. An FPL is a predetermined line along which grazing fire is placed to stop an enemy assault. If an FPL is assigned, the machine gun is sighted along it except when other targets are being engaged. An FPL becomes the machine gun's part of the unit's final protective fires. An FPL is fixed in direction and elevation; however, a small shift for search must be employed to prevent the enemy from crawling under the FPL and to compensate for irregularities in the terrain or the sinking of the tripod legs into soft soil during firing. Fire must be delivered during all conditions of visibility.

d. **Principal Direction of Fire**. A PDF is a direction of fire assigned priority to cover an area that has good fields of fire or has a likely dismounted avenue of approach. It also provides mutual support to an adjacent unit. Machine guns are sighted using the PDF if an FPL has not been assigned. If a PDF is assigned and other targets are not being engaged, machine guns remain on the PDF. A PDF has the following characteristics.

(1) It is used only if an FPL is not assigned; it then becomes the machine gun's part of the unit's final protective fires.

(2) When the target has width, direction is determined by aiming on one edge of the target area and noting the amount of traverse necessary to cover the entire target.

(3) The gunner is responsible for the entire wedge-shaped area from the muzzle of the weapon to the target, but elevation may be fixed for a priority portion of the target.

e. **Grazing Fire**. A good FPL covers the maximum area with grazing fire. Grazing fire can be obtained over various types of terrain out to a maximum of 600 meters. To obtain the maximum extent of grazing fire over level or uniformly sloping terrain, the gunner sets the rear sight at 600 meters. He then selects a point on the ground that he estimates to be 600 meters from the machine gun, and he aims, fires, and adjusts on that point. To prevent enemy soldiers from crawling under grazing fire, he searches (downward) by lowering the muzzle of the weapon. To do this, the gunner separates his elbows.

f. **Dead Space.** The extent of grazing fire and the extent of dead space may be determined in two ways. In the preferred method, the machine gun is adjusted for elevation and direction. A member of the squad then walks along the FPL while the gunner aims through the sights. In places where the soldier's waist (midsection) falls below the gunner's point of aim, dead space exists. Arm-and-hand signals must be used to control the soldier who is walking and to obtain an accurate account of the dead space and its location. Another method is to observe the flight of tracer ammunition from a position behind and to the flank of the weapon.

g. **Fire Control.** Predetermined targets, including the FPL or PDF, are engaged on order or by SOP. The signal for calling for these fires is normally stated in the defense order. Control these predetermined targets by using arm-and-hand signals, voice commands, or pyrotechnic devices. Gunners fire the FPL or PDF at the sustained rate of fire unless the situation calls for a higher rate. When engaging other predetermined targets, the sustained rate of fire is also used unless a different rate is ordered.

h. **Primary Sector of Fire.** The primary sector of fire is the area to be covered by an individual or unit.

i. **Secondary Sector of Fire.** The secondary sector of fire is the same area covered by the same individual or unit after it has moved to a different location.

5-16. RANGE CARD

The standard range card (DA Form 5517-R) provides a record of firing data and aids defensive fire planning. (See FM 7-8 for a reproducible copy of this form.) Its use enhances fire control and rapid engagement of predetermined targets. It is also used in estimating ranges to other targets within the sector of fire. Each gunner makes two copies—one for his position and one for the squad leader. The squad leader uses his copy to prepare his sector sketch. The range card is prepared immediately upon occupation and is constantly revised. Each range card contains the following:

- Weapon symbol (Figure 5-21, page 5-18).
- Sector of fire.
- PDF or FPL.
- Range, azimuth, and number label to predetermined targets.
- Dead space.
- Distance and azimuth from a known point or eight-digit grid coordinate (reference point).
- Magnetic north arrow.
- Data section.

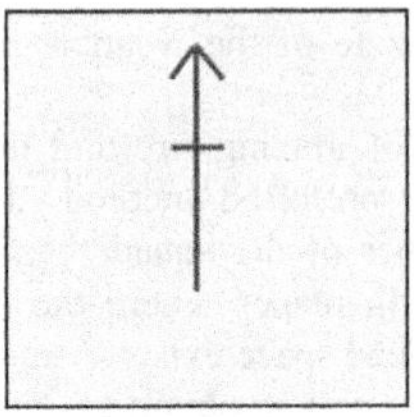

Figure 5-21. M249/M60 and M240B machine gun symbol.

a. **Procedures**. The machine gun is placed in the tripod-supported mode in the position it will be fired. The machine gun symbol is sketched on the range card pointing toward the most dangerous target in the sector.

(1) If using the FPL, the gunner aims the machine gun along the FPL. This procedure will also be either the left or right limit of the sector of fire. To set the limit, he slides the T&E mechanism all the way to the left or right end of the traversing bar. Then, he moves the tripod until the barrel lines up on the FPL. The sector of fire with the FPL along one limit is now prepared. The FPL is always labeled target number 1.

(2) To determine the range for all targets in the sector, the gunner ensures each circle, except the first one, represents 100 meters. Since the lowest setting on the M249 and M60 is 300 meters and the M240B is 200 meters, the first circle represents 200 or 300 meters. He indicates this on the range card in the data section just below the circles. On the top half of the range card, the gunner draws the left or right limits from the weapon position to the maximum effective range of the machine gun.

(3) If the FPL is assigned, the machine gun symbol is drawn along that line (left or right limit) (Figure 5-22). The extent of grazing fire is determined. A shaded blade is sketched on the inside of the FPL to represent the extent of the grazing fire. If there is dead space along the FPL, it is shown by breaks in the shaded area. The ranges to the *near* and *far* edges of the dead space are recorded above the FPL, and the *extent* of the grazing fire is recorded along the FPL. The magnetic azimuth of the FPL is determined and recorded below the shaded blade representing the FPL. The elevation reading and other data are recorded in the data section.

(4) If an FPL is not assigned, the gunner locks the T&E mechanism on 0 on the traversing bar scale and shifts the tripod until the muzzle points to the PDF. The machine gun symbol is sketched in the center of the left and right limits pointing in the direction of the PDF (Figure 5-23, page 5-20).

(5) The opposite primary sector limit is drawn. If a target is along this line, the target information is added to the data section. If the opposite side of the traversing bar cannot be used to mark the opposite side of the primary sector, a direction reading must be recorded in the sketch section.

(6) Next, the left and right limits of the secondary sector are drawn using a broken line. The area between the primary and secondary sector is labeled dead space.

(7) An arrow is drawn in the magnetic north block (upper right hand corner) pointing in the direction of magnetic north.

STANDARD RANGE CARD

For use of this form see FM 7-8 The proponent agency is TRADOC.

SQD 2
PLT 1
CO C

May be used for all types of direct fire weapons.

MAGNETIC NORTH

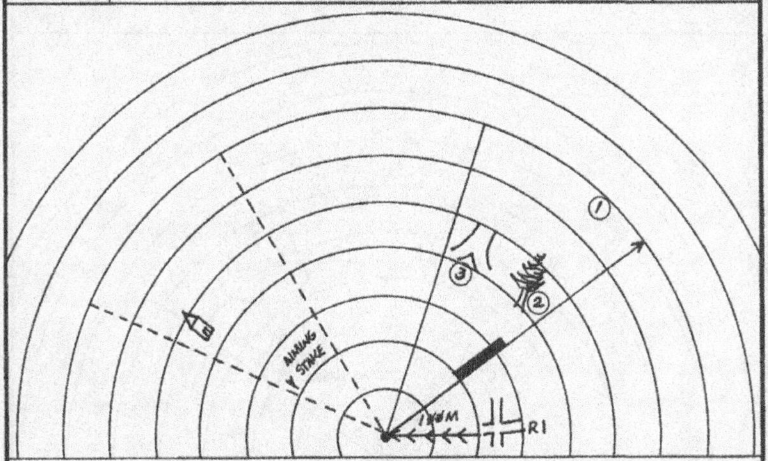

DATA SECTION

POSITION IDENTIFICATION

DATE 8 AUG 94

WEAPON M60 #1

EACH CIRCLE EQUALS 100 METERS METERS

NO.	DIRECTION/ DEFLECTION	ELEVATION	RANGE	AMMO	DESCRIPTION
1		+5Ø/3	6ØØ		FPL
2	R35Ø°	+5Ø/45	6ØØ		LONE TREE
3	L3ØØ°	Ø/28	65Ø		TRAIL JUNCTION

REMARKS:

DA Form 5517-R, FEB 86

Figure 5-22. Final protective line.

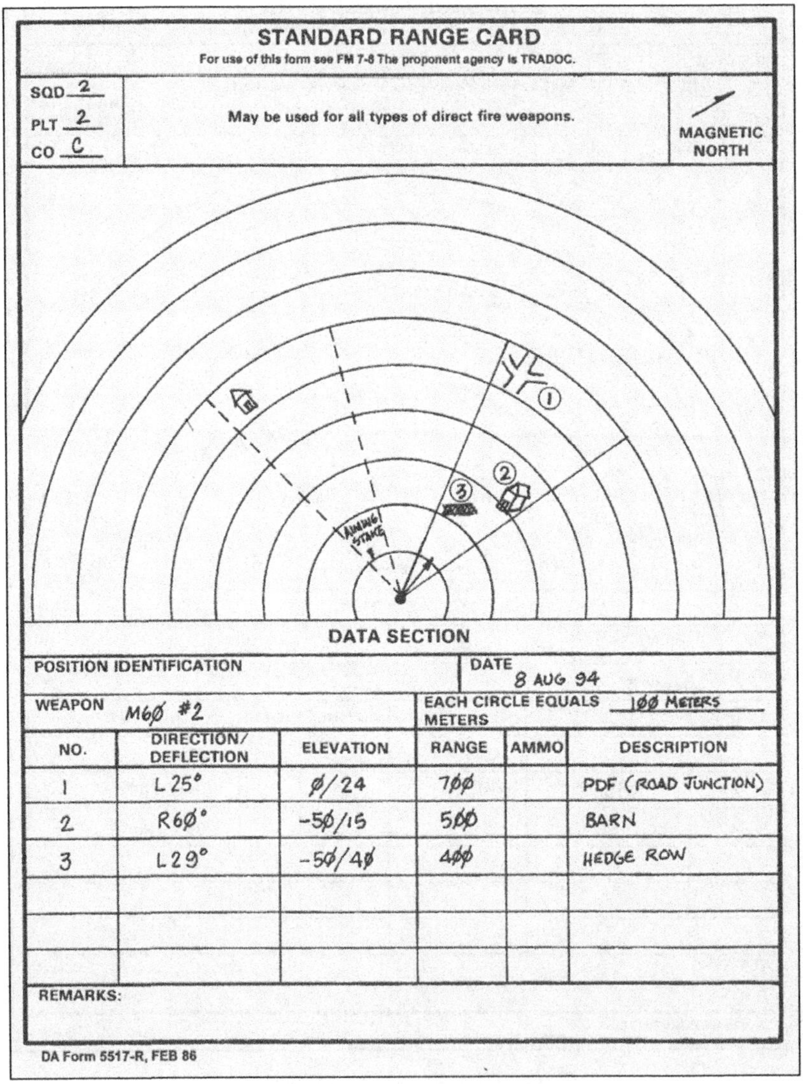

STANDARD RANGE CARD

For use of this form see FM 7-8 The proponent agency is TRADOC.

SQD 2
PLT 2
CO C

May be used for all types of direct fire weapons.

MAGNETIC NORTH

DATA SECTION

POSITION IDENTIFICATION

DATE 8 AUG 94

WEAPON M6Ø #2

EACH CIRCLE EQUALS 1ØØ METERS METERS

NO.	DIRECTION/ DEFLECTION	ELEVATION	RANGE	AMMO	DESCRIPTION
1	L 25°	Ø/24	7ØØ		PDF (ROAD JUNCTION)
2	R6Ø°	-5Ø/15	5ØØ		BARN
3	L29°	-5Ø/4Ø	4ØØ		HEDGE ROW

REMARKS:

DA Form 5517-R, FEB 86

Figure 5-23. Principal direction of fire.

(8) The position is oriented with a prominent terrain feature (recognizable on a map) by obtaining a magnetic azimuth to or from the terrain feature to the position. A line is

drawn between these two points. Arrow barbs are drawn along this line pointing in the direction the magnetic azimuth was taken. The magnetic azimuth is recorded in mils or degrees below the line. If a prominent terrain feature is not available, identify the position by using an eight-digit grid coordinate. The grid coordinate is recorded below the position on the range card.

(9) The gunner's number, unit designation (SQD, PLT, CO), and date are recorded in the upper left-hand corner. For security, do not use a unit designation higher than a company.

(10) Targets within the sector are identified. A symbol is drawn to represent the target in the appropriate place within the sector of fire. Targets in the primary sector are shown by numbers and enclosed in circles. An FPL, when assigned, is always labeled target Number 1. Other targets are assigned subsequent numbers in order of tactical importance.

(a) Wide targets in the primary sector are usually engaged in the center; however, the initial burst can be positioned anywhere the leader designates. The gunner measures the target width and records it in the data section; for example, TW-20 (target width is 20 mils). The gunner lays on the point on the target where the initial burst will be placed, and traverses to one edge of the target, while counting the clicks. He records the number of clicks he traverses and the direction he moves the muzzle; for example, TW-20/R7 (target width, 20 mils; right 7 clicks). After the initial burst, the gunner traverses 7 clicks to the right edge of the target and back to the left 20 clicks to cover the target area. To lay on the left edge of the target, the gunner records TW-20/R20.

(b) When field expedients are used with the machine gun to engage targets, they are sketched above the drawing of the target. Predetermined targets in the secondary sector are sketched on the range card and ranges to these targets are recorded below the targets but not in the data section. Field expedients should be used for targets in the secondary sector.

b. **Field Expedients.** When laying the machine gun for predetermined targets, the gunner can use field expedients as a means of engaging targets when other sources are not available. These methods are not as effective as the traversing bar and T&E mechanism method.

(1) *Base Stake Technique.* A base stake is used to define sector limits and may provide the lay for the FPL or predetermined targets along a primary or secondary sector limit. This technique is effective in all visibility conditions. The gunner uses the following steps:

(a) Defines the sector limits by laying the gun for direction along one sector limit and by emplacing a stake along the outer edge of the folded bipod legs. Rotates the legs slightly on the receiver, so that the gunner takes up the "play". Uses the same procedure for placing a stake along the opposite sector limit.

(b) Lays the machine gun along the FPL by moving the muzzle of the machine gun to a sector limit. Adjusts for elevation by driving a stake into the ground, so that the top of the stake is under the gas cylinder extension, allowing a few mils of depression to cover irregularities in the terrain.

(c) Lays the machine gun to engage other targets within a sector limit, in a primary sector by using the procedure described previously, except keeps the elevation fixed.

(2) *Notched-Stake or Tree-Crotch Technique.* The gunner uses the notched-stake or tree-crotch technique (Figure 5-24, page 5-22) with the bipod mount to engage

predetermined targets within a sector or to define sector limits. This technique is effective during all conditions of visibility, and it requires little additional material. The gunner uses the following steps:

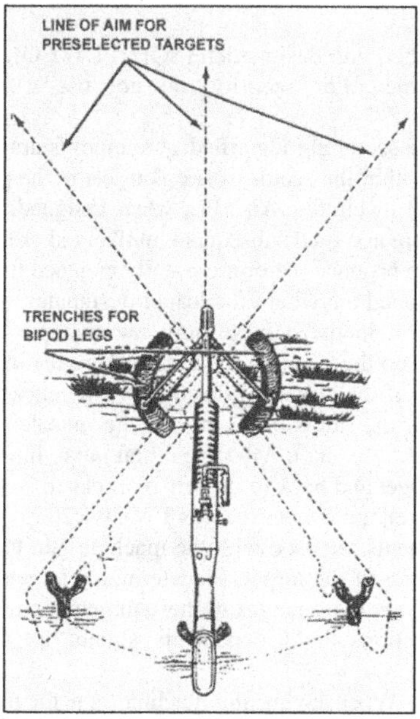

Figure 5-24. Notched-stake or tree-crotch technique.

(a) Drives either a notched stake or tree crotch into the ground where selected targets are anticipated. Places the stock of the machine gun in the nest of the stake or crotch and adjusts the weapon to hit the selected targets and to define his sector limits.

NOTE: If notched stakes and crotches are not available, tent poles can be used. It requires four poles for the left and right limits and additional poles for target areas. The gunner drives two poles in the ground in the shape of an X and then places the stock within that X as described with stakes and crotches.

(b) Digs shallow, curved trenches or grooves for the bipod feet. (These trenches allow for rotation of the bipod feet as the gunner moves the stock from one crotch or stake to another.)

(3) *Horizontal Log or Board Technique*. This technique is used with the bipod or tripod mount to mark sector limits and engage wide targets. This technique is good for all visibility conditions. It is best suited for flat, level terrain. The gunner uses the following steps:

(a) Using a bipod-mounted machine gun, places a log or board beneath the stock of the weapon, so that the stock can slide across it freely. Digs shallow, curved trenches or grooves for the bipod feet to allow rotation of the feet as he moves the stock along the log or board. (The gunner may mark the sector limits by notching or placing stops on the log or board. The gunner uses the bipod firing position and grip.)

(b) Using a tripod-mounted machine gun, places a log or board beneath the barrel, positioning it so that the barrel, when resting on the log or board is at the proper elevation to obtain grazing fire. Marks the sector limits, when appropriate, as described for the bipod in the preceding paragraph. (This technique is used only if a T&E mechanism is not available.)

Section V. FIRE CONTROL

Fire control includes all actions of the leader and soldiers in planning, preparing, and applying fire on a target. The leader selects and designates targets. He also designates the midpoint and flanks or ends of a target, unless they are obvious to the gunner. The gunner fires at the instant desired. He then adjusts fire, regulates the rate of fire, shifts from one target to another, and ceases fire. When firing, the gunner should continue to fire until the target is neutralized or until signaled to do otherwise by the leader.

5-17. METHODS OF FIRE CONTROL

The noise and confusion of battle may limit the use of some of these methods; therefore, the leader must select a method or combination of methods that will accomplish the mission.

a. **Oral**. This can be an effective method of control, but sometimes the leader may be too far away from the gunner, or the noise of the battle may make it impossible for him to hear. The primary means of the oral fire control method is the issuance of a fire command.

b. **Arm-and-Hand Signals**. This is an effective method when the gunner can see the leader. All gunners must know the standard arm-and-hand signals. The leader gets the gunner's attention and then points to the target. When the gunner returns the READY signal, the leader commands FIRE.

c. **Prearranged Signals**. These are either visual or sound signals such as casualty-producing devices, pyrotechnics, whistle blasts, or tracers. These signals should be included in SOPs. If the leader wants to shift fire at a certain time, he gives a prearranged signal such as smoke or pyrotechnics. Upon seeing the signal, the gunner shifts his fire to a prearranged point.

d. **Personal Contact**. In many situations, the leader must issue orders directly to individual soldiers and is used more than any other method by small-unit leaders. The leader must use maximum cover and concealment to keep from disclosing the position or himself.

e. **Range Cards**. When using this method of fire control, the leader must ensure all range cards are current and accurate. Once this is accomplished, the leader may designate certain targets for certain weapons with the use of limiting stakes or with fire commands. He should also designate no-fire zones or restricted fire areas to others. The key factor, in this method of fire control, is that gunners must be well-disciplined and pay attention to detail.

f. **Standing Operating Procedures**. SOPs are actions to be executed without command that are developed during the training of the squads. Their use eliminates many commands and simplifies the leader's fire control. SOPs for certain actions and commands can be developed to make gunners more effective. Some examples follow:

(1) *Observation*. The gunners continuously observe their sectors.

(2) *Fire*. Gunners open fire without command on appropriate targets that appear within their sectors.

(3) *Check*. While firing, the gunners periodically check with the leader for instructions.

(4) *Return Fire*. The gunners return enemy fire without order, concentrating on enemy automatic weapons.

(5) *Shift Fire*. Gunners shift their fires without command when more dangerous targets appear.

(6) *Rate of Fire*. When gunners engage a target, they initially fire at the rate necessary to gain and maintain fire superiority.

(7) *Mutual Support*. When two or more gunners are engaging the same target and one stops firing, the other increases the rate of fire and covers the entire target. When only one gunner is required to engage a target and the leader has alerted two or more, the gunner not firing aims on the target and follows the movements of the target. This is so that he can fire instantly should the other machine gun malfunction or cease fire before the target has been eliminated.

5-18. FIRE COMMANDS

A fire command is given to deliver effective fire on a target quickly and without confusion. When the leader decides to engage a target that is not obvious to the squad, he must provide them with the information they need to effectively engage the target. He must alert the soldiers; give a target direction, description, and range; name the method of fire; and give the command to fire. There are initial fire commands and subsequent fire commands.

a. **Initial Fire Commands.** Initial fire commands are given to adjust onto the target, change the rate of fire after a fire mission is in progress, interrupt fire, or terminate the alert.

b. **Elements**. Fire commands for all direct-fire weapons follow a pattern that includes similar elements. There are six elements in the fire command for the machine gun: alert, direction, description, range, method of fire, and command to open fire. The gunners repeat each element of fire command as it is given.

(1) *Alert*. This element prepares the gunners for further instructions. The leader may alert both gunners in the squad and may have only one fire, depending upon the situation. To alert and have both gunners fire, the leader announces, "Fire mission." If he desires to alert both gunners but have only one fire, he announces,"Gun number one, fire mission." In all cases, upon receiving the alert, the gunners load their machine guns and place them on FIRE.

(2) *Direction*. This element indicates the general direction to the target and may be given in one or a combination of the following methods.

(a) Orally. The leader orally gives the direction to the target in relation to the position of the gunner (for example, FRONT, LEFT FRONT, RIGHT FRONT).

(b) Pointing. The leader designates a small or obscure target by pointing with his finger or aiming with a weapon. When he points with his finger, a soldier standing behind him should be able to look over his shoulder and sight along his arm and index finger to locate the target. When aiming his weapon at a target, a soldier looking through the sights should be able to see the target.

(c) Tracer Ammunition. Tracer ammunition is a quick and sure method of designating a target that is not clearly visible. When using this method, the leader should first give the general direction to direct the gunner's attention to the target area. To prevent the loss of surprise when using tracer ammunition, the leader does not fire until he has given all elements of the fire command except the command to fire. The leader may fire his individual weapon. The firing of the tracer(s) then becomes the last element of the fire command, and it is the signal to open fire.

NOTE: Soldiers must be aware that with the night vision device, temporary blindness "white out" may occur when firing tracer ammunition at night or when exposed to other external light sources. Lens covers may reduce this effect.

EXAMPLE

FIRE MISSION
FRONT
FIVE HUNDRED
WATCH MY TRACER(S)

(d) Reference Points. Another way to designate obscure targets is to use easy-to-recognize reference points. All leaders and gunners must know terrain features and the terminology used to describe them (FM 21-26). When using a reference point, the word "reference" precedes its description. This is done to avoid confusion. The general direction to the reference point should be given.

EXAMPLE

FIRE MISSION
FRONT
REFERENCE: BUNKER, CENTER MASS
TARGET: TROOPS EXTENDING SHORT ONE HUNDRED, OVER ONE HUNDRED
FOUR HUNDRED
FIRE

(Sometimes the reference point may be outside the target area).

EXAMPLE

FIRE MISSION
FRONT
REFERENCE: BUNKER, RIGHT FOUR FINGERS, CENTER MASS
TARGET: TROOPS EXTENDING SHORT ONE HUNDRED, OVER ONE HUNDRED
THREE HUNDRED
SEARCH
AT MY COMMAND
FIRE

(Sometimes a target must be designated by using successive reference points).

EXAMPLE

GUN NUMBER ONE, FIRE MISSION
RIGHT FRONT
REFERENCE: RED-ROOFED HOUSE, LEFT TO HAYSTACK, LEFT TO BARN

(Finger measurements can be used to direct the gunner's attention to the right or left of reference points).

EXAMPLE

FIRE MISSION
LEFT FRONT
REFERENCE: CROSSROADS, RIGHT FOUR FINGERS

(3) *Description.* The target description creates a picture of the target in the minds of the gunners. To properly apply their fire, the soldiers must know the type of target they are to engage. The leader should describe it briefly. If the target is obvious, no description is necessary.

(4) *Range.* The leader always announces the estimated range to the target. The range is given, so the gunner knows how far to look for the target and what range setting to put on the rear sight. Range is announced in meters; however, since the meter is the standard unit of range measurement, the word "meters" is not used. With machine gun's, the range is determined and announced to the nearest hundred or thousand (in other words, THREE HUNDRED, or ONE THOUSAND).

EXAMPLE

FIRE MISSION
FRONT
REFERENCE: KNOCKED-OUT TANK, LEFT TWO FINGERS
TARGET: TROOPS
THREE HUNDRED

(5) Method of Fire. This element includes manipulation and rate of fire. Manipulation prescribes the class of fire with respect to the weapon. It is announced as

FIXED, TRAVERSE, SEARCH, or TRAVERSE AND SEARCH. Rate controls the volume of fire (sustained, rapid, and cyclic). Normally, the gunner uses the sustained rate of fire. The rate of fire is omitted from the fire command. The method of fire for the machine gun is usually 3- to 5-round bursts (M249) and 6- to 9-round bursts (M60/M240B).

EXAMPLE

FIRE MISSION
FRONT
REFERENCE: KNOCKED-OUT TANK, LEFT TWO FINGERS
TARGET: TROOPS
THREE HUNDRED
TRAVERSE

(6) *Command to Open Fire.* When fire is to be withheld so that surprise fire can be delivered on a target or to ensure that both gunners open fire at the same time, the leader may preface the command to commence firing with AT MY COMMAND or AT MY SIGNAL. When the gunners are ready to engage the target, they report READY to the leader. The leader then gives the command FIRE at the specific time desired.

EXAMPLE

FIRE MISSION
FRONT
TROOPS
FOUR HUNDRED
AT MY COMMAND or AT MY SIGNAL (The leader pauses until the gunners are ready and fire is desired.)
FIRE (The gunners fire on prearranged signal.)

If immediate fire is required, the command FIRE is given without pause and the gunners fire as soon as they are ready.

c. **Subsequent Fire Commands**. These fire commands are used to make adjustments in direction and elevation, to change rates of fire after a fire mission is in progress, to interrupt fires, or to terminate the alert. If the gunner fails to properly engage a target, the leader must promptly correct him by announcing or signaling the desired changes. When these changes are given, the gunner makes the corrections and resumes firing without further command.

(1) Adjustments in direction and elevation with the machine gun are always given in meters; one finger is used to indicate 1 meter and so on. Adjustment for direction is given first. For example: RIGHT ONE ZERO METERS or LEFT FIVE METERS. Adjustment for elevation is given next. For example: ADD FIVE METERS or DROP ONE FIVE METERS. These changes may be given orally or with arm-and-hand signals.

(2) Changes in the rate of fire are given orally or by arm-and-hand signals.

(3) To interrupt firing, the leader announces "Cease fire," or he signals to cease fire. The gunners remain on the alert. They resume firing when given the command FIRE.

(4) To terminate the alert, the leader announces "Cease fire, end of mission."

d. **Doubtful Elements and Corrections**. When the gunner is in doubt about any element of the fire command, he replies "Say again range, target." The leader then announces "The command was," repeats the element in question, and continues with the fire command.

(1) When the leader makes an error in the initial fire command, he corrects it by announcing "Correction," and then gives the corrected element.

EXAMPLE
FIRE MISSION
FRONT
TROOPS
SIX HUNDRED
CORRECTION
THREE HUNDRED
TRAVERSE
AT MY COMMAND

(2) When the leader makes an error in the subsequent fire command, he may correct it by announcing "Correction," and then repeating the entire subsequent fire command.

EXAMPLE
LEFT FIVE METERS, DROP ONE METER
CORRECTION
LEFT FIVE METERS, DROP ONE HUNDRED METERS

e. **Abbreviated Fire Commands**. Fire commands need not be complete to be effective. In combat, the leader gives only the elements necessary to place fire on a target quickly and without confusion. During training, however, he should use all of the elements to get gunners in the habit of thinking and reacting properly when a target is to be engaged. After the gunner's initial training in fire commands, he should be taught to react to abbreviated fire commands, using one of the following methods.

(1) *Oral*. The leader may want to place the fire of one machine gun on an enemy machine gun.

EXAMPLE
GUN NUMBER ONE, FIRE MISSION
MACHINE GUN
FOUR HUNDRED
FIRE

(2) *Arm-and-Hand Signals*. Battlefield noise and the distance between the gunner and the leader often make it necessary to use arm-and-hand signals to control fire (Figure 5-25). When an action or movement is to be executed by only one of the gunners, a preliminary signal is given to that gunner only. The following are commonly used signals for fire control.

Figure 5-25. Arm-and-hand signals.

(a) Ready. The gunner indicates that he is ready to fire by yelling "Up" or having the assistant gunner raise his hand above his head toward the leader.

(b) Commence Firing or Change Rate of Firing. The leader brings his hand (palm down) to the front of his body about waist level, and moves it horizontally in front of his body. To signal an increase in the rate of fire, he increases the speed of the hand movement to signal slower fire, and he decreases the speed of the hand movement.

(c) Change Direction or Elevation. The leader extends his arm and hand in the new direction and indicates the amount of change necessary by the number of fingers extended. The fingers must be spread, so the gunner can easily see the number of fingers extended. Each finger indicates 1 meter of change for the weapon. If the desired change is more than 5 meters, the leader extends his hand the number of times necessary to indicate the total amount of change. For example, *right nine* would be indicated by extending the hand once with five fingers showing and a second time with four fingers showing for a total of nine fingers.

(d) Interrupt or Cease Firing. The leader raises his arm and hand (palm outward) in front of his forehead and brings it downward sharply.

(e) Other Signals. The leader can devise other signals to control his weapons. A detailed description of arm-and-hand signals is given in FM 21-60.

Section VI. RANGE DETERMINATION

During combat, ranges are seldom known. Poor visibility and damp ground often make adjustment of fire by observation difficult if not impossible. Therefore, correct range determination is critical for accurate effective fire. Range estimation and lateral distance measurements are two methods used to determine the range to the target.

5-19. RANGE ESTIMATION

Range estimation is determining the distance between two points. In most situations, one of these points is the gunner's own position; the other point may be a target or prominent terrain feature. The gunner must accurately determine the range to set the sights and effectively fire on a target with the first burst.

a. Not only does the accurate estimation of range affect marksmanship, but it is also required in the reporting of information and the adjustment of artillery and mortar fire (Table 5-1).

FACTORS AFFECTING RANGE ESTIMATION	FACTORS CAUSING UNDERESTIMATION	FACTORS CAUSING OVERESTIMATION
The clearness of outline and details of the target.	When most of the target is visible and offers	When only a small portion of the target is small in relation to its surroundings.
Nature of terrain or position of the gunner.	When looking across a depression that is mostly hidden from view. When looking downward form high ground. When looking down a straight, open road or along a railroad. When looking over uniform surfaces like water, snow, desert, or grain fields.	When looking across a depres-sion that is totally visible. When looking from low ground toward high ground. When vision is narrowly confined as in streets, draws, or forest trails.

Table 5-1. Factors of range estimation.

FACTORS AFFECTING RANGE ESTIMATION	FACTORS CAUSING UNDERESTIMATION	FACTORS CAUSING OVERESTIMATION
Light and atmosphere.	In bright light or when the sun is shining from behind the gunner. When the target is in sharp contrast with the silhouette because of its size, shape, or color. When seen in the clear air of high altitudes.	In poor light such as dawn and dusk; in rain, snow, fog; or when the sun is in the gunner's eyes. When the target blends into the background or terrain.

Table 5-1. Factors of range estimation (continued).

b. There are several methods of estimating range. They include measuring distance on a map, pacing the distance between two points, and using an optical range finder. The gunner does not usually have a map and rarely has access to an optical range finder. He can pace the distance between two points if the enemy is not within range. Firing rounds to determine the range is not desirable, since it may reveal your position to the enemy. Most of the time, the gunner must use techniques that do not require equipment and can be used without exposing himself or revealing his position. There are two methods that meet these requirements: the appearance-of-objects and the 100-meter-unit-of-measure.

(1) *Appearance-of-Objects Method.* This method is a means of estimating range by the size and other characteristic details of the object.

(a) This is a common method of determining distances and is used most often. For example, a motorist trying to pass another car must judge the distance of oncoming vehicles based on his knowledge of how vehicles appear at various distances. In this example, the motorist is not interested in precise distances but only in having enough road space to safely pass the car. Suppose, however, the motorist knew that at a distance of 1 kilometer, an oncoming vehicle appeared to be 1 centimeter between headlights. Then, anytime he saw other oncoming vehicles that fit these dimensions, he would know they were about 1 kilometer away. This technique can be used by a gunner to estimate ranges on the battlefield. If the gunner knows the characteristic size and detail of men and equipment at known ranges, he can compare these characteristics to similar objects at unknown ranges. When characteristics match, so does the range.

(b) To use the appearance-of-objects method with any degree of accuracy, the gunner must know the characteristic details of objects as they appear at various ranges. For example, the gunner should study the appearance of a man standing at a range of 100 meters. He fixes the man's appearance firmly in his mind, carefully noting details of size and the characteristics of uniform and equipment. Next, he studies the same man in a kneeling position and then in a prone position. By comparing the appearance of the man at known ranges from 100 to 500 meters, the gunner can establish a series of mental images that will help determine range on unfamiliar terrain. Training should also be conducted in the appearance of other familiar objects such as weapons or vehicles. Because the successful use of this method depends upon visibility, anything that limits visibility (such as weather, smoke, or darkness) will also limit the effectiveness of this method.

(2) *100-Meter-Unit-of-Measure Method.* To use this method, the gunner visualizes a distance of 100-meters on the ground. For ranges up to 500-meters (Figure 5-26), he determines the number of 100-meter increments between the two points he wishes to measure. Beyond 500-meters (Figure 5-27), he selects a point halfway to the target, determines the number of 100-meter increments to the halfway point, and then doubles it to find the range to the target.

(a) During training, the gunner must become familiar with the effect that sloping terrain has on the appearance of a 100-meter increment. Terrain that slopes upward gives the illusion of longer distance, and observers have a tendency to overestimate the 100-meter increment. Terrain that slopes downward gives the illusion of shorter distance. In this case, the gunner's tendency is to underestimate the 100-meter increment and thus underestimate the range.

(b) Proficiency in the 100-meter-unit-of-measure method requires constant practice. When training in this technique, the gunner should make frequent comparisons between the range as determined by the himself and by pacing or other accurate means of measurement. The best training technique is to pace the range after he has visually determined it. In this way, he discovers the actual range for himself, which makes a much greater impression than if he is told the correct range.

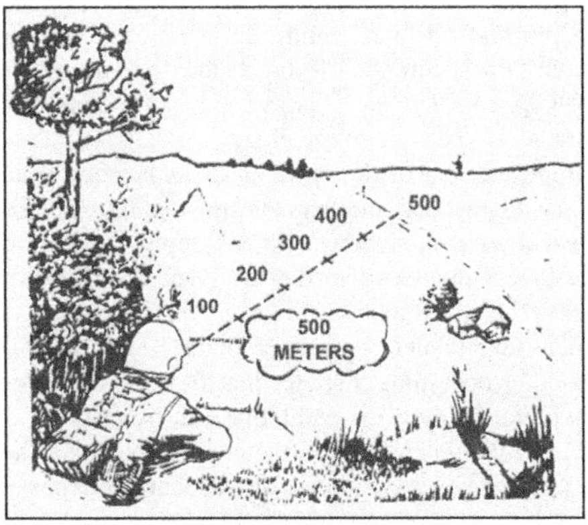

Figure 5-26. Applying the 100-meter-unit-of-measure method for ranges up to 500 meters.

Figure 5-27. Applying the 100-meter-unit-of-measure method for ranges up to 800 meters.

(c) A limitation of the 100-meter-unit-of-measure method is that its accuracy is directly related to the amount of terrain visible to the gunner. This is particularly true at greater ranges. If a target appears at a range of 500 meters or more and the gunner can only see a portion of the ground between himself and the target, it becomes difficult to use the 100-meter-unit-of-measure method of range estimation with any degree of accuracy.

(3) *Combination of Methods*. Under ideal conditions, either the appearance-of-objects method or 100-meter-unit-of-measure method is an effective method of estimating range. However, ideal conditions rarely exist on the battlefield, so the gunner must use a combination of methods. The terrain might limit the use of the appearance-of-objects method. For example, a gunner may not be able to see all the terrain out to the target; however, he may see enough to get a general idea of the distance. A slight haze may obscure many of the target details, but the gunner should still be able to judge its size. By carefully considering the ranges estimated by both methods, an experienced gunner should arrive at a figure close to the true range. The best way to reduce errors using these two methods is to train often.

5-20. LATERAL DISTANCE MEASUREMENT

In addition to estimating range accurately, the gunner needs a quick method of measuring lateral distance (right or left) from a reference point to a target. When the machine gun is tripod-mounted, width can be measured by aiming on a point, manipulating the traversing handwheel, and counting the clicks from one point of aim to another point of aim. Each click equals 1 mil and is equivalent to 1 meter at 1,000 meters, or half a meter at 500 meters. When the machine gun is bipod-mounted, the gunner can use his fingers to measure the lateral distance between a reference point and a target. He extends his arm with his palm outward, lowers his fingers, and locks his elbow. Then, he closes one eye, raises his index finger, and sights along its edge, placing the edge of his finger so that it appears to be along the flank of the target or reference point. The gunner fills the space remaining between the points by raising his fingers until the space is covered. He states the measurement from the reference point to the target as being one or more fingers, depending upon how many fingers are raised to cover this distance.

Section VII. ADVANCED GUNNERY

Once the gunner masters the four fundamentals of marksmanship in the prone position and fighting position, he needs practice in engaging targets that will most likely replicate the battlefield. The advanced gunnery field firing exercise for the gunner exposes him to different types of targets at various ranges to simulate combat conditions.

5-21. OBJECTIVES

The objectives of this training are to reinforce the fundamentals and increase the effectiveness of the gunner by building his confidence (*not for qualification*). He must acquire targets quickly and deliver an accurate volume of fire.

5-22. ORGANIZATION

The unit is assembled in the bleachers, given the training objectives, a range briefing, and a safety briefing. Gunners are then organized into firing orders with a gunner and an assistant gunner. (Concurrent training stations must be set up for those soldiers not on the firing line.)

5-23. AMMUNITION

This exercise requires 392 rounds of 7.62-mm linked ammunition (zero is included). The gunner is allotted two 7-round bursts for each target and fires twice.

5-24. FIRING SEQUENCE

The sequence of firing is to be conducted IAW Firing Table IV (Table 5-2). Commanders may score their soldiers to determine their most proficient gunners, to assess the marksmanship program, and to encourage competition.

a. **Task 1, Field Zeroing the 500-Meter, Single E-Type Silhouette**. The gunner is allocated 28 rounds of ammunition.

b. **Task 2, Engage Single E-Type Silhouettes From the Prone and Fighting Position, Bipod Supported (Point Targets) at Various Ranges.** The gunner will use his NBC equipment (Mask and Gloves). Targets are at 100, 200, 250, 300, and 400 meters. The gunner is allocated 70 rounds of ammunition.

c. **Task 3, Engage Double E-Type Silhouettes (Automatic Weapon Positions) at Various Ranges**. Targets are at 250, 300, 400, 500, and 600 meters. The gunner is allocated 70 rounds of ammunition.

d. **Task 4, Engage Linear E-Type Silhouettes (Troops on Line) at Various Ranges.** The gunner uses his NBC equipment (Mask and Gloves). Targets are at 300 and 600 meters. The gunner is allocated 28 rounds of ammunition.

WARNING

The M240B is carried loaded with the bolt locked to the *rear* in *tactical situations* where noise discipline is critical to the success of the mission. Trained gun crews are the only personnel authorized to load the M240B and only when command directs the crew to do so. During *normal training exercises*, the M240B is loaded and carried with the bolt in the *forward position*.

TASK	RANGE (M)	TIME	TOTAL ROUNDS PER INDIVIDUAL	TARGET	AMMO	TYPE FIRE
1	500	No limit	28	E-type silhouette	4:1	Zeroing 7-round bursts
2*	100 200 250 300 400	60 seconds	70	Single E-type silhouette	4:1	7-round bursts
3	250 300 400 500 600	120 seconds	70	Single E-type silhouette	4:1	7-round bursts
4*	300 600	120 seconds	28	Linear target Single E-type 1 meter apart	4:1	7-round bursts
* Indicates tasks fired with protective mask and gloves as a minimum.						

Table 5-2. Firing Table II.

5-25. ALTERNATE FIRING POSITIONS

All gunners must master the bipod-supported prone firing fighting position, and tripod-supported prone firing position to be effective. But it is also equally important that they know other positions. Each gunner must be trained to assume different positions quickly during various combat conditions. The situation determines the appropriate position. The gunner must establish his position so that he can effectively observe and engage the target yet minimize his exposure to enemy fire.

a. **Positions**. The underarm firing position and the hip firing position are used.

(1) *Underarm Firing Position*. This position is used almost exclusively when moving in and around the objective during the assault (Figure 5-28). To assume this position, the gunner—

(a) Puts the bipod legs and rear sight down for instant use in the prone position if necessary.

(b) Faces the target with his feet spread about shoulder width apart.

(c) Places his left foot in front of the right with most of his weight on his left foot.

(d) Bends both legs at the knees and leans forward at the waist.

(e) With his right hand, firmly grasps the pistol grip, and with his right forearm holds the stock firmly against the side of his body at a point between his armpit and waist.

(f) With his left hand, grasps the handguard firmly.

(g) Points his left foot in the direction of the target while his right foot provides stability.

(h) Depresses the muzzle of the machine gun slightly so the strike of rounds can be observed. (This reduces shooting high and takes advantage of ricochets.)

(i) Leans toward the target before and during firing.

Figure 5-28. Underarm firing position.

(2) *Hip Firing Position.* This position is used when closing with the enemy, when a heavy volume of fire in the target area is required, and when rapid movement is not necessary (Figure 5-29). The only differences between this position and the underarm position are that the gunner—

(a) Holds the rear of the stock firmly against the forward position of his right thigh.

(b) Extends his arms fully downward.

Figure 5-29. Hip firing position.

b. **Alternate Firing Position Exercises.** The assault fire exercise challenges the gunner. It consists of point and area targets during a variety of conditions replicating the battlefield. These exercises, which involve fire and maneuver, must be carefully controlled for safety purposes.

(1) *Objectives.* This exercise gives the gunner practice on engaging targets as quickly as possible, using any of the alternate firing positions.

(2) *Organization.* The unit is assembled in the bleachers, given instructions, and briefed on training that will be conducted while they are on the range. After the briefing, they are organized into firing orders and moved to firing lanes. Lanes are conducted and used IAW local range policies.

(3) *Ammunition.* This exercise requires a total of 168 rounds of 7.62-mm linked ammunition. The gunner is allowed two bursts per exposure, and he is also required to conduct at least one rapid reload during the exercise and the gunner will fire this twice. The commander has the option of when the rapid reload may take place. Ammunition is configured into two belts of any size that requires the gunner to reload.

(4) *Firing Sequence.* The sequence of firing is conducted IAW Firing Table V (Table 5-3). The suggested sequence of firing is as follows.

(a) Task 1, Dry Fire Walk-Through. Upon his arrival at the firing position, the gunner walks through his respective lane to become familiar with the targets. No ammunition is fired at this time. When he returns, he draws his ammunition.

NOTE: Commanders ensure that ammunition is used in such a manner that the gunner is required to rapidly reload sometime during his movement phases.

(b) Task 2, Engage Single E-Type Silhouette from the Hip-Firing Position. After being issued the ammunition, the gunner begins his movement. When targets are exposed, he uses the hip-firing technique. He is given a single exposed target at a distance of 25 meters, with an exposure time of 5 seconds.

(c) Task 3, Engage Single E-Type Silhouettes from the Underarm-Firing Position. As the gunner continues to move through the course, he is given two single exposed E-type silhouettes at distances of 50 and 25 meters, where he engages each silhouette using the underarm-firing position. The exposure time for each target is 5 seconds.

(d) Task 4, Engage Single E-Type Silhouettes from the Underarm or Hip-Firing Position. Once the gunner reaches this point, he is be given three single exposed E-type silhouettes at distances of 25, 50, and 75 meters. The exposure time for each target is 5 seconds.

NOTE: The commander may integrate firing under NBC conditions for selected tasks while negotiating the course, or he may conduct the course during limited visibility unaided.

TASK	RANGE (M)	TIME	TOTAL ROUNDS PER INDIVIDUAL	TARGET	AMMO	TYPE FIRE
1	NA	No limit	NA	NA	NA	NA
2	25	5 seconds	14	Single E-type silhouette	4:1	7-round bursts
3	50	10 seconds	28	Single E-type silhouette	4:1	7-round bursts
4	25 50 75	15 seconds	42	Single E-type silhouette	4:1	7-round bursts

Table 5-3. Firing Table VIII.

5-26. MOVEMENT, SPEED, AND ALIGNMENT

The gunner must keep up with the other soldiers of the assaulting element through individual movement techniques. To do this, he moves as rapidly as possible, consistent with his ability to fire accurately and maintain alignment.

5-27. RELOADING

The gunner must reload rapidly to avoid lulls in the firing. This can be achieved by practicing and by applying the following techniques.

a. Before the assault, the gunner conducts prefire checks on the machine gun. He inspects ammunition to ensure that it is clean and serviceable, and he checks the box for serviceability.

b. During the assault, the gunner must continue moving forward and reload as rapidly as possible. The sling allows the gunner to reload using both hands.

CHAPTER 6
TRAIN-THE-TRAINER PROGRAM

This chapter helps the leader develop a good train-the-trainer program. The goal of this program is to certify qualified machine gun trainers and achieve a high state of combat readiness. Knowledgeable, small-unit leaders and trainers are the key to any successful marksmanship training program; however, the entire leadership must be involved in the execution of training to standard. An effective train-the-trainer program reflects the priority, emphasis, and interest of the leaders and trainers to see that execution of training to standard is achieved.

The objectives of the train-the-trainer program are to develop in every machine gun trainer the confidence, willingness, knowledge, and skills required to consistently train their soldiers to be effective in combat.

6-1. MISSION-ESSENTIAL TASK LIST

The unit's combat mission is considered when establishing training priorities. This not only applies to the tasks selected but to the conditions under which the tasks are to be preformed. The tasks for the METL are developed for both defensive and offensive operations. Machine gun marksmanship will be a critical individual, crew and leader task. Each commander should develop a METL and organize a training program that devotes adequate time to conduct marksmanship training.

6-2. TRAINER ASSESSMENT

The leaders are also involved in determining the proficiency of potential trainers by reviewing the following information.

a. **Selection.** Trainers should be selected from the most highly qualified soldiers available within the unit. These soldiers should be knowledgeable of the machine guns, a high degree of proficiency in applying the fundamentals, and demonstrate a motivated attitude for marksmanship training. The leaders must ensure that a high level of proficiency is maintained. Knowledgeable trainers are the key to any successful training program.

b. **Trainer Course.** Once the leaders have identified these soldiers possessing the required knowledge, skills, and motivation in machine gun marksmanship, they must then ensure this knowledge can be effectively taught to other soldiers.

c. **Training**. There are several available means that may be used in the progression of trainer training or that can easily be tailored to the certain needs of the command. The more time and training initially invested, the better the trainer will be. The leaders should periodically evaluate each trainer and replace any that loses his motivation or desire. To maintain interest in the program, leaders should have a way of promoting competition and awards the best trainer.

6-3. ASSISTANT TRAINERS AND CADRE COACHES

Assisting the trainer and coaching a soldier to fire the machine gun are highly technical jobs that must be done well. The most valuable soldiers in the program are those who not only have obtained a high standard, but those who can effectively teach this knowledge to others. Once the individual is consistent in this train-the-trainer program, he can then develop into a competent assistant trainer. It is worth the effort to train these individuals to become a successful assistant trainer, because experience has shown that such training also develops leadership ability.

a. The primary responsibility of assistant trainers is to train individuals in the effective use of the machine gun. In addition, assistant trainers are responsible for enforcing safety regulations. They must maintain strict discipline on the firing lane at all times and constantly enforce compliance with the range regulations and training guidance.

b. To be an assistant trainer, a soldier must know the principles of accurate firing and coaching techniques, and he must have the following qualifications as well.

(1) *Knowledge.* The assistant trainer must know this manual. He must be prepared to accurately answer any question on the subject of marksmanship. Then, he must develop his ability to observe the actions of the soldier quickly and to correct them with sound recommendations.

(2) *Patience.* Gunners can be persuaded to accept the principles and procedures by patient repetition and demonstration.

(3) *Understanding.* The assistant trainers that have good "firing lane manners" enhance the success of training. Training new gunners is stressful to the soldiers and the trainers. The soldier may be sensitive to abruptness, impatience, or a lack of sympathy with his difficulties; and he will immediately react unfavorably to evidence of such attitude on the part of the coach.

(4) *Consideration.* Most soldiers, even those who do not fire well, enjoy firing and begin with a positive interest in their performance on the range. If the assistant trainer is considerate of his pupils' feelings from the beginning and encourages them throughout their training, he will find training a pleasant and rewarding duty.

(5) *Ability to Maintain Respect.* When a soldier is assigned the duties of assistant trainer, he too is an expert and should receive the same respect as the primary trainer. The assistant trainer must retain that respect throughout his contact with students by showing a thorough knowledge of his subject in a dignified manner.

(6) *Alertness.* Even the most capable student may forget or neglect some essential point in his instruction in the excitement of firing on the range. The assistant trainer must always be alert and patiently correct him as often as necessary. He must keep the gunners encouraged throughout the instruction by making the most of all progress. The assistant trainer must not allow his students to become discouraged or to lose interest.

(7) *Helpful Attitude.* When assisting on the range, as in most other lines of instruction, a combative, hard-boiled attitude is rarely effective.

(8) *Ability to Provide Encouragement.* The assistant trainer can encourage his soldiers by convincing them there is no mystery about good firing: The weapon and ammunition are mechanically developed for accuracy. The assistant trainer is there to assist the gunner and to help him gain the experience that produces a good marksman.

6-4. PROGRAM PHASES

Since firing is a learning process, certain prerequisites must be satisfied before a trainer passes from one phase of marksmanship to another. The trainer must qualify with the machine gun (the trainer must pass all tasks and qualify on the 10-meter and transition fire) that he will be teaching to the machine gunners in his unit. To obtain maximum results on the battlefield, the machine gunners are trained in the fundamentals before they engage a combat target. The phases of the train-the-trainer program are to develop this structure in the most progressive manner. They are sequenced to train-the-trainer in teaching tasks necessary to produce a quality machine gunner and a machine gun team.

a. **Preliminary Gunnery.** The machine gunners must receive this training before live firing. It is during this phase that sound foundations of good firing principles are constructed, reviewed, and reinforced. The degree of proficiency obtained or retained by the machine gunners depends on the foundation built during this phase. Correct firing and good safety habits must become natural. Drilling of the fundamentals and continued leader emphasis will bring the greatest return in the shortest time. Proper firing is a physical skill, which must be learned. When practiced, the process becomes a learned skill that will be retained. However, good firing is a perishable skill. All machine gunners must periodically familiarize themselves with the fundamentals regardless of their years of marksmanship experience. Even experienced machine gunners will develop a deficiency in applying certain fundamentals (for example, overconfidence).

b. **Basic Gunnery.** This training teaches the trainers how to qualify, plan, set up, and conduct 10-meter and transition firing exercises on the available ranges.

c. **Advanced Gunnery.** This training teaches the trainer how to develop teamwork among the machine gun teams. This training also gives them confidence in their ability to deliver a large volume of accurate fire against targets. During this phase, the trainer is responsible for the conduct of assault firing exercises. These exercises consist of assault fire, NBC assault fire, and field fire on available ranges.

6-5. TRAINING TASKS

This paragraph assists trainers in effectively training soldiers assigned to any machine gun team. It explains the tasks, organization, equipment needed, and instruction sequence for the three phases of gunnery. However, unit SOPs or post regulations may direct increases or decreases in these prescribed requirements.

a. **Phase I, Preliminary Gunnery Training**. This phase covers the basics that each trainer must know to teach the general care and maintenance of the machine gun.

(1) *Task 1: (071-312-3025) Disassemble the Machine Gun.* The trainer stresses that this task is not done hurriedly, because the soldiers may damage parts of the machine gun.

(a) Equipment Needed. A table is needed for placement of the machine gun so that the soldiers may better see the removal of parts of the gun. Nomenclature charts or mats are material aids in explaining mechanical training, and they help the gunners learn the nomenclature of parts.

(b) Class Organization. One assistant trainer is assigned for each group if possible; otherwise, assistant trainers are placed where they can supervise assigned groups.

(c) Sequence of Training. The trainer presents a brief history of the machine guns. He stresses the combat role it has played and the mission it is assigned. He emphasizes the purpose, scope, and importance of the instruction to be presented. He gives a brief description of the operation, general data, and exterior nomenclature of the machine gun. Assistant trainers should clear and disassemble the machine gun as the trainer explains the procedures. The gunners then practice until they become skilled in disassembly and can demonstrate this task to standards to either the trainer or assistant trainers. This practice often encourages the gunners to practice during free time and develops their individual skill and initiative.

(2) *Task 2: (071-025-0001) Perform Operator Maintenance on the Machine Gun.* The subtasks are inspect, clean, and lubricate.

(a) Equipment Needed. The trainer needs one cleaning rod and one dummy round for each machine gun, bore cleaner, (one cleaning reamer, one combination regulator scraper, one combination scraper and extractor tool), lubricating oil, patches, and rags. The trainer displays all the available cleaning materials, lubricants, and rust preventatives.

(b) Class Organization. One assistant trainer is assigned for each group if possible; otherwise, the assistant trainers are placed where they can supervise assigned groups.

(c) Sequence of Training. The trainer emphasizes meticulous cleaning, lubrication, inspection, and preventive maintenance to ensure performance. The care and cleaning period is used to further the gunner's knowledge of the nomenclature and skill in disassembly. The trainer emphasizes inspection, care, and preventive maintenance during combat conditions, which is the final test of the weapon maintenance program. Practical work is conducted. The trainer points out the differences in care and cleaning following an NBC attack. He emphasizes the importance of frequent inspection as a means of ensuring proper maintenance of the machine guns.

(3) *Task 3: (071-312-4025) Assemble the Machine Gun.*

(a) Equipment Needed. A table is needed for placement of the machine gun so that the soldiers may better see the removal of the parts. Nomenclature charts or mats are material aids in explaining mechanical training, and they help the gunners to learn the nomenclature of parts.

(b) Class Organization. One assistant trainer is assigned for each group if possible; otherwise, assistant trainers are placed where they can supervise assigned groups.

(c) Sequence of Training. Assistant trainers should assemble the machine gun as the trainer explains the procedures. The gunners practice until they become proficient in assembly and can demonstrate this task to standards to either the trainer or assistant trainer.

(4) *Task 4: (071-312-4026) Explain the Operation of the Machine Gun.*

(a) Equipment Needed. One machine gun placed on a table; one belt of six dummy rounds (5.56-mm or 7.62-mm linked), and a cleaning rod for each gunner.

(b) Class Organization. One assistant trainer is assigned for each group if possible; otherwise, assistant trainers are placed where they can supervise assigned groups.

(c) Sequence of Training. The trainer explains and the assistant trainer demonstrates loading, unloading, and clearing the machine gun. The trainer stresses the safety factors involved. The assistant trainer demonstrates letting the bolt go forward when the barrel is out of the machine gun. (Damage could be done to the chamber or the face of the bolt if the barrel is left in.)

(5) *Task 5: (071-312-3026) Explain the Functioning of the Machine Gun.*

(a) Equipment Needed. One machine gun for each gunner as in previous mechanical training instruction. Graphic training aids are useful if the class is about platoon size; otherwise, GTAs may be made available for study and discussion during breaks.

(b) Class Organization. One assistant trainer is assigned for each group if possible; otherwise, assistant trainers are placed where they can supervise assigned groups.

(c) Sequence of Training. Functioning is divided into eight steps—feeding, chambering, locking, firing, unlocking, extracting, ejecting, and cocking. The assistant trainer duplicates each demonstration on the machine gun with each group. Functioning is taught by seeing how the parts work, rather than by memorizing the text. The trainer tests retention of the training by asking questions concerning the steps of functioning.

(6) *Task 6: (071-312-3029) Explain Malfunction, Stoppage, Immediate Action, and Remedial Action.*

(a) Equipment Needed. A table is needed for placement of the machine gun so that the soldiers may better see the removal of parts. Nomenclature charts or mats are material aids in explaining mechanical training, and they help the gunners learn the nomenclature of parts.

(b) Class Organization. One assistant trainer is assigned for each group if possible; otherwise, assistant trainers are placed where they can supervise assigned groups.

(c) Sequence of Training. Malfunction and stoppages charts may be used as a guide in presenting instruction. These charts may be available from the local Training Support Center or the trainer can use tables of the appropriate chapters for the machine gun. The trainer stresses precision in detecting the cause and reducing stoppages. As the gunners progress, the trainer has them concentrate on speed in applying immediate action and other methods of reducing a stoppage. He stresses safety precautions in connection with a hangfire or cook off.

(7) *Task 7: (071-312-3030) Explain the Procedures for Sight Adjustments and Mechanical Zero.*

(a) Equipment Needed. A table is needed for placement of the machine gun so that the soldiers may better see the removal of parts. Nomenclature charts or mats are material aids in explaining mechanical training, and they help the gunners learn the nomenclature of parts, plus tripod. (Front sight adjustment tool for the M240B.)

(b) Class Organization. One assistant trainer is assigned for each group if possible; otherwise, assistant trainers are placed where they can supervise assigned groups.

(c) Sequence of Training. Assistant trainers demonstrate, while trainers describe the proper techniques to mechanically zero the machine gun. The trainer then describes sight adjustments. The trainer emphasizes the number of clicks in relation to the targets for M240B only.

(8) *Task 8: (071-020-0006) Explain the M122 or M122A1 Tripod with Appropriate Mounting Equipment.*

(a) Equipment Needed. One machine gun mounted on an M122 or M122A1 tripod complete with appropriate mounting equipment placed on a table. GTAs and film strips if available.

(b) Class Organization. One assistant trainer is assigned for each group if possible; otherwise, assistant trainers are placed where they can supervise assigned groups.

(c) Sequence of Training. The assistant trainers demonstrate, while the trainers describe the general nomenclature, data, functioning, and operation of the tripod with the appropriate mounting equipment.

(9) *Task 9: (071-020-0006) Place the Machine Gun Into Operation on the M122 or M122A1 Tripod with Appropriate Mounting Equipment Using Crew Exercises.*

(a) *Equipment Needed.* Basic TOE and individual equipment.

(b) Class Organization. Gunners are divided into three-man crews. To aid control and supervision, equipment is aligned with engineer tape or wire with the gunners in files behind the equipment.

(c) Sequence of Training. Assistant trainers demonstrate, while trainers explain how to place the machine gun into operation on the tripod with the appropriate mounting equipment. The crew drill is the first step in developing teamwork. Precision is a must for each crewmember. Crewmembers practice their duties with precision. Speed is increased as precision is attained. In advanced training, speed drills create interest and stimulate competition, while developing teamwork and testing individual crew performance. A well-trained crew can place the machine gun into action or take it out of action in 25 seconds.

(10) *Task 10: (071-312-3025) Demonstrate Fundamentals of Marksmanship.*

(a) Equipment Needed. One machine gun and a basic machine gun target for each firing lane on a range of suitable training area equipped with a prone and fighting position.

(b) Class Organization. The maximum number of required gunners is on the firing line or suitable training area.

(c) Sequence of Training. The trainer conducts conference, demonstration, and practical work on the fundamentals of marksmanship (steady position, aim, breath control, trigger control).

(11) *Task 11: (071-312-3026) Demonstrate Fundamentals of Firing Positions.*

(1) *Equipment Needed.* One machine gun and a basic machine gun target for each firing lane on a range of suitable training area equipped with a prone and fighting position.

(2) *Class Organization.* The maximum number of required gunners is on the firing line or suitable training area.

(3) *Sequence of Training.* The trainer conducts conference, demonstration, and practical work on the fundamentals of marksmanship (steady position, aim, breath control, trigger control).

(12) *Task 12: (071-312-3027) Demonstrate Fundamentals of Traverse and Search.*

(a) Equipment Needed. One machine gun and a basic machine gun target for each firing lane on a range of suitable training area equipped with a prone and fighting position.

(b) Class Organization. The maximum number of required gunners is on the firing line or suitable training area.

(c) Sequence of Training. The trainer conducts conference, demonstration, and practical work on the fundamentals of marksmanship (steady position, aim, breath control, trigger control).

(13) *Task13: (071-312-3031) Demonstrate Fundamentals of Engaging Night, NBC, and Moving Targets.*

(a) Equipment Needed. One machine gun, one night vision sight, one protective mask with gloves for each gunner, and a basic machine gun target for each firing lane on a range of suitable training area equipped with a prone and fighting position.

(b) Class Organization. The maximum number of required gunners is on the firing line or suitable training area.

(c) Sequence of Training. The trainer conducts conference, demonstration, and practical work on the fundamentals of marksmanship (steady position, aim, breath control, trigger control).

(14) *Task 14: (071-025-0007) Demonstrate Fire Commands.*

(a) Equipment Needed. One machine gun and a basic machine gun target for each firing lane on a range of suitable training area equipped with a prone and fighting position. GTAs and film strips if available.

(b) Class Organization. The maximum number of required gunners is on the firing line or suitable training area.

(c) Sequence of Training. The trainer conducts conference, demonstration, and practical work on the fundamentals of marksmanship (steady position, aim, breath control, trigger control), plus, the trainer gives the fire commands. (See Chapter 5.)

(15) *Task 15: (071-025-0007) Execute Dry-Fire Exercises.*

(a) Equipment Needed. One machine gun and a basic machine gun target for each firing lane on a range of suitable training area equipped with a prone and fighting position.

(b) Class Organization. The maximum number of required gunners is on the firing line or suitable training area.

(c) Sequence of Training. The trainer conducts conference, demonstration, and practical work on the fundamentals of marksmanship (steady position, aim, breath control, trigger control), plus, the trainer give the fire commands. (See Chapter 4.)

b. **Phase II, Basic Gunnery Training**. The information learned in this phase is essential to the development of the trainer who is to conduct the 10-meter firing, day transition firing, day NBC firing, and night transition instructional firing for the machine gun. The trainer must be qualified or certified on Phase I before moving to Phase II.

(1) *Task 16: (071-312-3031) Conduct 10-Meter Firing.*

(a) Equipment Needed. One machine gun and basic machine gun target (10-meter) for each firing lane; stopwatches; patches; and cleaning rods. Sound equipment is desirable during firing.

(b) Class Organization. The maximum number of required gunners is on the firing line with the remainder receiving concurrent instruction in the rear training area.

(c) Sequence of Instruction. The unit is assembled, given instructions, and briefed on the training that will be conducted while they are on the range. The trainer conducts conference, demonstration, and practical work on emplacing the machine gun on the firing lane. He places emphasis on preparation of the machine gun for firing, including safety checks. The interval between the two portions of the control command should be sufficient to permit the execution of the command and to allow the assistant gunner to perform his duties. Assistant gunners are required to perform their duties quickly and precisely. During practice, the trainer must ensure that the gunner is in the proper position before he is permitted to shoot. The OIC controls all firing. If space permits, all personnel are placed on the firing line. After the briefing, they are organized into firing orders and moved to firing lanes. Lanes are conducted IAW local range policies. Firing is conducted as described in Chapter 4. Concurrent training stations:

- Mechanical training.
- Care and cleaning.
- Any other machine gun subjects in which additional training is needed.

(2) *Task 17: (071-312-3031) Conduct Daytime Transition Fire on the Multipurpose Machine Gun Range.*

(a) Equipment Needed. Requires one machine gun for each firing lane, single and double E-type silhouette targets, and cleaning rods. Sound equipment is desirable during firing.

(b) Class Organization. Preferably, one gunner and one assistant gunner for each firing lane. Personnel not required in the operation of the range should receive concurrent training in rear area.

(c) Sequence of Instruction. The unit is assembled, given instructions, and briefed on the training that will be conducted while they are on the range. Before the conference and demonstration of firing, the trainer briefly reviews range estimation and techniques of adjustment. He also explains the characteristics of machine gun fire and their effect on field targets. The machine guns are zeroed at a known distance on the transition range. The gunner receiving the instruction should do the zeroing. The lane NCO requires the gunner to be in the correct position before letting him fire. The assist gunner helps the gunner in locating the targets and aids him in hitting the targets. The acting safety NCO may assist the assist gunner in locating the targets, but he is not permitted to aid in range estimation or fire adjustment. After the gunners complete this exercise, the gunner and assist gunner rotates duties. After completion of both gunner and assist gunner, the next gun team moves up. The OIC controls all firing. If space permits, all personnel are placed on the firing line. After the briefing, they are organized into firing orders and moved to firing lanes. Lanes are conducted IAW local range policies. Firing is conducted as described in Chapter 4. Concurrent training stations:

- Mechanical training.
- Care and cleaning.
- Any other machine gun subjects in which additional training is needed.

(3) *Task 18: (071-010-0006) Demonstrate Fundamentals of Engaging NBC Targets.*

(a) Equipment Needed. One machine gun, protective mask and gloves for every gunner, and a basic machine gun target (10-meter) for each firing lane on a range of suitable training area equipped with a prone and fighting position.

(b) Class Organization. The maximum number of required teams is on the firing line or suitable training area.

(c) Sequence of Training. The unit is assembled, given instructions, and briefed on the training that will be conducted while they are on the range. The trainer conducts conference, demonstration, and practical work on the fundamentals of marksmanship (steady position, aim, breath control, trigger control) while wearing the protective mask and gloves. (Same as Task 16.) After the briefing, they are organized into firing orders and moved to firing lanes. Lanes are conducted IAW local range policies. Firing is conducted as described in Chapter 4, Firing Tables I or IV. Concurrent training stations, using sand tables, charts, diagrams, or terrain, are set up to review rapid reloading techniques while wearing a protective mask and gloves:

- Mechanical training.
- Care and cleaning.
- Any other machine gun subjects in which additional training is needed.

(4) *Task 19: (071-025-0007) Conduct NBC Familiarization Transition Fire.*

(a) Equipment Needed. Requires protective mask and gloves, one machine gun, and appropriate ammunition for each firing lane, single and double E-type silhouette targets, and cleaning rods. Sound equipment is desirable during firing.

(b) Class Organization. Preferably, one gunner and one assistant gunner for each firing lane. Personnel not required in the operation of the range should receive concurrent training in rear area.

(c) Sequence of Instruction. The unit is assembled, given instructions and briefed on the training that will be conducted while they are on the range. The trainer conducts conference, demonstration, and practical work on the fundamentals of marksmanship (steady position, aim, breath control, trigger control) while wearing the protective mask and gloves. (Same as Task 16.) After the briefing, they are organized into firing orders and moved to firing lanes. Lanes are conducted IAW local range policies. Firing is conducted as described in Chapter 4, Firing Tables II or V. Concurrent training stations, using sand tables, charts, diagrams, or terrain, are set up to review rapid reloading techniques while wearing a protective mask and gloves:

- Mechanical training.
- Care and cleaning.
- Any other machine gun subjects in which additional training is needed.

(5) *Task 20: (071-010-0001) Conduct Nighttime Transition Fire.*

(a) Equipment Needed. One machine gun for each firing lane with appropriate night vision sight, and appropriate ammunition for each firing lane, single and double E-type silhouette targets, and cleaning rods. Sound equipment is desirable during firing.

(b) Class Organization. The firing area should have seating for the entire group during conferences. After the conference, the group is divided into small groups for practical work under the control of the assistant trainers. The OIC controls all firing. If space permits, all personnel are placed on the firing line. Requirements for the various exercises should be simple and progressive. If possible, the trainer selects terrain for the subject; otherwise, he applies the subject to the terrain.

(c) Sequence of Instruction. The unit is assembled, given instructions, and briefed on the training that will be conducted while they are on the range. Before the conference and demonstration, the trainer conducts the preparatory exercises of mounting the night vision sight and seating, boresighting, and zeroing procedures for the sight. He should also review and discuss range estimation, techniques of fire, adjustment, and characteristics of the machine gun fire. The machine gun should be zeroed to the sight using the night fire procedures in Chapter 4, Firing Tables III or VI. The acting safety NCO for each lane requires the gunner to be in the correct position before letting him fire. The OIC controls all firing. If space permits, all personnel are placed on the firing line. At the completion of the exercise, the next gunner moves up. After the briefing, they are organized into firing orders and moved to firing lanes. Lanes are conducted IAW local range policies. Firing is conducted as described in Chapter 4. Firing Tables III or VI. Concurrent training stations:

- Mechanical training.
- Care and cleaning
- Any other machine gun subjects in which additional training is needed.

c. **Phase III, Advanced Gunnery Training**. This training phase enables the trainer to develop his advanced skills. (See Chapter 5.)

(1) *Task 21: (071-025-0007) Conduct Day Assault Fire.*

(a) Equipment Needed. One machine gun and appropriate ammunition for each firing lane.

(b) Class Organization. The multipurpose machine gun transition range is used if the installation has one. *If need be, the lanes are modified so that the gunner has a trail of at least 150 meters in front of the weapon position.* The training area should have seating for the entire group during conferences. After the conferences, the group is divided into firing orders. If the group is large, two firing orders are placed on the firing line and all other personnel go to the concurrent training area for practical work under the control of the assistant gunners. The OIC controls all firing. If the group is small, all personnel go to the firing line at once. Requirements for the various exercises should be simple and progressive. If possible, the trainer selects terrain for the subject; otherwise, he applies the subject to the terrain.

(c) Sequence of Instruction. The unit is assembled, given instructions, and briefed on the training that will be conducted while they are on the range. After the briefing, they are organized into firing orders and moved to firing lanes. Lanes are conducted IAW local range policies. Firing is conducted as described in Chapter 5. Concurrent training stations, using sand tables, charts, diagrams, or terrain, are set up to review rapid reloading techniques. The OIC controls all firing. If space permits, all personnel are placed on the firing line.

- Underarm-firing position.
- Hip-firing position.
- Rapid reloading techniques.
- Any other machine gun subjects in which additional training is needed.

(2) *Task 22: (071-025-0009) Conduct NBC Assault Fire With the Machine Gun.*

(a) Equipment Needed. Requires protective mask and gloves, one machine gun, and appropriate ammunition for each firing lane.

(b) Class Organization. The maximum number of required gunners is on the firing line with the remainder receiving concurrent instruction in the rear training area.

(c) Sequence of Instruction. The unit is assembled, given instructions, and briefed on the training that will be conducted while they are on the range. After the briefing, they are organized into firing orders and moved to firing lanes. Lanes are conducted IAW local range policies. Firing is conducted as described in Chapter 5. Concurrent training stations, using sand tables, charts, diagrams, or terrain, are set up to review rapid reloading techniques while wearing a protective mask and gloves:

- Underarm-firing position while wearing a protective mask and gloves.
- Hip-firing position while wearing a protective mask and gloves.
- Rapid reloading techniques while wearing a protective mask and gloves.
- Any other machine gun subjects in which additional training is needed.

6-6. TRAINER CERTIFICATION PROGRAM

The certification program standardizes procedures for certifying and sustaining the proficiency of trainers. Their technical expertise must be continuously refreshed and updated, and leaders must manage it closely. One of the goals of the program is for the trainer to know the training mission.

a. **Training Base**. The training base can expect the same personnel changes as any other organization. Soldiers assigned as machine gun trainers have varying experience and knowledge of training procedures and methods. Therefore, the trainer certification program must address these variables. As a minimum, formal records document the trainer's progress in the certification program. All machine gun trainers must complete the three phases of machine gun training, and they must update their training quarterly.

b. **Certification Program Outline**. All trainers must attend, then conduct, all phases of the train-the-trainer program. Certified trainers have demonstrated the ability to train soldiers, to diagnose and correct problems, and to achieve standards. Those who fail to attend or fail any phase of the diagnostic examination will be assigned to subsequent training. The personnel designated to present instruction must complete the phases of the program in the prescribed sequence.

(1) *Phase I, Program Orientation*. In order for leaders to certify trainers, the trainers must visit training sites and ranges and demonstrate an understanding of—

- The certification program concept.
- The unit's marksmanship training outline and strategy.
- Issued reference materials and when to use which.

(2) *Phase II, Preliminary Gunnery*. During Phase II, the trainer must demonstrate his mastery of the fundamentals of marksmanship. He should complete within two weeks after completing Phase I. Leaders should review the following fundamentals. They record and maintain the results of this review on a trainer's progression sheet IAW the unit SOP:

- Characteristics.
- Capabilities.
- Disassembly.
- Cleaning, lubrication, and inspection.
- Assembly.
- Range determination and estimation.
- Classes of fire.
- Application of fire.
- Fire commands.
- Loading.
- Unloading.
- Immediate actions.
- Sight manipulations.
- Traverse and search.

(3) *Phase III, Basic Gunnery*. During this phase, the trainer must first qualify with the machine gun, then he must set up and conduct firing on the various ranges. He must brief leaders, explaining the targets as well as zeroing and scoring procedures. He explains the purpose of transition firing, field-zero procedures, range layout, and the conduct of training on the transition range. This briefing validates that he has the knowledge

necessary to conduct training. Leaders add the results of this interview to the trainer's progression sheet.

(4) *Phase IV, Advanced Gunnery*. This is the final test of the trainer's ability to train. He must set up a range and train at least one person. If ammunition is available, he conducts a firing exercise. If ammunition is not available, he is judged based on the quality of training.

APPENDIX A
UNIT TRAINING PROGRAM

This appendix provides guidance for conducting unit training as part of preliminary marksmanship training for each machine gun. The training program prepares the unit for war by enabling leaders and soldiers to develop and sustain proficiency in machine gun tasks. It does this by integrating individual training and evaluation with battle drills and other collective tasks.

A-1. FOCUS

An effective unit training program focuses on three battlefield variables: nature of the target (moving or stationary, single or multiple); nature of the firer (stationary or moving); and conditions (full or limited visibility, with or without protective mask, day or night).

A-2. STRUCTURE

This proposed training program is subdivided into the following periods:

a. Introduction.
b. Preliminary marksmanship training and dry fire.
c. Proficiency (performance) examination.
d. 10-meter zero practice and qualification.
e. Transition range, field zero, and practice fire.
f. Transition range qualification fire.
g. Night zero and instructional fire.
h. Night qualification firee.

A-3. PERIODS

After a brief description of the machine gun, soldiers receive the following instruction:

a. **Period 1--Introduction.**

DISASSEMBLY AND ASSEMBLY	TIME ALLOWED	
	HRS	MIN
Disassembly.		25
Operator maintenance:		
Inspection		10
Cleaning		10
Lubrication		10
Assembly		25
Disassembly and assembly (practice exercise--optional)		50
SUBTOTAL (including practice exercise):	2	10
CHARACTERISTICS OF FIRE	TIME ALLOWED	
	HRS	MIN
Burst fire		15
Trajectory and beaten zone		15
Engagements of targets:		
Point		10
Area		10
SUBTOTAL:		50
TOTAL:	3	0

b. **Period 2--Preliminary Marksmanship Training and Dry Fire**. After a brief description of the training to be conducted, soldiers receive the following instructions on the bipod, tripod, and mounting equipment:

TECHNIQUES OF FIRE	TIME ALLOWED	
	HRS	MIN
Position and grip, aiming, trigger manipulation, and T&E manipulation.		40
Loading ammunition.		15
Reducing stoppages and clearing the weapon.		5
SUBTOTAL:	1	10
DRY-FIRE PRACTICE	TIME ALLOWED	
	HRS	MIN
Aiming and firing sequence.		30
Sight setting and sight changes.		30
Zeroing procedures.	1	
SUBTOTAL:	2	
TOTAL:	3	10

c. **Period 3--Proficiency (Performance) Examination.** During this period, leaders use the proficiency examination to test and evaluate tasks learned during Periods 1 and 2.

PROFICIENCY (PERFORMANCE) EXAMINATION	TIME ALLOWED	
	HRS	MIN
TOTAL:	1	

d. **Period 4--10-Meter Zero Practice and Qualification.**

INSTRUCTIONAL SUBPERIODS	TIME ALLOWED	
	HRS	MIN
Preliminary marksmanship training including the aiming, sight picture, trigger control, bipod and tripod positions, and T&E manipulation.	1	
Function checks, loading, immediate action, clearing, and range safety.		15
Introduction to the 10-meter firing (includes zeroing and practice and qualification on the 10-meter target, and scoring of the target).		15
Zero firing.	1	
Practice day Table I.	1	
Qualification day Table I.	1	
TOTAL:	4	30

e. **Period 5—Transition Range, Field Zero, and Practice Fire.**

TRANSITION RANGE PRACTICE FIRE	TIME ALLOWED	
	HRS	MIN
Introduction to field firing (includes the transition range organization and operation).		15
Fire standard qualification course for practice (IAW with procedures in Chapter 4).	3	45
TOTAL:	4	

f. **Period 6—Transition Range Qualification Fire.** Soldiers fire the standard qualification course for record IAW procedures in Chapter 4.

TRANSITION RANGE QUALIFICATION FIRE	TIME ALLOWED	
	HRS	MIN
TOTAL:	4	

g. **Period 7—Night Zero, Instructional Fire, and Night Qualification Fire.** Soldiers receive instruction on mounting, placing the night vision device into operation, boresighting, and zeroing the device once it is seated. They fire the standard night course as Period 4, except a night vision device is mounted on the weapon.

NIGHT ZERO, INSTRUCTIONAL FIRE, NIGHT QUALIFICATION FIRE	TIME ALLOWED	
	HRS	MIN
TOTAL:	4	

A-4. HOURS

The unit training program requires a total of 23 hours and 40 minutes:

SUMMARY OF HOURS	TIME ALLOWED	
	HRS	MIN
Period 1	3	
Period 2	3	10
Period 3	1	
Period 4	4	30
Period 5	4	
Period 6	4	
Period 7	4	
TOTAL:	23	40

A-5. AMMUNITION

This paragraph summarizes ammunition required for the unit training program. Leaders should check STRAC requirements in DA Pamphlet 350-38.

a. **M249 Machine Gun**. See Table A-1.

TABLE	ROUNDS	TYPE
I Practice	185	Ball
I Record	91	Ball
II Practice	182	4:1
II Record	154	4:1
III Practice	371	4:1
III Record	154	4:1

Table A-1. M249 ammunition requirements.

b. **M60 and M240B Machine Guns**. See Table A-2.

TABLE	ROUNDS	TYPE
I Practice	231	Ball
I Record	117	Ball
II Practice	236	4:1
II Record	200	4:1
III Instructional	460	4:1
III Record	200	4:1

Table A-2. M60 and M240B ammunition requirements.

APPENDIX B

PROFICIENCY (PERFORMANCE) EXAMINATION

A proficiency examination determines if a soldier can perform all the tasks taught in the dry-fire training. This appendix is a guide for administering the examination to gunners.

B-1. DRY-FIRE PROFICIENCY EXAMINATION

The examination is a practical nonfiring exercise given during the last period of instruction on the machine gun before range firing. It may be held indoors if there are available facilities. The soldier must demonstrate techniques for the following tasks:

- General disassembly and assembly.
- Placement of direction and elevation readings on the T&E mechanism.
- Immediate action.
- Field zeroing.
- Engagement of a linear and a deep target.

B-2. CONDUCT OF THE EXAMINATION

Leaders can use the following schedule or modify it to fit their unit's training.

- Orientation, instructions, breakdown, and movement—15 minutes.
- Five stations—30 minutes each.
- Two breaks—10 minutes each.
- Four movement periods—5 minutes each.
- Total time—3 hours, 25 minutes.

B-3. STATION 1: PERFORM GENERAL DISASSEMBLY AND ASSEMBLY

This station normally has 11 setups. Each has one tripod-mounted machine gun with cover raised, bolt forward, and safety on the "F" position. This is placed on a mat to keep the parts free of dirt.

a. The following statement should be read at this station:

"DURING THIS PERIOD, YOU WILL BE ORGANIZED INTO THREE GROUPS AND REQUIRED TO DISASSEMBLE AND ASSEMBLE THE MACHINE GUN. THERE WILL BE ONE MACHINE GUN AND ONE GRADER FOR EVERY TWO GUNNERS. EACH GROUP WILL BE ALLOWED EIGHT MINUTES TO COMPLETE GENERAL DISASSEMBLY AND ASSEMBLY. IF YOU HAVE ANY TROUBLE, RAISE YOUR HAND AND THE GRADER WILL ASSIST YOU. THE TWO GROUPS NOT BEING TESTED WILL REMAIN TO THE REAR OF THE STATION WITH THEIR BACKS TOWARDS THE WORKING AREA UNTIL THEY ARE CALLED."

b. A scoresheet, like the example in Figure B-1, should be used in grading individual performance.

c. As each group completes this station, each grader should assemble the individuals he graded and gives them a thorough critique (6 minutes).

STATION 1: CHECKLIST
PERFORM GENERAL DISASSEMBLY AND ASSEMBLY

1. Cleared the machine gun using the procedures prescribed.
2. Disassembled the machine gun as prescribed for that machine gun.
   a. Removed the buttstock and buffer assembly.
   b. Removed the driving spring rod assembly.
   c. Removed the bolt and operating rod assembly. (Bolt is not separated from operating rod)
   d. Removed the trigger housing assembly.
   e. Removed the cover assembly.
   f. Removed the feed tray.
   g. Removed the barrel assembly hand heat shield.
   h. Removed the handguard from the receiver.
3. Assembled the machine gun as prescribed for that machine gun.
   a. Replaced the handguard on the receiver.
   b. Replaced the barrel assembly and heat shield.
   c. Replaced the cover assembly and feed tray.
   d. Replaced the trigger assembly.
   e. Replaced the bolt and operating rod assembly.
   f. Replaced the driving spring rod assembly.
   g. Replaced the buttstock, pulled the bolt to the rear, closed the cover, and pulled the trigger.
4. Completed task in 8 minutes.

TASK: GENERAL DISASSEMBLY AND ASSEMBLY	GO	NO GO
1. Cleared the machine gun using the procedures prescribed.		
2. Disassembled the machine gun as prescribed for that machine gun.		
a. Removed the buttstock and buffer assembly.		
b. Removed the driving spring rod assembly.		
c. Removed the bolt and operating rod assembly. (Bolt is not separated from operating rod.)		
d. Removed the trigger housing assembly		
e. Removed the cover assembly.		
f. Removed the feed tray.		
g. Removed the barrel assembly hand heat shield.		
h. Removed the handguard from the receiver.		
3. Assembled the machine gun as prescribed for that machine gun.		
a. Replaced the handguard on the receiver.		
b. Replaced the barrel assembly with the heat shield.		
c. Replaced the cover assembly and feed tray.		
d. Replaced the trigger assembly.		
e. Replaced the bolt and operating rod assembly.		
f. Replaced the driving spring rod assembly.		
g. Replaced the buttstock, pulled the bolt to the rear, closed the cover, and pulled the trigger.		
4. Completed task in 8 minutes.		

Figure B-1. Example format for Station 1 scoresheet.

B-4. STATION 2: PLACE DIRECTION AND ELEVATION READINGS ON THE T&E MECHANISM

This station normally has 11 setups. Each has one tripod-mounted machine gun complete with pintle and platform group and traversing-and-elevating mechanism.

a. For the first direction reading, the grader should ensure that the traversing slide is an even 5-mil graduation on the traversing bar. The gunner is required to place 1 to 4 mils on the traversing handwheel; for example, L242. This requires the gunner to center the traversing mechanism before he can place the next direction reading on it. The second direction reading should be in the opposite direction; for example, R240. The second elevation reading should also be a major change: +50/32 to -50/17.

b. The following statement should be read at this station:

> "DURING THIS PERIOD, YOU WILL BE ORGANIZED INTO THREE GROUPS AND REQUIRED TO PLACE TWO SETS OF READINGS ON THE TRIPOD-MOUNTED MACHINE GUN. A GRADER WILL CHECK YOUR FIRST SET OF READINGS BEFORE YOU PLACE THE SECOND SET ON THE TRIPOD-MOUNT MACHINE GUN. YOU WILL BE ALLOWED EIGHT MINUTES AT THIS STATION. IF YOU HAVE ANY TROUBLE, RAISE YOUR HAND AND THE GRADER WILL ASSIST YOU. THE TWO GROUPS NOT BEING TESTED WILL REMAIN TO THE REAR OF THE STATION WITH THEIR BACKS TOWARDS THE WORK AREA UNTIL THEY ARE CALLED."

c. A scoresheet, like the example in Figure B-2, should be used in grading individual performance.

d. As each group completes this station, each grader should assemble the individuals he graded and gives them a thorough critique (6 minutes).

STATION 2: CHECKLIST
PLACE DIRECTION AND ELEVATION READINGS ON THE T&E MECHANISM

1. Placed first set of readings on the T&E mechanism.
   a. Centered the traversing handwheel.
   b. Placed direction reading on T&E mechanism.
   c. Placed elevation reading on T&E mechanism.
2. Placed second set of readings on the T&E mechanism.
   a. Placed direction reading on T&E mechanism.
   b. Placed elevation reading on T&E mechanism.
3. Completed task in 8 minutes.

TASK: PLACE DIRECTION AND ELEVATION READINGS ON THE T&E MECHANISM	GO	NO GO
1. Placed first set of readings on the T&E mechanism.		
a. Centered the traversing handwheel.		
b. Placed direction reading on T&E mechanism.		
c. Placed elevation reading on T&E mechanism.		
2. Placed second set of readings on T&E mechanism.		
a. Placed directing reading on T&E mechanism.		
b. Placed elevation reading on T&E mechanism.		
3. Completed task in 8 minutes.		

Figure B-2. Example format for Station 2 scoresheet.

B-5. STATION 3: PERFORM IMMEDIATE ACTION

This station normally has six setups. Each has a cleaning rod and a bipod-mounted machine gun with bolt forward, cover closed, and safety on "F" position.

a. The grader should ask the individual to perform immediate action as he would if a round were in the chamber and would not fire providing the barrel was not hot enough to cause a cook off.

b. The following statement should be read at this station:

"DURING THIS PERIOD, YOU WILL BE ORGANIZED INTO FIVE GROUPS AND REQUIRED TO GO THROUGH THE STEPS OF IMMEDIATE ACTION AND REMEDIAL ACTION WITH THE MACHINE GUN. YOU WILL BE ALLOWED FIVE MINUTES. IF YOU HAVE ANY TROUBLE, RAISE YOUR HAND AND A GRADER WILL ASSIST YOU. THE GROUPS NOT BEING TESTED WILL REMAIN TO THE REAR OF THE STATION WITH THEIR BACKS TOWARDS THE WORK AREA UNTIL THEY ARE CALLED."

c. A scoresheet, like the example in Figure B-3, should be used in grading individual performance.

d. As each group completes this station, each grader assembles the individuals that he graded and gives them a thorough critique (5 minutes).

STATION 3: CHECKLIST
PERFORM IMMEDIATE ACTION

1. Pulled the cocking handle to the rear.
2. Observed a round, cartridge or linked being extracted from the machine gun.
3. Returned cocking handle to the forward position.
4. Attempted to fire the machine gun.
5. If the machine gun did not fire, pulled the cocking handle to the rear, locking the bolt assembly to the rear.
6. Place the safety on "S" and returned cocking handle to the forward position.
7. If machine gun was hot, kept it pointed downrange and waited 15 minutes, then performed remedial action. If machine gun was cold, performed remedial action.
8. Completed the task in 5 minutes.

TASK: IMMEDIATE ACTION	GO	NO GO
1. Pulled the cocking handle to the rear.		
2. Observed a round, cartridge or linked being extracted from the machine gun.		
3. Returned cocking handle to the forward position.		
4. Attempted to fire the machine gun.		
5. If the machine gun did not fire, pulled the cocking handle to the rear, locking the bolt assembly to the rear.		
6. Placed the safety on "S" and returned the cocking handle to the forward position.		
7. When machine gun was hot, kept it pointed downrange, waited 15 minutes, then performed remedial action. When machine gun was cold, performed remedial action.		
8. Completed the task in 5 minutes.		

Figure B-3. Example format for Station 3 scoresheet.

B-6. STATION 4: PERFORM FIELD ZEROING

This station normally has six setups. Each has a bipod-mounted machine gun with tools needed.

a. The grader tells each individual the range (500 meters) to the target and has him simulate the firing of a 7-round burst (M249) or 9-round burst (M60/M240B). The grader then tells the individual what corrections for deflection and elevation (in meters) are needed to hit the target. The gunner is then graded on his actions.

b. The following statement should be read at this station:

"DURING THIS PERIOD, YOU WILL BE ORGANIZED INTO FIVE GROUPS AND REQUIRED TO GO THROUGH THE STEPS OF FIELD ZEROING. YOU WILL BE ALLOWED FIVE MINUTES TO PERFORM AND EXPLAIN YOUR ACTIONS TO THE GRADER. IF YOU HAVE ANY PROBLEM, ASK YOUR GRADER. THE GROUPS NOT BEING TESTED WILL REMAIN TO THE REAR OF THE STATION WITH THEIR BACKS TOWARDS THE WORK AREA UNTIL THEY ARE CALLED."

c. A scoresheet, like the example in Figure B-4, should be used in grading individual performance.

d. As each group completes this station, each grader assembles the individuals that he graded and gives them a thorough critique (5 minutes).

STATION 4: CHECKLIST
PERFORM FIELD ZEROING

1. Placed the range setting on 500 meters.
2. Made corrections for deflection.
3. Made corrections for elevation.
4. Simulated firing the second burst and hitting the target.
5. Completed task in 5 minutes.

TASK: FIELD ZEROING	GO	NO GO
1. Placed the range setting on 500 meters.		
2. Made corrections for deflection.		
3. Made corrections for elevation.		
4. Simulated firing the second burst and hitting the target.		
5. Completed task in 5 minutes.		

Figure B-4. Example format for Station 4 scoresheet.

B-7. STATION 5: ENGAGE A LINEAR AND A DEEP TARGET

This station normally has six setups. Each should have a punchboard-type training aid or a blackboard and chalk.

a. The individual is required to show his point of initial lay, direction of manipulation, and extent of manipulation for a linear (single machine gun) and a deep target (pair of machine guns). The individual is also asked what rate of fire he would use to engage these targets if the rate were not stated in the fire command.

b. The following statement should be read at this station:

"DURING THIS PERIOD, YOU WILL BE ORGANIZED INTO FIVE GROUPS AND REQUIRED TO EXPLAIN HOW TO ENGAGE DIFFERENT TARGETS WITH THE MACHINE GUN EMPLOYED SINGLY AND IN PAIRS. YOU WILL BE ALLOWED FIVE MINUTES FOR SIMULATED ENGAGEMENT OF TWO TYPES OF TARGETS. IF YOU HAVE ANY QUESTIONS, ASK YOUR GRADER. THE GROUPS NOT BEING TESTED WILL REMAIN TO THE REAR OF THE STATION WITH BACKS TOWARDS THE WORK AREA UNTIL THEY ARE CALLED."

c. A scoresheet, like the example in Figure B-5, should be used in grading individual performance.

d. As each group completes this station, each grader assembles the individuals that he graded and gives them a thorough critique (5 minutes).

STATION 5: CHECKLIST
ENGAGE A LINEAR AND A DEEP TARGET

1. Engaged a linear target (single machine gun).
   a. Used initial point of aim.
   b. Manipulated T&E mechanism.
   c. Used sustained rate of fire.
2. Engaged a deep target (pair of machine guns).
   a. Used initial point of aim.
   b. Manipulated T&E mechanism.
   c. Used sustained rate of fire.
3. Completed task in 5 minutes.

TASK: ENGAGE A LINEAR AND A DEEP TARGET	GO	NO GO
1. Engaged a linear target (single machine gun).		
a. Used initial point of aim.		
b. Manipulated T&E mechanism.		
c. Used sustained rate of fire.		
2. Engaged a deep target (pair of machine guns).		
a. Used initial point of aim.		
b. Manipulated T&E mechanism.		
c. Used sustained rate of fire.		
3. Completed task in 5 minutes.		

Figure B-5. Example format for Station 5 scoresheet.

APPENDIX C

AERIAL DEFENSE

This appendix describes the use of machine guns in an air defense role, including the concept and two techniques for applying lead. Also discussed are the rules of engagement and firing positions.

C-1. PASSIVE AND ACTIVE MEASURES

A unit can take passive and active measures to defend itself against enemy air attack. Although volume fire is the key, there is a need to coordinate fires.

a. Passive measures are those that help the unit identify enemy aircraft before the aircraft locates the unit, make the unit difficult to locate, and make the unit less vulnerable when attacked. The unit must develop and practice camouflage as a passive measure. Concealment from the air must be considered when selecting routes, transportation means, or defensive positions. The use of air guards is important to give the unit time to react. Air guards should cover interlocking sectors of visible airspace.

b. Active measures for appropriate reactions to an air attack should be prescribed in unit SOPs. Each of the two techniques for applying lead is based on delivering a heavy volume of fire ahead of the target. The idea is to have every soldier in the unit engage the target. To achieve volume fire, soldiers armed with machine guns should fire at the cyclic rate.

c. If an aircraft is attacking his position, the soldier sees the aircraft in a head-on or diving view. To engage this aircraft, the soldier would fire slightly above the nose of the aircraft. Adjacent positions would see the aircraft in a crossing view. To engage the aircraft, these units would have to apply a proper lead. The method of applying lead depends on the technique used.

(1) *Football-Field Technique.* When engaging high-performance aircraft (those flying in excess of 200 mph), gunners should apply a one-football-field lead in front of the target and fire at the rapid rate until the target passes through the tracer stream. If the target is a low-performance aircraft, such as a helicopter, with a speed of 200 mph or less, gunners should apply half a football-field lead in front of the target, firing the cyclic rate. With all soldiers firing, a curtain of fire is formed because of slight differences in each soldier's estimate of the distance and lead.

(2) *Reference-Point Technique.* The unit leader designates terrain features as reference points. Upon spotting enemy aircraft, the leader commands, ENEMY AIR, REFERENCE POINT 1. At this time, the gunner points his weapon at reference point 1, elevates the weapon to about 45 degrees above the ground, and fires on command. Once he sights the target, he can make minor adjustments to align his fire on the target.

C-2. USE OF TRACERS

When planning for air defense, the leader should consider the use of tracers so that the gunner can observe the tracer stream and better align his fire on the target. A unit may engage an attacking aircraft without command. If an aircraft is not attacking, the unit may not fire on it unless ordered to do so.

C-3. FIRING POSITION

When firing the machine gun in an air defense role, the gunner should fire from a protected position, if possible. When not in a fighting position, he must position the weapon so that he has some type of support. In an emergency, another soldier can provide a firing support.

APPENDIX D

RANGE SAFETY

This appendix recommends safety precautions for the ranges described in this manual, but it does not replace AR 385-63 or local regulations. Range safety requirements vary because of the varied requirements of the courses of fire.

D-1. SAFETY PRECAUTIONS

The following safety precautions must be observed during all marksmanship training:

a. Display a red flag (red light for night firing) at the entrance to the range or in some other prominent location on the range during firing.

b. Always assume that weapons are loaded until they have been thoroughly examined and found to contain no ammunition and barrel is clear.

c. Mark firing limits with red-and-white-striped poles that are visible to all firers.

d. Never place obstructions in the muzzles of weapons about to be fired.

e. Keep all weapons in a prescribed area and proper safeguards when not in use.

f. Do not permit smoking near ammunition, explosives, or flammables.

g. Wear hearing-protection devices during firing.

D-2. RANGE PROCEDURES

The range can be a dangerous place, especially if safety procedures are not followed. Everyone must stay alert and adhere to the following precautions:

a. **Before Firing.**

(1) Close all prescribed roadblocks and barriers, and post necessary guards.

(2) Check all weapons to ensure that they are clear of ammunition and obstructions, and that the cover-feed mechanism assemblies are *up* to show they are cleared.

(3) Brief all personnel on the firing limits of the range and firing lanes.

(4) Obtain range clearance from the installation range-control office.

(5) Check the downrange area before firing to ensure that all personnel and equipment are clear of the area.

(6) Keep a complete first-aid kit on the range.

(7) Locate medical personnel on or near the range where they can be contacted quickly.

(8) Have all weapons checked by an officer or noncommissioned officer to ensure that they are operational.

(9) Do not handle weapons except on command from the tower operator or the officer in charge.

(10) Draw ammunition and issue it only on command of the officer in charge. When two or more lots of ammunition are used for firing, the officer in charge must ensure that the lots are separated and properly marked so that identification can be made by lot numbers in case of an accident or malfunction.

(11) Protect all ammunition from the direct rays of the sun.

(12) Do not allow anyone to move forward of the firing line without permission of the tower operator, safety officer, or officer in charge.

b. **During Firing.**

(1) Immediately order "CEASE FIRE" if an unsafe condition is noted during firing. Do not resume firing until directed to by the officer in charge.

(2) During firing, all personnel on the range must be aware of the danger in moving forward of the firing line to score their targets. Before the firing line is clear and anyone is allowed forward, the officer in charge or the safety officer will clear all weapons.

(3) The safety officer or NCO inspects each weapon that was fired on the firing line by making sure the bolt is locked to the rear and the safety is on. He makes sure each barrel is clear by running a cleaning rod through the barrel until he can see the end of the rod in the receiver. He performs the safety check.

c. **When Firing During Darkness.**

(1) Check the downrange area before firing to ensure that all personnel and equipment are clear of the area. This is accomplished by asking three times over a public address system, "IS THERE ANYONE DOWNRANGE?" Always pause each time to permit a response.

(2) Use a blinking red light in addition to the red flag. Display it at the entrance to the range or at some other prominent location.

(3) Mount two red lights on the striped poles marking the right and left limits of fire. They must be visible to all firers.

(4) Do not allow anyone to move from his position until told to do so by the officer in charge.

d. **After Firing.**

(1) Have safety personnel inspect all weapons to ensure that they are clear, and check to determine if the soldiers have any brass, links, or live ammunition.

(2) When weapons have been cleared, keep them in a prescribed area with the bolt forward and safety on.

APPENDIX E

EMPLOYMENT

Despite their Post-Civil War development, modern machine guns did not exhibit their full potential in battle until World War I. Although the machine gun has changed, the role of the machine gunner has not. The mission of machine guns in battle is to deliver fires when and where the leader wants them in both the offense and defense. Machine guns rarely, if ever, have independent missions. Instead, they provide their unit with accurate, heavy fires to accomplish the mission. This appendix provides information on three machine guns: the M249 light machine gun and the M60 and M240B general-purpose machine guns.

E-1. TACTICAL ORGANIZATION OF THE MACHINE GUN

The accomplishment of the mission demands an efficient machine gun crew who can fulfill all assigned missions. Each leader considers the mission and organizes machine guns to deliver firepower and fire support to any area or point needed to accomplish the assigned mission. Such organization takes time to achieve and reduces the flexibility of the unit.

E-2. MACHINE GUN IN THE ATTACK

The potential increase of air and ground attacks on the unit demands every possible precaution for maximum security while on the move. Where this situation exists, the machine gun crew must be thoroughly trained in the hasty delivery of antiaircraft fire and of counterfires against enemy ground forces. The distribution of the machine guns in the formation is critical. The machine gun crew is constantly on the alert, particularly at halts, ready to deliver fire as soon as possible. If the leader expects a halt to exceed a brief period, he carefully chooses machine gun positions to avoid unduly tiring the machine gun crew. If he expects the halt to extend for a long period, he can have the machine gun crew take up positions in support of the unit. They cover the direction from which he expects enemy activity as well as the direction the unit came from. He selects positions that permit the delivery of fire in the most probable direction of attack, such as into valleys, draws, ridges, and spurs. He chooses positions that offer obstructed fire.

a. **Machine Gun in the Offense.**

(1) *General Discussion.* Successful offensive operations result from the employment of fire and maneuver. Each is essential and greatly depends upon the other. Without the support of covering fires, maneuvering in the presence of enemy fire can result in disastrous losses. Covering fires, especially those that provide fire superiority, allow maneuvering in the offense. However, fire superiority alone rarely wins battles. The primary objective of the offense is to advance, occupy, and hold the enemy position.

(2) *Machine Gun.* The machine gun delivers an accurate, high-volume rate of fire on fairly large areas in a brief time. It is a great power to have on any offensive operation. When accurately placed on the enemy position, machine gun fires secure the essential element of fire superiority for the duration of the firing. Troops advancing in the attack should take full advantage of this period to maneuver to a favorable position from which

to facilitate the last push against the enemy. In addition to casualties, machine gun fire destroys the enemy's confidence and neutralizes his [actions in the?] defense.

(3) *Early Entry*. The early entry of machine guns in the offense is, with rare exceptions, highly desirable. Their continued action up to the moment of the assault enhances the probability of success. One desirable feature for employment of machine guns in the offense requires a proper handling of the ammunition for each machine gun. The other feature is to determine the actions of the machine gun crew to handle their weapon on the battlefield in order to deliver fire with the objective to support the maneuver unit at the time it is needed, regardless of physical difficulties encountered

(a) The machine guns seldom accompany the maneuver element. The gun's primary mission is to provide covering fire. The machine guns are only employed with the maneuver element when the area or zone of action assigned to the assault or company is too narrow to permit proper control of the guns. The machine guns are then moved with the unit and readied to employ on order from the leader and in the direction needing the supporting fire.

(b) Where the area or zone of action is too wide to allow proper coverage by the machine guns, the unit is assigned additional machine guns or personnel from within the battalion to permit the unit to accomplish its assigned mission. The machine guns are assigned a zone or a sector to cover and they move with the maneuver element.

(c) Under certain terrain conditions and for proper control, the machine guns move with the unit and are assigned a zone or sector to cover.

(d) When machine guns move with the unit undertaking the assault, the unit brings its machine guns to provide additional firepower. These weapons are fired either from the bipod or in an assault mode, from the hip or underarm position. They target enemy automatic weapons anywhere on the unit's objective. Once the enemy's automatic weapons have been destroyed, or if there are none, the gunners distribute their fire over their assigned zone or sector. In terms of engagement ranges, the machine gun in the assault engages within 300 meters of its target and frequently at point blank ranges.

c. **Machine Gun in a Base-of-Fire Element**. Machine guns organic to the company can help battalion machine guns lay the base of fire. In this case, the leader positions and controls the fires of all machine guns in the element. Machine gun targets include key enemy weapons or groups of enemy targets either on the objective or attempting to reinforce or counterattack. In terms of engagement ranges, machine guns in the base-of-fire element may find themselves firing at targets within 800 meters of the target. These ranges are simply a practical average. The nature of the terrain and desire to achieve some standoff, leads the leader to the correct tactical positioning of the base-of-fire element.

E-3. MACHINE GUN IN THE DEFENSE

Machine gun fire is distributed in width and depth in a defensive position. The leader can use machine guns to subject the enemy to increasingly devastating fire from the initial phases of his attack, and to neutralize any partial successes the enemy might attain by delivering intense fires in support of counterattacks. The machine gun's tremendous firepower is what enables the unit to hold ground. This is what makes them the backbone or framework of the defense.

a. The units' defense centers around the platoon's machine guns. The platoon leader sites the rifle squad to protect the machine guns against the assault of a dismounted enemy formation. The machine gun provides the requisite range and volume of fire to cover the squad front in the defense.

b. The primary requirement of a suitable machine gun position in the defense is that the machine gun be able to accomplish its specific missions. Secondarily, the position should be accessible and afford cover and concealment. Machine guns are sited to protect the front, flanks, and rear of occupied portions of the defensive position, and to be mutually supporting. Attacking troops usually seek easily traveled ground that provides cover from fire. This is not to say that they will avoid marshes, rough grounds, wooded areas, or any other type of terrain. Every machine gun should have three positions: primary, alternate, and supplementary. All of these positions should be chosen by the leader to ensure his sector is covered and that the machine guns are protected on their flanks.

c. The leader sites his machine gun to cover the entire sector or to overlap sectors with the other machine gun. The engagement range of a leader's weapon may extend from the last 300 meters where the enemy begins his assault to point-blank range. Machine gun targets include enemy automatic weapons and command and control elements.

E-4. MACHINE GUN ON A SECURITY MISSION

Security includes all command measures to protect against surprise, observation, and annoyance by the enemy. The principal security measures against ground forces include employment of security patrols and detachments covering the front flanks and rear of the units' most vulnerable areas. The composition and strength of these detachments depends on the size of the main body, its mission, and the nature of the opposition expected. The presence of machine guns with security detachments augments their firepower to effectively delay, attack, and defend, by virtue of their inherent firepower. When the machine guns are used as part of the security detachments in battalion trains or larger, the proportion of machine guns in such a detachment varies according to the situation. The main mission of the machine gun is to protect and defend through both defensive and offensive missions. For defense, the unit's main mission is to position the machine guns throughout the assigned area. For offense, the second mission, after a successful delay against an enemy attack, the unit expands its security outpost as needed to prevent another enemy attack.

APPENDIX F

TRAINING AIDS AND DEVICES

Training aids and devices must be included in a marksmanship program. This appendix lists those that units can make from materials most units have on-hand.

F-1. TRAINING DEVICES AND EXERCISES

The marksmanship training devices in this appendix are available to aid in sustainment training when used with the appropriate training strategies. These devices are beneficial when ammunition is limited for training or practice. Some training devices are complex and costly, but others are relatively simple and cheap to make. Devices and aids can be used alone or in combination. Individuals or squads can sustain or practice basic marksmanship skills and fundamentals with devices and aids.

F-2. FIRST SIGHTING AND AIMING EXERCISE

The purpose of the first sighting and aiming exercise is to teach the correct alignment of the sights on a target.

a. **Equipment Needed**:

One—sighting bar complete for each machine gun crew.

One—sighting target that measures 1 inch X 2 inches X 48 inches long. The sighting target is secured 5 1/2 inches away from one end (it should be moveable). The eyepiece is secured on the other end. The peep sight is secured 20 1/2 inches from the sighting target and 22 inches from the eyepiece (Figure F-1).

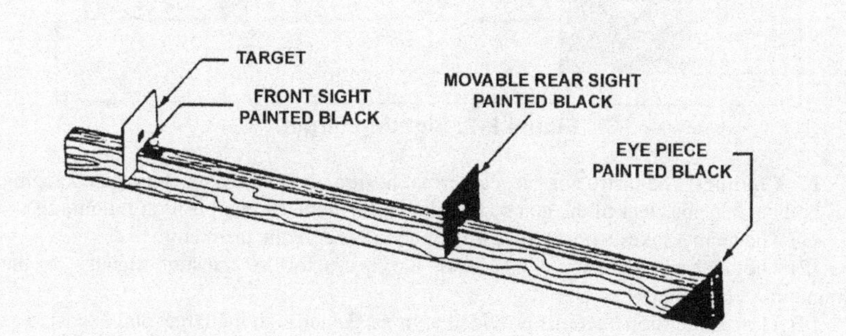

Figure F-1. Sighting bar.

b. **Conduct**. The instructor shows a sighting bar to the machine gun crews, points out its parts, and explains its use as follows:

(1) The sighting bar is used to assist in detecting small errors and in explaining them to the crews undergoing instruction.

(2) The front and rear sights on the sighting bar represent enlarged machine gun sights.

(3) The gunner looks through the eyepiece in such a position that he sees the sights in exactly the same alignment as the instructor does. Although there is no eyepiece on the machine gun, the use of an eyepiece on the sighting bar assists the gunners in learning how to align the sights properly when using the machine gun.

(4) The attachment of the removable target to the end of the sighting bar provides a simple method of readily aligning the sights on the target.

(5) Using a blackboard or a chart, the instructor explains and illustrates the correct sight alignment.

(6) A gunner from each crew adjusts the sights of the sighting bar and movable target to correctly align the sights on the target.

F-3. SECOND SIGHTING AND AIMING EXERCISE

The purpose of the second sighting and aiming exercise is to apply the preceding lesson to actual alignment of the machine gun sights on a target 25 meters away.

a. **Equipment Needed:**

One—Machine gun.

One—Basic machine gun target placed 25 meters away, with the reverse (blank) side showing.

One—Sighting target, 24 inches long 1 inch X 1 inch, a 3-inch square piece of wood painted black with a small 1/4-inch hole in the center (Figure F-2).

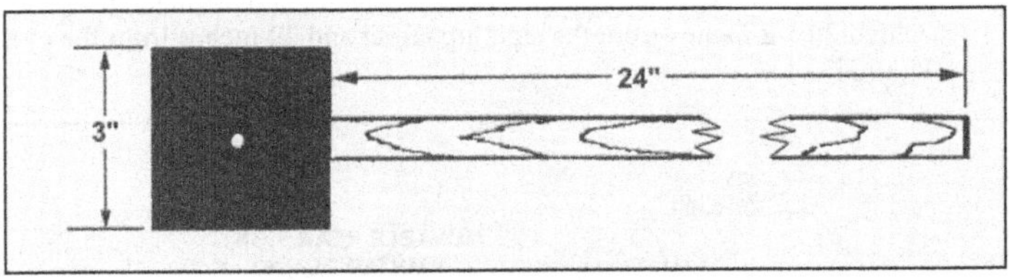

Figure F-2. Sighting target.

b. **Conduct**. The instructor takes a normal position behind the machine gun, keeping his body and hands clear of the gun so that the eye is in the correct position for aiming.

(1) The gunner takes a position near enough to observe the instructor.

(2) The assistant gunner stands near the instructor to transmit signals to the ammunition bearer.

(3) The ammunition bearer is provided with the 3-inch sighting target and is posted as the marker at the blank target, which is 25 meters away from the gun.

(4) The instructor, through improvising signals transmitted through the signalman, directs the marker to move the sighting target until it correctly aligns with the sights, and then commands: HOLD.

(5) The marker complies, holding the sighting target in place on the blank target. The gunner and assistant gunner then look through the sights.

(6) The instructor explains that, in aiming, the gunner's eye should first focus on the target, to ensure that he is aiming at the proper target. His eye then focuses on the top of

the front sight, to ensure that the line-of-sight established passes through the center of the peep sight and over the top of the center of the front sight.

(7) After the gunner and assistant gunner look through the sight, the instructor directs the marker (ammunition bearer) to move the sighting target out of alignment. He then requires the gunner to direct the marker to move the sighting target until it is in correct alignment with the sights.

(8) The instructor checks the alignment made by the gunner and points out any errors. When the instructor believes the gunner is proficient, the assistant gunner begins his exercise.

F-4. THIRD SIGHTING AND AIMING EXERCISE

The purpose of the third sighting and aiming exercise is to show the importance of uniform and correct aiming and to instill in the gunners a sense of exactness. This exercise can be used to check the consistency of the aiming and placement of a three-round shot group in a dry-fire environment.

a. **Equipment Needed**:

One—Machine gun.

One—Basic machine gun target placed at a distance of 25 meters away with the reverse (blank) side showing.

One—Sighting target (24 inches long 1 inch X 1 inch, a 3-inch square piece of wood painted black and a small 1/4-inch hole in the center) (Figure F-2).

b. **Conduct**. The instructor takes a normal position behind the machine gun, keeping his body and hands clear of the gun so that the eye is in the correct position for aiming.

(1) The gunner takes a position near the instructor to observe.

(2) The assistant gunner stands near the instructor to transmit signals to the ammunition bearer.

(3) The ammunition bearer is provided with a 3-inch sighting target and is posted as the marker at the blank target, which is 25 meters away from the gun.

(4) The instructor, improvising signals, transmits through the signalman to direct the marker to move the sighting target until it is in correct alignment with the sights and then commands: MARK.

(5) The marker complies, holding the sighting target in place on the blank target. The marker marks the position by inserting the tip of a pencil through the 1/4-inch hole in the center of the target. The marker marks the first dot with a number 1, second with a 2 and the third with a 3, ensuring the marker moves the target after every mark. All three dots should fit inside a 4-cm circle.

(6) The instructor explains that, in aiming, the gunner's eye should first be focused on the target to ascertain that he is aiming at the proper target. His eye is then focused on the top of the front sight to ensure the line of sighting established is a line through the center of the peep sight and over the top of the center of the front sight.

(7) The gunner and assistant gunner, without touching the gun, repeat the operation until three dots have been made and numbered.

(8) The instructor now explains the errors noted in the three sight alignments and the probable shape of the shot group formed by joining the three dots. Repeat as many times as needed until the gunner and assistant gunner can put seven rounds inside a 4-cm circle. The grader marks the first dot with a number 1, the second with a number 2 and the third

with a number 3. He makes sure the grader moves the target after every mark. When the instructor believes the gunner is proficient, the assistant gunner begins his exercise.

(9) The marker traces the three dots on a sheet of paper and connects them with lines. He writes the gunner's name at the bottom of the sheet and gives the paper to the squad leader.

F-5. MACHINE GUN T&E MANIPULATION DRILLS

The purpose of these exercises is to teach and instill confidence in the gunner in properly using his T&E. These exercises are conducted on a 25-meter line.

a. **Equipment Needed:**

One—Machine gun.

One—4-foot X 8-foot sheet of plywood. Paint the plywood white with black ruled lines of 1/8-inch width horizontally and vertically. The horizontal lines are 2 inches apart, starting 1 inch from the top and the bottom. The vertical lines are 3 inches apart, with 1 inch on either side.

(1) Label the horizontal lines starting from the top with 1, 3 and so on skipping every other line. Label the vertical lines along the middle horizontal line starting at the left; for example, A, B, C and so on (Figure F-3).

(2) When using the night vision sight, cut out a 4-cm-square circle from every four squares across and down. The squares or circles can be cut in any order the unit leader desires.

(3) Set up the machine 25 meters away from the target. Center the machine gun on the target.

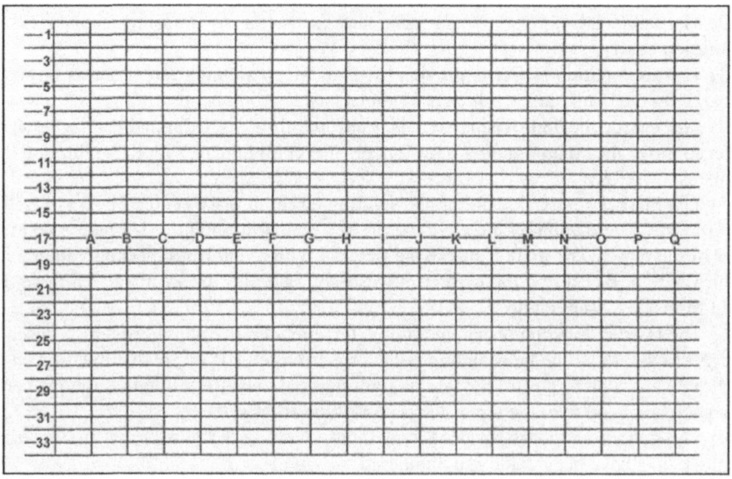

Figure F-3. Manipulator for T&E drills.

b. **Conduct.**

(1) The gunner is given the command to aim at vertical line B and horizontal line 4. The next command is RIGHT 150 mils; ADD 50 mils. The instructor then asks the gunner where he is aiming.

(2) Lay on the horizontal line 15 and vertical line O. The next command is, LEFT 100 mils; ADD 10 mils. The instructor then asks the gunner where he is aiming now.

(3) The leader or instructor can have as many of these types of questions set up, until he feels the gunner is proficient at this task.

F-6. TRAVERSE AND SEARCH EXERCISE

The purpose of the traverse and search exercise is to teach the gunner how to aim and point on a silhouette target and maintain that aim while the target is moving.

a. **Equipment Needed:**

One—Machine gun.

One—Basic machine gun target placed at a distance of 25 meters away with the reverse (blank) side showing.

One—Marking silhouette measures 1 inch X 1/2 inch X 22 inches long (wood).

One—F-type silhouette (reduced by half its size) measuring 7 1/2 inches long X 3 3/4 high (Chapter 4).

b. **Conduct**. The instructor shows the target to the machine gun crews. He points out its parts and explains its use as follows:

(1) The gunner takes a normal position behind the machine gun. The gunner is required to aim at a prescribed point on the silhouette target and to maintain that aim during the uniform movement of the target.

(2) The assistant gunner takes a normal position next to the gunner. The assistant gunner assists the gunner in maintaining the point of aim.

(3) The ammunition bearer is located 25 meters away behind the basic machine gun target with the target. The bearer moves the target back and fourth, up and down in any direction he wants. The movement should be consistent and not uniform, so that the gunner and assistant gunner can get proper training. The speed that the ammunition bearer moves depends on the instructor.

(4) The instructor watches the gunner and determines if the gunner or assistant gunner is properly doing the exercise by telling the ammunition bearer to stop and hold. He informs the gunner or assistant gunner to stand.

(5) The instructor positions himself behind the gun and looks through the sights to see where the gunner or assistant gunner was aiming. The instructor now explains the errors noted. When the instructor believes the gunner is proficient, the assistant gunner begins his exercise.

F-7. ENGAGEMENT SKILLS TRAINER

The engagement skills trainer (EST) 2000 (Figure F-4, page F-6) is a home station, indoor, multipurpose, multilane, small arms simulator. The EST augments and substitutes individual, crew, and static-squad collective training. Using projected imagery and laser-light technologies, the EST 2000 provides weaponry that simulates the same physical, functional, operational characteristics, and capabilities of service weapons. Weaponry for the EST 2000 includes the M16A2 rifle, M4 carbine, M9 pistol, M249 automatic rifle/light machine gun, M60, M240B and M2 machine guns, MK 19 grenade machine gun, M136 AT4, M1200 shotgun, and M203 grenade launcher.

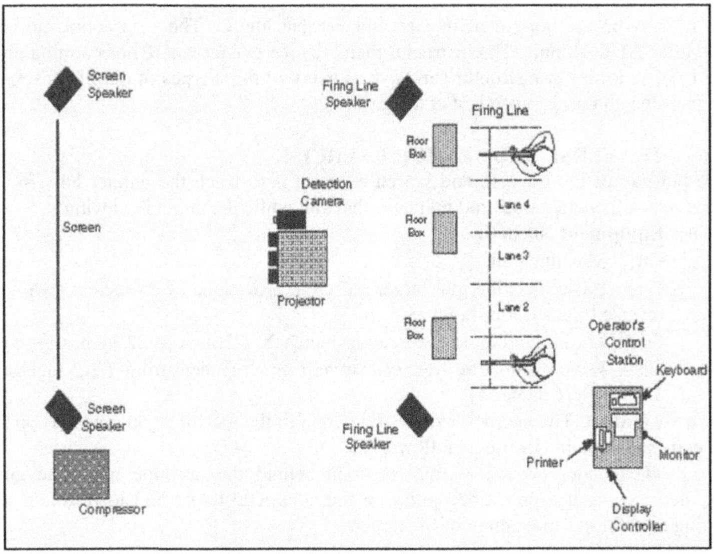

Figure F-4. EST 2000.

a. The EST 2000 is a modular system with an operational size of 35 feet by 35 feet by 8 feet high for a 10-lane system. Training programs include standard Army courses of fire, "shoot-don't shoot" decision training, and static-squad or element collective defensive training for infantry, scouts, engineers, military police, and CS or CSS. Basis of issue is one 15-lane EST 2,000 for each OSUT or BCT battalion supporting Initial Entry Training (IET) and one 10-lane EST 2,000 for each brigade-size element for Active and Reserve Component units for sustainment training.

b. The following tables (Tables F-1, F-2, and F-3) show the capabilities and limitations for each training program:

MARKSMANSHIP TRAINING PROGRAM	
209 marksmanship exercises, from zeroing through record-fire courses.	
Capabilities	**Limitations**
Train and sustain M16A2 rifle skills.	No qualification. No weather effects.
Train and sustain M4 carbine skills.	No qualification. No weather effects.
Train and sustain shotgun skills.	No qualification. No nonlethal munitions training.
Train and sustain M9 pistol skills.	No qualification. No weather effects.
Train and sustain light MG skills.	No maneuver training.
Train and sustain AR skills.	No maneuver training.
Train and sustain MMG skills.	No maneuver training.
Train and sustain M2 HB MG skills.	No maneuver training.
Train and sustain M203 GL skills.	No maneuver training. No nonlethal munitions.
Train and sustain MK 19 grenade MG skills.	No maneuver training.
Train antiarmor weapons employment.	No maneuver training.

Table F-1. Marksmanship training capabilities and limitations.

COLLECTIVE TRAINING PROGRAM	
178 squad tactical exercises against varying enemy targets, in terrain that varies from woodlands, desert, urban, and mountains that are typical of terrain found throughout the world.	
Capabilities	**Limitations**
Train and sustain defensive operations.	No offensive operations.
Train and sustain overwatch operations.	Nontactical interaction.
Train and sustain passage of lines.	Overwatch only.
Train and sustain ambush operations.	Limited eye points. Restricted interaction with on-screen CGI forces.
NOTE: No effects of weather and firing position distances under squad leader and fire-team leader control.	

Table F-2. Collective training capabilities and limitations.

"SHOOT-DON'T SHOOT" TRAINING PROGRAM	
40 "shoot-don't shoot" judgmental exercises.	
Capabilities	**Limitations**
Train and test rules of engagement.	No shoot back. No nonlethal training mode.
Train target selection under stress.	No shoot back.
Train target engagement under stress.	No shoot back.
Train verbal skills to deescalate.	No direct adversary reactions.

Table F-3. "Shoot-don't shoot" training capabilities and limitations.

c. The EST trains many skills, but it is not intended to replace live qualification or MILES force-on-force tactical training. Its many capabilities are clearly definable, but its limitations preclude total elimination of training ammunition resources. Training on the EST is planned to save ammunition resources, both at the assistant gunner and secondary firer positions and in peripheral training exercises such as protective mask and night-fire small-arms training. The EST 2000 enhances moving target training exercises. Sometimes shortages of range facilities or environmentally safe range resources prevent training in night, moving, and protective mask. The EST 2000 is the only three-mode training device capable of conducting realistic range operations while sustaining STRAC standards for individual weapons training, sustaining collective tactical skills and training, and sustaining judgmental engagements for the peace-keeping role of today's forces.

APPENDIX G

ADVANCED OPTICS AND LASERS

The newest systems greatly increase the night firing accuracy of all crew-served machine guns. They primarily enhance the gunner's ability to engage targets with great effectiveness to ranges of or beyond 1,000 meters. This appendix describes the AN/PAQ-4C, AN/PEQ-2A, laser systems. It also describes the AN/PAS-13 thermal sight and the M145 (MGO) telescope. Information for mounting, boresighting, and zeroing procedures for each weapon is also given.

Section I. ZERO PROCEDURES

An established day/night advanced marksmanship program equipped with training strategies and proposed qualification standards has been developed. Before beginning a night marksmanship program, soldiers must qualify on their assigned weapons during daylight conditions as outlined in the previous chapters of this manual and TMs. However, this appendix implements the new night qualification standards to compliment current Army training strategies. Commanders should follow these training strategies and abide by the qualification standards set forth to the best of their abilities. Although some courses of fire at first may seem redundant or inappropriate, numerous tests show that these training strategies work, and the qualification standards are achievable, if the strategy is followed.

G-1. BORE LIGHT

The bore light is an accurate means of zeroing weapons and the most aided vision equipment without the use of ammunition. Time and effort are applied to ensure a precise boresight, which in turn saves time and ammunition. Table G-1 outlines weapon/aided vision device combinations that can be zeroed using the bore light with the M249/M60/M240B weapons. With optics, a 25-meter zero must follow boresighting. The precise boresighting of a laser allows the soldier to directly engage targets without a 25-meter zero.

	M249	M60	M240B
MGO	X	X	X
AN/PAQ-4C	X	X	X
AN/PEQ-2A	X	X	X
AN/PAS-13	X	X	X

Table G-1. Weapon/aided vision device combinations.

a. **Intent**. Align the bore of the weapon to the optic, laser, or iron sight being fired to reduce or eliminate the time and ammunition it currently requires to live fire zero.

b. **Special Instructions**. Here are some special instructions the gunner should follow:

- Zero the bore light.
- Use only approved 10-meter offsets from Picatinny Arsenal (noted on offset).

- Ensure the proper 10-meter offset is used for the weapon configuration.
- Ensure boresighting is conducted 10 meters from the end of the barrel.
- Ensure weapon and offset is stabilized or the boresight will not be accurate.
- Ensure filters for aiming lasers are installed to reduce blooming.

b. **Observables.** The gunner should do the following:

- Confirm that the bore light spins on itself when zeroed at 10 meters.
- Boresight using the official and proper offsets only.
- Confirm that the weapon and offset are not moving during the boresighting procedure.
- Confirm that the bore light is centered on the circle of the offset.
- Confirm that the aiming device is aimed at the center of the crosshair on the offset.

DANGERS

PERMANENT INJURY COULD OCCUR WHEN DOING THE FOLLOWING:

1. STARE INTO THE VISIBLE LASER BEAM.

2. LOOK INTO THE VISIBLE LASER BEAM THROUGH BINOCULARS OR TELESCOPE.

3. POINT THE VISIBLE LASER BEAM AT MIRROR-LIKE SURFACES.

4. SHINE THE VISIBLE LASER BEAM INTO OTHER INDIVIDUALS' EYES.

WARNINGS

1. **Make sure the weapon is CLEAR and on SAFE before using the bore light.**
2. **Ensure that the bolt is locked in the forward position.**
3. **When rotating the bore light to zero it, ensure the mandrel is turning counterclockwise (from the gunner's point of view) to avoid loosening the bore light from the mandrel.**

G-2. CONCEPT

Boresighting is a simple procedure that can save time and ammunition only if the procedures outlined in this section are strictly followed. The bore light is a visible laser that is aligned with the barrel of the designated weapon. Then using a 10-meter offset, boresight the weapon with any optic, laser, or iron sight that you are assigned to fire. This procedure is accomplished by simply ensuring that the visible laser of the bore light is in line with the barrel by zeroing the bore light to the weapon, and then placing the visible laser of the bore light in a designated spot on the 10-meter offset. Once this procedure is accomplished, simply move the aiming point of the aiming device to the cross hair on the 10-meter offset. The weapon system is now boresighted and you are ready to engage targets or conduct a 25-meter zero. With optics, such as the MGO, TWS, and PVS-4, the bore light will put you on paper at 25 meters, thus reducing time and ammunition trying to locate rounds during the 25-meter zeroing. With lasers, the bore light allows you to boresight and then engage targets, eliminating the 25-meter zeroing procedures all together.

G-3. ZERO of BORE LIGHT TO WEAPON

Before boresighting the weapon system, make sure the bore light is first zeroed to the weapon. Take a flashlight and shine it through the barrel of the weapon and at 10 meters mark the spot that the flashlight is shining on. Then, without moving the weapon, fire one round. The impact of the round and the mark made with the flashlight should be the same. This procedure is the same principle used when zeroing the bore light to the weapon. This procedure is nothing more than aligning the visible laser with the barrel of the weapon, which would then hit the same area that the flashlight hit.

CAUTION
Do not over adjust the laser at soldiers.

G-4. STABILIZE

Stabilization of the weapon is crucial. The weapon can be stabilized on a tripod or using the bipod as a field-expedient method. The weapon does not have to be perfectly level with the ground when boresighting the weapon. Every barrel is different; therefore, with every weapon, steps (d) through (f) must be done to ensure that the bore light is zeroed to that barrel. If the bore light is zeroed, then go directly to the boresighting procedures.

a. Attach the 5.56-mm or 7.62-mm mandrel to the bore light.

b. Insert the mandrel into the muzzle of the weapon. The bore light is seated properly when no further travel of the mandrel into the muzzle is permitted and the mandrel spins freely.

c. Stabilize the weapon so that it will not move. The commands of *Start Point* and *Half Turn* are given to ensure clear communication between the soldier by the weapon and the soldier by the offset.

(1) *10 Meters.* 10 meters can be measured with the 10-meter cord that comes with the bore light or simply pace off eleven paces.

(2) *Zeroing Mark.* The zeroing mark (Figure G-1) is nothing more than a small dot drawn on a piece of paper, tree bark, or the bore light reference point on the 10-meter offset.

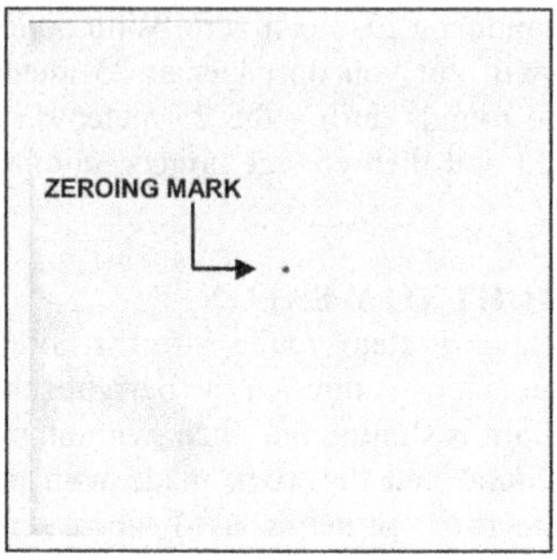

Figure G-1. Example of a zeroing mark.

(3) *Start Point.* Rotate the bore light until the battery compartment is facing upward and the adjusters are on the bottom. This position of the bore light and where the visible laser is pointing is identified as the start point (Figure G-2).

Figure G-2. Bore light in the START-POINT position.

(4) *Half Turn.* Rotate the bore light until the battery compartment is down and the adjusters are on top. This technique allows for easy access to the adjusters and helps with communication and stabilization of the weapon while conducting the boresighting procedures. This position of the bore light and where the visible laser is pointing is identified as the half turn position (Figure G-3).

Figure G-3. Bore light in the HALF-TURN position.

(5) *Reference Point.* The reference point is the point that is about half way between the *start point* and the *half-turn* point (Figure G-4).

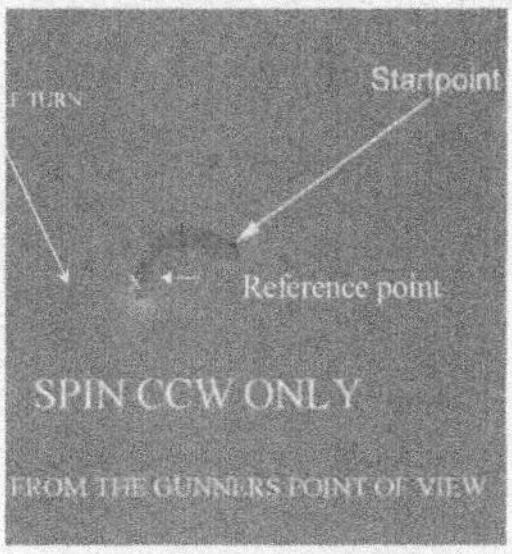

Figure G-4. Example of a start point, half-turn point, and reference point.

d. Turn the bore light on and spin it until it is in the ***start point*** position. Get a zeroing mark and mark a dot on it. Place the zeroing mark about 10 meters from the end of the barrel so that the visible laser strikes the zeroing mark.

e. Slowly rotate the bore light 360 degrees, while watching the visible laser made by the bore light. If the visible laser stops on the zeroing mark, the bore light is zeroed to the weapon.

f. If the bore light does not stop on the zeroing mark, then elevation and windage adjustments must be made to the bore light.

g. From the ***start point***, realign the zeroing mark with the visible laser and rotate the bore light 180 degrees into the *half turn* position. Then identify the *reference point*. Using the adjusters on the bore light move the visible laser to the reference point. Rotate the bore light back to the *start point*; move the zeroing mark to the visible laser. If the visible laser cannot be located once the bore light has been spun into the *half turn* position, start this procedure again at 2 meters instead of 10 meters. Once the visible laser is adjusted to the *reference point* at 2 meters, then start the procedure again at 10 meters.

h. Repeat step g until the visible laser spins on itself.

G-5. BORESIGHT TARGET OFFSET

Appendix H shows the most current 10-meter boresight target offsets. The 10-meter boresight offset grids are 1-centimeter squares, unlike the 25-meter zero targets. Some boresight target offsets are not available at this print. For more information, contact the infantry proponent at C co, 2/29 IN at Fort Benning, GA. Stability of the weapon is *crucial* in boresighting. Place the weapon in the *Bolt Forward* position during boresighting procedures. Do not cant the weapon left or right during this process. Two soldiers (a firer and a target holder) are required to properly boresight a weapon.

a. **Firer.** The firer's primary duty is to zero the bore light and make all adjustments on the aided vision device being used.

b. **Target Holder.** The target holder secures the 10-meter boresight target straight up and down 10 meters from the bore light and directs the firer in making the necessary adjustments to the aiming device. The target holder *must* wear night vision goggles when boresighting infrared aiming lasers.

NOTE: Weapon stabilization is crucial, orientation is irrelevant.

Section II. AN/PAQ-4C AIMING LIGHT

The AN/PAQ-4 aiming light projects an infrared laser beam, which cannot be seen with the naked eye but can be seen with night vision devices. This aiming light works with the AN/PVS-7B/C/D goggles and AN/PVS-14. The AN/PAQ-4C mounts on various weapons with mounting brackets and adapters.

G-6. EQUIPMENT DATA AND DESCRIPTION

The AN/PAQ-4 aiming light (Figure G-5) is activated by rotating the ON/OFF switch lever or the button on the optional cable switch. Either switch connects power from two AA batteries to an internal electronic circuit, which produces the infrared laser. Internal lenses focus the infrared light into a narrow beam. Rotating the mechanical adjusters with click detents controls the direction of the beam. These adjusters are used to zero the aiming light to the weapon. Once zeroed to the weapon, the aiming light projects the beam along the line of fire of the weapon. The optical baffle prevents off-axis viewing of the aiming light beam by the enemy.

a. **Data**. The AN/PAQ-4 aiming light includes the following equipment data:

Optics:	100 percent parallax free, anti-reflective coated lens system.
Length (sight):	14 centimeters (5.5 inches)
Weight:	164 grams (5.78 ounces)
Height:	3 centimeters (1.2 inches)
Width:	6.5 centimeters (2.5 inches)
Range:	Beyond 600 meters. (Actual range depends on light level and night vision device used for observation.)
Battery life:	100-hour operating (ON) time for AA batteries in temperatures above 0 degrees centigrade (32 degrees Fahrenheit); 36-hours for temperatures below 0 degrees centigrade (32 degrees Fahrenheit).

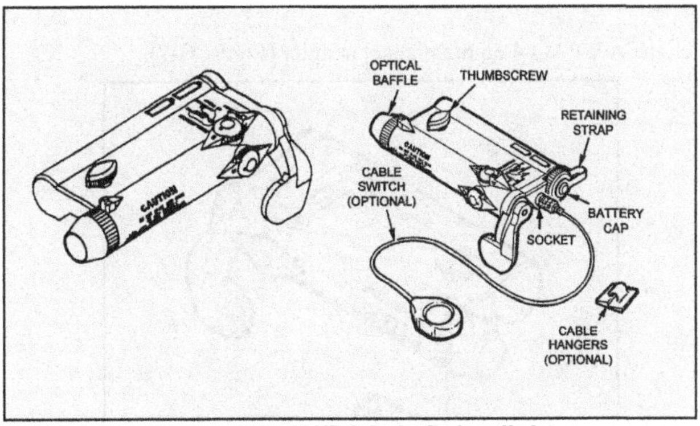

Figure G-5. AN/PAQ-4 aiming light.

c. **M249 Machine Gun.**

(1) *Mounting Procedures.* The gunner uses the AN/PVS-4 mounting bracket when mounting the AN/PAQ-4 on the M249 machine gun.

CAUTION

When mounting an AN/PAQ-4 to the mounting bracket, make sure that the hole for the screw in the AN/PAQ-4 is aligned and flush against the bracket screw. If not, the screw will strip the threads in the screw hole of the AN/PAQ-4 and prevent use with the M249.

(a) Install the AN/PVS-4 mounting bracket by first placing the mounting bracket on top of the feed cover mechanism assembly so that the two forked ends are secured around the headless pins. Next, remove the screw cover behind the rear sight assembly, and screw the bracket knob in until it is tight.

(b) Install the bracket adapter (Figure G-6).

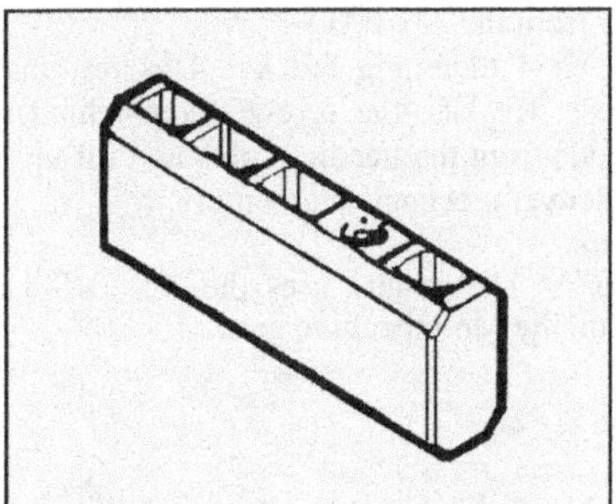

Figure G-6. Bracket adapter.

(c) Attach the AN/PAQ-4 on the bracket adapter (Figure G-7).

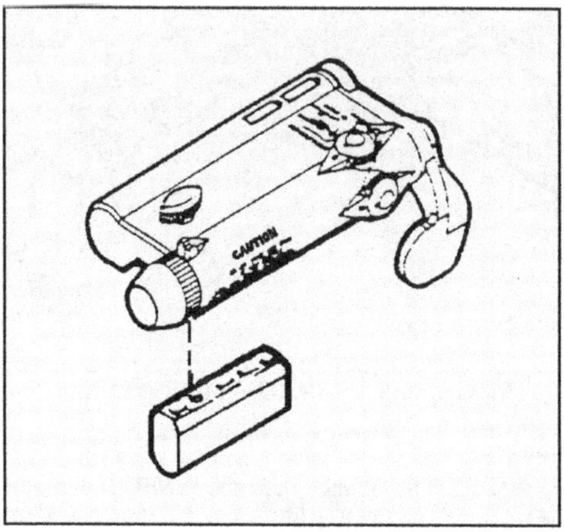

Figure G-7. Attaching bracket adapter to aiming light.

(d) Attach the AN/PAQ-4 to the AN/PVS-4 mounting bracket by first positioning the AN/PAQ-4 on top of the bracket so the spacer of the AN/PAQ-4 is aligned with the mounting knob of the bracket. Then, turn the mounting knob clockwise until the AN/PAQ-4 is tight.

WARNING
Do not store the AN/PAQ-4 with batteries installed.

(2) *Dismounting Procedures.*

(a) Detach the AN/PAQ-4 from the AN/PVS-4 mounting bracket. Turn the mounting knob counterclockwise until the AN/PAQ-4 is loose. Remove the AN/PAQ-4 and spacer from the bracket.

(b) Remove the spacer from the AN/PAQ-4.

(c) Remove the AN/PVS-4 mounting bracket. Unscrew the bracket knob until the rear of the bracket is loose. Replace the screw cover behind the rear sight assembly. Remove the two forked ends from the headless pins and lift up on the mounting bracket to remove it from the feed cover mechanism assembly.

c. **M60 Machine Gun.**

(1) *Mounting Procedures.* The gunner uses the AN/PVS-4 mounting bracket when mounting the AN/PAQ-4 on the M60 machine gun.

CAUTION

When mounting an AN/PAQ-4 to the mounting bracket, make sure that the hole for the screw in the AN/PAQ-4 is aligned and flush against the bracket screw. If not, the screw will strip the threads in the screw hole of the AN/PAQ-4 and prevent use with the machine gun.

(a) On the M60 machine gun bracket (Figure G-8), remove the M60 hinge pin latch and hinge pin from the over assembly by pressing on the latch (open end of pin) with an empty cartridge case, and separates the latch and pin. Place the pin and latch in the aiming guides on the left side of the mounting bracket and press together.

(b) Position the mounting bracket assembly on top of the machine gun cover so that the holes in the front of the bracket align with the cover assembly pinholes.

(c) Insert the longer hinge pin supplied with the bracket through the bracket and cover assembly, and secure it by inserting the hinge pin latch.

(d) Loosen the wing nuts on both leg clamps and positions the clamps under the cover assembly. Secure the mounting bracket by tightening the wing nuts firmly. Place the split washer is placed next to the wing nut and the flat washer is placed next to the bracket.

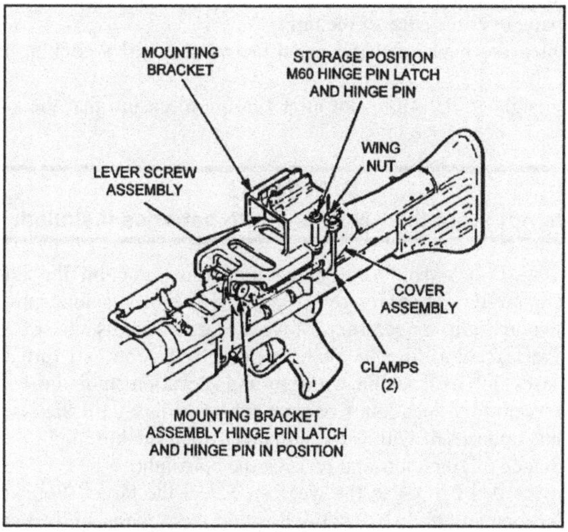

Figure G-8. Installation of M60 mounting bracket.

(f) Install the sight on the M60 mounting bracket assembly by positioning it in the groove on top of the bracket so that the scribe line on the bracket is aligned with the scribe line on the sight-mounting adapter. Tighten the lever screw assembly to secure the sight to the bracket. Use an empty cartridge case placed over the lever arm to increase leverage as the screw is tightened.

(2) *Dismounting Procedures.*

(a) Unscrew the lever screw assembly by turning it counterclockwise and remove the AN/PAQ-4 from the mounting bracket. Use an empty cartridge case placed over the lever arm for increased leverage if necessary.

(b) Loosen the wing nuts on both leg clamps.

(c) Push down on the longer hinge pin latch with an empty cartridge case and remove the hinge pin and latch.

(d) Remove the mounting bracket from the machine gun.

(e) Remove the M60 hinge pin and latch from the left side of the bracket by pushing on the hinge pin latch and pulling out the hinge pin.

(f) Replace the M60 hinge pin and latch.

(g) Remove the batteries when storing the sight.

d. **M240B Machine Gun.**

(1) *Mounting Procedures.* Before zeroing and qualifying with the AN/PAQ-4, the gunner must mount the sight onto his weapon. The M240B machine gun has a rail mount already attached to the cover assembly.

(a) Install the sight on the M240B rail mount by loosening the mounting knob located on the left side.

(b) Position the sight in a slot on the rail mount. Any slot can be used as long as the mount does not hang over the edge of the rail.

(c) Hand tighten the knob (clockwise) on the mount until a clicking noise is heard (two clicks).

(d) Places the sight in the same slot after zeroing to ensure that the sight retains its zero.

WARNING

Do not store the AN/PAQ-4 with batteries installed.

(2) *Dismounting Procedures.* Loosen the mounting knob on the left side of the bracket, then lift up on the sight to remove it from the M240B machine gun.

(a) Adjust the bore light, if necessary. Move the target to a distance of 2 meters. Mark the location of the laser dot. Slowly rotate the bore light one half turn. Note the new location of the laser dot. Adjust the windage and elevation until the laser dot moves halfway back to its original location. Continue the procedure until the laser dot remains stationary (or spins upon itself within 1 centimeter) when the bore light is rotated. Move the target to a distance of 10 meters and recheck the boresight.

(b) Boresight the AN/PAQ-4 to the weapon. Select the target (boresight offset) for the appropriate weapon and the AN/PAQ-4. Position the weapon so the bore light strikes the small dot on the boresight target. Zero the AN/PAQ-4 by turning both windage and elevation knobs (for the pointer and illuminator) fully clockwise until they stop. Rotate counterclockwise three turns and align the white dot on the adjuster with the center of the front adjuster flange (Figure G-9).

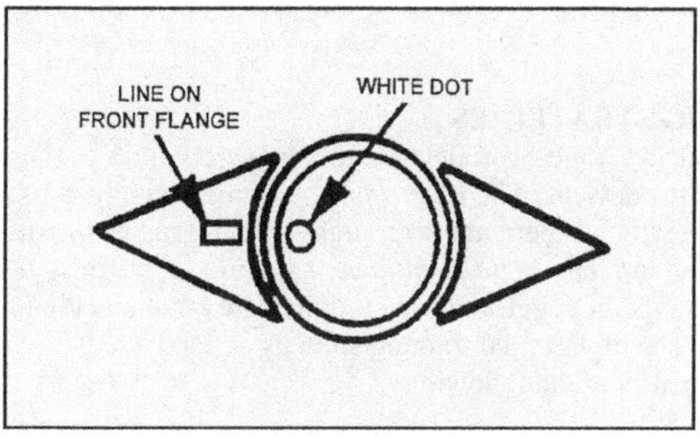

Figure G-9. Adjuster alignment.

G-7. BORESIGHT AND ZERO PROCEDURES

The following procedures are used to boresight and zero the M249, M60, and M240B machine guns using the AN/PAQ-4C aiming light.

a. **Boresight**. Check the alignment of the bore light.

(1) Place the appropriate mandrel with the bore light attached in the muzzle of the weapon.

(2) Turn on the bore light so that the laser dot strikes the target (offset) 10 meters away.

(3) Slowly rotate the bore light one-half turn (180 degrees) while watching the dot made by the laser on the target area.

(4) If the dot remains stationary, the laser is boresighted. Go to step 3 and use the appropriate boresight target for the weapon being boresighted.

(5) If the dot rotates in a circle, adjust the windage, elevation, or both until the dot remains stationary or rotates on itself no more than 1 cm. Go to step 2.

b. **25-Meter Zero**. The gunner chooses the appropriate strike zone from TM 11-5855-309-12&P, and draws it on the 25-meter zero target. He follows the procedures above and zeroes the weapon with the AN/PAQ-4. He fires a total of three rounds (one at a time) before triangulating the shot group. Once the weapon is zeroed on the 25-meter zero range, the gunner follows the procedures outlined below for field zeroing to obtain an accurate zero.

c. **Field Zero**. If the adjustment screws are turned too far, they will break. If the laser is side-mounted, the elevation and windage knobs will switch roles.

(1) Assume a good prone bipod-supported position. Fire a seven-round burst at the target. Note the burst pattern on the ground in relation to the target.

(2) Correct for windage (if rounds are impacting to the right or left of the target). Note that each click of the windage knob sight moves the group on the target mil (1/2 mil equals 6 inches at 300 meters or 10 inches at 500 meters).

(3) Correct for elevation (if rounds are impacting over or short of the target).

After making corrections on the sight, fire a confirming burst. If the target is not hit with the confirming burst, repeat the above procedures. Treat each subsequent burst as if it were the initial burst.

(2) Prepare the sight for field zeroing and check the sight. The preferred range is 300 meters.

G-8. TRAINING STRATEGIES

Two training strategies have been devised to adequately train soldiers in the use of the infrared aiming laser devices. The *night initial* training strategy is used for soldiers who have little or no previous experience with night vision goggles or with units beginning a night training program. The *night sustainment* training strategy is for soldiers who are familiar with night vision goggles and with units who have already implemented a night training program. The night initial training strategy is conducted over a two- to three-day time frame and consists of the following:

a. **Day 1**.

(1) *Night Vision Goggles Training With Terrain Walk When Firing Aiming Lasers*. Soldiers are given in-depth instruction on the proper use and fit of night vision goggles, to include characteristics and capabilities, maintenance, and mounting procedures. At night, soldiers conduct a terrain walk to become more familiar and build confidence using the night vision goggles.

(2) *Aiming Device Training*. Soldiers receive familiarization training on the aiming device. This training covers operation and characteristics, maintenance procedures, and mounting procedures.

(3) *Fundamentals of Firing*. This includes body position, foxhole, and prone. Soldiers review and practice firing positions and fundamentals of marksmanship and any changes that occur when using this aiming device.

b. **Day 2 to 3**.

(1) *Weapon and Equipment*. Inspect the weapon and aiming device to ensure the aiming device is mounted securely to the weapon. A review of clearing or misfire procedures is important. Let the soldiers use dummy rounds to practice correcting malfunctions with their eyes closed or during limited visibility.

(2) *Boresight Aiming Device to Weapon*. Each soldier boresights the weapon and aiming device at a range of 10 meters. If a bore light is not available, then 25-meter zero the weapon and sight by using the specified zeroing procedure for that weapon/sight combination. For course of fire, refer to Table 4-4 in Chapter 4 for a description of the aiming laser and AN/PVS-4 firing table. MGO and TWS will fire the day qualification table.

c. **Field Fire I**. Perform dry-fire exercise with no ammunition:

(1) Assume a stable, prone firing position. Fire the weapon maintaining sight alignment and sight picture. Pull the trigger correctly by pulling straight to the rear and releasing the trigger. Pull the trigger to the rear and say to yourself, "*fire a burst of seven*." After the word "*seven*," let go of the trigger.

(2) Apply correct traversing and searching techniques. Traversing means moving the muzzle of the weapon to the left or to the right to distribute fire laterally. Shift the shoulder slightly to the right or left for minor changes. For major changes, move the elbows and align the body directly behind the weapon. Searching means moving the muzzle of the weapon up or down to distribute fire in depth. To make changes in elevation, move the elbows closer or farther apart.

(3) The following targets are used for field fire I:

- 200-meter target single E-type silhouettes.
- 300-meter target single E-type silhouettes.
- 400-meter target double E-type silhouettes.
- 500-meter target double E-type silhouettes.
- 200-meter single E-type and 500-meter double E-type silhouettes.
- 300-meter single E-type and 400-meter double E-type silhouettes.
- 200-meter single E-type silhouettes.
- 400-meter double E-type silhouettes.
- 600-meter double E-type silhouettes.

d. **Field Fire II**. Perform a timed, dry-fire exercise on a transition range. This familiarizes the gunner with acquiring targets on a timed exercise. It also allows the soldier and the point safety time to make any corrections on-sight, or the body position and grip before the qualification table.

(1) Assume a stable, prone firing position. Fire the weapon maintaining sight alignment and sight picture. Pull the trigger correctly. Pull straight to the rear and release. Pull the trigger to the rear and say to yourself, "*fire a burst of seven*." After the word "*seven*," let go of the trigger.

(2) Apply correct traversing and searching techniques. Traversing means moving the muzzle of the weapon to the left or to the right to distribute fire laterally. Shift the shoulder slightly to the right or left for minor changes. For major changes, move the elbows and align the body directly behind the weapon. Searching means moving the muzzle of the weapon up or down to distribute fire in depth. To make changes in elevation move the elbows closer or farther apart.

(b) The following targets are used for field fire II:

- 200-meter target single E-type silhouettes (15 seconds).
- 300-meter target single E-type silhouettes (20 seconds).
- 400-meter target double E-type silhouettes (25 seconds).
- 500-meter target double E-type silhouettes (30 seconds).
- 200-meter single E-type silhouettes (40 seconds).
- 500-meter double E-type silhouettes (40 seconds).
- 300-meter single E-type silhouettes (35 seconds).
- 400-meter double E-type silhouettes (35 seconds).
- 200-meter single E-type silhouettes (50 seconds).
- 400-meter double E-type silhouettes (50 seconds).
- 600-meter double E-type silhouettes (50 seconds).

e. **Night Record Qualification**. Perform live-fire exercises with ammunition. Engage targets on a transition range timed. This exercise measures the gunner's ability to engage targets during a timed exercise (154 rounds of A064 4x1 mix 5.56-mm linked M249, 154 rounds of A131 4x1 mix 7.62-mm linked M60 and M240B):

(1) Assume a stable, prone firing position. Fire the weapon maintaining sight alignment and sight picture. Pull the trigger correctly. Pull straight to the rear and release. Pull the trigger to the rear and say to yourself, "*fire a burst of seven*." After the word "*seven*," let go of the trigger.

(2) Apply correct traversing and searching techniques. Traversing means moving the muzzle of the weapon to the left or to the right to distribute fire laterally. Shift the

shoulder slightly to the right or left for minor changes. For major changes, move the elbows and align the body directly behind the weapon. Searching means moving the muzzle of the weapon up or down to distribute fire in depth. To make changes in elevation move the elbows closer or farther apart.

(3) Use observation and adjustment of fire. Observe machine gun fire by noting the strike of the projectiles in the target area. When firing the bipod-mounted gun, adjust fire by changing body position.

(4) Apply the adjusted aiming point method. The adjusted aiming point method is a means of rapidly and accurately adjusting fires without making sight adjustment. If the initial burst misses the target, select a new aiming point on the ground. The aiming point should be the same distance from the target as the initial burst's center of impact.

(5) The following targets are used for night record qualification with the AN/PVS-4 and the aiming lasers. TWS and MGO will fire the day record fire table:

- 200-meter target single E-type silhouettes (10 seconds).
- 300-meter target single E-type silhouettes (15 seconds).
- 400-meter target double E-type silhouettes (15 seconds).
- 500-meter target double E-type silhouettes (25 seconds).
- 200-meter single E-type silhouettes (35 seconds).
- 500-meter double E-type silhouettes (35 seconds).
- 300-meter single E-type silhouettes (30 seconds).
- 400-meter double E-type silhouettes (30 seconds).
- 200-meter single E-type silhouettes (45 seconds).
- 400-meter double E-type silhouettes (45 seconds).
- 600-meter double E-type silhouettes (45 seconds).

Section III. AN/PEQ-2A TARGET POINTER/ ILLUMINATOR/AIMING LIGHT

The AN/PEQ-2A target pointer/illuminator/aiming light (TPIAL) (Figure G-11) is a Class IIIB laser that emits a highly collimated beam of infrared light for precise aiming of the weapon as well as a separate infrared illuminating beam with adjustable focus. A safety block is provided for training purposes (blue side), which limits the operator from selecting the high power modes (black side).

G-9. DESCRIPTION

This paragraph gives a description of the AN/PEQ-2A along with its accessories (Figure G-10). This paragraph also describes how to install and operate the AN/PEQ-2A. The TPIAL projects an infrared laser beam that cannot be seen with the eye but can be seen with night vision devices. It is capable of projecting a much wider infrared illuminating beam from an integral illuminator. The TPIAL works with night vision goggles and mounts on various weapons with mounting brackets and adapters. Leaders use the AN/PEQ-2A in the hand-held mode to illuminate and designate targets.

a. **Data**.

Weight:	7.5 ounce (with 2 AA batteries)
Length:	16.26 centimeters (6.4 inches)
Width:	7.12 centimeters (2.8 inches)
Height:	3.05 centimeters (1.2 inches)
Range:	600 meters in low power (eye safe)
	2,000 meters in high power (non-eye safe)
Output Power:	
Aiming laser	25 meters wide (+- 10 percent)
Illuminator	30 meters wide (+50, -20 percent)
Beam Divergence:	
Aiming laser	5 meter rad
Illuminator	1.0 to 115 meter rad
Batteries:	Two 1.5-volt AA batteries

b. **Components**. (See Figure G-10.)

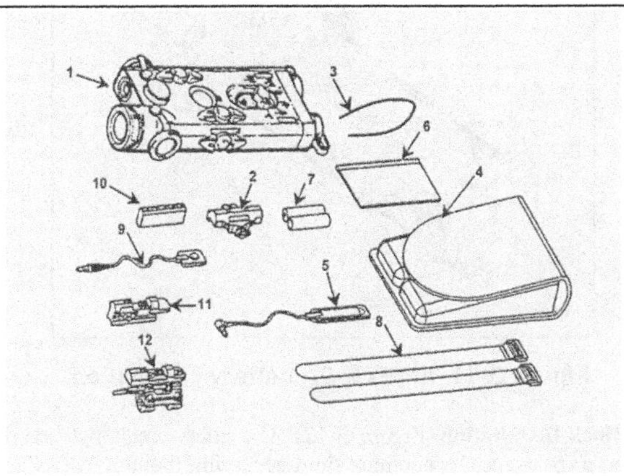

Figure G-10. AN/PEQ-2A with accessories.

(1) TPIAL assembly with safety block.
(2) Rail.
(3) Neck cord.
(4) Textile bag.
(5) Cable switch, 12-inch, membrane.
(6) Operator's manual.

(7) Batteries 1.5V AA.
(8) Strap, retention.
(9) Cable Switch, 20-inch, button.
(10) Bracket Adapter.
(11) Training extender (Army only).
(12) M4/M16A2 bracket assembly.

G-10. OPERATION

This paragraph describes how to operate the AN/PEQ-2A, to include battery installation, safety block installation, the button switch, the cable switch, the mode switch, and boresight adjusters.

a. **Battery Installation** (Figure G-11). Unscrew the battery caps and install two AA batteries. Orient the batteries as indicated by the markings on the AN/PEQ-2A body.

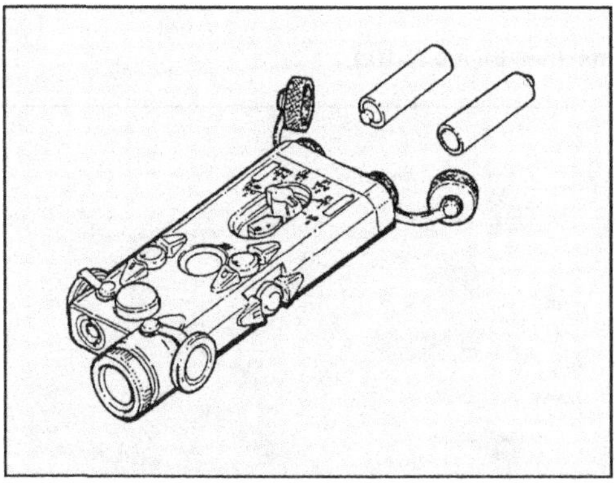

Figure G-11. AN/PEQ-2A battery installation.

b. **Safety Block Installation** (Figure G-12). The safety block installed in the training mode (blue side up) prevents the operator from accessing the non-eye safe modes (AIM HI, DUAL LO/HI, DUAL HI/HI). A .050 hex head Allen wrench is needed to unscrew the block from the body and reinstall it in the tactical mode (black side up).

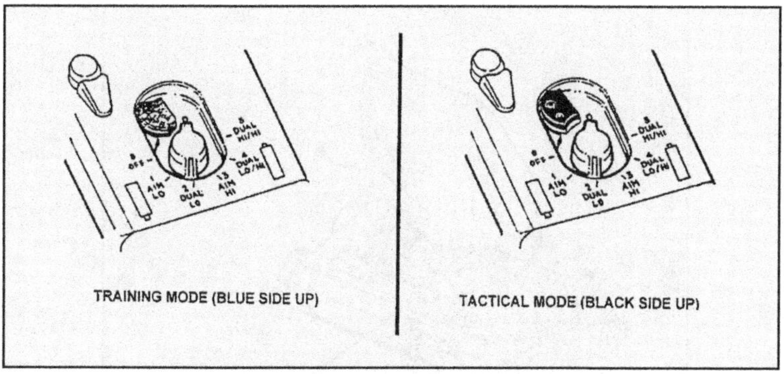

Figure G-12. Safety block installation.

c. **Mode Selector**. The mode selector is used to set the mode in which the AN/PEQ-2A operates when the cable-switch button or the push button is depressed. The mode selector has six positions (Table G-2).

KNOB POSITION	OPERATION
0 OFF	The AN/PEQ-2A does not operate.
1 AIM LO	The aiming beam operates at low power.
2 DUAL LO	The aiming beam operates at low power and the illuminating beam operates at low power.
3 AIM HI	The aiming beam operates at high power.
4 DUAL LO/HI	The aiming beam operates at low power and the illuminating beam operates at full power.
5 DUAL HI/HI	The aiming beam operates at high power and the illuminating beam operates at full power.

Table G-2. Mode selector positions.

d. **Button Switch** (Figure G-13, page G-18). The button switch is used when the AN/PEQ-2A is hand held. Pressing the button switch operates the AN/PEQ-2A in the operational mode set by the mode selector. When the button is released, the AN/PEQ-2A turns off. A green LED is incorporated into the body of the AN/PEQ-2A to indicate that the AN/PEQ-2A is ON. Whenever the AN/PEQ-2A is activated, the green LED lights and stays lit until the unit is turned OFF. If continuous operation of the AN/PEQ-2A is desired, pressing the button switch twice in rapid succession latches the AN/PEQ-2A to ON. The AN/PEQ-2A remains on until the push button is pressed a third time.

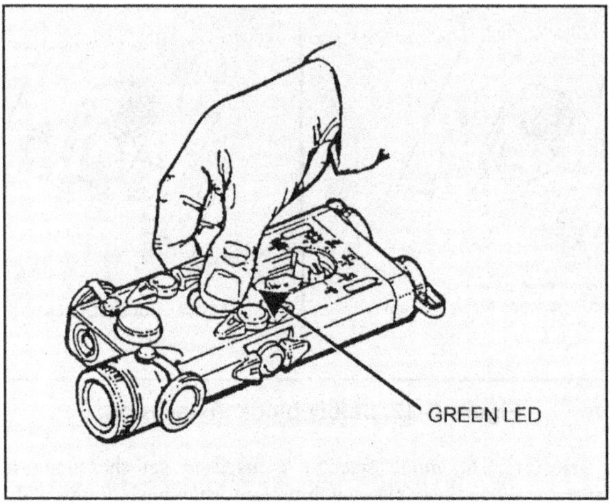

Figure G-13. Operation of the button switch.

e. **Cable Switch**. The cable switch (Figure G-14) is used when the AN/PEQ-2A is mounted on a weapon. The switch plugs into the back of the AN/PEQ-2A assembly. Pressing the button or pad at the end of the cable switch causes the AN/PEQ-2A to turn on in the operational mode selected by the mode select switch. When the button is released, the AN/PEQ-2A turns off. If continuous operation of the AN/PEQ-2A is desired, pressing the cable switch twice in rapid succession latches the AN/PEQ-2A to ON. The AN/PEQ-2A remains on until the push button is pressed a third time. When the cable switch plug is installed in the AN/PEQ-2A, it automatically locks into place. To remove the switch, pull back on the plug sleeve and pull the plug out. DO NOT TRY TO REMOVE THE PLUG BY PULLING ON THE CABLE.

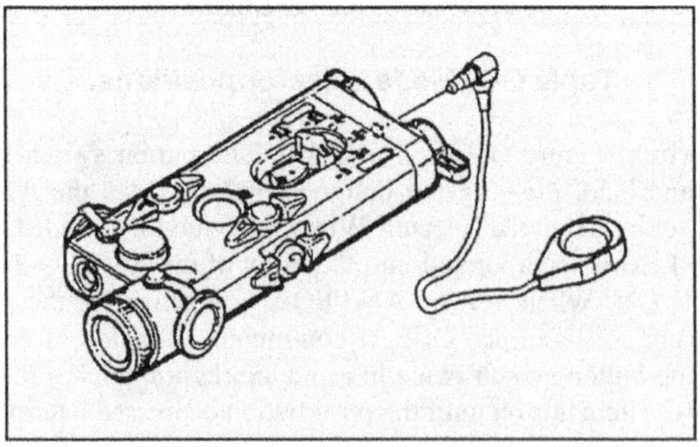

Figure G-14. Installation of the cable switch.

f. **Focus Knob** (Figure G-15). The focus knob is used to vary the spread of the illumination beam based on the range and size of the area to be illuminated.

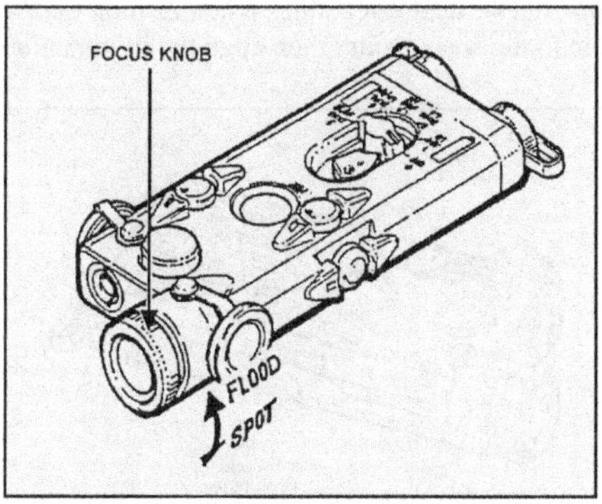

Figure G-15. Using the focus knob.

g. **Lens Cap** (Figure G-16).

(1) The *black lens cap* blocks the AN/PEQ-2A illuminator or aiming laser beam should the device be activated. To use the black lens cap, pull it from its stored location on the side of the AN/PEQ-2A and stretch it over the front of the focus knob or aiming beam so that it fits snugly in place.

(2) The *diffuser lens cap* enables the illuminator or aiming laser to emit in a 45-degree cone (10 feet at 10 feet). To use the diffuser lens cap, pull it from its stored location on the side of the AN/PEQ-2A and stretch it over the front of the focus knob or aiming beam so that it fits snugly in place.

(3) The *neutral density lens cap* enables the AN/PEQ-2A illuminator or aiming laser to be operated in low power. To use the neutral density lens cap, pull it from its stored location on the side of the AN/PEQ-2A and stretch it over the front of the focus knob or aiming beam so that it fits snugly in place.

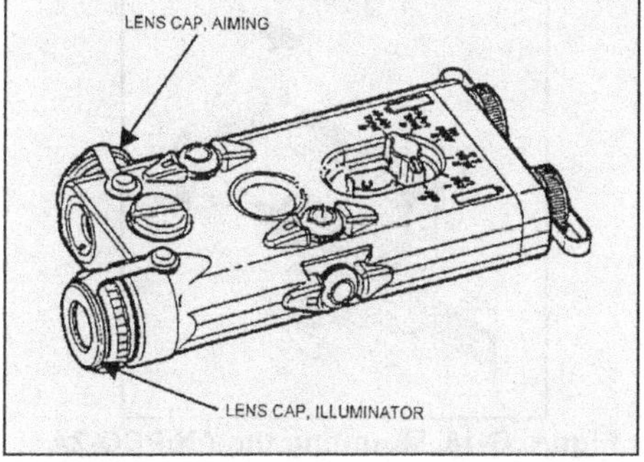

Figure G-16. Installing the lens caps.

h. **Adjusters** (Figure G-17). The AN/PEQ-2A is equipped with boresight adjusters for zeroing the aiming beam and illumination beam. The AN/PEQ-2A adjusters move the beams in true horizontal and vertical directions. When zeroing the AN/PEQ-2A, it is best to zero the aiming beam to the weapon and then align the illumination beam to the aiming beam.

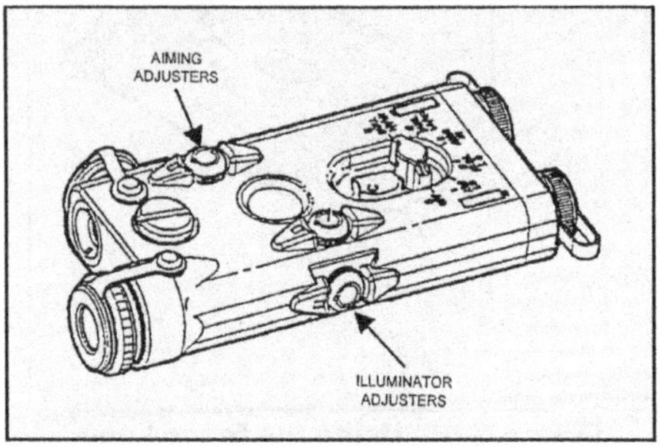

Figure G-17. Boresight adjusters for both aiming and illumination beams.

G-11. MOUNT/DISMOUNT PROCEDURES

This paragraph describes the mounting procedures for the AN/PEQ-2A (M249, M60, M240B). Reverse the procedures to dismount the AN/PEQ-2A.

a. **TWS Mounting Bracket.** (See Figure G-18.)

(1) Attach the rail grabber (3) (P/N 12598120) to the AN/PEQ-2A (4).

(2) Loosen the knob on the AN/PEQ-2A rail grabber.

(3) Select the slot on rail (5) for mounting. Any slot may be used as long as the mount does not hang over the edge of the rail.

(4) Place the bar of the rail grabber (3) in the slot of the rail (5) and hand tighten the knob on mount until a clicking noise can be heard.

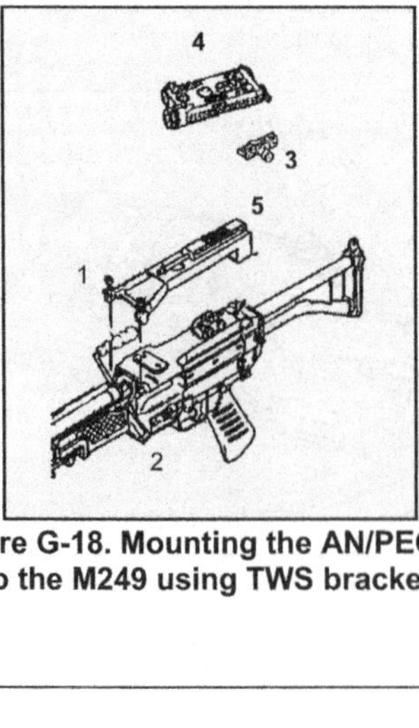

Figure G-18. Mounting the AN/PEQ-2A to the M249 using TWS bracket.

b. **M60 Machine Gun.**

(1) Remove the M60 hinge pin latch and hinge pin from the cover assembly. Place the pin latch in the aiming guides on the left side mounting bracket and press together.

(2) Place the mounting bracket on top of the machine gun cover so that the holes in the front of the bracket align with the cover assembly pinholes.

(3) Insert the longer hinge pin supplied with the bracket through the bracket and cover assembly and secure by inserting the hinge pin latch.

(4) Turning counterclockwise loosen the wing nuts on both leg clamps and position the leg clamps under the cover assembly. Secure the mounting bracket by tightening the wing nuts firmly. The split washer should be next to the wing nut and the flat washer next to the bracket.

(5) Place the bracket adapter in the AN/PEQ-2A mounting groove flush with the front of the AN/PEQ-2A. Tighten the thumbscrew clockwise.

(6) Position the AN/PEQ-2A with the bracket adapter on the M60 mounting bracket mounting groove. Align the front edge of the bracket adapter and the front edge of the groove. Tighten the lever screw assembly (Figure G-19).

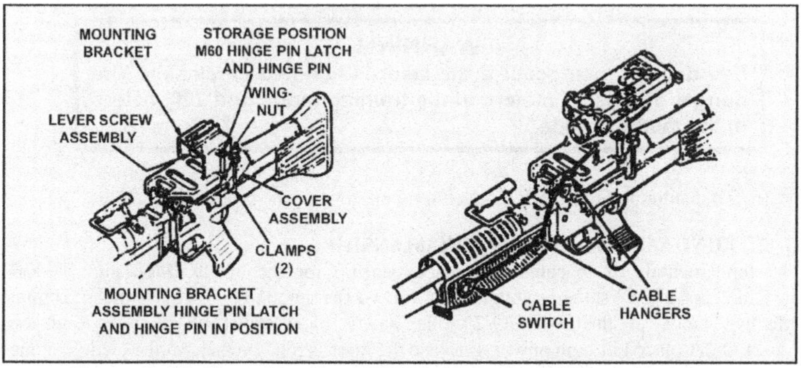

Figure G-19. Mounting the AN/PEQ-2A to the M60 machine gun.

c. **M240B Machine Gun.**

(1) Mount the AN/PEQ-2A 1 on the top cover rail (see arrow) using the AN/PEQ-2A bracket adapter and a standard rail grabber. The unit armorer preassembles the rail grabber and the bracket adapter.

(2) Mount the bracket adapter to the AN/PEQ-2A using the thumbscrew on the AN/PEQ-2A.

(3) Loosen the clamping knob until the rail grabber has sufficient space to fit over the top cover rail. Tighten the clamping knob until two clicks are heard.

(4) Place the AN/PEQ-2A at a position on the rail that is the most convenient for the operator. If however the AN/PEQ-2A is removed from the rail, the operator must take note of the position at which the device was zeroed, and then return the device to the same position to ensure the zero is retained.

(5) Install the remote switch in a convenient location using the provided cable hangers (Figure G-20).

Figure G-20. Mounting the AN/PEQ-2A to the M240B machine gun.

> **WARNING**
> **Eye damage can occur if the laser is handled carelessly. The danger area is 15 meters in the training mode and 220 meters in the tactical mode.**

(6) To dismount the devices for each weapon, reverse the procedures.

G-12. FUNDAMENTALS OF MARKSMANSHIP

The fundamentals of machine gun marksmanship for the M249, M60 and M240B machine guns are the same as with the AN/PAQ-4 (paragraph G-7). To use the maximum effective range of the AN/PEQ-2A, the safety block must be removed and the AN/PEQ-2A placed in high power. Because the laser is non-eye safe in the tactical mode (high power), batteries for the AN/PEQ-2A are issued to the firers on the firing line only when the weapon is oriented down range.

a. **Boresight Procedures**. Follow the procedures outlined in paragraph G-6 to boresight the machine gun and the AN/PEQ-2A.

b. **Zero Procedures**. Follow the procedures outlined in paragraph G-6 to zero and field zero the machine gun and the AN/PEQ-2A. (Refer to TM 11-5855-308-12&P for further information on zeroing procedures and zeroing targets with designated strike zones.)

c. **Sight Adjustments**. Sight adjustments for the aiming light and illuminator beam are the elevation adjustment screw (1 click at 25 meters equals 4 millimeters) (clockwise equals up) and the windage adjustment screw (1 click at 25 meters equals 4 millimeters) (clockwise equals right).

(1) The adjustment screws will break if they are turned too far.

(2) If the laser is side-mounted, the elevation and windage knobs will switch roles.

G-13. TRAINING STRATEGIES

Training strategies are the same for all weapons that use the AN/PAQ-4 except for the use of the illuminating beam. The illuminating beam allows soldiers to detect targets at ranges up to 2,000 meters (dependent upon terrain and weather constraints) in the high-power mode (tactical mode). Remember that the laser is not eye-safe in the tactical mode. In the low-power mode, targets can be engaged out to 600 meters during ideal limited visibility conditions (for example, 75 percent illumination, no rain or fog). The tactical mode should only be used on the M60 machine gun, and M240B machine gun.

Section IV. AN/PAS-13 (V2) MEDIUM WEAPON THERMAL SIGHT

The AN/PAS-13 (V2) medium-weapon, thermal sight (MWTS) (Figure G-21) is silent, lightweight, compact, and durable battery-powered thermal imaging sensors that operate with low battery consumption.

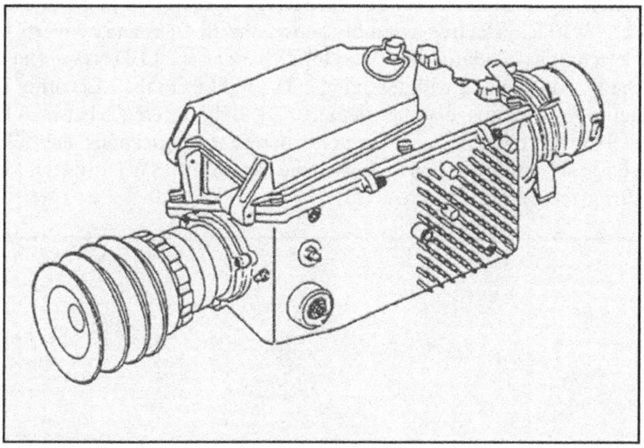

Figure G-21. Model of the medium-weapon thermal sight.

G-14. DESCRIPTION

The MWTS is capable of target acquisition under conditions of limited visibility such as darkness, smoke, fog, dust, and haze. The MWTS operates at night and during the daytime. Infrared light is received through the telescope, detected by an IR sensor, converted to digital data, processed, and then displayed for the user. The MWTS is composed of two functional groups: the telescope and the basic sensor.

a. **Telescope**. The telescope receives IR light emitting from an intended target and its surroundings. The telescope magnifies and projects the IR light on the scanner on the basic sensor (Figure G-22, page G-24).

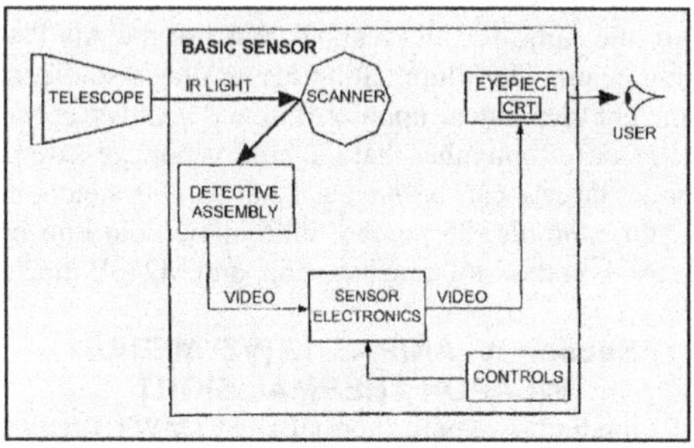

Figure G-22. MWTS configurations.

b. **Basic Sensor**. The scanner reflects the IR light received from the telescope on the detective assembly. The detective assembly senses the IR light and converts it to video. The sensor's electronics condition the video for display on the LED array. The LED array illuminates the IR image along with the reticle. The light from the LED array is reflected off the scanner to form an image at the eyepiece. The only difference between the MWTS and the HWTS are the telescopes, which are different magnifications and hold different reticles. The basic sensor on the two models is the same. The MWTS fits the M249, M60, and the M240B machine gun. (Figure G-23).

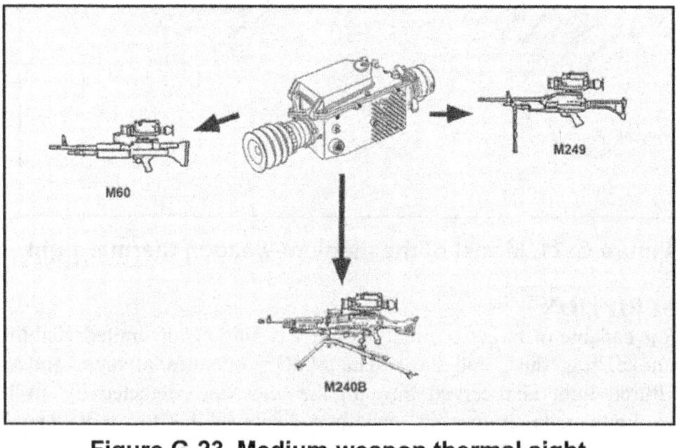

Figure G-23. Medium-weapon thermal sight.

c. **Data**. The following is the equipment data for the medium-weapon thermal sight.

- Field of view (FOV):
  Narrow—9 degrees
  Wide—15 degrees
- Telescope magnification:
  Narrow FOV—3.3X

Wide FOV—2X
- Length—15.5 inches
- Width—6.25 inches
- Height—6.25 inches
- Weight—4.1 pounds
- Power consumption:
  ON mode—12.0 watts
  STANDBY mode—5.5 watts
- Battery life:
  71 percent STANDBY mode—10 hours
  EMERGENCY mode—3.5 hours

G-15. OPERATION

The MWTS has three modes of operation: STANDBY, ON, and EMERGENCY. Figure G-24, page G-26, shows controls and indicators.

a. **Modes**.

(1) *STANDBY Mode*. When the system is first turned on, the MWTS begins a cool-down period of about 2 minutes. After the cool-down period, the MWTS enters the STANDBY mode. During the STANDBY mode, power is not applied to the scanner or display to extend the life of the battery.

(2) *ON Mode*. When the MWTS is in the STANDBY mode and pressure is applied to the eyecup, the MWTS switches to the ON mode, and a switch engages to provide power to the scanner and display. After a three-second delay, the system is fully operational.

(c) EMERGENCY mode. When switched to the EMERGENCY mode, the MWTS continuously applies power to the entire system. This mode allows the operator to bypass the 3-second delay experienced when switching from the STANDBY to the ON mode. Since power is applied to the entire system while in the emergency mode, battery life is greatly reduced.

b. **Controls and Switches**.

(1) The CONTRAST CONTROL (1) adjusts the contrast of the thermal scene. When turned fully clockwise to AUTO, contrast is automatically set.

(2) The EMERGENCY CONTROL (2) overrides the eyecup switch and turns entire system on, which places the MWTS in the EMERGENCY mode.

(3) The BRIGHTNESS CONTROL (3) is a nine-position rotary switch with an off detent position (turned fully counterclockwise). The purpose is to turn the system on or off and adjust the brightness of the eyepiece display.

(4) The FOCUS RING (4) adjusts the telescope focus from 20 meters to infinity. It requires a manual adjustment and affects both the wide and narrow fields of view.

(5) The FIELD OF VIEW RING (5) is located on the telescope. It has a wide and a narrow field of view. The wide FOV is for using low magnification during target detection, and the narrow FOV is for using high magnification during recognition and engagement.

(6) The RETICLE SELECT SWITCH (6) selects one of the available reticles depending on the MWTS model (medium or heavy). It must be held for 2 seconds to enable reticle changes. After 2 seconds, release the switch to cycle to the next reticle. This control is disabled after 10 seconds of inactivity.

(7) The RETICLE ADJUST SWITCH (7) adjusts the reticle aiming features in azimuth and elevation. It is used during zeroing, and it must be held for 2 seconds to allow changes to be made. After 2 seconds, each press moves the reticle aiming features one increment. This control is also disabled after 10 seconds of inactivity.

(8) The BLACK/WHITE POLARITY SWITCH (8) selects the polarity of the thermal image displayed on the raster. The initial setting is WHITE HOT. The polarity switch affects the appearance of the target.

(9) The EYECUP (9) controls STANDBY/ON when the system is on. When forward pressure is applied to the eyecup, the system is in the ON mode. When the pressure is removed for longer than 30 seconds, the system returns to the standby mode.

(10) The DIOPTER FOCUS RING (10) adjusts the focus of the raster and indicators to the operator's eye. It ranges from +2 to -6 diopters.

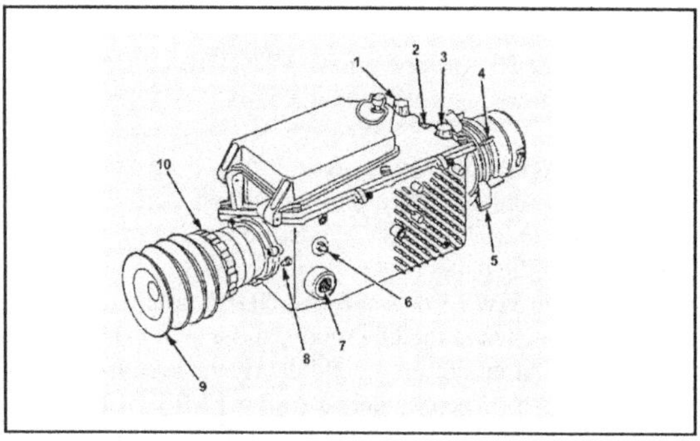

Figure G-24. MWTS controls and indicators.

c. **Eyepiece Indicators** (Figure G-25). Eyepiece indicators illuminate as follows:

(1) NOT COOL (11) when the detectors are not cool enough for proper operation.

(2) WHT HOT/BLK HOT (12) polarity.

(3) EMER (13) during emergency mode operation.

(4) LOW (14) when battery power has about 15 minutes of useful power left with go-to-war batteries.

(5) RETICLE SELECT/FOV (15) identifies which reticle is currently selected and indicates wide FOV or narrow FOV as selected by the FOV.

(6) ELEVATION INDICATOR (16) indicates the elevation zeroing adjustment of the reticle. It also displays the number of increments the reticle is UP (U) or (D) DOWN from the center zero position.

(7) NFOV/WFOV (17) indicates wide FOV or narrow FOV.

(8) ZOOM (18) indicates the zoom mode is selected.

(9) RET ADJ (19) indicates the reticle adjustment mode is selected.

(10) RET SEL (20) indicates the reticle select mode is selected.

(11) AZIMUTH INDICATOR (21) indicates the azimuth zeroing adjustment of the reticle. It also displays the number of increments the reticle is left (L) or right (R) from the center zero position.

(12) SENSOR COOLING (22) is displayed when the power is turned on, the indicator displays about a 2-minute cool-down period. After cool-down, the current reticle is displayed.

(13) The COOLDOWN PERIOD INDICATORS blink the first 10 seconds after the system is turned on. After 10 seconds, the NOT COOL indicator is lit and the POLARITY, EMERGENCY, and LOW BATTERY indicators function normally. The DISPLAY RASTER displays the thermal image with a superimposed reticle.

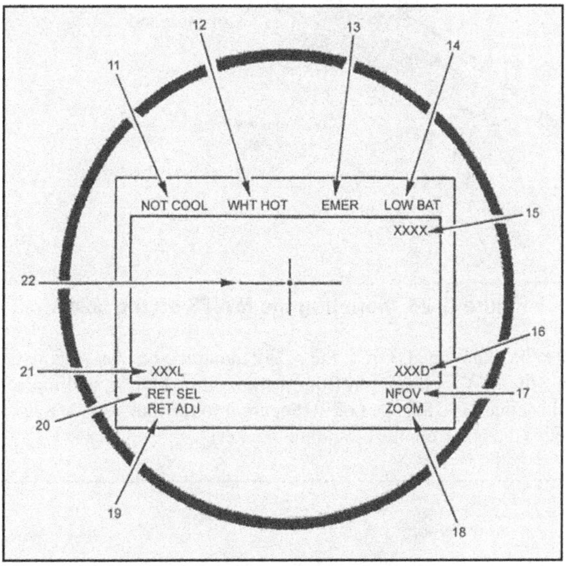

Figure G-25. Eyepiece indicators.

CAUTION

When raising or lowering the cover with the MWTS installed, avoid slamming the MWTS into the heat shield or slamming the cover closed.

d. **M249 Machine Gun.**

(1) *Mounting Procedures.* Before zeroing and qualifying with the AN/PAS-13, the gunner must mount the sight on the weapon (Figure G-26, page G-28).

(a) Open the cover of the weapon.

(b) Rotate the hook-retaining pin downward (over the top of the hinge pin) and remove the retaining clip from the left side of the hinge pin.

(c) Remove the weapon's hinge pin. **Do not lose the retaining clip.**

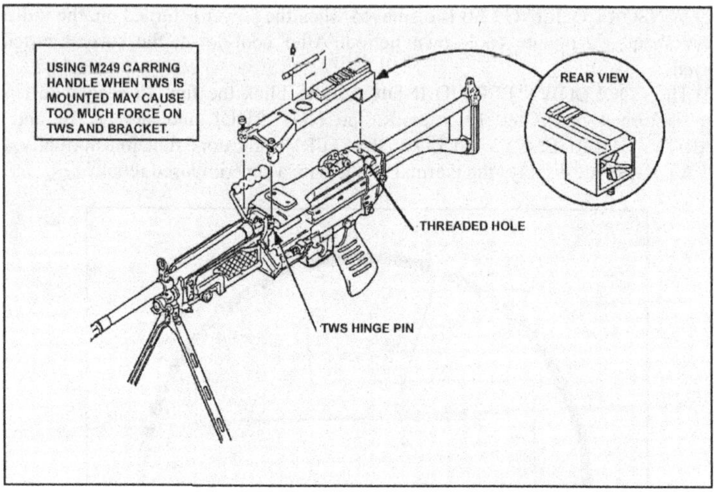

Figure G-26. Mounting the MWTS on the M249.

(d) Remove the retaining clip from the M249 mounting bracket (Figure G-27).

(e) Remove the MWTS hinge pin from the mounting bracket and replace it with the M249 hinge pin (Figure G-28, page G-29). Secure it to the mounting bracket by replacing the retaining clip (the one from the mounting bracket).

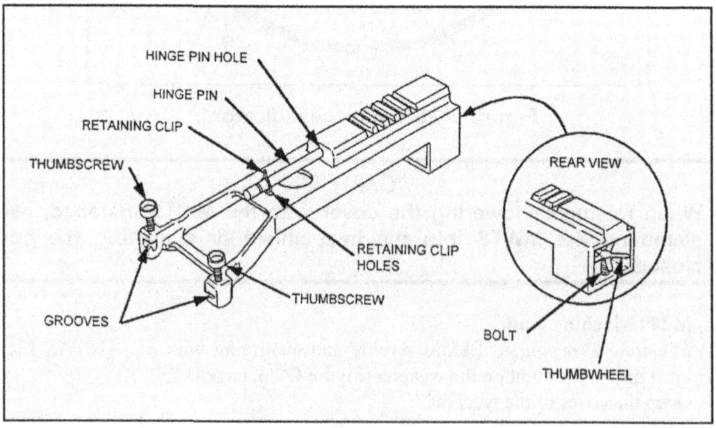

Figure G-27. M249 bracket.

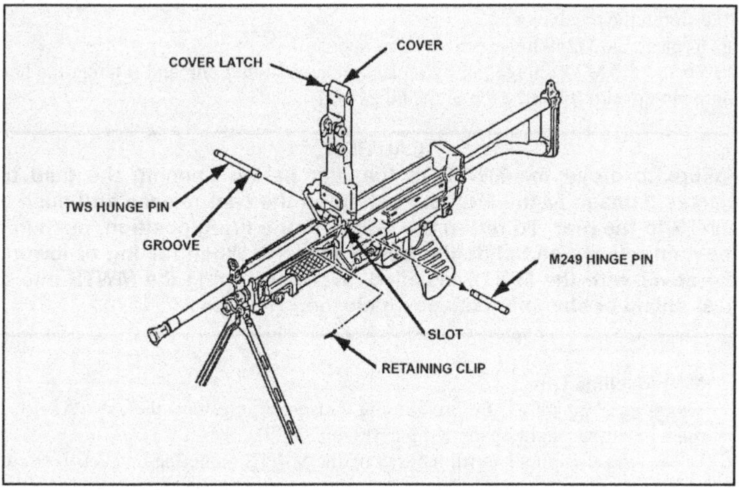

Figure G-28. M249 hinge pin replacement.

(f) Replace the cover of the M249 so that the slot in the cover lines up with the holes in the receiver group.

(g) Push the MWTS hinge pin through the slot and the spring.

(h) Secure the MWTS hinge pin by rotating the hook-retaining pin upward (over the hinge pin) and replacing the retaining clip on the left side of the pin. Close the cover of the weapon.

(i) Place the grooves on the front of the bracket over the edges of the MWTS hinge pin and rotate the bracket downward onto the cover. Ensure that the thumb wheel on the rear of the mounting bracket lines up with the threaded hole on the cover of the weapon.

(j) Secure the mounting bracket to the cover by tightening the thumbscrews and the thumb wheel.

(k) Select a slot on the rail of the bracket and place the bar of the MWTS mount in the slot. Hand-tighten the knob on the MWTS mount until it clicks at least twice.

(2) *Dismounting Procedures.*

(a) Loosen the knob on the MWTS mount until it is free of the rail and remove the MWTS.

(b) Unscrew the thumb wheel and thumbscrews and remove the mounting bracket from the weapon.

(c) Open the cover of the weapon and remove the retaining clip from the MWTS hinge pin.

(d) Rotate the hook-retaining pin downward (over the top of the hinge pin) and remove the MWTS hinge pin.

(e) Remove the M249 hinge pin from the mounting bracket and replace it with the MWTS hinge pin. Secure it with the MWTS retaining clip.

(f) Place the cover of the M249 on top of the weapon, aligning the holes in the cover with the slot in the receiver.

(g) Replace the M249 hinge pin to the weapon.

(h) Secure the M249 hinge pin by replacing the retaining clip and rotating the hook-retaining pin upward (over the top of the hinge pin).

CAUTION

Failure to move the MWTS to the rear before opening the feed tray causes damage to the MWTS. Push in on the cam release and push the MWTS to the rear. To return the MWTS to the firing position, push in on the cam release and slide the MWTS forward. When raising or lowering the cover with the MWTS installed, avoid slamming the MWTS into the heat shield or slamming the cover closed.

e. **M60 Machine Gun.**

(1) *Mounting Procedures.* Before zeroing and qualifying with the AN/PAS-13, the gunner must mount the sight on the weapon (Figure G-29).

(a) Loosen the clamping bars all the way on the MWTS mounting brackets by turning the thumb wheels.

(b) Loosen the locking nuts on the front of the MWTS mounting bracket (Figure G-30, page G-31).

(c) Open the feed tray cover of the M60 and remove the M60 hinge pin and hinge pin latch (Figure G-31, page G-31).

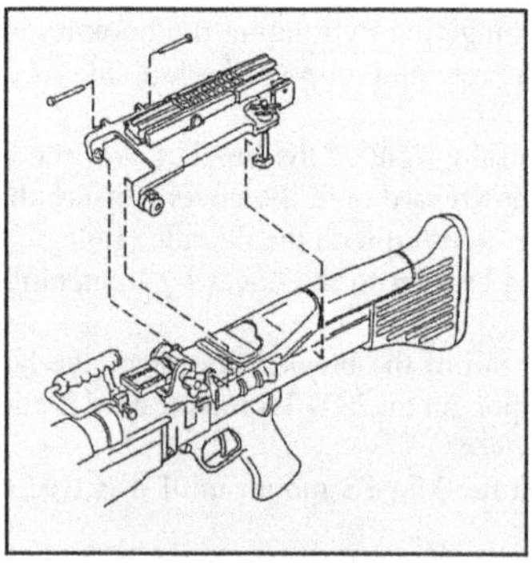

Figure G-29. Mounting the MWTS on the M60.

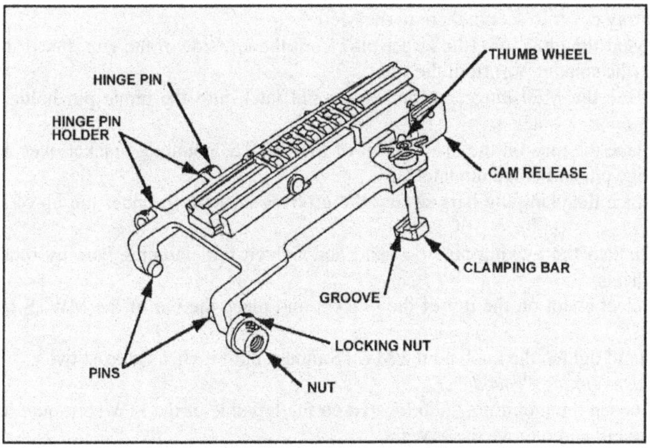

Figure G-30. M60 bracket.

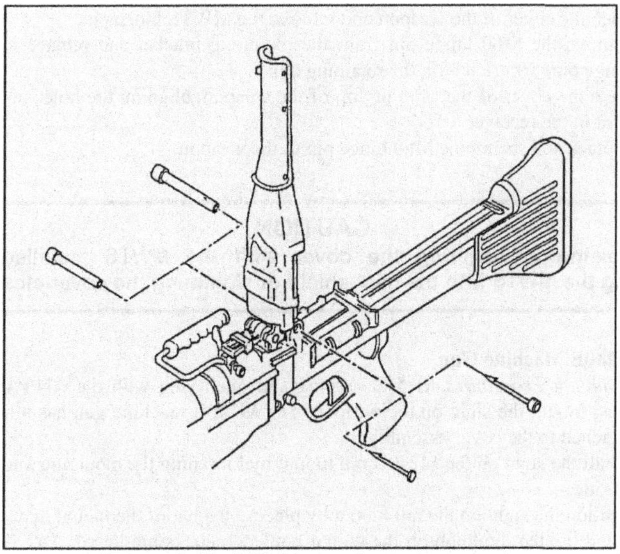

Figure G-31. M60 hinge pin replacement.

(d) Remove the MWTS hinge pin and hinge pin latch from the M60 mounting bracket.

(e) Place the MWTS hinge pin and hinge pin latch through the holes in the feed tray and feed tray cover to secure them to the M60.

(f) Insert the hinge pin (the larger pin) from the left side of the gun. Insert the hinge pin latch (the smaller pin) from the right.

(g) Place the M60 hinge pin and hinge pin latch into the hinge pin holder on the MWTS mounting bracket.

(h) Place the pins on the inside front of the MWTS mounting bracket over the holes in the hinge pin and hinge pin latch.

(i) Place the clamping bars so that the grooves in them fit under the lip of the feed tray cover.

(j) Tighten the locking nuts by hand and tighten the clamping bars by rotating the thumb wheels.

(k) Select a slot on the rail of the bracket and place the bar of the MWTS mount in the slot.

(l) Hand tighten the knob on the MWTS mount until it clicks at least twice.

(2) *Dismounting Procedures.*

(a) Loosen the mounting knob located on the left side of the MWTS mount until it is free of the rail and remove the MWTS.

(b) Unscrew the thumb wheels and locking nuts and remove the mounting bracket from the weapon.

(c) Open the cover of the weapon and remove the MWTS hinge pin.

(d) Remove the M60 hinge pin from the mounting bracket and replace it with the MWTS hinge pin. Secure it with the retaining clip.

(e) Place the cover of the M60 on top of the weapon, aligning the holes in the cover with the slot in the receiver.

(f) Replace and secure the M60 hinge pin to the weapon.

CAUTION

When raising or lowering the cover with the MWTS installed, avoid slamming the MWTS into the heat shield or slamming the cover closed.

f. **M240B Machine Gun.**

(1) *Mounting Procedures.* Before zeroing and qualifying with the AN/PAS-13, the gunner must mount the sight on the weapon. The M240B machine gun has a rail mount already attached to the cover assembly.

(a) Install the sight on the M240B rail mount by loosening the mounting knob located on the left side.

(b) Position the sight on the rail mount by placing the bar of the mount in a slot on the rail and hand tighten the knob on the mount until two clicks are heard. Any slot can be used as long as the mount does not hang over the edge of the rail (Figure G-32).

WARNING

Do not store the AN/PAS-13 with batteries installed.

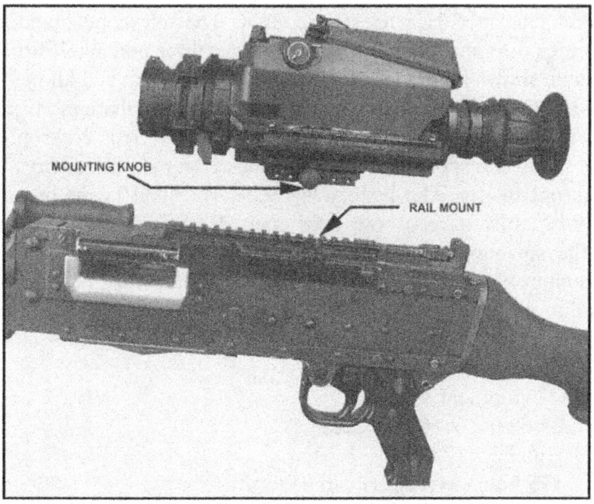

Figure G-32. Mounting the ANPAS-13 to the M240B.

(2) *Dismounting Procedures.*

(a) Unscrew the mounting knob on the left side.

(b) Lift up on the sight to remove it from the M240B machine gun.

g. **Boresighting Procedures**. Follow the instructions to boresight the weapon as outlined in TC 23-AIMSS. Choose the correct reticle for the weapon by pressing the reticle select switch for 2 seconds. Then scroll through the reticles until the correct one is found. Follow the instructions for boresighting the MWTS on the boresight target offset. Adjust the aiming point until the red dot is centered on the cross hair on the offset. You must boresight in both the narrow FOV and wide FOV.

h. **Zeroing Procedures**. Refer to TM 11-5855-309-12&P for zeroing procedures for the M249, M60, and M240B machine guns.

G-16. TRAINING STRATEGIES

The training strategies for the 5.56-mm and 7.62-mm caliber weapons with the MWTS are the same as with the AN/PVS-4 (as outlined in paragraph G-8) for both the night initial and night sustainment training strategies.

Section V. M145 STRAIGHT TELESCOPE (MGO)

The M145 telescope is a fixed 3.4-power, 28-mm optical sight that has been designed to accurately engage targets out to 1,200 meters. The optical sight weighs 24 ounces (681 grams) and is extremely rugged for rough field conditions. The sight has an 8.2-mm diameter exit pupil, which provides excellent vision in low light levels; for example, dawn and dusk, and also for rapid target acquisition.

G-17. DESCRIPTION

The M145 straight telescope is a telescopic sight. The telescope magnifies targets by three and a half times or it appears to bring the shooter three and a half times closer to the target. The telescope shows more clearly the strike of the round and allows more accurate shooting. In low-light conditions, removal of the laser filter enhances target detection far better than that visible to the naked eye (Figure G-33). The lens cover protects the lens when the sight is being transport or stored. The lens cover should always be kept closed when the sight is not in use. The battery has an O-ring that keeps out moisture. With practice you may be able to keep both eyes open. With both eyes open, the gunner is more aware of the surroundings and feels less strain on his eyes. The M145 straight telescope must remain matched with the same weapon, attached at the same slot in the rail system, or if not, it must be rezeroed.

G-18. DATA

The following is the equipment's data.

Weight:	24.0 ounces (681 grams)
Length:	7 inches (175mm)
Battery Life:	175 hours average (fresh battery) Sight is packed with a new battery.
Optics:	Antireflective coated lens system, (28mm) clear objective, x3.4 magnification.

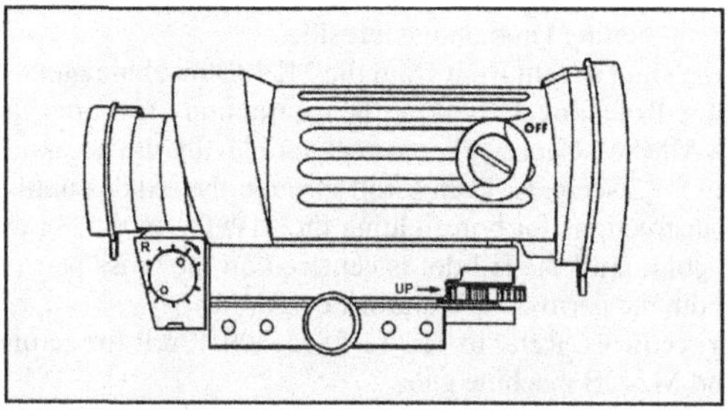

Figure G-33. M145 Telescope (MGO).

G-19. CONTROLS

The following is a brief description of the operation controls and its indicators.

a. **Elevation Adjustment Dial**. The elevation adjustment dial is used for zeroing the telescope to the weapon. The dial can only be rotated when the silver lock (1) is moved to the UP position. Turning the elevation adjustment dial (2) counterclockwise in the direction of the arrows one click moves the point of impact up 2.5 mm at 10 meters. Turning the elevation adjustment dial (2) clockwise (opposite direction to the arrow) one click moves the point of impact down 2.5 mm at 10 meters. Ensure that the silver lock (1) is moved down to prevent any further movement of the elevation adjustment dial (Figure G-34).

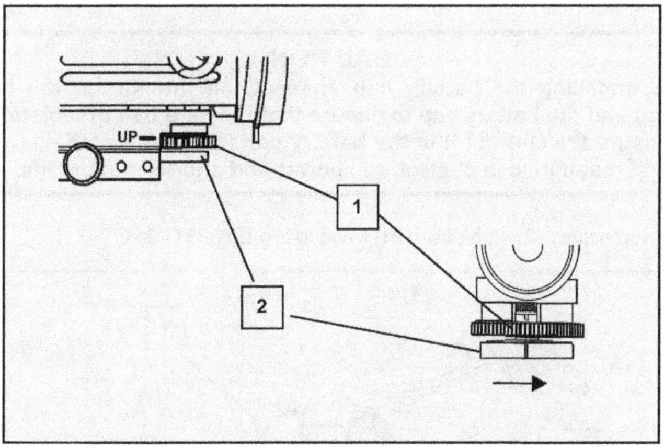

Figure G-34. Rotation of the elevation adjustment dial.

b. **Windage Adjustment Screw**. The windage adjustment screw is used when zeroing the weapon. Turning the windage adjustment screw (3) clockwise one click moves the point of impact left 2.5 mm at 10 meters. Turning the windage adjustment screw (3) counterclockwise, one click moves the point of impact right 2.5 mm at 10 meters (Figure G-35).

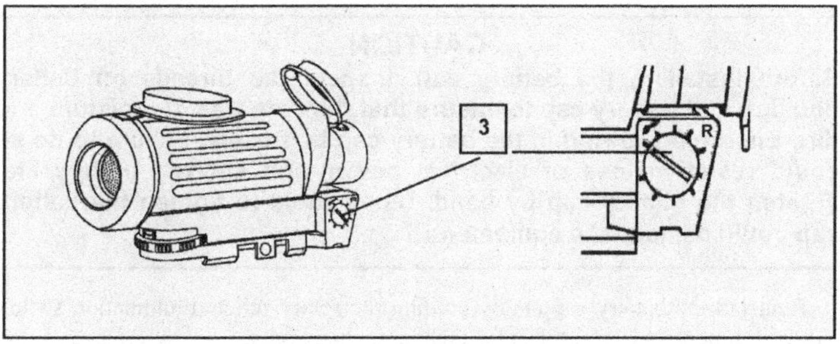

Figure G-35. Rotation of the windage adjustment screw.

G-20. BATTERY

This paragraph explains how to install and check the battery.

a. Remove battery cap (1) by turning it counterclockwise and holding the rotary reticle illumination switch (3) stationary.

CAUTION

Before installing the battery cap, inspect the threads on the battery housing and the battery cap to ensure that they are free of moisture and dirt. Ensure the O-ring (4) in the battery cap (1) is present. Failure to do so could result in loss of electrical power and shorten battery life.

b. Insert battery (2) with positive (+) end to cap (Figure G-36).

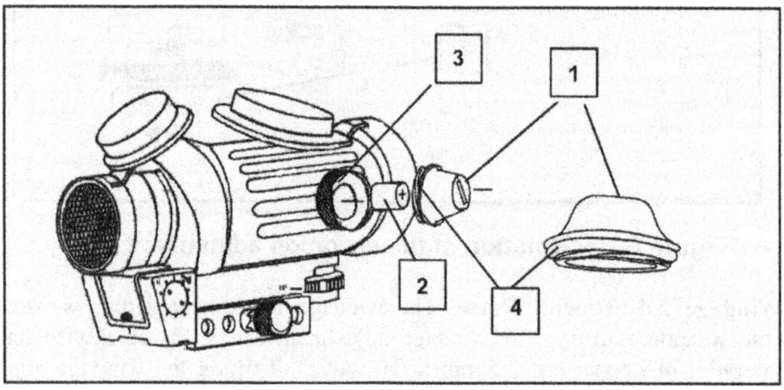

Figure G-36. Installing and checking batteries.

CAUTION

Before installing the battery cap, inspect the threads on battery housing and battery cap to ensure that they are free of moisture and dirt. Ensure the O-ring in the battery cap is present. Failure to do so could result in loss of electrical power and shorten battery life. Tighten the battery cap by hand. Using tools to tighten the battery cap could damage the equipment.

c. Reinstall the battery cap (1) by holding the rotary reticle illumination switch (3) stationary turning clockwise until snug. Tighten by hand only.

d. Open the rear lens cover (4). Turn the rotary reticle illumination switch (3) and look through rear lens. Verify that the reticle is illuminated. If not, replace the battery. When finished, turn the rotary switch to the OFF position, and then replace the rear lens cover (Figure G-37).

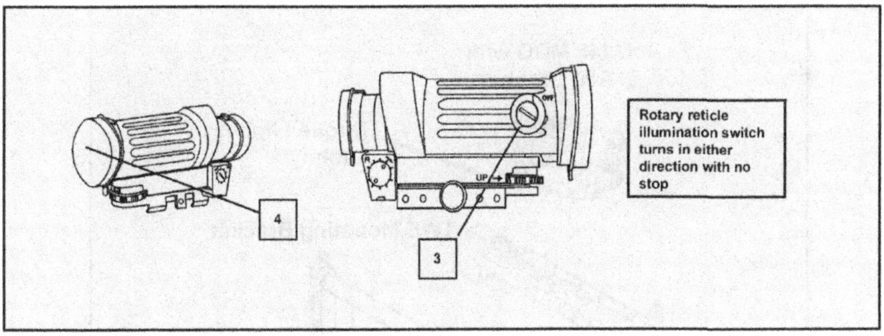

Figure G-37. Rotary reticle illumination switch.

G-21. INSTALLTION

The M145 straight telescope mounts directly to the mounting rail on the M249, and M240B machine guns. It is necessary that you adjust the position of the M145 telescope either backwards or forward on the rail in order to achieve the correct eye relief (distance of the eye from the back of the telescope). If the same sight is installed in the same position slot on the rail on the same weapon, rezeroing is not required.

CAUTION

By hand, tighten the torque limiting knob until you hear two clicks. Using tools to tighten the mounting hardware could damage equipment.

a. Back off on the torque limiting knob just enough for the rail grabber to go over the rail. Do not force the torque limiting knob past its intended stop. Mount the M145 straight telescope firmly over the rail. Ensure that the mount is seated squarely over the rail. Tighten the torque limiting knob (clockwise) until it rotates with two clicks. Ensure that the mount is securely fastened before commencing eye relief adjustment. The sight is now mounted to the weapon about 3 inches (70 mm) in front of the firing eye (Figure G-38 [M249] and Figure G-39 [M240B], page G-38). Assume a comfortable firing position and achieve a good stockweld (at trigger pull length) with both eyes closed. Open the sighting eye and compare the view through the scope with the following figures:

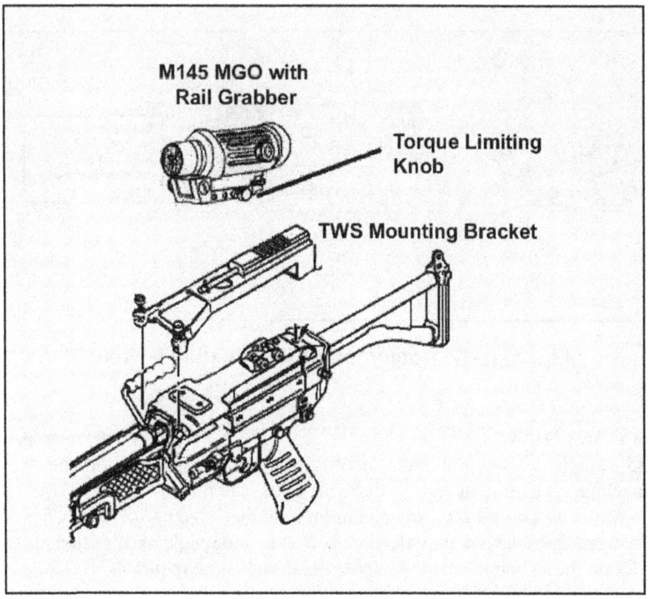

Figure G-38. Mounting the MGO to the M249.

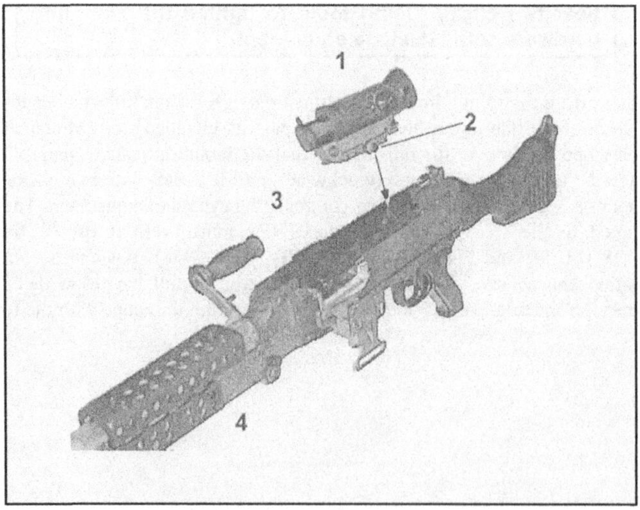

Figure G-39. Mounting the MGO the M240B.

(1) If the target scene fills the scope to provide the maximum field of view, the correct eye relief has been attained (Figure G-40). No further repositioning of the M145 on the mounting rail is required.

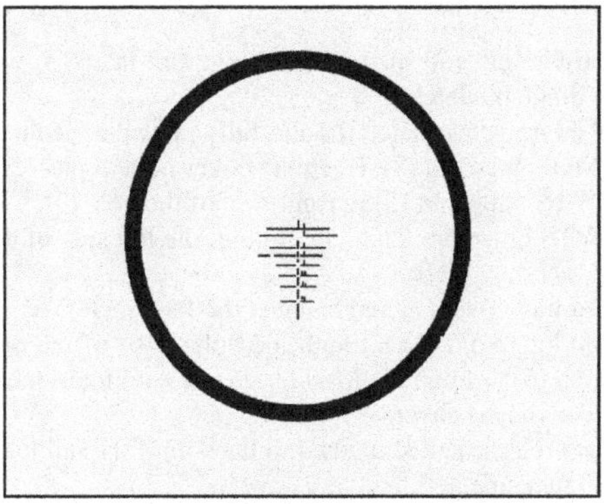

Figure G-40. Correct eye relief.

(2) If the target scene does not fill the sight's field of view, the optical sight must be repositioned on the rail either forwards or backwards (Figure G-41).

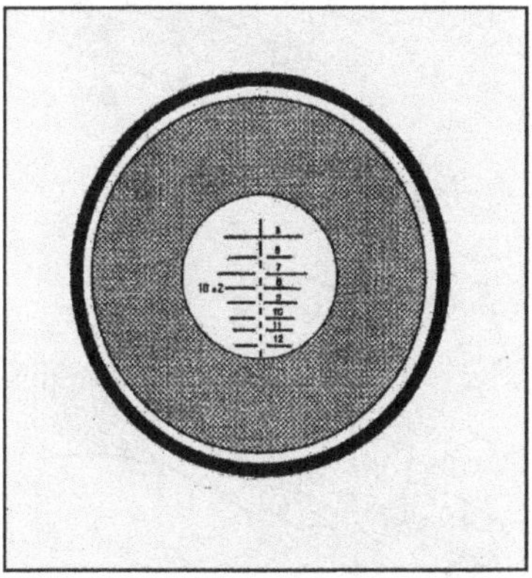

Figure G-41. Incorrect eye relief.

b. The optical sight must be repositioned for correct eye relief. Loosen the torque limiting knob and move the optical sight in the appropriate direction (forward or backward), which provides the full field of view.

c. Repeat a until the correct sight picture is obtained as shown in Figure G-41.

G-22. MOUNTING PROCEDURES (M60 ONLY)

Refer to Figures G-42a and G-42b when mounting MWTS on M60 machine gun.

a. Press on the end of the MWTS hinge pin latch (1) and remove the MWTS hinge pin latch and the MWTS hinge pin (2) from the M60 bracket (3).

b. Install the MWTS hinge pin (2) and the MWTS hinge pin latch (1) on the M60 machine gun (4).

c. Press on the right side end of the M60 hinge pin latch (5) with a pointed object and remove the M60 hinge pin latch.

d. Release the feed tray cover latch (6) and fully raise the feed tray cover (7).

e. Remove the M60 hinge pin (8). Keep the cover held in place.

f. Install the MWTS hinge pin (2) in right side of the hole (9).

g. Insert the MWTS hinge pin latch (1) through the left side of the MWTS hinge pin (2) until interlocked.

h. Lower the feed tray cover (7) and connect the feed tray cover latch (6).

i. Insert the M60 hinge pin (8) through the holes (10) of the bracket (3) and insert the M60 hinge pin latch (5) through the M60 hinge pin until interlocked.

j. Close the objective lens cover.

k. Push in the cam release and slide the rail back until the rail locks in place.

l. Open the M60 machine gun feed tray cover.

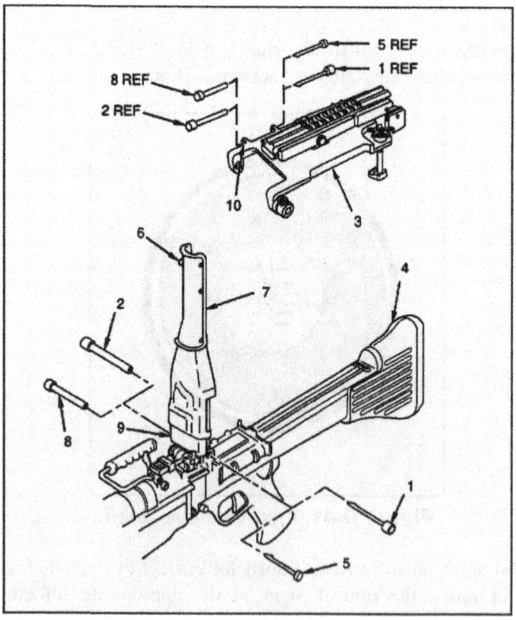

Figure G-42a. Mounting the MGO to the M60 machine gun.

m. Install the bracket (3) on the M60 machine gun (4). Fold the rear sight (11) forward. Place the pins (12) of the bracket (3) in the cups (13) of the MWTS hinge pin (2) and the MWTS hinge pin latch (1) and tighten the nut (14) by hand. Tighten the locking nut (15) by hand. Place the groove (16) of the two clamping bars (17) on the edge of the feed tray cover (4) and tighten the two thumbwheels (18).

n. Install the MGO on the rail. Loosen the clamping knob on the mount. Select the slot on the rail for mounting. Any slot may be used as long as the mount does not hang over edge of rail. Place the bar of the mount in the slot of the rail and tighten the knob on the mount by hand until two clicks are heard.

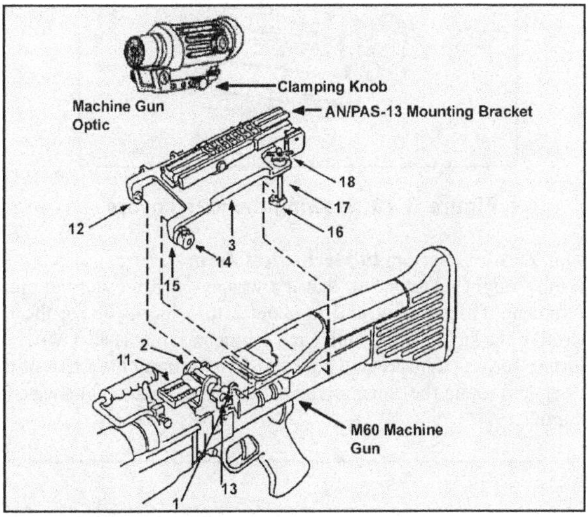

Figure G-42b. Mounting the MGO to the M60 (continued).

G-23. ZEROE PROCEDURES

The following paragraphs explain the different ways to zero the M145 straight telescope on the M240B, M60, and M249 machine guns.

a. **Zeroing to Weapon**. Zeroing the M145 straight telescope aligns the sight to the barrel of the machine gun so that the point of aim equals the point of impact. Adjustment of the M145 straight telescope is centered at the factory.

(1) Open the front (1) and the rear (2) lens covers.

(2) Turn each cover inside out to stow the lens covers while the sight is being used (Figure G-43).

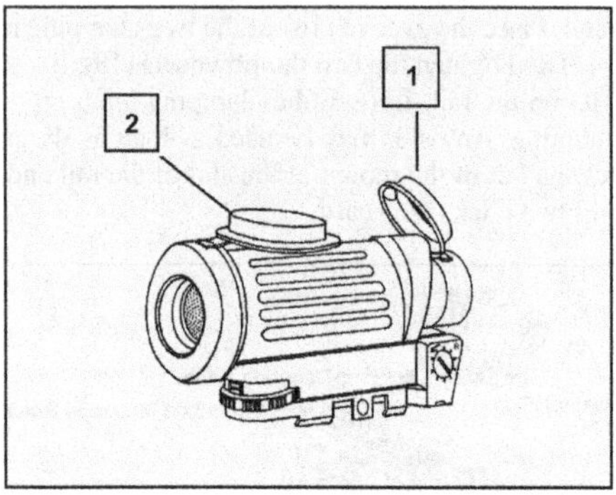

Figure G-43. Stowing the lens covers.

b. **10-Meter Zeroing, Setting to Mechanical Zero.**

(1) Adjust the straight telescope so that the weapon's barrel and the optical sighting axis are in alignment. The sighting axis is about 2 to 3 inches above the machine gun barrel, and therefore the strike of the bullet at a 10-meter range is also about 2 to 3 inches low without further zeroing adjustment (Figure G-44). To bring the strike of the bullet up, lift the silver lock and rotate the elevation adjustment dial counterclockwise (to the right) about one full turn.

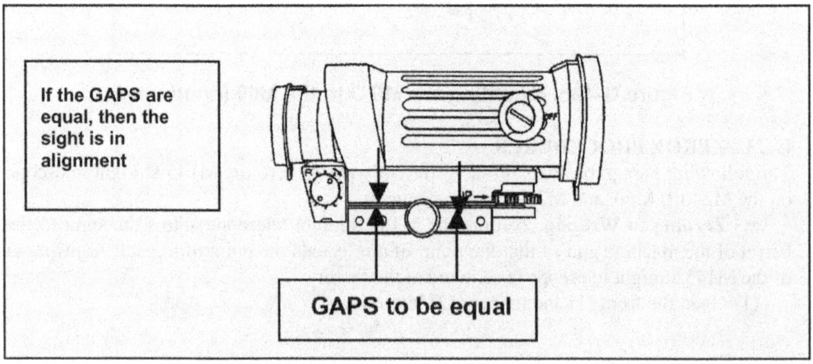

Figure G-44. GAPS to be equal.

(2) Adjust the windage dial to center the markings on the front of the sight. This adjustment brings the bullet's point of impact to the middle of the point of aim

(Figure G-45). Each click of zeroing adjustments makes a 2.5 mm movement of the point of impact at 10 meters.

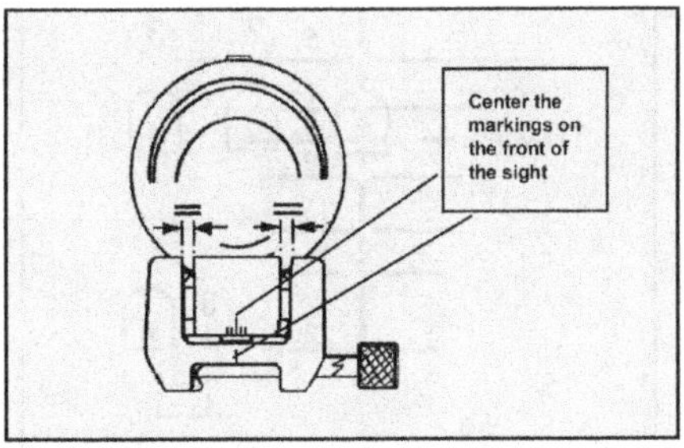

Figure G-45. Centering the markings.

(3) To move the point of impact to the right, turn the windage adjustment screw (2) counterclockwise with the arrow marked on the dial.

(4) To move the point of impact to the left, turn the windage adjustment screw (2) clockwise opposite to the arrow.

(5) To move point of impact up, turn the elevation adjustment screw (1) counterclockwise (right) with the direction of the arrow marked and "UP."

(6) To move the point of impact down, turn the elevation adjustment screw (1) clockwise (left) opposite to the arrow (Figure G-46).

Figure G-46. Adjusting point of impact.

c. **10-Meter Range Zeroing.** In the zeroing process, groups of three-single shot rounds are fired at a target. After each three rounds, the center of the group has to be determined.

(1) Look through the telescope and align the reticle's 10-meter zeroing mark on the center base of the aiming points on the basic machine gun marksmanship target (Figure G-47, page G-44).

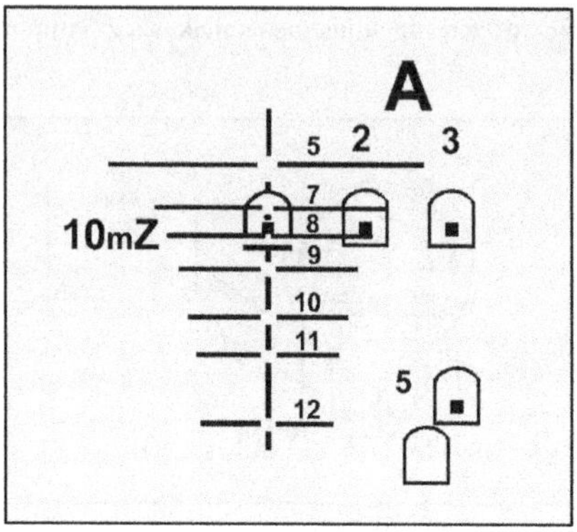

Figure G-47. 10-meter reticle aiming point.

(2) Fire three-single rounds loaded individually without making any sight adjustments.

(3) The three-round shot group should be within a 4-cm circle to establish the center of the shot group in relation to the center base of the aiming paster.

(4) Measure the amount of movement that is required left or right (windage) and either up or down (elevation) to move the three-round shot group onto the center of the aiming paster.

(5) Upon completion, return to the firing line to make corrections to the weapon and refire a three-round shot group to confirm zero (Figure G-48).

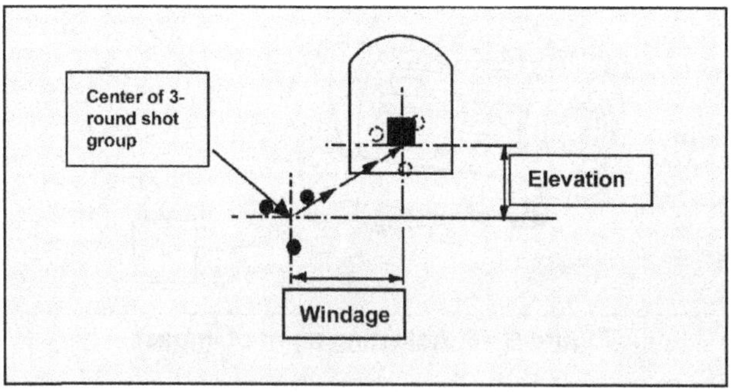

Figure G-48. Three-round shot group with adjustments.

(6) Repeat the above steps until the strike of the round is coincident with the center of the target. Close the silver lock down to prevent any further movement of the elevation zeroing adjustment dial.

d. **Field Zeroing at 500-Meter Range**. Look through the telescope and align the reticle's 500-meter mark on the center of mass of the double "E" silhouette target (Figure G-49):

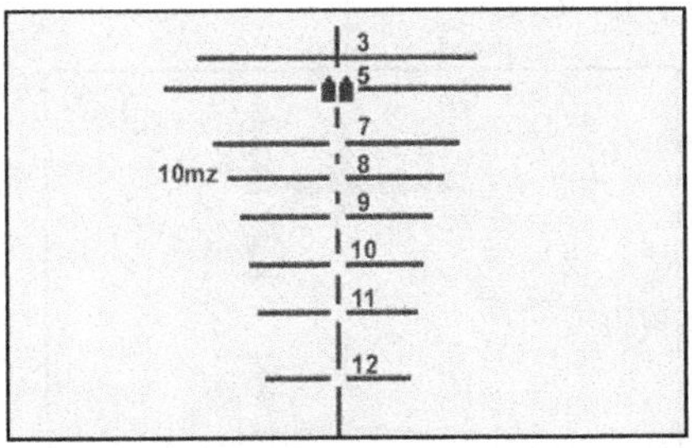

Figure G-49. 500-meter reticle aiming point.

(1) M240B, M60, and M249 machine guns: Fire a 5- to 7-round burst.

(2) Observe the impact of the rounds.

(3) Determine the direction of impact to be moved (up or down, left or right).

(4) Estimate or measure the amount of movement required to move the strike of the round to the center of the target (at 500 meters; 5 inches equals one click of adjustment in both windage and elevations). Repeat these steps until the strike of the round coincides with the center of the target.

NOTE: Close the silver lock down to prevent any further movement of the elevation zeroing adjustment dial. The M145 straight telescope is now zeroed and ready for operational shooting.

e. **Using Reticle to Estimate Range**. The vertical gap in the stadia lines is for estimating ranges. The height of gaps in the stadia lines represents a 60-inch high target at the range noted; for example, 5, 7, 8,, 9, 10, 11, or 1,200 meters (Figure G-50).

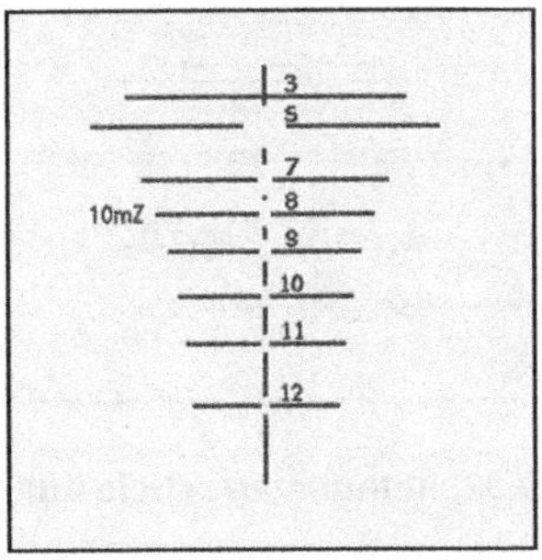

Figure G-50. Reticle stadia lines.

f. **Illuminating the Reticle.** For low light operations, the reticle can be illuminated to show the 300-, 500-, 700-, and 800-meter aiming marks (Figure G-51). The sight is equipped with variable intensity LED illumination of the reticle. It has 10 positions: one OFF position and 9 positions for different reticle intensity settings.

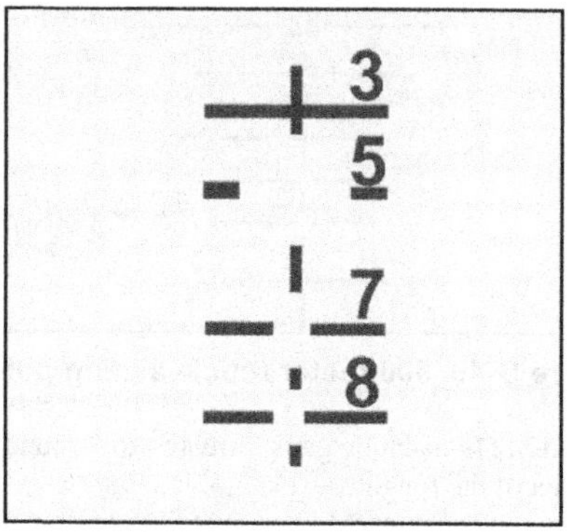

Figure G-51. Illuminated reticle.

(1) To make reticle illumination adjustments, turn the rotary switch (1) clockwise. The intensity of the illumination increases the further the switch is turned (Figure G-52).

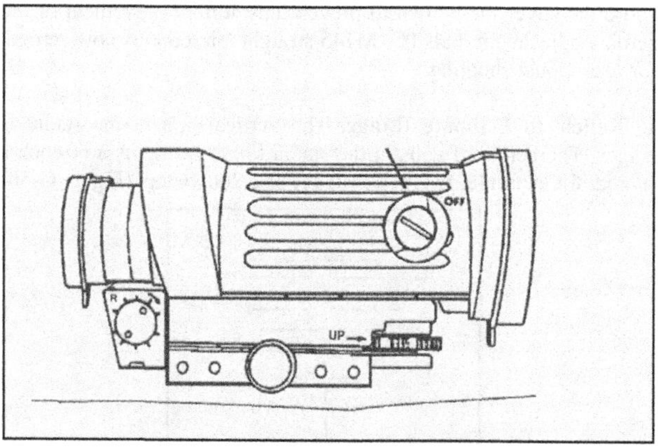

Figure G-52. Illuminated reticle adjustments.

(2) Turn the rotary switch to the OFF position when the telescope is being used during normal daylight or when illumination is not required (Figure G-53). Ensure the reticle illumination switch is turned to the OFF position when not required.

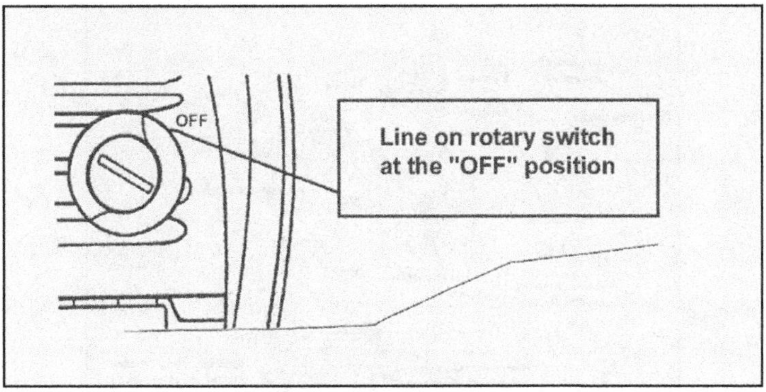

Figure G-53. OFF switch.

G-21. MOUNTING AND ZEROING THE AN/PVS-4

Zeroing aligns the AN/PVS-4 to the M249. The sight may be zeroed during daylight or darkness. (See TM 11-5855-213-10.) If done during daylight, the daylight cover must be used. To obtain a precise zero, it is best done at 300 meters and at night. Once an AN/PVS-4 has been zeroed on an M249 machine gun, any soldier who knows how to use the reticle should fire the weapon effectively. However, there may be some changes in zero when the objective focus is adjusted to engage targets at various ranges and when the diopter focus is adjusted for the vision of different firers. A metal target is excellent for zeroing purposes, because the strike of the round can be easily observed with an AN/PVS-4. The procedures to zero are as follows.

a. **Mounting the Bracket and Device.** Before zeroing and qualifying with the AN/PVS-4, the gunner must mount the bracket and sight onto his weapon.

> **CAUTION**
>
> When mounting an AN/PVS-4 to the mounting bracket, make sure that the hole for the screw in the AN/PVS-4 is aligned and flush against the bracket screw. If not, the screw will strip the threads in the screw hole of the AN/PVS-4 and prevent use with the M249 machine gun.

(1) Place the mounting bracket on top of the feed cover mechanism assembly so that the two forked ends are secured around the headless pins.

(2) Remove the screw cover behind the rear sight assembly, and screw the bracket knob in until it is tight.

(3) Position the AN/PVS-4 on top of the bracket so that the mount of the AN/PVS-4 is aligned with the mounting knob of the bracket.

(4) Turn the mounting knob clockwise until the AN/PVS-4 is tight (Figure G-54).

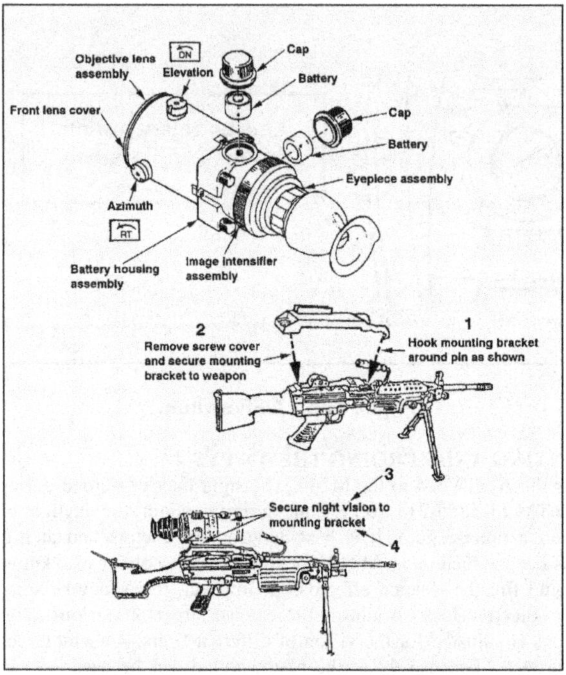

Figure G-54. Mounting the AN/PVS-4 to the M249 light machine gun.

b. **Seating the Device**. Once the device is mounted, the gunner fires a 3-round burst to seat the device, checks and tightens the mounting knob, and then fires another 3-round burst. He checks the device to ensure it is settled and securely fastened and tightens the mounting knob if necessary. He does not fire at the boresight target during this procedure.

c. **Centering the Reticle in the Field of View**. The gunner turns the device on and centers the reticle pattern in the field of view by using the azimuth and elevation actuators. To be accurate, he does this by rotating the elevation and azimuth actuators from one side to the other and from top to bottom, while counting the number of clicks. The elevation actuator has the down direction marked DN with an arrow. This moves strike of the round. The azimuth actuator has the right direction marked with RT with an arrow. This also moves the strike of the round. He divides the number of clicks for each by two and moves the elevation and azimuth actuators that number of clicks. This manually centers the reticle in the field of view horizontally and vertically. This enables the gunner to reach an accurate boresight between the point of aim (reticle) and the center of the bore (Figure G-55).

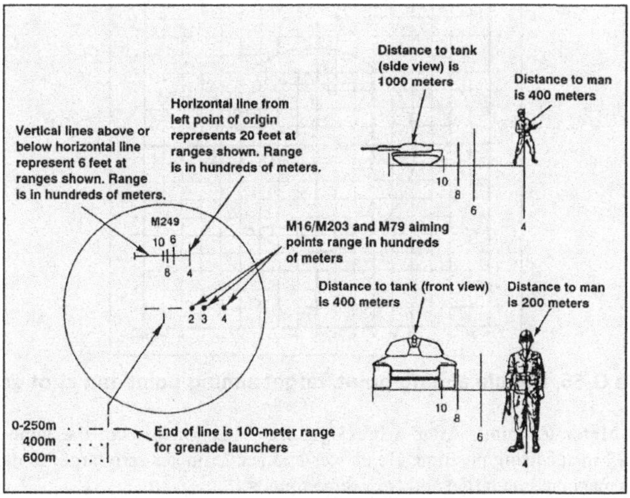

Figure G-55. Centered reticle pattern.

d. **Confirming the Boresight**. To do this, the gunner centers and affixes a 25-meter (M16A2) zero target to the back of a basic machine gun paster target. This provides a large, clear surface for identifying the strike of the round. Then, he emplaces the target 10 meters from the firing position. The gunner places the reticle aiming point on the 25-meter zero target aiming point (Figure G-56, page G-50) and fires a single round. If the round impacts anywhere near the aiming point, he fires two more rounds to establish his group.

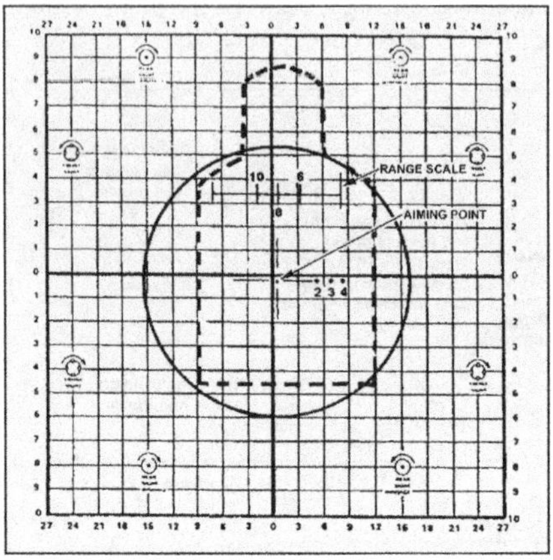

Figure G-56. Reticle aiming point, target aiming point and shot group.

e. **25-Meter Zeroing**. After a boresight has been established, the gunner moves back to a 25-meter firing position. He affixes another 25-meter zero target to the back of a 10-meter machine gun target and fires three rounds.

(1) Locate and triangulate the center of the shot group. From the center of the shot group, adjust the reticle to move the center of the shot group to a point 8 cm below and 2 cm right of the target aiming point (Figure G-57). This location on the 25-meter zero target is 9 squares below (8 cm divided by .9) and 2 squares right (2 cm divided by .9) of the target aiming point. Make the adjustment using the AN/PVS-4 azimuth and elevation adjustment actuators. Each square on the 25-meter zero target is .9 cm. Each click of the actuators moves the strike of the round .25 cm (or .1 inch) at 10-meters. Therefore, 4 clicks on either the elevation or azimuth actuator moves the strike of the round one square.

(2) After making the adjustments, assume a stable position, place the reticle aiming point on the target aiming point, and fire three more single rounds. Repeat the process until the rounds impact within the desired location (9 squares below and 2 squares right).

(3) If you miss the 25-meter zero target with the first round but strike the 10-meter machine gun paster target, make a large adjustment with the elevation and azimuth actuators. Continue this process with three-round groups and adjustments until the rounds strike the desired location.

(4) Do not record the zero because when the AN/PVS-4 is dismounted and remounted on the same M249, some changes may occur. It is best to zero each time.

(5) Once the AN/PVS-4 is mounted and boresighted, fire a 3-round burst at the center base of the target and note the strike of the rounds. While maintaining the reticle aiming point on the target, move the reticle aiming point to the strike of the rounds by manipulating the elevation and azimuth actuators.

(6) Acquire a good sight picture on the target with the reticle aiming point once again and fire another 3-round burst. Note the strike of the rounds and repeat the process until the rounds impact on target.

f. **Mounting the AN/PVS-4 on the M60**. Before zeroing and qualifying with the AN/PVS-4, the gunner must mount the bracket and sight onto his weapon. The M60 machine gun requires a mounting bracket.

> **CAUTION**
> **When mounting an AN/PVS-4 to the mounting bracket, make sure that the hole for the screw in the AN/PVS-4 is aligned and flush against the bracket screw. If not, the screw strips the threads in the screw hole of the AN/PVS-4 and prevents its use with the machine gun.**

(1) The gunner removes the M60 hinge pin latch and hinge pin from the over assembly by pressing on the latch (open end of pin) with an empty cartridge case, and separates the latch and pin. He places the pin and latch in the aiming guides on the left side of the mounting bracket and presses together.

(2) The gunner positions the mounting bracket assembly on top of the machine gun cover so that the holes in the front of the bracket align with the cover assembly pin holes.

(3) The gunner inserts the longer hinge pin supplied with the bracket through the bracket-and-cover assembly, and secures by inserting the hinge pin latch.

(4) The gunner loosens the wing nuts on both leg clamps and positions the clamps under the cover assembly. He secures the mounting bracket by tightening the wing nuts firmly (Figure G-57).

(5) The split washer is placed next to the wing nut and the flat washer is placed next to the bracket.

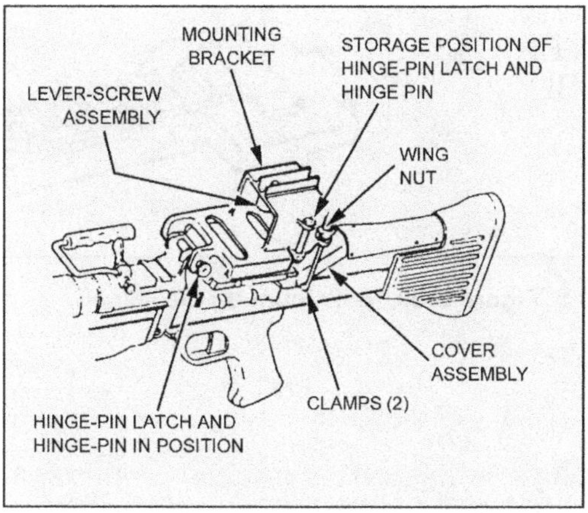

Figure G-57. Installation of mounting bracket.

(5) The gunner installs the sight on the M60 mounting bracket assembly by positioning it in the groove on top of the bracket so that the scribe line on the bracket is aligned with the scribe line on the sight mounting adapter. He tightens the lever screw assembly to secure the sight to the bracket. He uses an empty cartridge case placed over the lever arm to increase leverage as the screw is tightened. (See Figure G-58.)

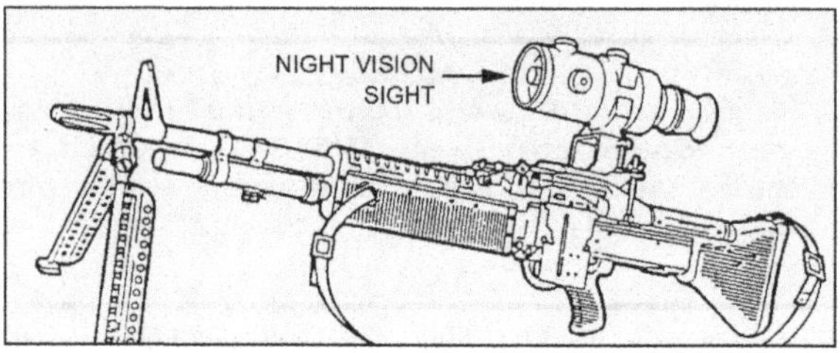

Figure G-58. Mounting the AN/PVS-4.

g. **Mounting the AN/PVS-4 on the M240B.**
Before zeroing and qualifying with the AN/PVS-4, the gunner must mount the sight onto his weapon. The M240B machine gun has a rail mount already attached to the cover assembly (Figure G-59). The gunner installs the sight on the M240B rail mount by loosen the mounting knob located on the left side. Positioning it in a slot on the rail mount. Any slot may be used as long as the mount does not hang over the edge of the rail. By placing the bar of the mount in a slot on the rail and hand tighten knob on mount until clicking noise is heard (2 clicks). As long as the gunner places the sight in the same slot after zeroing the sight will retain its zero to a degree. (See Figure G-60.)

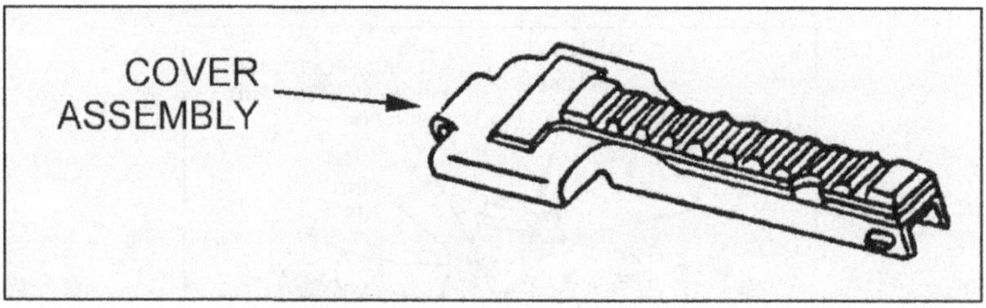

Figure G-59. Rail mount on the M240B.

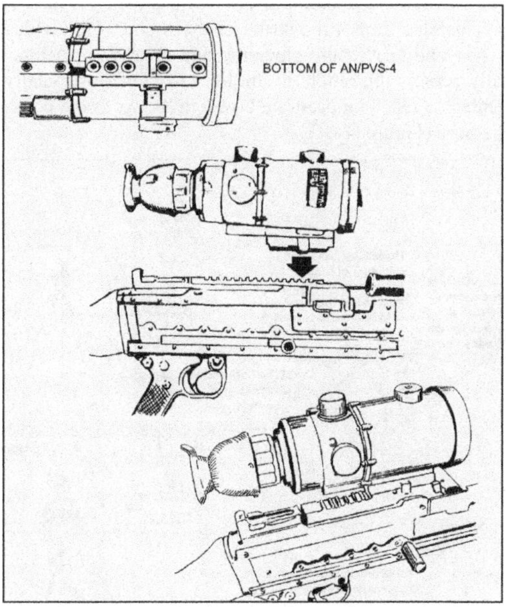

Figure G-60. Mounting the AN/PVS-4.

h. **Zeroing the AN/PVS-4 to the M60/M240B**. Zeroing aligns the AN/PVS-4 to the machine gun. The sight may be zeroed during daylight or darkness. (AN/PVS-4 TM 11-5855-213-10) When zeroing during daylight, the daylight cover must be used. To obtain a precise zero, it is best done at 300 meters and at night. Once the AN/PVS-4 is zeroed on the machine gun, any soldier who knows how to use the reticle should fire the weapon effectively. However, there may be some changes in zero when the objective focus is adjusted to engage targets at various ranges and when the diopter focus is adjusted for the vision of different firers. A metal target is excellent for zeroing purposes, because the strike of the round can be easily observed with an AN/PVS-4. The procedures to zero are as follows:

(1) *Seating the Device.* Once the device is mounted, the gunner fires a seven-round burst to seat the device, checks and tightens the mounting knob, and then fires another three-round burst. He checks the device to ensure it is settled and securely fastened and tightens the mounting knob if necessary. He does not fire at the boresight target during this procedure.

(2) *Centering the Reticle in the Field of View of the AN/PVS-4.* The gunner turns the device on and centers the reticle pattern in the field of view by using the azimuth and elevation actuators. To be accurate, he does this by rotating the elevation and azimuth actuators from one side to the other and from top to bottom, while counting the number of clicks. (The elevation actuator has the down direction marked DN with an arrow.

This moves strike of the round. The azimuth actuator has the right direction marked with RT with an arrow. This also moves the strike of the round.) He divides the number of clicks for each by two and moves the elevation and azimuth actuators that number of clicks. This manually centers the reticle in the field of view horizontally and vertically. This enables the gunner to reach an accurate boresight between the point of aim (reticle) and the center of the bore (Figure G-61).

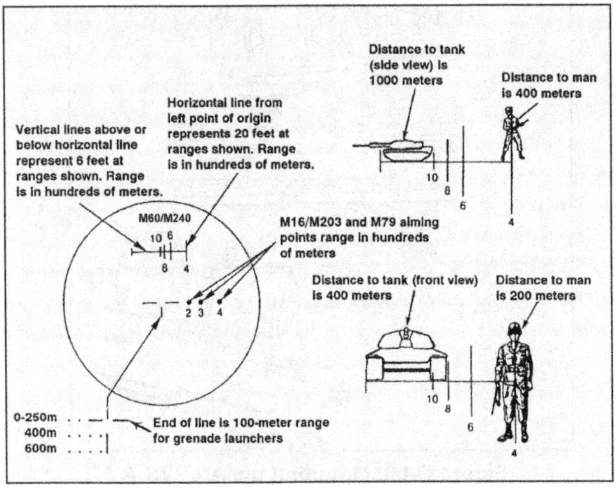

Figure G-61. Centered reticle pattern.

(3) *Confirming the Boresight.* To confirm the boresight, the gunner centers and affixes a 25-meter (M16A2) zero target to the back of a basic machine gun paster target. This provides a large, clear surface for identifying the strike of the round. Then, he emplaces the target 10-meters from the firing position. The gunner places the reticle aiming point on the 25-meter zero target aiming point (Figure G-62) and fires a single round. If the round impacts anywhere near the aiming point, he fires two more rounds to establish his group.

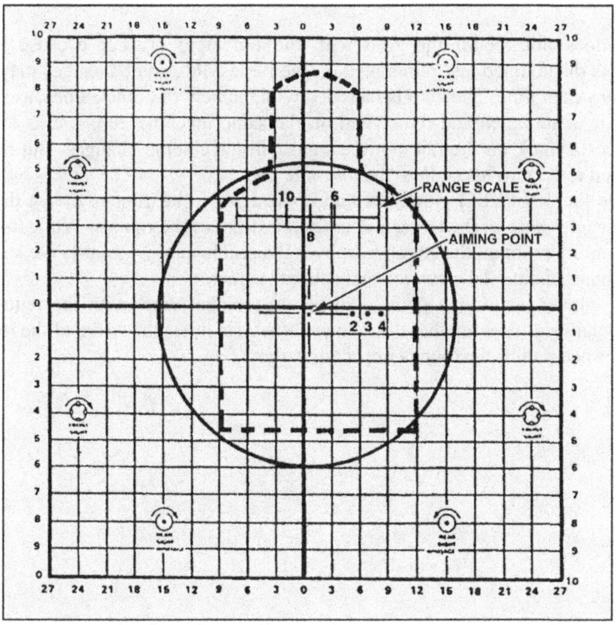

Figure G-62. Reticle aiming point, target aiming point and shot group.

(4) *25-Meter Zeroing AN/PVS-4.* After a boresight has been established, the gunner moves back to a 25-meter firing position. He affixes another 25-meter zero target to the back of a 10-meter machine gun target and fires three rounds.

(a) AN/PVS-4 (M60 and M240B) The gunner locates and triangulates the center of the shot group. From the center of the shot group, he adjusts the reticle to move the center of the shot group to a point 8 cm below and 2 cm right of the target aiming point. This location on the 25-meter zero target is 9 squares below (8 cm divided by .9) and 2 squares right (2 cm divided by .9) of the target aiming point. He makes the adjustment using the AN/PVS-4 azimuth and elevation adjustment actuators. Each square on the 25-meter zero target is .9 cm. Each click of the actuators moves the strike of the round .25 cm (or .1 inch) at 10 meters. Therefore, 4 clicks on either the elevation or azimuth actuator moves the strike of the round one square.

(b) After making the adjustments, the gunner assumes a stable position, places the reticle aiming point on the target aiming point, and fires three more single rounds. He repeats the process until the rounds impact within the desired location (9 squares below and 2 squares right).

(c) If he misses the 25-meter zero target with the first round but strikes the 10-meter machine gun paster target, he makes a large adjustment with the elevation and azimuth actuators. He continues this process with three-round groups and adjustments until the rounds strike the desired location.

(d) He does not record the zero with the old style bracket because when the AN/PVS-4 is dismounted and remounted on the same M60, some changes may occur. It is best to zero each time. The zero is recorded with the new rail bracket because when the AN/PVS-4 is dismounted and remounted on the same machine gun, and as long as the gunner places a mark on the rail slot the sight will have some changes. But the gunner does not need to re-zero his sight to the machine gun.

(e) Once the AN/PVS-4 is mounted and boresighted, the gunner fires a three-round burst at the center base of the target and notes the strike of the rounds. While maintaining the reticle aiming point on the target, he moves the reticle aiming point to the strike of the rounds by manipulating the elevation and azimuth actuators.

(f) The gunner acquires a good sight picture on the target with the reticle aiming point once again and fires another three-round burst. He notes the strike of the rounds and repeats the process until the rounds impact on target.

APPENDIX H

10-METER BORE LIGHT/ 25-METER TARGET OFFSETS

This appendix provides all the 10-meter/25-meter target offsets for the M249/M240B weapons mounted with iron sight, optics, MILES, or aiming lasers. A blank reproducible 10-meter target offset is provided along with a table and example of every weapon configuration. The M16A2 300-meter zero target is used for 25-meter zeroing with all weapon configurations.

H-1. 10-METER TARGET OFFSET

To mark the proper 10-meter target offset, the gunner must do the following:

a. Find the correct template for your weapon configuration.

b. Count the number of squares starting from center of the bore light circle on the offset to the desired point of aim. Each template also provides you with a number formula for the proper offset. Example (L2.0, U2.4): Starting from the center of the bore light circle (0.0, 0.0) move LEFT 2 squares and UP 2.4 squares.

c. Place the appropriate symbol or mark (Figure H-1).

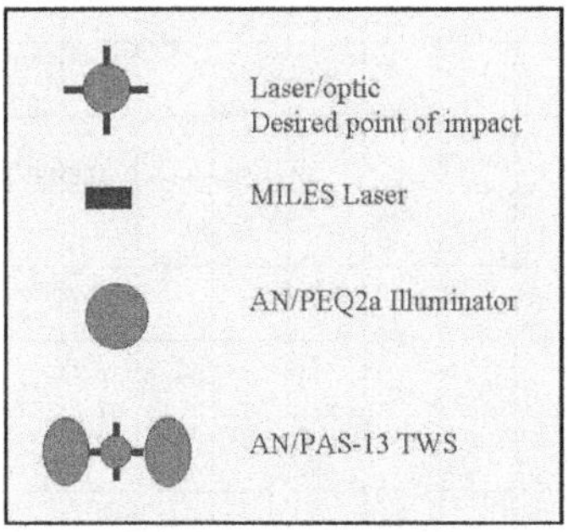

Figure H-1. 10-meter target offset symbols.

H-2. 25-METER TARGET OFFSET

To mark the proper 25-meter target offset do the following:

a. Use only M16A2 300-meter zero targets.

b. Find the correct target template for your weapon configuration.

c. Count the number of squares starting from the center of the 300-meter zeroing silhouette.

d. Mark the designated strike point by drawing a small circle at the appropriate number of squares from the center of the 300-meter zeroing silhouette.

e. Draw a 4-centimeter by 4-centimeter square keeping the designated strike point center mass of the 4-centimeter by 4-centimeter square.

H-3. REPRODUCE

Figure H-2 is an example completed 10-meter offset target. Reproduce onto 8 1/2" x 11" paper the blank target located in the back of this manual The quick reference card (Figure H-3) is located on the back of the target. The blank target can be laminated and used over and over again.

10-METER BORESIGHT OFFSET TARGET

For use of this form, see FM 3-22.68. Proponent agency is TRADOC.

See the back of this form for a quick reference card.

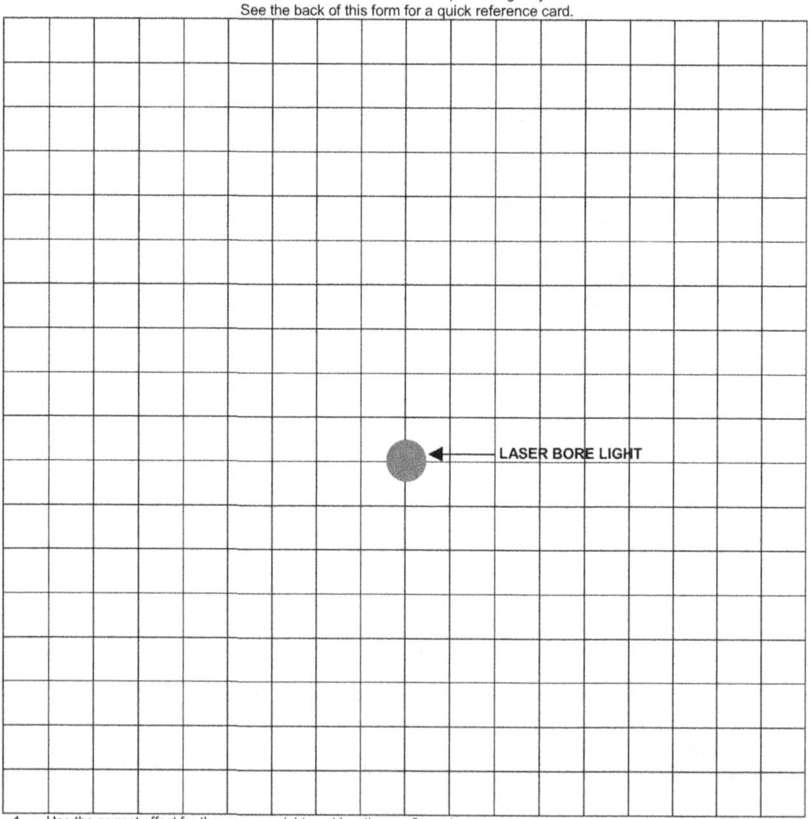

1. Use the correct offset for the weapon, sight, and location configured.
2. Stabilize the weapon and the offset.
3. Zero the bore light while inside the barrel of the weapon.
4. Align the laser of the bore light with the dot on the 10-meter offset.
5. Align the MILES laser with the MILES rectangle on the 10-meter offset (if applicable).
6. Adjust the aiming laser until it is centered on the cross hair.
7. Place the optic aim point centered on the cross hair and make adjustments on the optic until the bore light laser is aligned with the dot on the 10-meter offset.
8. Reconfirm that all devices are still aligned to their aiming mark.

DA FORM 7476-R, OCT 2002 | Page 1 of 1

Figure H-2. Example completed DA Form 7476-R.

WPN	ACCESSORY	MOUNT	RANGE ZEROED TO	ZERO TARGET OFFSET	BORESIGHT TARGET OFFSET	MILES OFFSET
M249	IRON SIGHTS	NA	400m	NA	TBD	1.9L/.5U
M249	MGO	IFTC RAIL	400m	TBD	0.0/7.9U	1.9L/.5U
M249	MGO	TWS BRACKET	400m	TBD	0.0/2.15U	1.9L/.6U
M249	AN/PAQ-4C	TWS BRACKET TOP	400m	0.5R/1.5U	1.85L/7.7U	1.9L/.5U
M249	AN/PAQ-4C	AN/PVS4 BRACKET	400m	2.5R/1.5D	4.1L/6.1U	1.9L/.5U
M249	AN/PAQ-4C	INSIGHT RAIL GRABBER WITH IFTC	400m	TBD	1.75L/4.69U	1.9L/.5U
M249	AN/PAQ-4C	PICATINNEY RAILGRABBER WITH IFTC	400m	1.75R/0.0	1.75L/5.39U	1.9L/.5U
M249	AN/PAQ-4C	INSIGHT RAIL GRABBER FORWARD RAILS RIGHT	400m	5.9R/9.6D	5.9R/4.0D	1.9L/.5U
M249	AN/PAQ-4C	INSIGHT RAIL GRABBER FORWARD RAILS LEFT	400m	6.0R/13.3D	6.OR/8.3D	1.9L/.5U
M249	AN/PAQ-4C	PICATINNY RAIL GRABBER WITH AL SPACER FORWARD RAILS RIGHT	400m	7.7R/9.6D	7.7R/4.0D	1.9L/.5U
M249	AN/PAQ4-C	PICATINNY RAIL GRABBER FORWARD RAILS LEFT	400m	7.6R/13.3D	7.6R/8.3D	1.9L/.5U
M249	AN/PEQ-2A	TWS BRACKET TOP	400m	1.8L/2.7D	1.8R/7.95U	1.9L/.5U
M249	AN/PEQ-2A	AN/PVS4 BRACKET WITH SPACER	400m	5.0R/4.0D	.45L/6.5U	1.9L/.5U
M249	AN/PEQ-2A	INSIGHT RAIL GRABBER WITH IFTC	400m	2.0L/1.5U	1.95R/4.79U	1.9L/.5U
M249	AN/PEQ-2A	PICATINNEY RAIL GRABBER WITH IFTC	400m	2.0L/0.5D	1.95R/6.49U	1.9L/.5U
M249	AN/PEQ-2A	INSIGHT RAIL GRABBER FORWARD RAILS RIGHT	400m	6.1R/13.2D	6.1R/7.6D	1.9L/.5U
M249	AN/PEQ-2A	INSIGHT RAIL GRABBER FORWARD RAILS LEFT	400m	6.0R/9.4D	6.0R/4.4D	1.9L/.5U
M249	AN/PEQ-2A	PICATINNY RAIL GRABBER WITH AL SPACER FORWARD RAILS RIGHT	400m	7.8R/13.2D	7.8R/7.6D	1.9L/.5U
M249	AN/PEQ-2A	PICATINNY RAIL GRABBER FORWARD RAILS LEFT	400m	7.6R/9.4D	7.6R/4.4D	1.9L/.5U
M249	AN/PVS-4	IFTC TOP WITH SPACER	400m	0.0/4.3D	0.0/10.00U	1.9L/.5U
M249	AN/PVS-4	AN/PVS-4 BRACKET	400m	2.5R/4.9D	2.25L/11.25U	1.9L/.5U
M249	AN/PAS-13	IFTC TOP	400m	0.0/2.75D	0.0/8.6U	1.9L/.5U
M249	AN/PAS-13	TWS BRACKET	400m	0.0/5.5D	0.0/10.05U	1.9L/.5U
M240B	IRON SIGHTS	NA	500m	TBD	TBD	1.9L/.5U
M240B	MGO	FEEDTRAY COVER RAIL	500m	NA	0.0/0.0	5.0R/4.1D
M240B	AN/PAQ-4C	PICATINNEY RAIL GRABBER TOP	500m	1.75R/2.2D	1.5L/3.5U	5.0R/4.1D
M240B	AN/PEQ-2A	INSIGHT RAIL GRABBER TOP	500m	2.0R/1.5D	1.7R/3.71U	5.0R/4.1D
M240B	AN/PAQ-4C	INSIGHT RAIL GRABBER FORWARD RAILS RIGHT	500m	TBD	TBD	5.0R/4.1D
M240B	AN/PAQ-4C	INSIGHT RAIL GRABBER FORWARD RAILS LEFT	500m	6.2R/16.8D	6.2R/8.1D	5.0R/4.1D
M240B	AN/PAQ-4C	PICATINNY RAIL GRABBER FORWARD RAILS RIGHT	500m	TBD	TBD	5.0R/4.1D
M240B	AN/PAQ-4C	PICATINNY RAIL GRABBER FORWARD RAILS LEFT	500m	7.9R/16.8D	7.9R/8.1D	5.0R/4.1D
M240B	AN/PEQ-2A	INSIGHT RAIL GRABBER FORWARD RAILS RIGHT	500m	TBD	TBD	5.0R/4.1D
M240B	AN/PEQ-2A	INSIGHT RAIL GRABBER FORWARD RAILS LEFT	500m	6.2R/12.8D	6.2R/4.1D	5.0R/4.1D
M240B	AN/PEQ-2A	PICATINNY RAIL GRABBER FORWARD RAILS RIGHT	500m	TBD	TBD	5.0R/4.1D
M240B	AN/PEQ-2A	PICATINNY RAIL GRABBER FORWARD RAILS LEFT	500m	7.9R/12.8D	7.9R/4.1D	5.0R/4.1D
M240B	AN/PVS-4	FEEDTRAY COVER RAIL PICATINNY RAIL GRABBER WITH SPACER	500m	0.0/6.2D	0.0/6.0U	5.0R/4.1D
M240B	AN/PAS-13	FEED TRAY COVER	500m	0.0/2.3U	0.0/8.0U	5.0R/4.1D

DA FORM 7476-R, OCT 2002 Page 1 of 2

Figure H-3. Quick reference card is included on back of DA Form 7476-R.

GLOSSARY

AAR	after-action report
AR	Army regulation
ARTEP	Army Training and Evaluation Program
BFA	blank firing attachment
CLP	cleaner, lubricant, preservative
cm	centimeters
CO	company
DA	Department of the Army
DODAC	Department of Defense Ammunition Code
EST	engagement skills trainer
FM	field manual
FPF	final protective fire
FPL	final protective line
FSN	Federal stock number
FTX	field training exercise
GTA	graphic training aid
HMMWV	high-mobility, multipurpose wheeled vehicle
IAW	in accordance with
IET	initial entry training
IOAC	Infantry officer's advanced course
IOBC	Infantry officer's basic course
instr	instruction
LAW	lubricating oil, arctic weather
LFX	live-fire exercise
LRA	local reproduction authorized
LSA	lubricating oil, semifluid, automatic weapons
LTA	local training area
m	meter
METL	mission-essential task list

MGO	machine gun optic
MILES	multiple-integrated laser engagement system
mm	millimeter
mph	miles per hour
MPRC	multipurpose range complex
MTA	major training area
MTP	mission training plan
MWTS	medium-weapon thermal sight
N/A	not applicable
NATO	North Atlantic Treaty Organization
NBC	nuclear, biological, and chemical
NCO	noncommissioned officer
NCOES	Noncommissioned Officer Education System
NCOIC	noncommissioned officer in charge
NSN	national stock number
NVD	night vision device
OIC	officer in charge
PAM	pamphlet
PDF	principal direction of fire
PLT	platoon
POPP	Pull, Observe, Push, and Press (memory aid for immediate action)
prac	practice
qual	qualification
RD	round
sec	second
SM	soldier's manual
SMCT	soldier's manual of common tasks
SOP	standing operating procedures
sqd	squad
STP	soldier's training publication
STRAC	standards in training commission
STX	situational training exercise
T&E	traversing and elevating
TC	training circular

TM	technical manual
TOE	table of organization and equipment
TPIAL	target pointer/illuminator/aiming light
TRADOC	US Army Training and Doctrine Command
US	United States

REFERENCES

SOURCES USED

These are the sources quoted or paraphrased in this publication.

ARTEP 7-8-MTP	Mission Training Plan for the Infantry Rifle Platoon and Squad. 30 September 1988.
ARTEP 7-8-DRILL	Drills for the Infantry Rifle Platoon and Squad. 3 November 1993.
FM 7-8	Infantry Platoon and Squad. 22 April 1992.
FM 21-26	Map Reading and Land Navigation. 7 May 1993.

DOCUMENTS NEEDED

These documents must be available to the intended uses of this publication.

AR 385-63	Policies and Procedures for Training, Target Practice and Combat. 15 October 1983.
DA Form 2404	Equipment Inspection and Maintenance Worksheet. April 1979.
DA Form 5517-R	Standard Range Card (LRA). February 1986.
DA Form 85-R	Scorecard for M249, M60/M240B Machine Guns. October 2002.
DA Form 7476-R	10-Meter Boresight Offset Target. October 2002.
DA Pam 25-30	Army Electronic Library. 1 July 2002.
DA Pam 350-38	Training Standards for Weapon Training. September 1990.
*FM 3-5	NBC Decontamination. 23 July 1992.
*FM 21-60	Visual Signals. 30 September 1987.
STP 21-24-SMCT	Soldier's Manual of Common Tasks (Skill Levels 2-4). 1 October 2001.

*This source was also used to develop this publication.

*TC 25-8	Training Ranges. 25 February 1992.
TM 9-1005-319-10	Operator's Manual List for Rifle 5.56-mm, M16A2. 1 October 1998.
*TM 9—1005-201-10	Operator's Manual Machine Gun, 5.56-mm, M249 w/Equip. 26 July 1991.
*TM 9-1005-224-10	Operator's Manual for Machine Gun, 7.62-mm, M60, W/E; Mount, Tripod, Machine Gun, 7.62-mm, M122. 2 April 1998.
*TM 9-1005-313-10	Operator's Manual for Machine Gun, 7.62-mm, M240B, W/E; Flex-mount, Tripod, Machine Gun, 7.62-mm, M122A1. 19 July 1996.
*TM 9-1300-200	Ammunition General. 3 October 1969.
*TM 11-5855-213-10	Operator's Manual Night Vision Sight, Individual Served Weapon, AN/PVS-4. 1 February 1993.

INTERNET WEB SITES

U.S. Army Publishing Agency
http://www.usapa.army.mil

Army Doctrine and Training Digital Library
http://www.adtdl.army.mil

*This source was also used to develop this publication.

INDEX

10-METER BORESIGHT OFFSET TARGET

For use of this form, see FM 3-22.68; the proponent agency is TRADOC.

See back of this form for quick reference card.

1. Use the correct offset for the weapon, sight, and location configured.
2. Stabilize the weapon and the offset.
3. Zero the bore light while inside the barrel of the weapon.
4. Align the laser of the bore light with the dot on the 10-meter offset.
5. Align the MILES laser with the MILES rectangle on the 10-meter offset (if applicable).
6. Adjust the aiming laser until it is centered on the cross hair.
7. Place the optic aim point centered on the cross hair and make adjustments on the optic until the bore light laser is aligned with the dot on the 10-meter offset.
8. Reconfirm that all devices are still aligned to their aiming mark.

WPN	ACCESSORY	MOUNT	RANGE ZEROED TO	ZERO TARGET OFFSET	BORESIGHT TARGET OFFSET	MILES OFFSET
M249	IRON SIGHTS	NA	400m	NA	TBD	1.9L/.5U
M249	MGO	IFTC RAIL	400m	TBD	0.0/7.9U	1.9L/.5U
M249	MGO	TWS BRACKET	400m	TBD	0.0/2.15U	1.9L/.5U
M249	AN/PAQ-4C	TWS BRACKET TOP	400m	0.5R/1.5U	1.85L/7.7U	1.9L/.5U
M249	AN/PAQ-4C	AN/PVS4 BRACKET	400m	2.5R/1.5D	4.1L/6.1U	1.9L/.5U
M249	AN/PAQ-4C	INSIGHT RAIL GRABBER WITH IFTC	400m	TBD	1.75L/4.69U	1.9L/.5U
M249	AN/PAQ-4C	PICATINNY RAIL GRABBER WITH IFTC	400m	1.75R/0.0	1.75L/5.39U	1.9L/.5U
M249	AN/PAQ-4C	INSIGHT RAIL GRABBER FORWARD RAILS RIGHT	400m	5.9R/9.6D	5.9R/4.0D	1.9L/.5U
M249	AN/PAQ-4C	INSIGHT RAIL GRABBER FORWARD RAILS LEFT	400m	6.0R/13.3D	6.0R/8.3D	1.9L/.5U
M249	AN/PAQ-4C	PICATINNY RAIL GRABBER WITH AL SPACER FORWARD RAILS RIGHT	400m	7.7R/9.6D	7.7R/4.0D	1.9L/.5U
M249	AN/PAQ-4C	PICATINNY RAIL GRABBER FORWARD RAILS LEFT	400m	7.6R/13.3D	7.6R/8.3D	1.9L/.5U
M249	AN/PEQ-2A	TWS BRACKET TOP	400m	1.8L/2.7D	1.8R/7.95U	1.9L/.5U
M249	AN/PEQ-2A	AN/PVS4 BRACKET WITH SPACER	400m	5.0R/4.0D	.45L/6.5U	1.9L/.5U
M249	AN/PEQ-2A	INSIGHT RAIL GRABBER WITH IFTC	400m	2.0L/1.5U	1.95R/4.79U	1.9L/.5U
M249	AN/PEQ-2A	PICATINNY RAIL GRABBER WITH IFTC	400m	2.0L/0.5D	1.95R/6.49U	1.9L/.5U
M249	AN/PEQ-2A	INSIGHT RAIL GRABBER FORWARD RAILS RIGHT	400m	6.1R/13.2D	6.1R/7.6D	1.9L/.5U
M249	AN/PEQ-2A	INSIGHT RAIL GRABBER FORWARD RAILS LEFT	400m	6.0R/9.4D	6.0R/4.4D	1.9L/.5U
M249	AN/PEQ-2A	PICATINNY RAIL GRABBER WITH AL SPACER FORWARD RAILS RIGHT	400m	7.8R/13.2D	7.8R/7.6D	1.9L/.5U
M249	AN/PEQ-2A	PICATINNY RAIL GRABBER FORWARD RAILS LEFT	400m	7.6R/9.4D	7.6R/4.4D	1.9L/.5U
M249	AN/PVS-4	IFTC TOP WITH SPACER	400m	0.0/4.3D	0.0/10.00U	1.9L/.5U
M249	AN/PVS-4	AN/PVS4 BRACKET	400m	2.5R/4.9D	2.25L/11.25U	1.9L/.5U
M249	AN/PVS-4	IFTC TOP	400m	0.0/2.75D	0.0/8.6U	1.9L/.5U
M249	AN/PVS-4	TWS BRACKET	400m	0.0/5.5D	0.010.05U	1.9L/.5U
M240B	IRON SIGHTS	NA	500m	TBD	TBD	1.9L/.5U
M240B	MGO	FEEDTRAY COVER RAIL	500m	NA	0.0/0.0	5.0R/4.1D
M240B	AN/PAQ-4C	PICATINNY RAIL GRABBER TOP	500m	1.75R/2.2D	1.5L/3.5U	5.0R/4.1D
M240B	AN/PEQ-2A	INSIGHT RAIL GRABBER TOP	500m	2.0R/1.5D	1.7R/3.71U	5.0R/4.1D
M240B	AN/PAQ-4C	INSIGHT RAIL GRABBER FORWARD RAILS RIGHT	500m	TBD	TBD	5.0R/4.1D
M240B	AN/PAQ-4C	INSIGHT RAIL GRABBER FORWARD RAILS LEFT	500m	6.2R/16.8D	6.2R/8.1D	5.0R/4.1D
M240B	AN/PAQ-4C	PICATINNY RAIL GRABBER FORWARD RAILS RIGHT	500m	TBD	TBD	5.0R/4.1D
M240B	AN/PAQ-4C	PICATINNY RAIL GRABBER FORWARD RAILS LEFT	500m	7.9R/16.8D	7.9R/8.1D	5.0R/4.1D
M240B	AN/PEQ-2A	INSIGHT RAIL GRABBER FORWARD RAILS RIGHT	500m	TBD	TBD	5.0R/4.1D
M240B	AN/PEQ-2A	INSIGHT RAIL GRABBER FORWARD RAILS LEFT	500m	6.2R/12.8D	6.2R/4.1D	5.0R/4.1D
M240B	AN/PEQ-2A	PICATINNY RAIL GRABBER FORWARD RAILS RIGHT	500m	TBD	TBD	5.0R/4.1D
M240B	AN/PEQ-2A	PICATINNY RAIL GRABBER FORWARD RAILS LEFT	500m	7.9R/12.8D	7.9R/4.1D	5.0R/4.1D
M240B	AN/PVA-4	FEEDTRAY COVER RAIL PICATINNY RAIL GRABBER WITH SPACER	500m	0.0/6.2D	0.0/6.0U	5.0R/4.1D
M240B	AN/PAS-13	FEEDTRAY COVER	500m	0.0/2.3U	0.0/8.0U	5.0R/4.1D

SCORECARD FOR M249, M60/M240B MACHINE GUNS

For use of this form, see FM 3-22.68; the proponent agency is TRADOC.
See back of this form for instructions.

PRIVACY ACT STATEMENT

AUTHORITY: 10 USC 30129(g) Executive Order 9397.

PRINCIPAL PURPOSE: Records individual's performance on record fire range.

ROUTINE USES: Evaluate individual's proficiency and basis for determination of award of proficiency badge; SSN is used for positive identification purposes only.

DISCLOSURE: Voluntary, individuals not providing information cannot be rated/scored on a mass basis.

NAME	SSN	UNIT	DATE *(YYYYMMDD)*	LANE

TABLE I (10-METERS)

TASK	RANGE (meters)	TIME	HITS
1*	10	N/A	N/A
2*	10	N/A	N/A
3*	10	N/A	N/A
4*	10	N/A	N/A
5*	10	N/A	N/A
6*	10	N/A	N/A
7*	10	N/A	N/A
8*	10	N/A	N/A
9	10	40 SEC	
10	10	50 SEC	

TOTAL HITS (POINTS) ______

TABLE II (DAY TRANSITION)

TASK	RANGE (meters)	TIME	*PRACTICE		**QUALIFY	
			HIT	MISS	HIT	MISS
1*	500	N/A	N/A	N/A	N/A	N/A
2**	400	10 SEC				
3**	500	15 SEC				
4**	600	20 SEC				
5**	800	30 SEC				
6**	400 600	30 SEC				
7**	700 800	45 SEC				
8**	400 500 600	45 SEC				

**TOTAL POINTS ______

TABLE III (LIMITED VISIBILITY)

TASK	RANGE (meters)	TIME	*PRACTICE		**QUALIFY	
			HIT	MISS	HIT	MISS
1*	10	N/A	N/A	N/A	N/A	N/A
2*	10	N/A	N/A	N/A	N/A	N/A
3*	500	N/A	N/A	N/A	N/A	N/A
4	200	10 SEC				
5	400	15 SEC				
6	100	10 SEC				
7	300	15 SEC				
8	200 400	25 SEC				
9	100 300	25 SEC				
10	100 200 400	30 SEC				

TOTAL HITS ______

TOTAL SCORE ______

* NONSCORED TASKS
**10 POINTS PER HIT

CHECK APPROPRIATE WEAPON

EXPERT		FIRST CLASS		SECOND CLASS	
☐ **M249 182-201**	☐ **M60/M240B 206-227**	☐ **M249 158-181**	☐ **M60/M240B 180-205**	☐ **M249 158-181**	☐ **M60/M240B 180-205**

OIC SIGNATURE	GRADER	RATING

Use the following procedures to fill out the M249, M60/M241B scorecard:

1. **NAME:** Enter the gunner's last name, first name, middle initial, and rank.

2. **SSN:** Enter the gunner's social security number.

3. **UNIT:** Enter the gunner's unit designation.

4. **DATE:** Enter the date of firing.

5. **LANE:** Enter the lane number for the gunner's firing point.

6. **RECORD:** Tasks used for record and qualification are Firing Table I, tasks 9 through 10; Firing Table II, tasks 2 through 8; and Firing Table III, tasks 4 through 10.

7. **HIT/MISS:** For Table I, tasks 9 through 10, enter the number of rounds impacting within target spaces (maximum of 7 [M249] or 9 [M60/M240B] per space).
For Table II, tasks 2 through 8, and Table III, tasks 4 through 10, enter an X for hit and an O for a miss (regardless of whether the target is hit on the first or second burst).

8. **TOTAL HITS/ POINTS:** For Table I, tasks 9 through 10, give 1 point for each round impacting within a scoring space.
For Table II, tasks 2 through 8, give 10 points for each target hit.
For Table III, tasks 4 through 10, enter the number of targets hit (no points awarded).

9. **TOTAL SCORE:** Add points from Tables IV and V. Use the following qualification levels*:

	M249	M60/M240B
EXPERT GUNNER	182-201	206-227
GUNNER 1ST CLASS	158-181	180-205
GUNNER 2D CLASS	133-157	151-179
UNQUALIFIED	0-132	0-150

* The gunner must score 63 points (M249), 81 points (M60/M240B) on Table I, 70 points on Table II, and 6 hits on Table III to meet the minimum score for each

Made in the USA
Coppell, TX
05 November 2026

11299179R00374